Computer-Aided Design:

An Integrated Approach

Computer-Aided Design:

An Integrated Approach

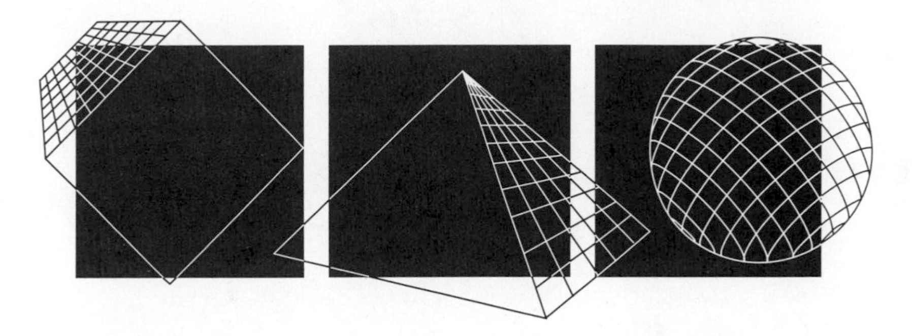

Tai-Ran Hsu
San Jose State University

Dipendra K. Sinha
San Francisco State University

▪ WEST PUBLISHING COMPANY ▪
St. Paul ▪ New York ▪ Los Angeles ▪ San Francisco

Copyediting ■ Sherry Goldbecker
Proofreading ■ Sylvia Dovner, Technical Texts, Inc.
Interior and Cover Design ■ Katherine Townes, TECH*arts*
Interior artwork ■ Katherine Townes, TECH*arts*
Composition ■ Polyglot
Cover art ■ Courtesy of WABCO Construction and Mining Equipment, Peoria, Illinois.

Printed in the United States of America

99 98 97 96 95 94 93 92 8 7 6 5 4 3 2 1 0

Library of Congress Cataloging-in-Publication Data

Hsu, Tai-Ran.
Computer-aided design: an integrated approach/Tai-Ran Hsu, Dipendra K. Sinha.
p. cm.
Includes bibliographical references (p.) and index.
ISBN 0-314-80781-0
1. Engineering design—Data processing. 2. Computer-aided design.
I. Sinha, Dipendra K. II. Title.
TA174.H77 1991
620′.0042′0285—dc20

90-46674
CIP

To our patient wives:

Grace Su-Yong Hsu,
Basanti Sinha

our children:

Jean, Euginette, and Leigh Hsu,
Priyamvada and Udayan K. Sinha

and our students

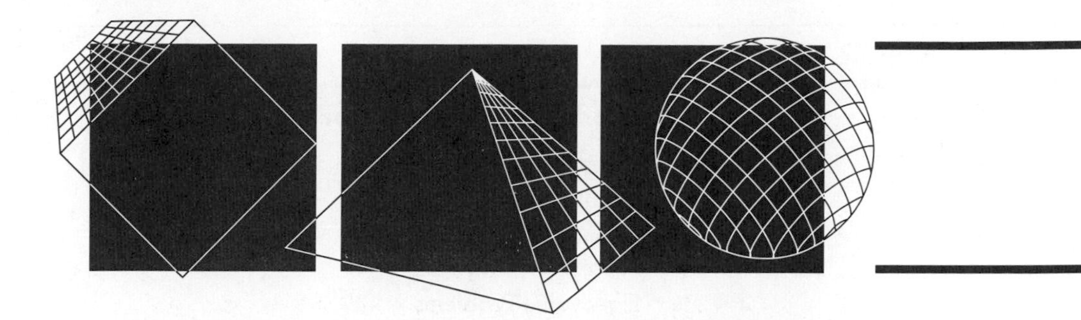

Contents

Preface xiii

About the Authors xvii

CHAPTER 1

An Overview of Computer-Aided Design and Analysis 1

1.1 ■ Introduction 1
1.2 ■ CAD and the CIMS 2
1.3 ■ CAD and Traditional Design 4
1.4 ■ Essential Hardware Requirements for CAD 8
1.5 ■ Interactive Computer Graphics 15
1.6 ■ General Procedure for Engineering Design 20
1.7 ■ Engineering Analysis 27
1.8 ■ Summary 29
■ References 29
■ Problems 30

CHAPTER 2

Review of Numerical Techniques for CAD 33

2.1 ■ Introduction 33
2.2 ■ Matrices and Determinants 34
2.3 ■ Solution of Simultaneous Linear Equations 35
2.4 ■ Eigenvalue Problems 42
2.5 ■ Numerical Differentiation 50
2.6 ■ Numerical Integration 58
2.7 ■ Summary 66
■ References 67
■ Problems 67

CHAPTER 3

Principles of Computer Graphics 73

3.1 ■ Introduction 73
3.2 ■ Mathematical Formulations for Graphics 74
3.3 ■ Basic Curve-Fitting Techniques 82
3.4 ■ Algorithms for Raster-Scan Graphics 91
3.5 ■ Algorithms for Scan Conversion 92
3.6 ■ Two-Dimensional Transformations: Homogeneous Coordinates 94
3.7 ■ Transformations of a Point 95
3.8 ■ Transformation of Plane Objects 101
3.9 ■ Image Manipulation About Arbitrary Axes 103
3.10 ■ Graphics Packages and Graphics Standards 106
3.11 ■ GKS 107
3.12 ■ Graphics Primitives in GKS 108
3.13 ■ Other Graphics Standards 112
3.14 ■ Principles of Three-Dimensional Graphics 114
3.15 ■ Three-Dimensional Transformations: Simple Cases 114
3.16 ■ Three-Dimensional Transformations: Complex Cases 116
3.17 ■ Summary 122
■ References 122
■ Problems 124

CHAPTER 4

Computer Graphics and Design 127

4.1 ■ Introduction 127
4.2 ■ Geometric Modeling 127
4.3 ■ Surface Modeling 128
4.4 ■ Solid Modeling 131
4.5 ■ Viewing in Three-Dimensions 135
4.6 ■ Principles of Projections 136
4.7 ■ Mathematics of Projections 140
4.8 ■ Hidden Line/Surface Removal Algorithms 147
4.9 ■ Surface Patches 158
4.10 ■ Geometric Properties of Graphics Models 161
4.11 ■ Computer Simulation and Animation 165
4.12 ■ Windows, Viewports, and Viewing Transformations 167
4.13 ■ Clipping 170
4.14 ■ Summary 172
■ References 172
■ Problems 173

CHAPTER 5

Introduction to Design Databases 177

5.1 ■ Introduction 177
5.2 ■ Database Management Systems 179
5.3 ■ Data Models 180
5.4 ■ Design Databases 192
5.5 ■ Geometric Databases for Two-Dimensional Objects 194
5.6 ■ Geometric Databases for Three-Dimensional Objects 199
5.7 ■ The IGES Standard 200
5.8 ■ Basic Requirements for Design Database Management Systems 201
5.9 ■ Databases for Integrated Engineering Systems 202
5.10 ■ Summary 204
■ References 205
■ Problems 206

CHAPTER 6

Overview of the Finite Element Method 209

6.1 ■ Introduction 209
6.2 ■ The Concept of Discretization 210
6.3 ■ Application of the Finite Element Method in Engineering Analysis 214
6.4 ■ Steps in the Finite Element Method 216
6.5 ■ Automatic Mesh Generation 229
6.6 ■ Integration of CAD and Finite Element Analysis 235
6.7 ■ Summary 240
■ References 240
■ Problems 243

CHAPTER 7

Elastic Stress Analysis by the Finite Element Method 247

7.1 ■ Introduction 247
7.2 ■ Review of Basic Formulations in Linear Elasticity Theory 247
7.3 ■ Finite Element Formulation 255
7.4 ■ One-Dimensional Stress Analysis of Solids 259
7.5 ■ Two-Dimensional Stress Analysis of Solids (Plane Stress Case) 269
7.6 ■ General-Purpose Finite Element Programs 279
7.7 ■ The ANSYS Program 281
7.8 ■ Summary 285
■ References 286
■ Problems 286

CHAPTER 8

Design Optimization 293

8.1 ■ Introduction 293
8.2 ■ Design Variables, Parameters, and Constraints 294

8.3 ■ Objective Function 295
8.4 ■ Constrained and Unconstrained Optimization 296
8.5 ■ Unconstrained Optimization with One Design Variable 299
8.6 ■ Constrained Optimization with One Design Variable 302
8.7 ■ Unconstrained Optimization with Two Variables 304
8.8 ■ Unconstrained Optimization with Multiple Design Variables 307
8.9 ■ Constrained Optimization: Indirect Methods 316
8.10 ■ Constrained Optimization: Direct Method 326
8.11 ■ Use of the Finite Element Method in Optimization 326
8.12 ■ Summary 334
■ References 335
■ Problems 336

CHAPTER 9

Linkage Synthesis on Microcomputers 339

9.1 ■ Introduction 339
9.2 ■ Mechanism Synthesis and Analysis 340
9.3 ■ Mechanism and Linkage 340
9.4 ■ Planar Mechanisms 344
9.5 ■ Linkage Synthesis 346
9.6 ■ Precision Points and Their Selection 347
9.7 ■ Geometric Method of Mechanism Synthesis 350
9.8 ■ Analytic Method of Linkage Synthesis 355
9.9 ■ Cognate Four-Bar Mechanisms 360
9.10 ■ Miscellaneous Synthesis Problems 362
9.11 ■ Summary 369
■ References 369
■ Problems 369

CHAPTER 10

Integrated CAD Systems 373

10.1 ■ Introduction 373
10.2 ■ Overview of Integrated CAD Systems 375
10.3 ■ Integrated CAD Systems 378
10.4 ■ Major Functions of Commercial CAD Systems 382
10.5 ■ The MICROCAD System 384
10.6 ■ Design Case Study Using the ANSYS Program 397
10.7 ■ Expert Systems for CAD 402
10.8 ■ CAD/CAM Interface 406
10.9 ■ Summary 408
■ References 409
■ Problems 410

APPENDIXES

I ■ **Review of Matrix Operations** 413
II ■ **The Graphical Kernal System (GKS)** 421
III ■ **PROFILE—A Program for a Two-Dimensional Graphic Database** 447
IV ■ **Computer Configuration of the MICROCAD System** 467

Glossary 473

Index 481

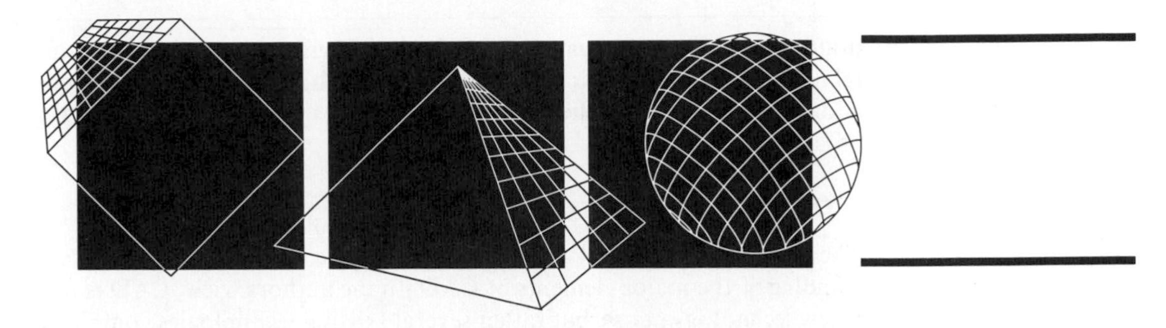

Preface

Computer-aided design (CAD) is an essential element of the emerging technology of the computer-integrated manufacturing system (CIMS). It is becoming increasingly evident that the competitiveness of an industrialized society will be strongly affected by the extent to which computers are used in its design and manufacturing sectors.

The use of computers in engineering practice is not new. However, the possibility of performing all design functions on a computer, as implied in CAD, has been available for little over a decade. The birth of CAD technology was obviously a result of the maturity attained in the areas of interactive graphics-related computer hardware and software. The pace at which CAD has been adopted by various sectors of industry is truly unparalleled by any other technology developed in the past. The popularity of CAD has made its instruction in engineering schools a high priority. Unfortunately, as of now, very few schools have cohesive CAD curricula. Students who graduate from these schools without a good background in this fast-growing technology may find themselves disadvantaged in the job market.

The noticeable lack of effective CAD instruction in many institutions may be attribtued to two main factors: (1) the shortage of funds to acquire necessary hardware and software and (2) the lack of proper textbooks. Fortunately, the first problem has been partially solved by the steady decline in the cost of computers. The second problem, however, remains unresolved. In past years, we tried to identify a textbook that has a balanced presentation of the various aspects of CAD. However, almost all CAD textbooks currently in print are dominated by computer graphics and drafting; many topics that are relevant to engineering design and the relationship of these topics to computer applications are not adequately addressed. It was this deficiency that prompted the writing of this book.

❑ OBJECTIVE

The principal objective of this book is to provide students with an understanding of the major elements of CAD. In the authors' view, CAD is not a new technology per se, but rather several existing technologies combined to create a new format. Therefore, a detailed description of the various components of CAD technology is not absolutely necessary in the book. The authors believe that an introduction to the fundamental principles underlying these relevant areas is required. Advanced topics should be covered in depth in separate courses (e.g., in computer graphics and finite element analysis).

Another important objective of the book is to give students an appreciation of not only the individual elements of CAD, but also the importance of the integrated nature of these elements. The authors believe this objective can be achieved by presenting students first with an overview of CAD followed by the principles of the essential elements that constitute CAD technology. Numerous examples and problems are included to provide the students with ample opportunity to practice various CAD functions.

❑ ORGANIZATION AND SCOPE

The textbook is organized as follows: Students are first introduced to the fundamental principles of major CAD elements in chapters on an overview of CAD, numerical techniques relevant to CAD, principles of computer graphics, the application of computer graphics to design; CAD databases, and the principle of design optimization. Relevant design analysis is also presented. Two chapters are devoted to the finite element method, covering an overview of this method and its application to elastic stress analysis. Formulations and algorithms for linkage synthesis are included in Chapter 9. These advanced topics are presented at a level that is suitable for undergraduate students.

An overview of the main features offered by typical commercial CAD packages is included in the last chapter. The purpose of this is to provide students with an opportunity to become familiar with the performance of commercial turnkey CAD packages. Also included in this chapter are introductory sections on expert systems in CAD and on CAD/CAM interaction. These two topics are presented to provide students with a sense of two ongoing efforts in the development of CAD technology.

Two design case studies are included in the last chapter. One case deals with the design of a gear tooth using a microcomputer-based MICROCAD system. The second case involves the optimum modal analysis of the rocker arm of an overhead engine using a commercial CAD system, the ANSYS program. The hardware and software configurations of MICROCAD are given in Appendix IV. A working diskette of this program is available on request from the authors.

❑ USING THIS TEXTBOOK

The content of this book is designed for one semester of instruction to senior students or graduate students at the master's level. For a junior year course, we suggest the instructor use his or her discretion regarding the depth of coverage of each topic. One suggestion is to treat Chapter 8 on design optimization and Chapter 9 on computer-aided linkage synthesis as optional reading materials. The instructor may also wish to skip Chapter 2 on the application of numerical techniques in CAD should he or she be satisfied with students' knowledge and skill in these areas.

❑ ACKNOWLEDGEMENTS

The task of preparing a textbook of this breadth certainly cannot be accomplished by the authors' effort alone. The authors have the good fortune and great pleasure to have many young and dedicated research associates and students working for them during their tenure on the academic staff at the University of Manitoba in Winnipeg, Canada. The following are just a few names identifiable with specific areas in the book: G. Pizey, F. De Luca, D.R. Young, and T. Malaher in computer graphics and the development of the MICROCAD; G. Chen, Z.L. Gong, and N.S. Sun in the development of the finite element method and part of the solutions manual; and P.A. Dutchak in design databases. The authors also wish to express their gratitude to R.J. March and J. Twerdok of Swanson Analysis Systems, Inc., for their cooperation and supply of information on the ANSYS program; and M. Gilberg of the Onan Corporation for providing the design case study in Section 10.6. Permission given by WABCO Construction and Mining Equipment and the Structural Dynamics Research Corporation for the use of their valuable photoplates is also gratefully acknowledged. The authors wish to thank V. Lee and G. Ferguson of the University of Manitoba for their excellent work in preparing a large portion of the manuscript.

The authors are grateful for the valuable comments and suggestions expressed by their colleagues in reviewing the manuscript for this text.

David Hoeltzel, *Columbia University*
Ampere Tseng, *Drexel University*
C. Alex Hou, *Howard University*
Rollin Dix, *Illinois Institute of Technology*

Martin Valderploeg, *Iowa State*
Mukesh Gandhi, *Michigan State*
Leo LaFrance, *New Mexico State*
Henry Busby, *Ohio State*
Y.J. Lin, *University of Akron*
Ara Arabyan, *University of Arizona*
Ali Manesh, *University of Arkansas*
S. Kota, *University of Michigan*
Bo Ping, *University of Texas, Arlington*
Pradip Sheth, *University of Virginia*
Alice Agogino, *University of California—Berkeley*
Charles Beadle, *University of California—Davis*
Leon Levine, *University of California—Los Angeles*
Geoffrey Shiflett, *University of Southern California*
Mohammed Khosrowjerdi, *Western New England University*

Last, but not the least, the authors wish to acknowledge the superb services provided by the editorial and production staff at the West Publishing Company. Able support and assistance by T. Michael Slaughter, Christine Hurney, and Beth Hatton are greatly appreciated.

TAI-RAN HSU
DIPENDRA K. SINHA

About the Authors

❑ TAI-RAN HSU

Tai-Ran Hsu is Professor and Chair of Mechanical Engineering at San Jose State University. Much of the work presented in this book, however, was carried out during his tenure as a Professor and Head of Mechanical Engineering at the University of Manitoba in Winnipeg, Canada.

Prior to joining the academic ranks, Professor Hsu worked extensively in the United States and Canada as a design engineer for heat exchangers, urban mass transit vehicles, steam power plants, large steam turbines, and nuclear reactors. He received his B.Sc. from the National Cheng-Kung University in China, his M.Sc. from the University of New Brunswick in Canada, and his Ph.D. from McGill University in Canada. All degrees were in mechanical engineering.

His current research interests lie in computer-aided design/computer-aided manufacturing, finite element analysis, thermofracture mechanics, and experimental stress analysis. He has authored three other books and over one hundred technical papers in several of these areas.

❑ DIPENDRA K. SINHA

Dipendra K. Sinha is currently Associate Professor of Mechanical Engineering at San Francisco State University, a position he has held since 1987. He received his B.Sc. from Patna University in India and his M.Sc. and Ph.D. from the University of Manchester Institute of Science and Technology in England.

Between 1968 and 1976 Dr. Sinha participated in the design and development of mechanical equipment at Tata Iron and Steel Company in India. From 1981 to 1987 he served on the faculties of the University of Manitoba, the Virginia Military Institute, and the University of Wisconsin–Platteville.

Dr. Sinha has been involved in the development of computer-aided design (CAD) technology since 1981. He has authored or coauthored three books on different aspects of CAD and has published frequently in refreed journals and conference proceedings. His current research interests lie in computer-aided manufacturing and expert mechanical design systems.

CHAPTER 1

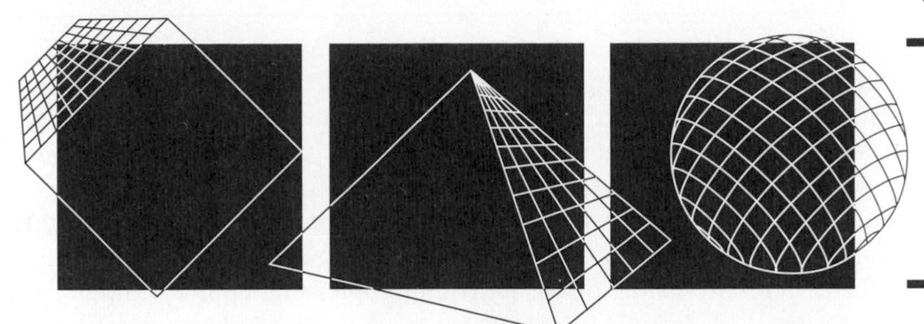

An Overview of Computer-Aided Design and Analysis

1.1 Introduction

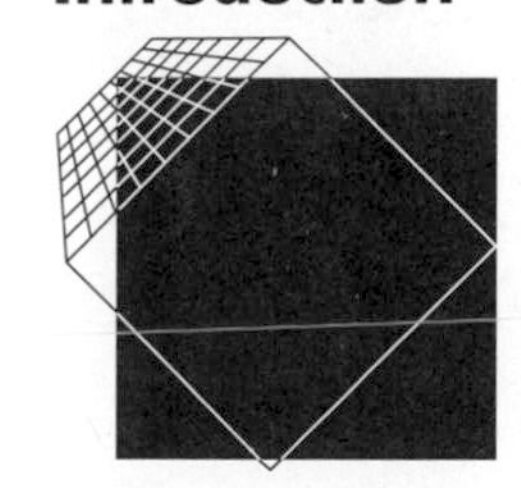

Computer-aided design (CAD) has been one of the most frequently mentioned technical terms in the engineering community in recent years. It is perceived as a major element of *computer-aided engineering* (CAE). CAD will also play a major role in future industrial production systems such as the *computer-integrated manufacturing system* (CIMS) as we move toward total industrial automation.

The rate at which CAE is spreading to various sectors of industry is truly phenomenal and has few, if any, parallels in technological history. The adoption of CAE systems has accelerated with recent advances in microcomputer technology. Computer-assisted design analysis and graphics are no longer a luxury enjoyed by only a handful of large companies that can afford mainframe computers and peripherals. Engineers everywhere have access to better and cheaper computer hardware and software to produce sophisticated computer graphics and drafting.

The technological capabilities of computer hardware and graphics are increasing so quickly that engineers are constantly finding new applications of this technology in various aspects of design work. The real power of CAD is no longer confined to just producing sophisticated solid geometries and engineering drawings. As a result, precise description of the scope of CAD is far from being unanimous. The computer buff looks at CAD as computer graphics, spectacular solid geometric modeling, and animation. The hardcore analyst, on the other hand, interprets CAD as sophisticated finite element analyses by a computer, with the results depicted graphically. To a large number of practising engineers, CAD is still not a well-known entity and is often viewed as a passing fad. The general feeling, however, is that CAD is the same as conventional design, except that calculations are carried out by computers. In the authors' view, CAD is all these and more. The intent of this chapter is to present to future and practising engineers an overview of CAD before touching on the individual topics that comprise it.

1.2 ▪ CAD and the CIMS

The term *computer-integrated manufacturing system* (CIMS) was formally recognized by the Institute of Industrial Engineering (IIE) at its 1983 spring meeting [1]. It was a natural evolution of CAD/CAM (computer-aided manufacturing) technology, which had been in existence in some industries for over a decade.

To many practicing engineers, the CIMS is an integrated CAD/CAM system that encompasses all activities from the planning and design of a product to its manufacturing and shipping. In a true sense, such a system covers an even broader scope and includes the management and control of an entire business. The ultimate goal of a CIMS is to optimize the total design/manufacturing/business operation, rather than its individual components. The major elements of the CIMS, according to Sadowski [1], are interactive computer graphics, artificial intelligence, computer-aided process planning, computer numerical control, database technology and management, flexible manufacturing systems, information flow, material requirements planning, process and adaptive control, and robotics. A detailed description of a CIM system is given in Figure 1.1. The CIM system consists of seven major elements, with a central database that serves as a common link. It is apparent that the first two components, Items 1 and 2, can be classified as CAD functions. Most of these functions have already been incorporated into many commercial CAD/CAM packages, such as McAuto (McDonnell-Douglas Corp), CALMA (General Electric Co.), and CADAM, CATIA (IBM)[2]. Recent advances in microcomputer technology have made personal computer (PC)–based CAD systems effective design and drafting tools for small and medium-sized operations. A detailed description of PC-based commerical CAD systems, such as AutoCAD, Cadkey, Versacad, and those major systems as mentioned above, will be presented in Chapter 10. These systems can give us an appreciation for the role that CAD will play in future industrial operations.

The scope of this book is limited primarily to the first four functions listed under Item 1, mechanical design—that is, geometric modeling,

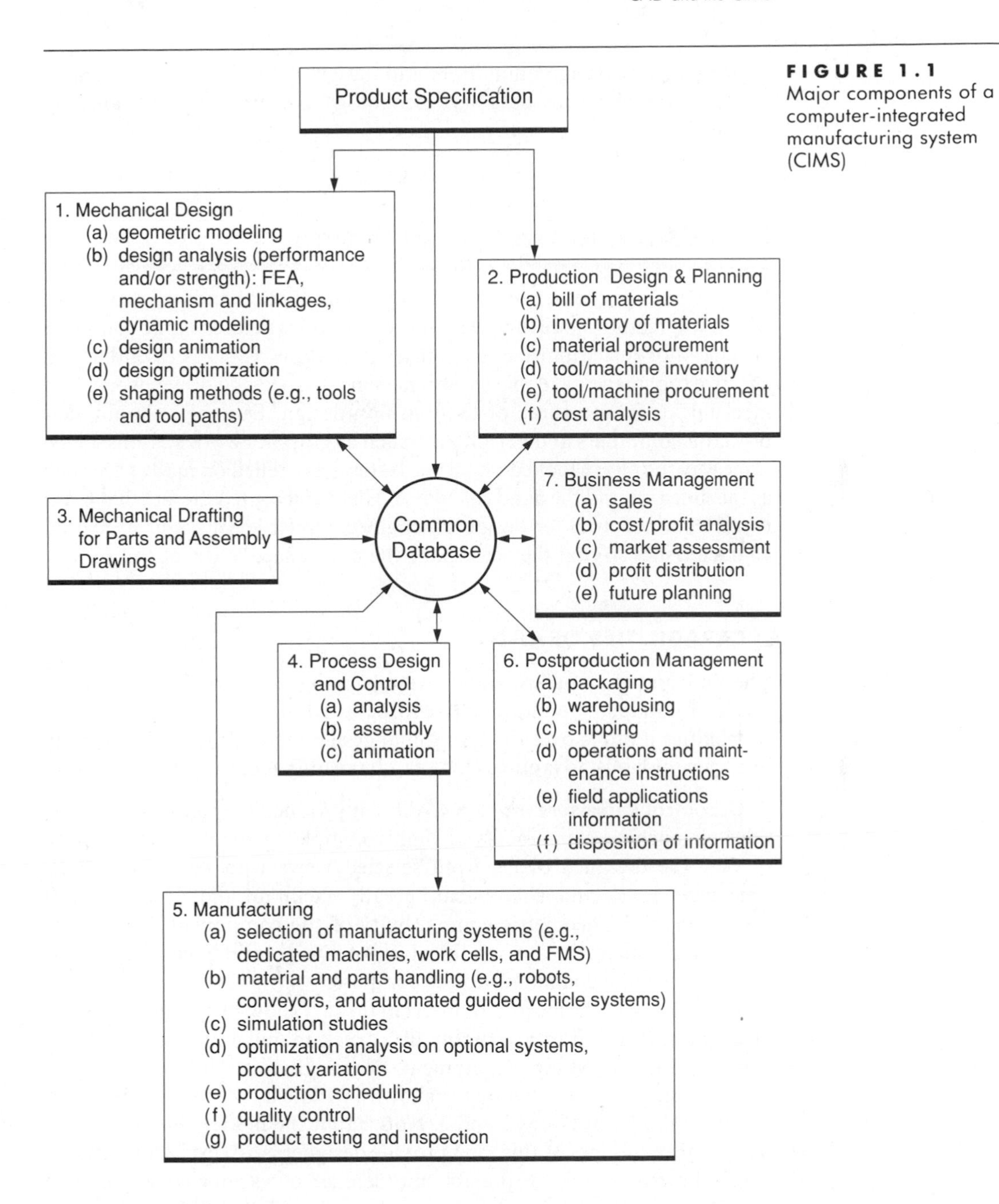

FIGURE 1.1 Major components of a computer-integrated manufacturing system (CIMS)

design analysis, design animation, and design optimization. Chapters on related topics, such as database technology and management, are also included.

1.3 ▪ CAD and Traditional Design

Many who have been exposed to CAD and have had hands-on experience with it will readily agree that the essence of CAD is neither exotic computer graphics nor just sophisticated computer analysis. CAD is essentially an effective tool made available to designers in all aspects of design work, be it mechanical, architectural, or even fashion design. Designers should view powerful computers and peripherals, such as graphics work stations, in the same way they have looked on the tools they have relied on in the past, such as the slide rule and the hand calculator. Having accepted the fact that CAD is merely a new tool for designers, engineers must learn about its special characteristics so that they can make the most effective use of this tool.

❑ CAPABILITIES OF CAD

The undisputed popularity and overwhelming adoption of CAD technology by industry undoubtedly can be attributed to the many unique capabilities it offers over the traditional design method. The following are just a few obvious advantages that can be readily identified with CAD.

Graphical Representations ▪ CAD can produce graphical representations of machine components of highly complex geometries. The ability to view the designed object from selected views with perspectives (e.g. isometric views) and shades can greatly facilitate the designer's appreciation of the geometric compatibility of the component. Interactive, or real-time, alteration or modification of the geometry can be accomplished by simple commands to the computer.

Plates 1 to 4 in the four-color insert in the text show the geometric model of a rear axle and frame assembly of a truck manufactured by WABCO Construction and Mining Equipment, Peoria, Illinois. The designer's perspective image of the assembly is shown in Plate 1, with its shaded solid geometry shown in Plates 2 and 3. Notice that the hidden lines in Plate 1 were removed. Views of this kind provide designers with excellent views by which they can check for possible interference of various components in the assembly. The interrelationship of the assembled components can be even better recognized in the solid geometric model with a desirable degree of transparency—50 percent in this case, as shown in Plates 2 and 3. It is also possible to get a close-up or zoomed view of a particular portion of the assembly, as shown in Plate 4, allowing an even more accurate check of the fitting of components. The shaded portion indicates the geometric interference during a certain portion of the duty cycle of the truck. These views give the same information as does a prototype in traditional design. An experienced CAD operator can make substantial changes to these models in a matter of a few minutes.

Through high-quality graphical depiction, CAD provides the designer with a much better perception of the assembly of design components, as demonstrated in Plates 1 to 4. Hence, the designer has the ability to specify more accurately the necessary clearances and tolerances of these interrelated components.

Animation ▪ CAD has the unique capability to animate on a graphical screen the kinematics of mechanical systems that involve complex linkages and mechanisms. Even slightly abnormal movements of any part of the system can be readily detected with the aid of such special features as zooming, clipping, and windowing, which are available in most commercial CAD packages. Plates 5 and 6 illustrate the frozen motion pictures of the suspension system of the rear axle and frame assembly system described in Plates 1 to 4. Plate 5 consists of eight frames that depict the relative positions of the rear axle of a truck at eight selected sequential instants. Optionally, these frames can be superimposed and shown at selected instants in an enlarged view, as shown in Plate 6. From these, the designer can visually check for desired clearances and/or interferences.

The ability to animate sequences of events by using CAD is of great value to the designer who must evaluate moving mechanical systems, process performance, and even flight simulation and urban traffic control.

Engineering Analysis ▪ CAD has the ability to perform sophisticated engineering analysis by such powerful techniques as the finite element method and to present the results in comprehensive graphical form. Critical areas in machine components with high stress or strain concentrations, excessive heat fluxes, or excessive deformations can be shown in detail with a distinct color code. A designer can detect such critical areas in a single glance. All these functions could not be done with traditional design. Following are two examples relating to color coding.

The first example involves the use of color coding in the design of an injection molding process for plastic computer keyboards, as shown in Plates 7 to 10. Plate 7 shows the assembly of the mold and the workpiece in purple. Plate 8 shows an enlarged view of the molded keyboard and simulates the ejection of the finished product, with the ejection pins shown in yellow. The distributions of stress and temperature in the keyboard being molded are assessed by finite element analysis. The finite element meshes for the special triaquamesh model used in this analysis are described in Plate 9. Here, the stress variations are also shown by color coding, which is correlated to the stress level scale shown at the top of this plate.

Similarly, the temperature variation in the keyboard, which can be observed from the color variations, is shown in Plate 10.

Color coding can also be used in the stress and modal analysis of rocker arms in a valve train with optimal geometry and mass. A graphical illustration on such an application produced by the ANSYS program is shown in Plates 11 to 13. A detailed description of this case will be presented in Chapter 10. The solid model of the rocker arm is shown in Plate 11, with the finite element mesh illustrated in Plate 12. The variation of stresses as well as stress concentrations can be readily observed by the color variation over the solid model, as depicted in Plate 13. The magnitude of the stress varies from the highest level, shown in red, to the lowest level, shown in blue.

Design Optimization ▪ Many commercial CAD packages can perform various degrees of design optimization involving hundreds of design variables and constraints, such as weight, volume or space, material specifications, and costs. They can also optimize the methods of production, such as the selection of tools, tool materials, and tool paths and the scheduling and control of production. Such an optimization process cannot be handled nearly as effectively by the traditional method as it can be by CAD technology.

❑ BENEFITS OF CAD

The unique capabilities of CAD outlined above provide substantial benefits to industrial users.

1. CAD can enhance the quality of engineering design with excellent graphical representation of product geometries (Plates 1, 2, and 3) and production drawings.

2. It enables engineers to visually inspect the geometry of the product from every conceivable viewing angle. The need to build prototypes of the product, as required in many traditional design processes, has thus diminished. Geometric definition of the product can be thoroughly studied and modified by CAD.

3. Graphic simulation and animation of a CAD system can be used to inspect the tolerance and interference of matching components of a product (Plates 7 and 8). They can minimize errors in specifying such tolerances and interferences of fittings.

4. Ever-improving and increasingly available computing power has further extended the use of CAD for design problems involving hundreds of variables. Many CAD systems can provide engineers with optimum solutions for these complex problems.

5. CAD offers great flexibility and economy in the management of design data, which include all pertinent design information and product geometries. It provides engineers with the means to easily modify existing design files and production drawings and saves storage space.

6. Many CAD systems can be interfaced directly to a CAM system that uses intelligent machines to carry out production without human intervention. Such an interface can reduce human error and result in a highly efficient, highly accurate production system.

❑ CHARACTERISTICS OF CAD

There are many obvious differences in the ways in which CAD and the traditional method are used in engineering design. It is useful for engineers to recognize the unique characteristics of CAD in order to use this tool effectively.

The two principal characteristics of CAD are its dependence on software programs and its strong basis on computer algorithms. Following is a brief description of each.

CAD and Software ▪ There are generally two types of software programs: system software and application software. System software used for the internal data management and operations of the components is normally supplied by the hardware manufacturer. On the other hand, application software must either be purchased from a commercial outlet or be developed in house. Design engineers are expected to be users, but not developers, of software. In most cases, they can operate the CAD system, which involves both hardware and software, as a "black box." They do not have access to the contents of the software as they would otherwise have with the handbooks, the design codes, and the mathematical formulas, as in the case of traditional design. However, it is essential that they be knowledgeable about the principles of the application softwares used for the job so that they can prepare the input and interpret the output correctly in an engineering sense.

CAD and Computer Algorithms ▪ All computations performed by a computer are based on the discrete sets concept. In other words, all physical variables and functions, including "smooth" curves, can only be handled in discrete or incremental piecewise linear forms. Algorithms with iterative procedures for numerical approximations, such as those described in Chapters 2, 6, and 7, are thus standard practice. The same concept applies to the graphics algorithms. For example, a circle appearing on a monitor screen may well be a polygon composed of many minute straight-line segments. Code developers require adequate knowledge in adopting the most effective algorithm for the problem at hand, and users must have sound engineering common sense in selecting the right sizes of increments in both time and space domains.

Overall, it appears that the most fundamental difference between traditional design and the CAD process lies not in the design methodology, but in the implementation of the design method. Computerization of the design process can produce quick, accurate design analyses and impressive, easily modifiable interactive graphics displays, but the fundamental design principles remain intact. It is inconceivable that an engineering designer can command a CAD system effectively and produce good designs without adequate knowledge of the traditional engineering disciplines.

1.4 ▪ Essential Hardware Requirements for CAD

At one time, hardware requirements for CAD included a large mainframe host computer and sophisticated peripheral equipment, and the cost of such an installation would be millions of dollars. Thus, the luxury of having one could be enjoyed only by a handful of large companies in the aerospace and auto industries in which laborious, but highly sophisticated, engineering design and manufacturing processes are standard practice. Such companies could justify a large investment in highly efficient computer-assisted operations, but the exorbitant cost, of necessity, kept most of the small and medium-sized industries from benefiting from this advanced technology.

The situation, however, has dramatically changed in recent years as a result of rapid advances in microcomputer technology. More and more powerful microcomputers have been made available in the marketplace at steadily falling prices. The same trend has occurred in the prices of peripheral equipment. This drastic reduction in the cost of computer hardware with enhanced capabilities has made CAD affordable to many medium and small-sized industries. The advance of microcomputer technology is so spectacular that one computer expert suggested that a Rolls Royce would cost only $2.75 and run 3 million miles on a gallon of gasoline if the automobile industry had developed the way the computer industry has.

Detailed descriptions of CAD hardware components are neither necessary nor practical because most commercially available hardware components are constantly being replaced by newer and better products. The complete, state-of-the-art CAD hardware system, shown in Figure 1.2, consists of the central processing unit, memory disks, the keyboard, the digitizer tablet, the monitor, the plotter, and the printer. Following is a brief discussion of the working principles of these essential components.

❑ CENTRAL PROCESSING UNIT

The central processing unit (CPU) is the essential component of a digital computer. The two main subcomponents of most CPUs are a microprocessor and memory boards. The former controls all activities related to the

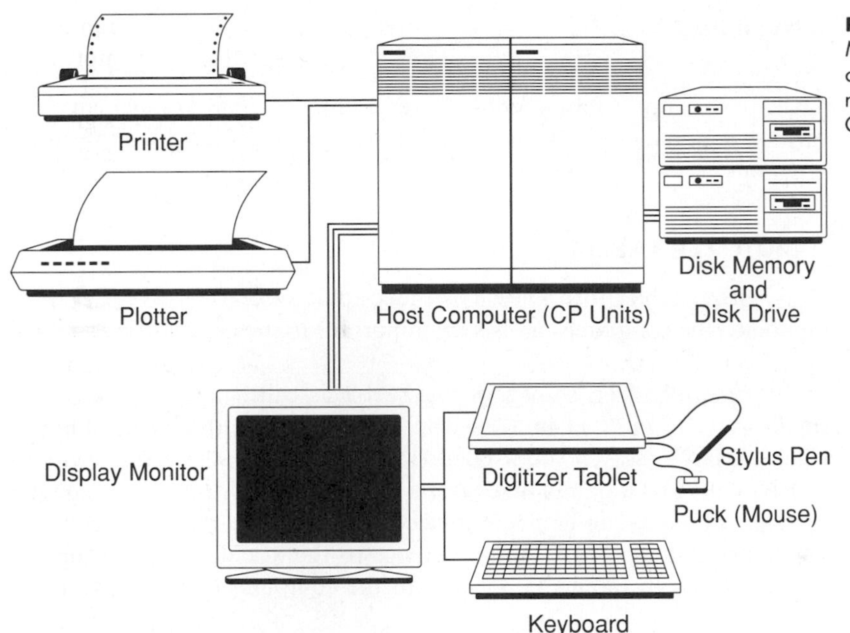

FIGURE 1.2
Major hardware components for a minicomputer-based CAD system

system operation and performs all arithmetic and logic functions. Most high-powered microcomputers are equipped with two types of memory functions, the read-only memory (ROM) and the random access memory (RAM). The main function of the ROM is to store the system software, as described in Section 1.3, whereas the RAM contains instructions for the particular job the user wants the computer to perform. These instructions are contained in the application software supplied by the user. In the early 1980s the size of the RAM of computers used for the CAD function varied from 512 K (Kilo-bytes) for many 16-bit microcomputers to 50 M (Mega-bytes) for 32-bit minicomputers. However, the amazing evolution of the very large systems integration (VLSI) and surface-mount technologies in recent years has made it possible for a lap-top computer weighing less than 20 pounds to have up to 20 M of RAM at a CPU speed of up to 35 MHz.

❑ MEMORY DISKS

Most application software programs require more memory storage than the RAM can provide. Thus, additional memory has to be provided. Magnetic tapes and disks are commonly used for this purpose. Most microcomputers use magnetic disks with capacities varying from 360 K for floppy disks to 30 M for hard disks, although 20- to 60-M hard disks are not uncommon. Hard disks with 450 M of capacity are common for 32-bit minicomputers.

❑ KEYBOARD

The keyboard is the principal input device for almost every type of computer. In addition to alphanumeric keys, function keys are available for various commands, such as the plot options included in application programs.

❑ DIGITIZER TABLET

The digitizer tablet, also known as the graphic tablet, is used for graphic input into the computer and is very important in the interactive graphical operation of CAD.

A typical digitizer, shown in Figure 1.2, is called a tablet when it is small—say, 11 in. × 11 in. (280 mm × 280 mm)—and a table or board when it is large—say, 40 in. × 60 in. (1 m × 1.52 m). A stylus, shaped like a pencil or pen, or a hand cursor (a puck or a mouse) moves over the flat surface. Digitizers can be made to accurately measure the position of the hand cursor or the stylus with reference to the x and y axes located on the tablet and to input the position to the computer directly for further processing.

Several types of digitizers are available commercially. Two of the most commonly used are the electromagnetic digitizer and the touch sensitive digitizer.

Electromagnetic Digitizers ▪ One popular variety of digitizers contains either a grid of fine wires embedded into the tablet surface or a printed board with a thin protective covering. To get a clear picture of a digitizer board, consider one that is 48 in. × 72 in. (1.22 m × 1.83 m) with a resolution of 0.05 in. (1270 microns). Such a board will contain 960 wires along its width and 1,440 wires along its length. The board may be provided with an electronic control unit that sends unique digitally coded signals down every wire in sequence. Inside the stylus is a sensitive amplifier that picks up the signal from the wires nearest to it in the x or y direction, amplifies it, and transmits it via coaxial cable to the decoding logic, which in turn finds the location of the stylus. The x and y coordinates of the position of the stylus are placed in the tablet buffer, which is then read by the computer.

An alternative technique for obtaining coordinates through electromagnetic means involves the use of a resistive plate. Digitizers based on this technique have a sheet of partially conductive material for the surface of the tablet. At regular time intervals, a potential is impressed on the sheet first in the x direction and next in the y direction. The stylus, kept lightly pressed against the sheet, senses potentials that are directly related to its distance from the reference axes.

Yet another device adopted to sense the position of the stylus electromagnetically replaces the wire grid with a moving coil underneath the surface of the digitizing tablet. The coil carriage is moved in the *x* and *y* directions by two separate motors. The stylus detects the field generated by the coil when it is directly over the coil and sends a signal to the digitizer control unit. The current position of the stylus is determined by the position of the coil when the signal is sent.

Touch-Sensitive Digitizers ▪ These contain two thin sheets arranged in layers with a small gap between them. One of the sheets has a thin coating of some conducting material while the other sheet has a resistive coating. If pressure is applied by a pencil or pen, the sheets come into contact with each other, thus closing an electric circuit. The voltage drop across the resistive substance in the *x* and *y* directions is then utilized to determine the position of the stylus or the point of contact.

A major advantage of a touch-sensitive digitizer is that it senses the position coordinates in an analog way; therefore, theoretically it has infinite resolution. Another advantage is that it has no stylus and no wire.

❑ MONITOR

The main function of the monitor is to display the user's input, as well as the computer's output, in either alphanumeric or graphic form. Although only limited information can be displayed on a monitor screen and this display is non permanent. The monitor is an essential piece of equipment for interactive graphic communication between the computer and the user. In many CAD processors, images are produced on the monitor screen for preliminary viewing. The keyboard or digitizer tablet is then used on a real-time basis to reformat or alter the display. Once the image has been perfected, it can be directed to a printer or a plotting device that produces a hard copy. For more sophisticated applicants, such as computer animation, display devices are required to produce and erase images several times a second.

In view of the similarity between the requirements of television broadcasting and those of interactive graphics, it is not surprising that the first display devices used were cathode ray tubes (CRTs), similar to TV monitors. CRTs are still the most prevalent display devices used in computer graphics today in spite of their unwieldy size and the high voltage required for their operation. In recent years, however, the marketplace has demanded thin, flat monitors that fit in portable (or lap-top) microcomputers. Two new types of image display devices have been developed: the liquid crystal display (LCD) and the gas-plasma/electro-luminescent display (ELD). Following are brief descriptions of the working principles of these various display devices.

Cathode-Ray Tube (CRT) Display ▪ The refresh cathode ray tube is the most commonly used type of monitor. It is basically a sealed tube of glass that is conical at the front and cylindrical at the back. The cylindrical portion of the tube houses an electron gun that emits a very thin beam of electron particles when energized. The electron beam is directed at a doubly curved surface of large radius, coated inside with phosphor. Phosphor particles glow when struck by the beam, and an image is produced from a large number of glowing particles.

A deflector system is provided at the junction of the cylindrical and conical portions of the CRT. Its function is to direct the electron beam to the appropriate portion of the screen where the image is to be produced. The deflector system works with magnetic or electrostatic fields, as will be described later.

There are two distinct methods of creating images: the stroke-writing method and the raster scan method. In the stroke-writing method, vectors or lines are drawn one at a time. These vectors may have any orientation or position on the screen. In the raster scan method, horizontal lines are scanned across the display screen in a sequential manner. The scan lines contain numerous closely packed glowing dots. Vector images are built by dots belonging to several consecutive scan lines.

A CRT may be either the refresh type, as described above, or the storage (condensing) type. In the storage tubes, an image can be retained indefinitely. In the refresh CRTs, on the other hand, the glow of the phosphor particles starts to fall off immediately; therefore, the image cannot last for more than a fraction of a second. In order to retain the image and avoid flickers, it is recreated (refreshed) frequently—say, 30 to 60 times per second.

Having just discussed the working principles of the image display devices, it is now appropriate to briefly describe two principal methods for creating images on the monitor screen: the vector display and the raster display.

Vector display on a refresh CRT. The vector or random display system contains a display processor or controller and a display memory buffer. A display program or list generated by the host computer is stored in the buffer. The display processor reads the information contained within the display buffer and produces analog voltages to drive the deflection system of the refresh CRT from the digital coordinate data provided. The display program contains point, line, and character generation commands in addition to the coordinate data.

For flicker-free images, a refresh rate of 30 Hz is desirable (i.e., the information contained in the buffer memory should be cycled through at least 30 times every second). Considering the fact that the length of a typical display program for an engineering drawing may be in excess of 3 K (Kilo-bytes), it is clear that a very fast display processor is required. The combination of a large buffer memory and a fast processor make a vector display system on a refresh CRT an expensive proposition.

Raster display on a refresh CRT. A raster screen contains a huge number of picture elements (pixels) or points. An average-quality raster screen may have more than a quarter of a million pixels (say, 512 lines in both directions with 512 pixels each). Images are created by lighting some of these pixels and leaving the rest unlit or by lighting some in one color and others in other colors. It is necessary, however, to supply information about the condition (on/off) or the color of each of these pixels. The list containing this information is known as a *bit map*.

Bit maps for most graphical displays are very large for obvious reasons. They require a large buffer memory (250 K for a 512 × 512 black-and-white screen) and fast processors. Fortunately, given the present state of solid-state electronics, neither the memory nor the processor requirements are difficult to meet. The simple working principle for raster displays, as described above, has made them popular for graphic terminals. However, the lengthy time required to store and convert the large amount of information in bit maps to the corresponding displays on the monitor screen has limited its applications to animated displays in real time.

Color images can be produced on CRT monitor screens with specially arranged pixels. By varying the combinations of intensities of the emission from each pixel at a specific optical spectrum (e.g., in the three basic colors of red, blue, and green), a variety of colors can be produced. The control of such an emission is carried out by the computer memory bit planes built in for each pixel. A *bit plane* is an assembly of bit maps. Each bit map contains binary information on either the on or off status of the pixel, or the two present levels of emission when the pixel is turned on. Thus, a bit plane comprised of two bit maps can provide $2^2 = 4$ levels of emission by the pixel. The number of bit planes can vary from 2 for some IBM-PCs, to 8 for some more sophisticated microcomputers, to 24 for advanced graphics work stations. The total number of available colors, in theory, is 2^n, where n is the number of bit planes. Thus, for high-level work stations, the user would have 2^{24}, or 16,777,216, different colors at his or her disposal. In reality, however, the total number of colors available to users is limited by the color definition and the speed or bandwidth of the display system. A total of 4,096 displayable colors appears to be common.

Liquid Crystal Display (LCD) ▪ Liquid crystals (LCs) are materials whose optical properties can be controlled by an applied electric field. The technology of using liquid crystals for the display of electronic images is more than 20 years old. However, the use of this technology was seriously hampered by its many shortcomings, such as slow response speed (fractional seconds for small panels or seconds for larger panels), narrow temperature range, limited viewing angles, and degree of multiplexibility. The use of LCDs was primarily limited to such simple electronic gadgets as hand-held calculators, watches, and telephone switches.

Despite the many shortcomings mentioned above, the LCD has two principal advantages over other known image display devices: (1) it requires very low power (e.g., a few watts) for the display, and (2) it can be used in the shape of thin layers of LC cells, with images displayed on a flat panel. These are the principal reasons for the use of LCDs as monitor devices for small, portable lap-top computers.

The working principle of LCDs is similar to that of the raster display on a refresh CRT, as described earlier. Here the LCD cells replace the pixels. These LCD cells are installed in the matrix form with a number of horizontal and vertical electrodes that apply the necessary driving voltage through the crystal cells. The change of optical properties of the electrified crystal cells will control the amount of light to be passed through these cells. Visual images can thus be created on the screen and viewed by the operator.

Most LC materials used in display devices are classified as nematic LCs. Readability of the LCD can be improved by the twisted nematic (TN) technology in which double layers of LCs are used. The molecule orientations of the crystals in each layer differ by 90°, thus enhancing their ability to efficiently block and transmit light. The viewing angle for a TN LCD monitor is limited to 10°–15° from the normal of the screen. Better readability can be achieved by using the double-twist nematic (DTN) technique in which an extra LCD layer is added between the two existing LC layers. Two polarizers are attached to these sandwiched LC layers, one at the front and the other at the rear. The compensating layer is made optically active. Its primary function is to reorient all light bands to the same polarity to get brighter bright and darker dark. The DTN monitor thus can provide sharp black images on a white background. The super twist nematic (STN) technique was developed to improve both readability and viewing angles. The twisting angles of the LC molecules in this case are from 180° to 270°, with higher tilt angles of the molecules at the alignment layers. The STN LCD can operate in three basic modes: yellow mode, blue mode, and neutral grey mode. One major advantage of STN is that its background color is sensitive only to temperatures very near the −20°C and 70°C extremes.

Most LCD monitors operate in a temperature range of −30°C to 85°C, with an achievable resolution of 960 × 240 or 640 × 480 dots. With rapid advances in the LC layer structures and materials, and strong demand in the marketplace for flat-panel monitors, it is entirely conceivable that high-resolution, durable-color LC monitors will soon replace the bulky CRT monitors in both the computer and the television industries.

Gas-Plasma/Electro-Luminescent Display (ELD) ▪ ELD is another flat-panel image display device that is gaining popularity in the portable microcomputer industry. The working principle of this device involves the light emission from gas-discharge elements arranged in a matrix. It works much like a fluorescent or neon light in which light at certain wavelengths (white in the case of fluorescent light and red or another color in the case of neon light) is emitted by ionizing gases entrapped between cathodes and anodes with the passage of a current at high voltage. In the case of a flat panel used for image display, the gas is trapped in a thin film between two thin glass plates. Embedded in one plate is a set of vertical conductors used as anodes, and another set of horizontal conductors is embedded in the second glass plate as cathodes. Emission of monochromatic light takes place in gas at the cross-point, which is electrically activated—a phenomenon similar to a lighted pixel in a raster display on a refresh CRT. Two parallel developments have taken place in ELD technology: the development of AC-driven panels, mainly for computer monitors, and that of DC-driven panels, directed at color displays for television receivers. The former type of panel is used in computers because it sustains images as long as the ignition pulse lasts at the cross-point.

Like the LCD, the ELD suffers from slow response time. Another major disadvantage is its requirement for high electric voltage to ionize the gas between the electrodes. This excludes the use of batteries with this type of display panel—thus eliminating a convenient feature that most portable computers provide.

❑ PLOTTER AND PRINTER

These devices are used to produce permanent records of the CAD output. Various types of plotters are available. Smaller configurations, typically A or B size drawings, can usually be handled by a flat-bed-type plotter, whereas larger drawings in the C, D, or E sizes are generally produced by a drum-type plotter, shown in Figure 1.2.

Different quality printouts are available from various types of printers. A dot matrix printer produces lower-quality characters, but at a higher speed. Letter-quality printout is commercially available on low-speed printers. Dot matrix printers are being rapidly replaced by newer and superior laser printers. The working principle of laser printing is similar to that of electrostatic particle deposition, but with a computer-controlled laser beam doing the scanning. A resolution of 300 dots per inch is achievable by this printing technique.

1.5 ▪ Interactive Computer Graphics

The term *interactive computer graphics* refers to the real-time or "instantaneous" communication with the computer by the designer or operator. This form of communication offers the operator much more freedom over the traditional input methods that use the card deck and the keyboard. An effective interactive computer graphic system allows sketching and drawing to be both input to and output from the computer. A fully interactive graphics system has almost instant feedback from the graphic input by the designer.

As described in Section 1.3, interactive graphics is a very important part of CAD. It enhances the designer's visual appreciation of the component's geometry. It also permits the designer to identify obvious errors and thus prompts immediate action by the designer to rectify these errors by adjusting proper design parameters.

Three general types of interactive graphics descriptions are used in computer-aided design and analysis.

❑ GEOMETRIC MODELING

The shape of a component at the initial (or conceptual) design stage can be constructed graphically on a monitor screen of the CAD system, as shown in Plates 1 to 4. Detailed descriptions of this practice will be given in Chapters 3 and 4. The construction of such geometry on the monitor screen usually can be accomplished either by a light pen or cursor on a digitizer tablet or by a mouse. Once the configuration of the initial geometry of the component has been finalized by visual inspection from various viewing angles, it may be converted into proper models for design analysis (e.g., in finite element meshes), which may then be stored in a common database for the subsequent design analysis.

Two types of geometric models are available for visual inspection from the monitor screen: the wire frame model and the solid model, illustrated in Figure 1.3 (a) and (b), respectively.

After having satisfied all the specified design criteria, the final shape of the component can be stored or retrieved from the common database in a CAD system, as will be described in Chapter 5. This geometric model can then be used as input to an automated drafting system to produce engineering drawings. In many advanced systems in which CAD is interfaced

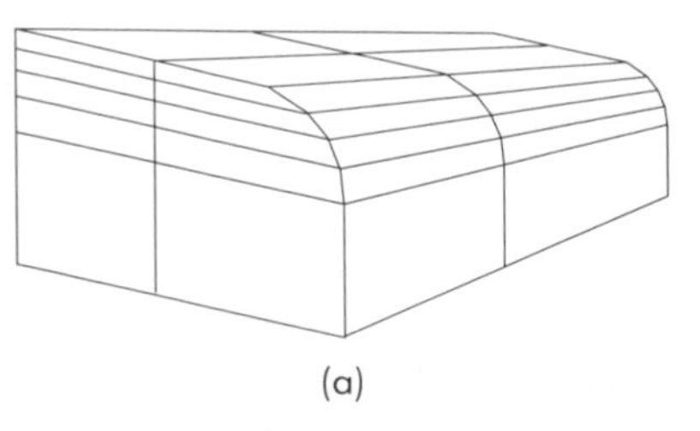
(a)

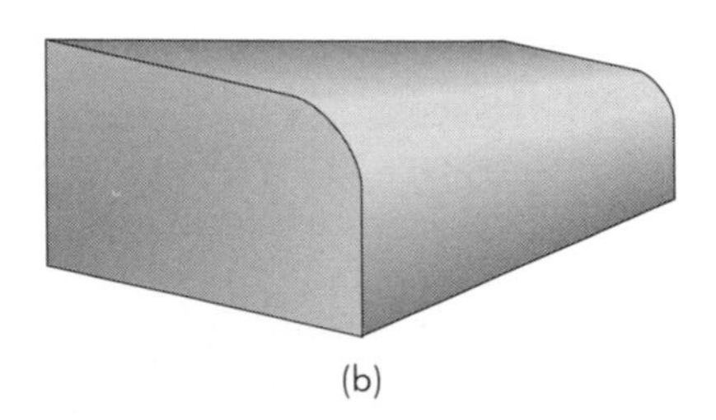
(b)

FIGURE 1.3 Simple illustration of geometric models: (a) wire frame with hidden lines removed; (b) solid geometry

with CAM, the geometric model will be used to produce numerically controlled (NC) tapes or to command computer-numerically-controlled (CNC) machines for production of the component. The logic used for the interface will be elaborated on in Chapter 10.

❑ GRAPHICAL REPRESENTATION OF ANALYTICAL RESULTS

Results from either traditional design formulas or computer-assisted methods (e.g., the finite element and the finite difference methods) are

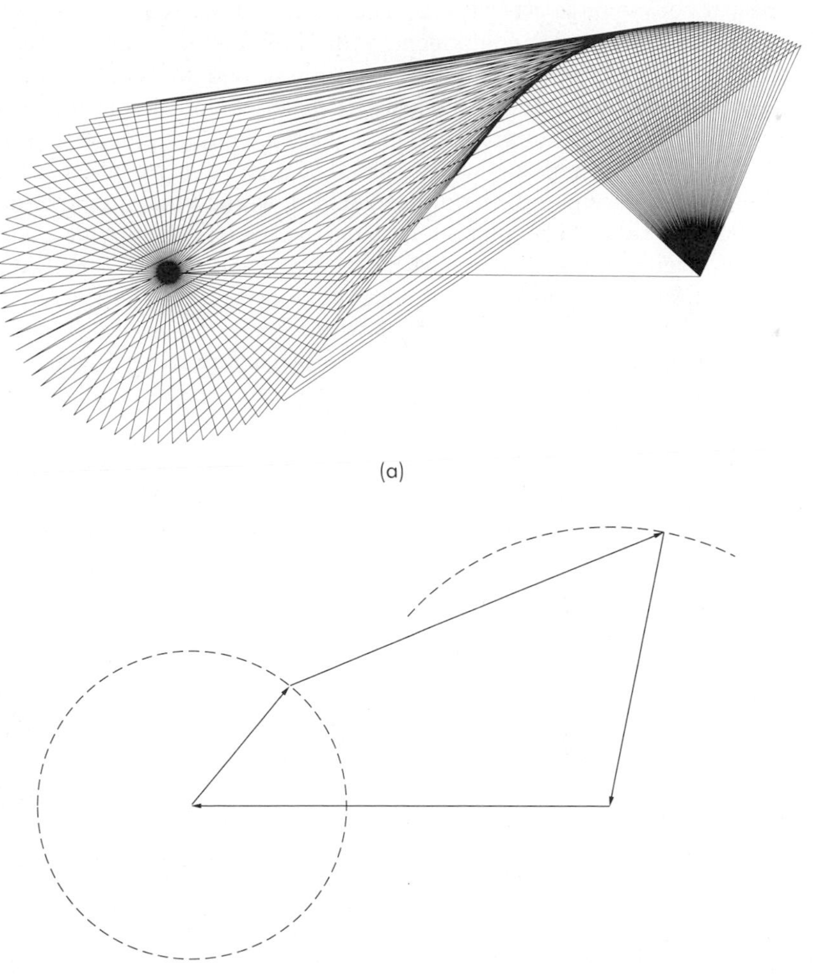

FIGURE 1.4
Animation of a four-bar linkage: (a) accumulated images; (b) trace of the follower

usually presented in the form of large arrays of numbers. Analyzing these numbers for the physical interpretation of the results is a time-consuming task. The graphical capability of a CAD system can convert these numeric results into (1) functional form, (2) contours (e.g., isothermal lines or stress contours, as shown in Plate 16), or (3) visible patterns illustrating design parameters in two- or three-dimensional geometries. Graphical expressions of areas with critical stresses or temperatures by means of color codes are depicted in Plates 9, 10, 13, and 16.

❑ KINEMATICS AND ANIMATION

Many CAD systems have kinematic features for depicting moving components in an animated manner. This feature is very useful in comparing the actual path of moving parts with the set designed path and in checking the actual passage of moving components with small tolerances in the traces. Animation is extensively used in linkage synthesis and analysis. The computer animation of a four-bar linkage system shown in Figure 1.4 can be produced by the program presented in "Kinematic Simulations of Planar Mechanisms" [3]. Animation has proved to be useful in many other applications, such as designing conduits for fluid flow and studying large deformations of solids. The animation of the motion of a front-end loader shown in Figure 1.5 was produced by the same program [3], allowing optimum design of the linkages through visual inspection. The animation of a truck suspension system is also shown in Plates 5 and 6. Graphical animation, if done properly, can replace expensive prototype construction of the component, still a standard process in traditonal design-manufacturing practice.

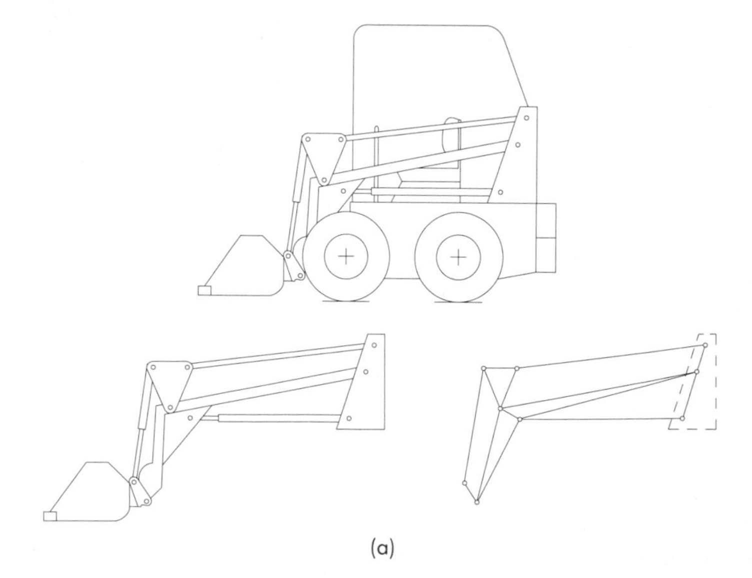

FIGURE 1.5
Animation of front-end loader movements: (a) typical mechanism—front-end loader; (b) trace paths; (c) accumulated images

FIGURE 1.5
(*Continued*)

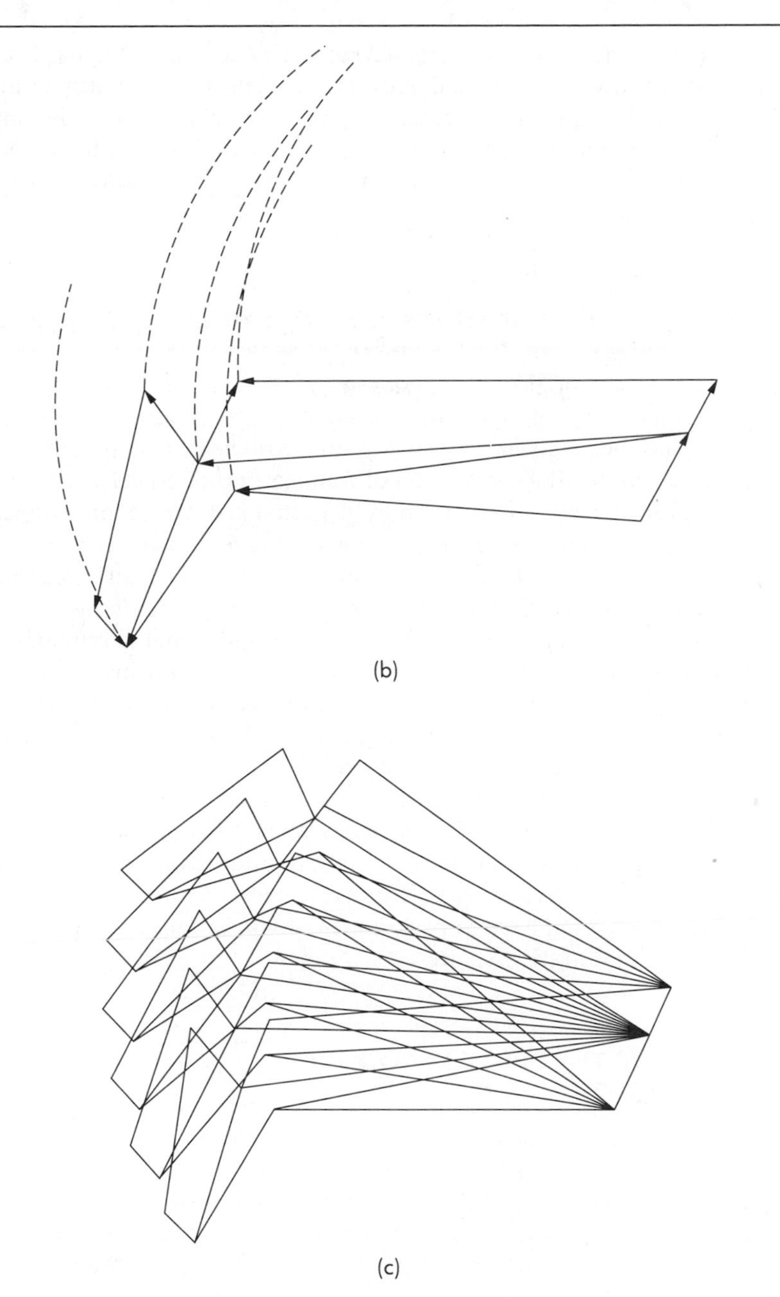

(b)

(c)

Kinematics and animation require sophisticated computer hardware and graphic software and could only be performed on mainframe and midicomputer systems until recently. However, recent advances in computer technology allow such animation to be carried out on microcomputers. The results shown in Figures 1.4 and 1.5 were produced by this class of computers.

1.6 ▪ General Procedure for Engineering Design

The interpretation of the term *design* may vary from person to person despite the fact that the design process is as old as civilization itself. A simple definition has been given by Shigley and Mitchell: "To design is to formulate a plan for the satisfaction of a human need" [4]. The scope of design can range from design of a purely mechanical nature for products, structures, devices, etc., to mechanical engineering design involving several disciplines of mechanical engineering, such as solid and fluid mechanics, heat transfer, and metallurgy. The scope of CAD, in the authors' view, covers an even wider scope, including such additional disciplines as production methods and control and optimization, as illustrated in Figure 1.1. However, in this textbook, we shall focus our attention only on the mechanical engineering design aspect of CAD and use the terminology to mean simply engineering design.

There are generally five distinct stages in engineering design: setting up the product specification, conducting a design synthesis, establishing conceptual geometry, performing design analysis, and finalizing the product geometry. These stages are by no means sequential in most design processes. Designers often have to perform lengthy iterations among various functions involved in these stages.

❑ SETTING UP THE PROPER PRODUCT SPECIFICATION

A product specification defines the nature of a single product or of a system to be produced. Obviously, it has to begin with the identification and recognition of some need either by self-motivation or through a thorough market survey in the public domain. Often the need for a particular product or system exists as a part of another larger system. For example, many military hardware and equipment needs originate from the government's overall defense strategy plans.

Depending on the nature of the product, a specification can be as brief as a few lines of description for a simple machine component or as lengthy as several hundred volumes for production and installation of major military hardware. Regardless of length, a typical specification would include such items as the intended applications of the design product, loading conditions (thermal, mechanical, wind, seismic, etc.), materials for essential components, space availability, environmental constraints, life expectancy, cost limits, quality assurance, and other relevant conditions. In some cases, methods of fabrication and product inspection are also specified. A good specification has to be both concise and comprehensive and requires that the writer possess sound technical knowledge and good engineering common sense as well as a high level of technical writing skill.

A product specification that leaves out essential technical conditions and constraints or that is open to misinterpretation not only will cause serious problems for the designer, but also can be potentially disastrous to the manufacturer and even to the public at large. It is not surprising that specification writing has become a subdiscipline within engineering.

❑ CONDUCTING A THOROUGH DESIGN SYNTHESIS

Once the product requirements are fully defined, the designer should immediately start the process of synthesis and optimization. According to Pahl and Beitz [5], *synthesis* is the putting together of parts or elements to produce new effects and to demonstrate that these effects create an overall order—the product. The purposes of this process are thus to conduct a detailed logical search and analysis of all the conditions and constraints described in the product specification and to derive all possible solutions and other considerations that will satisfy the specifications. An optimized, but not necessarily the best solution, as well as a number of alternative approaches, may emerge from this process.

Because a great variety of design problems has to be dealt with in the real world, it is not possible to present the reader with a fixed synthesis procedure. The following description can be considered only a general guideline for the designer:

Step 1 *Itemize the major expected performance capabilities of the design product.* This step encompasses listing, quantifying, and prioritizing the major requirements of the design product as described in the specification. Every attempt should be made to satisfy these requirements. For instance, major performance requirements for a mobile fork-lift to be used in a warehouse can be itemized as the design for maximum capacity of the lift in terms of the weight and volume of the articles to be lifted, the required maximum height to which an article must be lifted, the maximum and minimum speeds of the vehicle, the maneuverability in terms of the minimum steering angle, and the overall size of the vehicle and the fork-lift. Other requirements such as cost, life expectancy, and frequency of maintenance should also be considered.

Step 2 *Itemize the major design constraints.* Depending on the nature of the design product, there may be a great many design constraints that, in a real sense, tell the designer what not to do. Again, most of these constraints can be found in the specification. Some constraints are fairly explicit and can be translated into technical terms, whereas many others are nontechnical. Translating these constraints into proper technical terms and incorporating them into the design process may require significant industrial experience.

In the case of the fork-lift design, constraints such as the speed limit of the car, the space for maneuvering the vehicle, fuel cost, and exhaust pollution control are not uncommon.

Step 3 *Set the design criteria.* A designer can usually identify the relevant design criteria for the particular product by reviewing the special requirements described in the specification as well as by summarizing the assessments made in steps 1 and 2. However, it is important that those requirements be translated into technical terms and that they be met based on a set of analyses to be described later. A common design criterion is the failure criterion used in a strength analysis for a structure where certain stress components or combined stresses are to be kept below specified limits.

Step 4 *Itemize the major elements in the design consideration.* Having identified the critical requirements and constraints, the designer now can begin to put together a package of all major design considerations relevant to the product and the options that are available. It is also useful to outline all the advantages and disadvantages associated with each of these options. A logical optimum solution may then be derived from such an outline.

Table 1.1 illustrates a sample of design synthesis for a fork-lift design. The reader is reminded that the rationale presented in this example is by no means absolute. Uniformity of synthesis analyses carried out for a product by different designers is highly unlikely due to the fact that such an analysis is conducted largely on the basis of the designer's own professional judgment.

Other design considerations to be included in the synthesis process are the analytical methods (e.g., design handbooks, mathematical formulations, numerical analysis), the fabrication and assembling methods, and the human factor, such as workers' comfort and safety.

❑ ESTABLISHING CONCEPTUAL GEOMETRY

Once the product requirements are thoroughly defined and design synthesis is performed, the designer can then begin sketching a tentative

TABLE 1.1 Design synthesis analysis of a fork-lift

Major Design Considerations	Options	Main Advantages	Main Disadvantages	Decisions
Geometry of the fork	(1) 2 flat plates	Secure positioning and better balance for the payload. Requires small clearance to insert the fork.	Tapered cross-sections are required for better strength. Special tools are required to produce these forks.	Stability of payload is important for safe operation. The flat fork will provide much more stable positioning of the payload during transportation and lifting because wooden crates with flat bottoms are normally used to hold loads. Floor clearance, though, is of minor importance: it is desirable that it be kept to a minimum. Hence, the flat geometry is chosen.
	(2) 2 round pipes	Readily available and less expensive. Very easy to produce.	Payload may roll sideways. Requires a floor clearance equal to or more than the diameter of the pipe to insert the fork.	
Material	(1) Steel	High unit strength and durability. Common material, less expensive.	Heavy. Corrosive.	Steel has the higher unit strength; hence, less material is required. Aluminum alloy is lighter, but requires more to lift the same payload. No obvious solution on the weight advantage is available. However, the resistance to wear is important for warehouse application. Steel is chosen because of its high wear resistance.
	(2) Aluminum alloy	Light, so saves power.	Too soft so becomes worn easier. Lower unit strength may require bulkier forks.	
Power source for the lift	(1) Mechanical	Relatively simple.	Requires complex system to achieve high mechanical efficiency. Very slow lift speed. System is too heavy to be economical.	The hydraulic lift scheme is chosen for its high lift capacity and relatively simple control system. Pumping of the hydraulic cylinder may be supplied by the engine power of the vehicle.
	(2) Electrical	Simple. Light.	Limited lifting capacity. Power source will be a problem.	
	(3) Pneumatic	Relatively simple. Heavier.	Requires the vehicle to carry an air compressor and air cylinder, adding too much extra weight.	
	(4) Hydraulic	Has lighest lifting capacity of all options.	Requires the vehicle to carry a hydraulic pump and control system. Heavy in weight.	
	(5) Combination of the above.			

(*continues*)

TABLE 1.1 (*Continued*)

Major Design Considerations	Options	Main Advantages	Main Disadvantages	Decisions
Power source for the vehicle	(1) Gasoline	Higher engine performance. Fuel economy.	Emission causes pollution of air in warehouse.	Option (1) with tight emission control and option (2) are suitable for heavy-duty applications.
	(2) Propane gas	Lower engine performance. Clean emission.	Requires frequent change of propane gas cylinder.	
	(3) Electric	Cleanest form of power. Lightest engine weight and volume.	Requires the vehicle to carry batteries and undergo frequent recharging. Limited power.	

geometry with dimensions for the product. Sketches of the product at this preliminary design stage are usually penciled freehand, but with the convenient and readily available interactive graphic capability of CAD systems, these tentative geometries can be created on the monitor screen, and necessary modifications can be made with great ease at any time. An experienced designer will select stock materials as much as possible. For example, a designer will select pipes from available commercial products made with even-numbered outside diameters. The same may apply to the selections of beam cross-sections, gears, bearings, and many other machine elements that have been standardized in the marketplace. In many cases, the tentative geometry of the design product will probably be a combination of standard products and those of the designer's own creation.

PERFORMING APPROPRIATE DESIGN ANALYSIS

This stage is the most important part in the whole design process. There are two major elements involved in this stage: (1) engineering analysis and (2) interactive graphics.

Two common types of analysis (i.e., mechanical or strength analysis and performance analysis) are first performed on the conceptual (or initial) geometry of the component to ensure not only that the product will satisfy the specified conditions and criteria, but also that it is structurally sound. Analytical results can be derived from conventional design codes, formulas, mathematical expressions, and, of course, such numerical methods as the finite element and the finite difference methods. Interactive graphical facility will provide the designer with visual appreciation of the input data and the analytical results at various stages of the analysis, as well as in the graphical interpretation of the results. The animation of a moving component, as shown in Plates 5 and 6, and the color-coded stress and temperature variations in structures, as shown in Plates 9, 10, 13, and 16, are typical examples of the graphical interpretation of results by CAD systems.

Because there is no such thing as the "correct answer" to any design problem [4], what the designer may develop at this stage is just one of many possible solutions based on the adopted options, as illustrated in Table 1.1. It is important for a designer to be aware of alternative solutions based on other conditions that may appear to be equally relevant. In this case, an optimized solution will have to be sought by following an optimization procedure. Engineering design optimization is a special subject of its own and cannot be adequately covered in this text. However, an introduction to this topic is presented in Chapter 8. A comprehensive description of this technique is also available in *Optimization Methods for Engineering Design* [6] and in other sources cited in Chapter 8.

Another major task involved in this step is the determination of design factors (or allowances) for various aspects in the design process. One such factor is the frequently used safety factor in the mechanical or strength analysis. Like all other factors, the proper numerical value of the safety factor is assigned on the basis of the design conditions and confidence level established in the analysis. A safety factor of 4 is commonly used for the unfired pressure vessels, as specified in the *ASME Pressure Vessel Code* [7], whereas this factor becomes a variable corresponding to the stress classification for pressure vessels used in the nuclear industry [8]. It is a well-known fact that a safety factor as low as 1.25 is used in the design of some aircraft components. The low safety factor is allowed because of the highly sophisticated stress analysis, stringent material specifications, and frequent thorough inspections implemented in the aerospace industry.

❑ FINALIZING GEOMETRY

Once all the design criteria have been met through proper analyses, minor adjustments of the geometry will be made to comply with such other design requirements as space and allowances for environmental effects (corrosion, tolerances for machining, etc.). For problems involving a large number of independent variables and parameters, a separate optimization processs would be necessary, as illustrated in Figure 1.6. After such modifications and optimization, the geometry is finalized, and the formal design procedures have been completed. The final geometry is now ready to be transferred to engineering drawings used for production.

For most practical cases, the above procedure involves multiple iterations and refinements, as illustrated in Figure 1.6.

FIGURE 1.6 General design procedure*

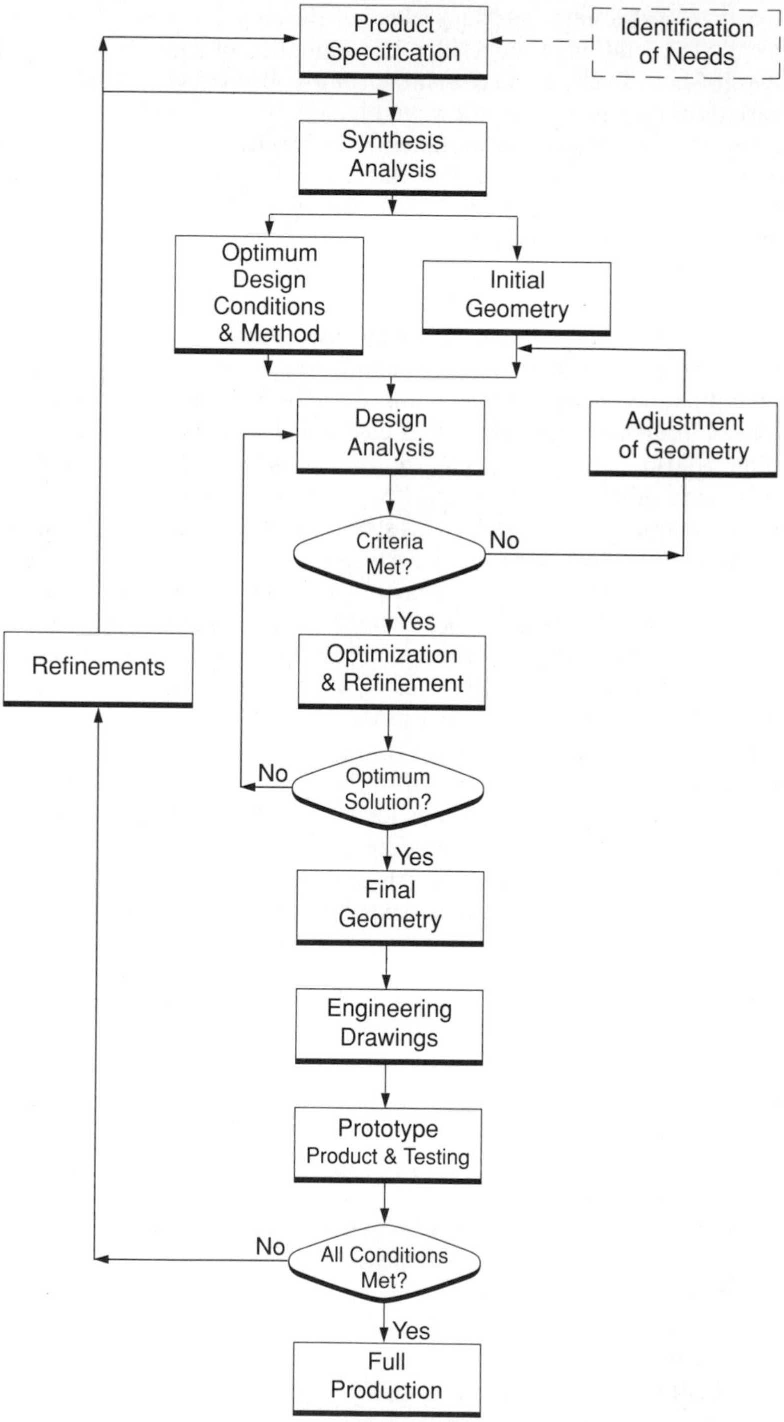

Product Specification
Identification of Needs
Synthesis Analysis
Optimum Design Conditions & Method
Initial Geometry
Design Analysis
Adjustment of Geometry
Criteria Met?
No
Yes
Optimization & Refinement
Refinements
No
Optimum Solution?
Yes
Final Geometry
Engineering Drawings
Prototype Product & Testing
No
All Conditions Met?
Yes
Full Production

1.7 ▪ Engineering Analysis

As described in the foregoing section, the function of engineering analysis in the design process is to provide a reliable basis on which to judge the suitability of the initial (or conceptual) geometry of the component with respect to the preset design conditions and criteria. Established design codes (e.g., [7] and [8]) and empirical charts and curves are frequently used for established products. For components that have no established design code to be followed and that involve complex geometries under complicated conditions, sophisticated mathematical models have to be derived and solved. These models may involve relatively simple formulations (e.g., simple beam theory for some beam structures or the theory of elasticity for more refined analyses).

The finite difference technique has been widely used in the past for problems that can be identified with certain forms of differential equations. In recent years, however, the finite element method, which will be described in Chapters 6 and 7, has become a commonly used technique for many complicated practical problems.

In addition to the various methods available for the engineering analysis, interactive graphical representation, as described in Section 1.5, has made CAD a great deal more effective as well as efficient in terms of saving the designer's time and effort. Thus, interactive graphics has become a very important element in the CAD operation.

Most engineering analyses involve four distinct stages: identification of physical conditions, mathematical modeling, model analysis, and interpretation of results.

❑ IDENTIFICATION OF PHYSICAL CONDITIONS

In a real sense, the main effort here is to identify particular conditions from the product specification that require analytical solutions. A thorough understanding of such conditions as loading and boundary conditions is particularly important.

❑ MATHEMATICAL MODELING

Once the problem is fully defined, the search begins for appropriate analytical tools and solutions. The designer will soon find out that the chance of identifying an analytical method that will satisfy all the specified physical conditions is extremely slim. It is therefore very important that engineers know how to idealize physical situations so that the problem can be handled by existing mathematical tools that are known to him or her.

The most viable approach may be to follow a synthesis analysis in choosing from among all possible analytical tools. Another major effort in formulating the mathematical model is to idealize complex physical conditions to a manageable level. Idealizations can be made in several areas—for example, in the geometries of the components and in the loading and boundary conditions. It should be emphasized, however, that intelligent idealizations can be made only by persons with sufficient mathematical ability and full awareness of the physical situations that are involved in the analyses.

❑ MODEL ANALYSIS

Work at this stage mainly involves the formulation of suitable mathematical models. One such modeling effort can be deriving equations (e.g., algebraic, differential, or integral equations) based on the assumptions made in the last stage of the analysis, as well as translating the physical loading and boundary conditions into the mathematical expressions. Solutions to these equations may be obtained by various techniques described in classical mathematical textbooks, (e.g., [9, 10, and 11]) or by numerical techniques such as the finite difference method [12, 13, and 14]. In recent years, finite element methods based on the variational principle and the Galerkin residual method [15, 16, and 17] have become an effective and versatile solution technique to many complex problems. Detailed description of the principle and relevant applications of this technique will be given in Chapters 6 and 7.

Whatever method the designer uses in the analysis, effort should be made at the end to evaluate its validity and reliability. It is apparent that little doubt would be raised by an analysis of a simple beam structure with clearly defined boundary or end conditions. On the other hand, for a complicated analysis such as the case described in Plate 1, validation of the analytical results may become a very important step in the analysis.

❑ INTERPRETATION OF RESULTS

Results obtained from the model analysis usually are in the form of numbers and do not provide direct answers to the physical problems. It is up to the engineers to interpret these numbers in light of engineering knowledge. Numerical values representing physical quantities, such as stresses, strains, velocities, and temperatures, should be translated into appropriate forms that can be related to the design criteria identified earlier in the design synthesis described in Section 1.6. In many cases, the results obtained from stages 3 and 4 are fed back to stage 2 to optimize the analysis.

1.8 ▪ Summary

The word *CAD* has become one of the most frequently mentioned technical terms in recent years. Despite its popularity, the interpretation of its real meaning as well as its relationship with traditional design is far from unanimous. This chapter provides the reader with an overview of this subject and its relationship with the computer-integrated manufacturing system.

Although CAD appears to be a new technology to many people, its real essence, in the authors' view, does not deviate significantly from traditional design philosophy and methodology. CAD gives designers much better and more effective tools with which to produce better-quality design work.

Despite the close similarity between CAD and the traditional design method, the reader should not underestimate the great value of the CAD in design technology. No one can dispute the fact that CAD, even at its present state-of-the-art, can provide designers with many unique capabilities, as described in Section 1.3, which would never have been thought possible with the traditional method. With these additional capabilities, designers can handle projects that involve a great many more variables and constraints as well as complexities and varieties in terms of product geometry and design conditions. It is a new challenge to the engineering profession, and success in facing this challenge lies in how effectively engineers can use this new tool.

High-level engineering analysis and interactive graphics are the two major ingredients of most CAD systems. In addition to the obvious requirements for the high quality of these two elements and the hardware facilities, the effective use of CAD systems also depends on the user's appreciation and knowledge of these elements. The remainder of the book will illustrate basic principles of these elements.

References

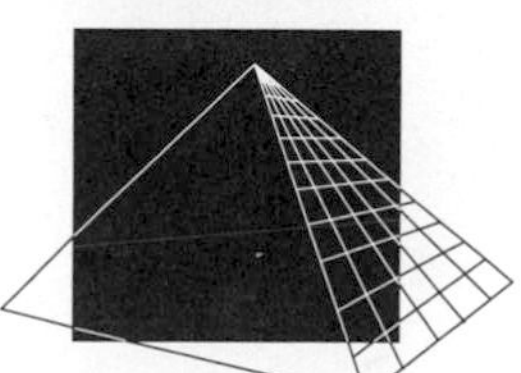

1. Sadowski, R.P. "Computer-Integrated Manufacturing Series Will Apply Systems Approach to Factory of Future." *Industrial Engineering*, Vol. 16, no. 1 (January 1984): 35–40.
2. Allan, III, J.J., ed. *A Survey of CAD CAM Systems.* 3d ed. Dallas: Productivity International Inc., 1983.
3. Han, P.S., R. Tsuyuki, and G.Y. Tu. "Kinematic Simulations of Planar Mechanisms." ISMM Institute Conference on Computer Applications in Design, Simulation and Analysis, New Orleans, March 1990.
4. Shigley, J.E., and LD. Mitchell. *Mechanical Engineering Design.* 4th ed. New York: McGraw-Hill, 1983.

5. Pahl, G., and W. Beitz. *Engineering Design, a Systematic Approach.* Translated into English by A. Pomerans and K. Wallace. London: The Design Council, 1988. Chapter 2.
6. Fox, R.L. *Optimization Methods for Engineering Design.* Reading, Mass.: Addison-Wesley, 1971.
7. "Pressure Vessels." *ASME Boiler and Pressure Vessel Code.* Section VIII—Div. 1: Rules for construction of pressure vessels; Div. 2: Alternative rules. New York: ASME, 1977.
8. "Nuclear Power Plant Components." *ASME Boiler and Pressure Vessel Code.* Section III. New York: ASME, 1977.
9. Sokolnikoff, I.S., and R.M. Redheffer. *Mathematics of Physics and Modern Engineering*, New York: McGraw-Hill, 1958.
10. Lebedev, N.N. *Special Functions and Their Applications.* Translated by R.A. Silverman. Englewood Cliffs, N.J.: Prentice-Hall, 1965.
11. Potter, M.C. *Mathematical Methods in the Physical Sciences.* Englewood Cliffs, N.J.: Prentice-Hall, 1978.
12. Ames, W.F. *Nonlinear Partial Differential Equations in Engineering.* New York: Academic Press, 1965.
13. Hildebrand, F.B. *Finite-Difference Equations and Simulations.* Englewood Cliffs, N.J.: Prentice-Hall, 1968.
14. James, M.L., G.M. Smith, and J.C. Wolford. *Applied Numerical Methods for Digital Computation.* New York: Harper & Row, 1977.
15. Desai, C.S. *Elementary Finite Element Method.* Englewood Cliffs, N.J.: Prentice-Hall, 1979.
16. Zienkiewicz, O.C. *The Finite Element Method in Engineering Science.* New York: McGraw-Hill, 1971.
17. Hsu, T.R. *The Finite Element Method in Thermomechanics.* London, Boston, Mass.: George Allen & Unwin, 1986.

Problems

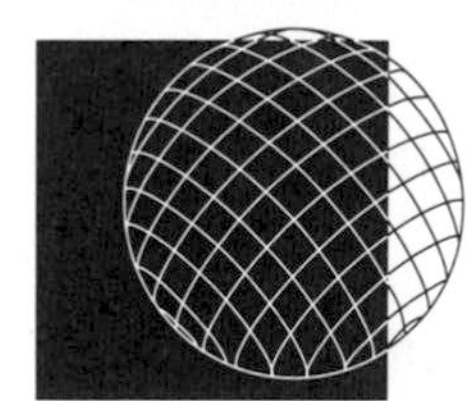

1.1 Give brief descriptions (less than 50 words) of each of the following common terms: CAD, CAM, CAE, and CIMS.

1.2 Given all the unique advantages of CAD described in Section 1.3, rank the effectiveness of this technique on a scale of 0 to 10 when it is applied to the design of the following products: a racing bicycle, a pencil sharpener, a utility trailer in a household, a pneumatic wrench, a golf club, a transmission gear train for a small car, and a hydraulic cylinder for dump trucks. Briefly describe your reasons for each ranking.

1.3 Define RAM and ROM, and describe their respective functions.

1.4 If the resolution of the monitor of a computer is 512×320 pixels in respective horizontal and vertical directions, how would a horizontal line appear on the screen when the coordinates of the two terminal points of this line are entered from a digitizer pad with 300 × 200 resolution? Also illustrate the shape of lines appearing on the screen in the vertical as well as in an inclined position with a 45° angle from the horizontal axis.

1.5 If safety of the driver is the only design constraint, what would be the synthesis factors to be considered in the example of a fork-lift, as described in Table 1.1. A stronger protection shroud, effective warning signals, and even a deadman's switch are among various options you may consider.

1.6 How would you synthesize the design of a surfboard?

1.7 Prepare product specifications for a snowmobile to be used for

(a) Leisure enjoyment
(b) Emergency rescue operations in urban and rural areas
(c) Hunting and fishing for residents in northern settlements near the Arctic.

1.8 Explain why a buffer memory of 256 K is needed in order to have a raster display on a refresh CRT with 512 × 512 resolution.

1.9 Relate two design cases that you have experienced in the past that can be categorized by procedures described in Section 1.6 and Figure 1.6.

1.10 Write an application software program to find the shortest length of aluminum tubes 50 mm O.D. × 45 mm I.D. with minimum induced stresses to be used in a two-bar truss. The two tubes are hinged to two supports that are 2 m apart, as illustrated below. Stress in the tubes should be kept below the yield strength of the material at 320 MPa.

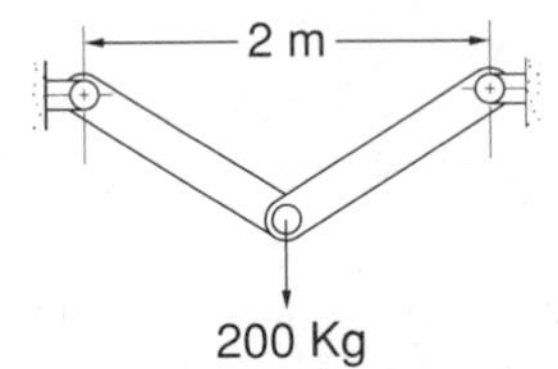

1.11 Evaluate the design of a short bridge that must withstand a truck load up to 20 tons. The geometries and dimensions of the I-beams selected by the designer are shown on the following page.

Consider the following options:

(a) The load is uniformly distributed between all four wheels, and the weight of the beams can be neglected.

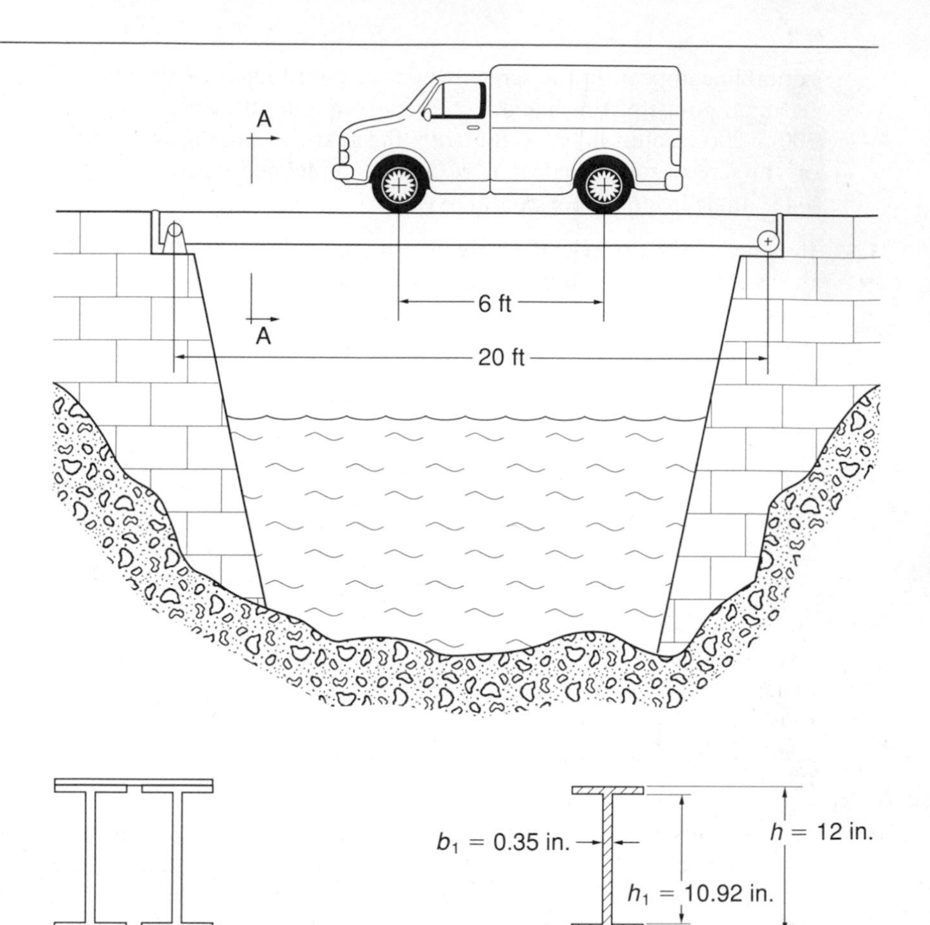

(b) Only one-quarter of the load is transmitted through the front wheels, and the weight of the beams is considered in the analysis.

1.12 Write a computer program to illustrate the shifting of maximum longitudinal stress in the beams as the truck crosses the bridge in both options described in problem 1.11.

CHAPTER 2

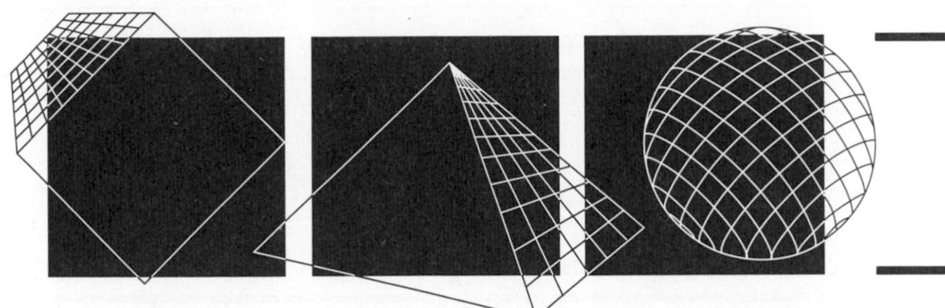

Review of Numerical Techniques for CAD

2.1 Introduction

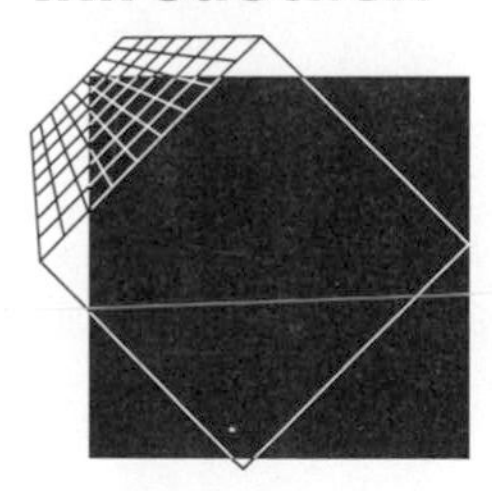

Digital computers play a very important role in modern engineering design and analysis. Engineers now use this effective tool to solve complex problems that would never have been thought possible only decades ago. However, the power of computers at the present level of technology lies in their incredible speed and memory capacity, but not in any intelligence or independent judgment. These machines essentially are capable of performing only simple arithmetical functions.

A logical question then is, how can we expect a digital computer to perform mathematical operations (e.g., evaluating everything from simple trigonometric functions to complex integrals, or even solving extremely complicated differential or integral equations)? The techniques used to convert all these complicated mathematical functions and equations into simple arithmetical operations that can be handled by computers are called *numerical techniques.* One can readily appreciate the important role that these techniques play in CAD, which relies heavily on the use of computers to perform design analysis and produce graphical representations of complex machine geometries.

Numerical techniques, sometimes referred to as *numerical analyses*, are used in wide spectrum of the physical sciences, and numerous books and monographs are specially devoted to this subject [1, 2, 3]. Because it is not possible to cover in this chapter all the topics involved in numerical analysis, the reader is referred to the above references for wide-ranging, rigorous treatment of such topics. In this chapter, we will present the principles of some key topics that are particularly relevant to the applications in computer-aided design. Most of the topics reviewed herein will be used in other parts of the book.

2.2 ▪ Matrices and Determinants

Among the most common numerical solution techniques currently used in engineering analysis are the finite difference method, the finite element method, and the boundary element method. In all these methods, the physical domain is subdivided into a number of smaller domains. The advantage of subdividing into smaller domains is that simple approximations can be adopted for each subdomain, but the discretized model does not deviate significantly from the real model. In the limit when the subdomains become infinitesimally small, the discretized model coincides with the real model. The fields of investigation involved in most design analyses, such as displacement, velocity, temperature, pressure, and other engineering quantities, can be mapped by the discretized values obtained for the individual subdomains. All these methods of analysis yield, in the final stage, a solution consisting of a set of simultaneous linear equations of the following form:

$$\begin{aligned} a_{11}x_1 + a_{12}x_2 + \cdots + a_{1n}x_n &= r_1 \\ a_{21}x_1 + a_{22}x_2 + \cdots + a_{2n}x_n &= r_2 \\ \vdots \qquad \vdots \qquad\quad \vdots \quad &\quad \vdots \\ a_{m1}x_1 + a_{m2}x_2 + \cdots + a_{mn}x_n &= r_m \end{aligned} \tag{2.1}$$

where $a_{11}, \ldots, a_{mn}$ comprise known coefficients and the set $x_1, x_2, \ldots, x_n$ comprise the n unknown discrete values of certain physical quantities that are to be solved with a given set of specified values of $r_1, r_2, \ldots, r_m$.

Our aim in this section is to illustrate the use of matrix algebra necessary to solve a set of equations such as (2.1), which can be expressed in a shorthand form:

$$[A]\{x\} = \{r\} \tag{2.2}$$

where

$$[A] = \begin{bmatrix} a_{11} & a_{12} & \cdots & a_{1n} \\ a_{21} & a_{22} & \cdots & a_{2n} \\ \vdots & & & \vdots \\ a_{m1} & a_{m2} & \cdots & a_{mn} \end{bmatrix} \tag{2.3}$$

is an $m \times n$ matrix (m rows and n columns) with the constant coefficients of (2.1) as its element. The other two matrices

$$\{x\} = \begin{Bmatrix} x_1 \\ x_2 \\ \vdots \\ x_n \end{Bmatrix} \quad \text{and} \quad \{r\} = \begin{Bmatrix} r_1 \\ r_2 \\ \vdots \\ r_m \end{Bmatrix} \tag{2.4}$$

are column matrices, which are often used to express vector quantities representing such physical quantities as force, displacement, and velocity.

As shown in (2.2) above, if $m = n$, the matrix $[A]$ is a square matrix. The diagonal elements, $a_{11}, a_{22}, \ldots, a_{nn}$, of a square matrix form the *main diagonal.*

If $m = 1$, the coefficient matrix in (2.3) is called a *row vector*. If $n = 1$, the matrix will have just one column, and, hence, the name *column vector* is used.

Matrices are also extensively used in the systematic handling of large arrays of numbers involved in transforming discrete sets of data points (e.g., geometric data on an object) from one coordinate system to another. Much of this type of transformation will be presented in Chapters 3 and 4. In view of the important role matrices play in modern numerical analyses, a review of certain concepts, definitions, and operations of matrices appears appropriate. Important matrix formulas and algebraic operations of matrices are presented in Appendix I.

2.3 ▪ Solution of Simultaneous Linear Equations

As mentioned in the foregoing section, most common numerical techniques for design analysis (e.g., the finite element method) require the solution of a set of simultaneous linear equations such as that shown in (2.2). The equations can be expressed in a compact form as

$$[A]\{x\} = \{r\}$$

where, in general, $\{x\}$ is the $n \times 1$ vector of unknown quantities, $[A]$ is the $n \times n$ coefficient matrix, and $\{r\}$ is an $n \times 1$ vector of specified values.

The number of such equations, which is identical to the number of rows in the above matrices, may be as few as two or may be as many as thousands. In the case of finite element analysis, the matrix $[A]$ is usually square and symmetric, and the number of rows or columns is identical to the total number of unknown variables required in the solution.

One may solve these simultaneous equations in a number of ways. For some simple cases (e.g., those with less than four unknowns and thus four equations), the well-known Cramer's rule can be used effectively with a hand-held calculator. However, when the number of unknowns is increased to hundreds or thousands, Cramer's rule becomes impractical. The matrix inversion technique described in section AI.4 may be marginally better in terms of computational efficiency, but it is not economically feasible for a large system of equations.

Alternative methods for finding the solution are thus desirable. A number of different methods, are, in fact, available for such a purpose, many of which have evolved from the basic Gaussian elimination method. The principle of this method will be illustrated below. Also presented in this section is Cholesky's decomposition method, which is widely used in the solution of eigenvalue problems related to the modal analysis of structures. Description of several other popular methods and their corresponding microcomputer programs are available in reference [4].

❑ GAUSSIAN ELIMINATION METHOD

Let us consider a set of simultaneous equations as follows:

$$\begin{aligned} 8x_1 + 4x_2 + x_3 &= 12 \\ 2x_1 + 6x_2 - x_3 &= 3 \\ x_1 - 2x_2 + x_3 &= 2 \end{aligned} \tag{2.5}$$

We can express the same equation in a matrix form simply as

$$[A]\{x\} = \{r\}$$

where

$$[A] = \begin{bmatrix} 8 & 4 & 1 \\ 2 & 6 & -1 \\ 1 & -2 & 1 \end{bmatrix} \tag{2.6}$$

and

$$\{r\} = \begin{Bmatrix} 12 \\ 3 \\ 2 \end{Bmatrix} \tag{2.7}$$

From the first equation in (2.5), we have $x_1 = 1.5 - 0.5x_2 - 0.125x_3$. Substituting this expression in the second and third equations of (2.5), we get

$$\begin{aligned} 8x_1 + 4x_2 + \quad x_3 &= 12 \\ 5x_2 - 1.25x_3 &= 0 \\ -2.5x_2 + 0.875x_3 &= 0.5 \end{aligned} \tag{2.8}$$

or

$$\begin{bmatrix} 8 & 4 & 1 \\ 0 & 5 & -1.25 \\ 0 & -2.5 & 0.875 \end{bmatrix} \{x\} = \{r\} \tag{2.9}$$

Again, the above equations can be expressed as

$$[A']\{x\} = \{r'\}$$

From the second equation of (2.8), we get $x_2 = 0.25x_3$, which when substituted in the third equation yields

$$\begin{aligned} 8x_1 + 4x_2 + \quad x_3 &= 12 \\ 5x_2 - 1.25x_3 &= 0 \\ 0.25x_3 &= 0.5 \end{aligned} \tag{2.10}$$

or

$$\begin{bmatrix} 8 & 4 & 1 \\ 0 & 5 & -1.25 \\ 0 & 0 & 0.25 \end{bmatrix} \{x\} = \begin{Bmatrix} 12 \\ 0 \\ 0.5 \end{Bmatrix} \tag{2.11}$$

or in a condensed form

$$[A'']\{x\} = \{r''\} \tag{2.12}$$

Thus, the matrix $[A]$ in (2.6) has been converted into a triangular matrix $[A'']$. For an obvious reason, this process is known as *triangularization.* Now, the equations in (2.11) are easily solved by first finding x_3 from the third equation, to be followed by the solutions for x_2 from the second equation and x_1 from the first. This process of solving for the unknowns in reverse order is known as the *backward substitution.* The reader can easily verify that $x_1 = 1$, $x_2 = 0.5$ and $x_3 = 2$.

We may observe from the above exercise that, starting from a square coefficient matrix in (2.6), we have obtained an upper triangular matrix in (2.11). This procedure for solving simultaneous equations is called the *Gaussian elimination method.* Because the method involves the reduction of the original square matrix to a triangular matrix, the Gaussian elimination method is also referred to as a *triangularization method.*

We will now attempt to obtain a general equation for the Gaussian elimination method by considering the following set of equations:

$$\begin{aligned} a_{11}x_1 + a_{12}x_2 + a_{13}x_3 &= r_1 \\ a_{21}x_1 + a_{22}x_2 + a_{23}x_3 &= r_2 \\ a_{31}x_1 + a_{32}x_2 + a_{33}x_3 &= r_3 \end{aligned} \tag{2.13}$$

Express x_1 in terms of x_2 and x_3 from the first of the above equations:

$$x_1 = \frac{r_1}{a_{11}} - \frac{a_{12}}{a_{11}} x_2 - \frac{a_{13}}{a_{11}} x_3 \tag{2.14}$$

Then substitute for x_1 into the second and third equations:

$$\begin{aligned} & a_{11}x_1 + a_{12}x_2 + a_{13}x_3 = r_1 \\ & \left(a_{22} - a_{21}\frac{a_{12}}{a_{11}}\right)x_2 + \left(a_{23} - a_{21}\frac{a_{13}}{a_{11}}\right)x_3 = r_2 - \frac{a_{21}}{a_{11}}r_1 \\ & \left(a_{32} - a_{31}\frac{a_{12}}{a_{11}}\right)x_2 + \left(a_{33} - a_{31}\frac{a_{13}}{a_{11}}\right)x_3 = r_3 - \frac{a_{31}}{a_{11}}r_1 \end{aligned} \tag{2.15}$$

or

$$\begin{bmatrix} a_{11} & a_{12} & a_{13} \\ 0 & a_{22}^1 & a_{23}^1 \\ 0 & a_{32}^1 & a_{33}^1 \end{bmatrix} \{x\} = \{r^1\} \tag{2.16}$$

where the superscript 1 indicates that this is the first step toward the triangularization and also that the coefficients of x_1 in the second and third equations have been eliminated. We can easily see that

$$\begin{aligned} a_{22}^1 &= a_{22} - a_{21}\frac{a_{12}}{a_{11}} & a_{32}^1 &= a_{32} - a_{31}\frac{a_{12}}{a_{11}} \\ a_{23}^1 &= a_{23} - a_{21}\frac{a_{13}}{a_{11}} & a_{33}^1 &= a_{33} - a_{31}\frac{a_{13}}{a_{11}} \\ r_2^1 &= r_2 - \frac{a_{21}}{a_{11}}r_1 & r_3^1 &= r_3 - \frac{a_{31}}{a_{11}}r_1 \end{aligned} \tag{2.17}$$

We can now write the general equations for the modified elements in each successive row in equation (2.1) as follows:

$$\begin{aligned} & a_{ij}^0 = a_{ij}, \qquad r_i^0 = r_i \\ & a_{ij}^k = a_{ij}^{k-1} - a_{ik}^{k-1}\frac{a_{kj}^{k-1}}{a_{kk}^{k-1}} \\ & r_i^k = r_{ij}^{k-1} - a_{ik}^k\frac{r_k^{k-1}}{a_{kk}^{k-1}} \end{aligned} \tag{2.18}$$

for $i > k$, $j > k$, and $k = 1, \ldots, m - 1$ where m is the number of equations.

The numerical illustration of the Gaussian elimination method presented above produced exact solutions as expected. This was possible due to the fact that all coefficients, as well as the unknown values, are integers. In reality, however, most of these values are fractions. The computer handles fractions in decimal form to a limited number of decimal places (e.g., 7 decimal places for a single precision arithmetic algorithm and 14 decimal places for double precision operations). Fractions are treated as nonterminating decimals in the computation. Errors are introduced during the computation by the dropping of numbers associated with those fractions beyond the limiting decimal place that the computer is designed to accommodate. This error is referred to as the *roundoff error.*

Roundoff errors may not substantially affect the accuracy of the results if the number of arithmetic operations involved in the computation is not very large. Otherwise, such errors can cumulate rapidly as the number of operations increases. In general, the number of simultaneous equations that can be solved satisfactorily by the Gaussian elimination method is limited to 15 to 20 with dense coefficient matrices [5]. Many numerical analyses that require the solution of a large number of simultaneous equations (e.g., in the finite element method described in Chapters 6 and 7) require double precision in the computation in order to reduce the roundoff error.

❑ CHOLESKY LU DECOMPOSITION

We have demonstrated in the previous section that a set of simultaneous equations such as that shown in (2.2) can be solved by converting the coefficient matrix $[A]$ to an upper triangular matrix using the Gaussian elimination method. Here, we will introduce another useful algorithm that can yield a triangularized coefficient matrix. This method, called the *Cholesky LU decomposition method*, is widely used in the solution of problems involving eigenvalues (e.g., the free vibration analysis of structures).

There are generally two ways to decompose a square matrix $[A]$. The first form of decomposition is expressed as follows:

$$[A] = [L][U] \tag{2.19}$$

in which $[L]$ and $[U]$ are respectively the lower and upper triangular matrices, as illustrated in equations (AI.7) and (AI.8) in Appendix I.

If one denotes l_{ij} and u_{ij} to be corresponding entries of the $[L]$ and $[U]$ matrices, then the following recurrence relations may be used [2, 4]:

$$l_{i1} = a_{i1} \tag{2.20}$$

$$u_{1j} = a_{ij}/a_{11} \tag{2.21}$$

$$l_{ij} = a_{ij} - \sum_{k=1}^{j-1} l_{ik}u_{kj} \qquad j \le i,\ i = 1, 2, \ldots, n \tag{2.22}$$

$$u_{ij} = \left(a_{ij} - \sum_{k=1}^{i-1} l_{ik}u_{kj} \right) \Big/ l_{ii} \qquad i \le j,\ j = 2, 3, \ldots, n \tag{2.23}$$

where n is the total number of rows or columns.

The solution of the simultaneous equations in (2.2) can be obtained by the substitution of (2.19) for the matrix $[A]$:

$$[L][U]\{x\} = \{r\}$$

By premultiplying both sides of the above equation by the inverse matrix of $[L]$, or $[L]^{-1}$, one obtains

$$[I][U]\{x\} = [U]\{x\} = [L]^{-1}\{r\} \tag{2.24}$$

where $[I]$ is the identity matrix. The product $[I][U]$ is an upper triangular matrix similar to the one derived in the previous section.

Another way to decompose the $[A]$ matrix is to express the matrix in the following form:

$$[A] = [L][L]^T \tag{2.25}$$

where the elements of the lower triangular matrix $[L]$ become

$$l_{ii} = \left(a_{ii} - \sum_{j=1}^{i-1} l_{ij}^2 \right)^{1/2} \qquad i = 1, 2, \ldots, n \tag{2.26}$$

$$l_{ki} = \frac{1}{l_{ii}} \left(a_{ki} - \sum_{j=1}^{i-1} l_{ij}l_{kj} \right) \qquad k = i + 1, \ldots, n \tag{2.27}$$

EXAMPLE ■ 2.1

Solve the set of simultaneous equations in (2.5) by the Cholesky decomposition method.

The coefficient matrix $[A]$ in (2.2) for (2.5) has the entries

$$\begin{array}{lll} a_{11} = 8 & a_{12} = 4 & a_{13} = 1 \\ a_{21} = 2 & a_{22} = 6 & a_{23} = -1 \\ a_{31} = 1 & a_{32} = -2 & a_{33} = 1 \end{array}$$

The entries of the decomposed matrices $[L]$ and $[U]$ can be computed using the recurrence relations given in (2.20) through (2.23):

For $i = 1$:

$$l_{11} = a_{11} = 8$$

$$u_{11} = a_{11}/l_{11} = 1;\ u_{12} = a_{12}/l_{11} = 0.5;\ u_{13} = a_{13}/l_{11} = 0.125$$

For $i = 2$:

$$l_{21} = a_{21} = 2;\ l_{22} = a_{22} - l_{21}u_{12} = 5$$

$$u_{22} = (a_{22} - l_{21}u_{12})/l_{22} = 1;\ u_{23} = (a_{23} - l_{21}u_{13})/l_{22} = -0.25$$

For $i = 3$:

$$l_{31} = a_{31} = 1;\ l_{32} = a_{32} - l_{31}u_{12} = -2.5;$$

$$l_{33} = a_{33} - l_{31}u_{13} - l_{32}u_{23} = 0.25$$

$$u_{33} = (a_{33} - l_{31}u_{13} - l_{32}u_{23})/l_{33} = 1$$

Hence, the two decomposed matrices take the form

$$[L] = \begin{bmatrix} 8 & 0 & 0 \\ 2 & 5 & 0 \\ 1 & -2.5 & 0.25 \end{bmatrix} \tag{2.28}$$

$$[U] = \begin{bmatrix} 1 & 0.5 & 0.125 \\ 0 & 1 & -0.25 \\ 0 & 0 & 1 \end{bmatrix} \tag{2.29}$$

By following the procedures for matrix inversion presented in Appendix I, the inverse of $[L]$ was readily computed and is shown below.

$$[L]^{-1} = \begin{bmatrix} 0.125 & 0 & 0 \\ -0.05 & 0.2 & 0 \\ -1 & 2 & 4 \end{bmatrix} \tag{2.30}$$

Substituting for $[U]$ from (2.29) and for $[L]^{-1}$ from (2.30) into (2.24) results in

$$\begin{bmatrix} 1 & 0 & 0 \\ 0 & 1 & 0 \\ 0 & 0 & 1 \end{bmatrix} \begin{bmatrix} 1 & 0.5 & 0.125 \\ 0 & 1 & -0.25 \\ 0 & 0 & 1 \end{bmatrix} \begin{Bmatrix} x_1 \\ x_2 \\ x_3 \end{Bmatrix} = \begin{bmatrix} 0.125 & 0 & 0 \\ -0.05 & 0.2 & 0 \\ -1 & 2 & 4 \end{bmatrix} \begin{Bmatrix} 12 \\ 3 \\ 2 \end{Bmatrix}$$

or in a compact form

$$\begin{bmatrix} 1 & 0.5 & 0.125 \\ 0 & 1 & -0.25 \\ 0 & 0 & 1 \end{bmatrix} \begin{Bmatrix} x_1 \\ x_2 \\ x_3 \end{Bmatrix} = \begin{Bmatrix} 1.5 \\ 0 \\ 2 \end{Bmatrix} \tag{2.31}$$

It is readily seen from the third row of (2.31) that

$$x_3 = 2$$

from which one may obtain by the back substitution process

$$x_2 = 0 - (-0.25x_3) = 0.5$$

and

$$x_1 = 1.5 - 0.5x_2 - 0.125x_3 = 1.0$$

The numerical values of x_1, x_2, and x_3 are identical to those calculated by the Gaussian elimination method as described above.

❑

2.4 ▪ Eigenvalue Problems

One type of design analysis that engineers often encounter is eigenvalue problems. The solution of eigenvalue problems also requires the solution of simultaneous algebraic equations. Unique solutions to these equations are possible only if the value of a parameter involved in these equations is known. This parameter is referred to as an *eigenvalue*.* The solution to the equations corresponding to each eigenvalue is called the *eigenvector* of that eigenvalue. Eigenvalues are associated with the natural frequencies of a structure in a modal analysis, and eigenvectors characterize the shapes of corresponding modes of vibration. Eigenvalues also determine the critical loads for stability analyses of structures, in which the eigenvalues are used to compute the critical loads that cause the instability of the structure.

The general formulation of an eigenvalue problem is

$$([A] - \lambda[I])\{x\} = 0 \tag{2.32}$$

where $[A]$ is an $n \times n$ matrix, $[I]$ is a unit matrix of order n, and λ and $\{x\}$ are the respective n eigenvalues and eigenvectors.

Note that equation (2.32) has a nonzero solution only if the determinant of the coefficient matrix $([A] - \lambda[I])$ is zero; that is,

$$|A - \lambda I| = 0 \tag{2.33}$$

where A and I represent elements of the respective matrices $[A]$ and $[I]$.

One may visualize that the above relation will yield a polynomial equation of order n, from which the values of the n unknown parameters, λ, can be solved.

---

* The word *eigen* is German, meaning "characteristic." Eigenvalue, -vector, and -function are mixtures of German and English, meaning "characteristic value," "characteristic vector," and "characteristic function," respectively.

EXAMPLE ■ 2.2

Find the natural frequencies of the two-degrees-of-freedom mass-spring system illustrated in Figure 2.1.

Before we actually begin to solve this problem, let us review the physical meaning of natural frequencies of a dynamic system and why they are important in engineering design.

Natural frequencies of a dynamic system are associated with the *free vibration* of the system. One may imagine that the mass of the familiar mass-spring system can vibrate from its equilibrium position as a result of a small disturbance. The word "small" is used here in the mathematical sense, meaning the absence of the force function in the equation of motion. Free vibration takes place when a system oscillates under the influence inherent in the system itself and when external forces are not present. Free vibration analysis is a very important part of the design of machine components or structures that are vulnerable to vibration. This analysis will result in finding the *natural frequencies* of the structures at various modes of vibration. *Resonant vibration* of the structure can occur when the frequency of the externally excited vibration of the structure approaches any of these natural frequencies. Resonant vibration will not only cause extreme discomfort to operators or riders of the machines, but can also be detrimental to the integrity of the structures.

Let us now derive the equations of motion for mass M_1 and M_2 in the system illustrated in Figure 2.1. These equations can be readily established by using Newton's second law with respect to the free body force diagram in the figure:

$$M_1\ddot{x}_1 = -k_1x_1 - k(x_1 - x_2) \tag{2.34}$$

$$M_2\ddot{x}_2 = k(x_1 - x_2) - k_2x_2 \tag{2.35}$$

where x_1 and x_2 are the respective instantaneous displacements of M_1 and M_2 from their equilibrium positions, and k, k_1, and k_2 are the spring constants of sections of springs illustrated in Figure 2.1. The "double dot" above x_1 and x_2 denotes accelerations of M_1 and M_2.

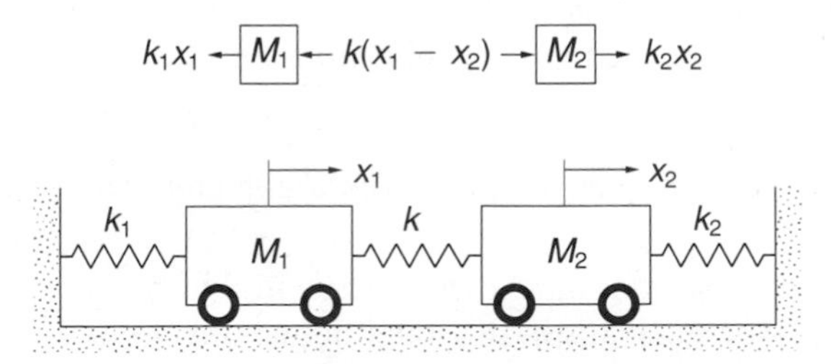

FIGURE 2.1
A two-degree-of-freedom mass spring-system

Assume harmonic functions for the displacement function; that is, $x_1 = A_1 \sin \omega t$ and $x_2 = A_2 \sin \omega t$, in which ω represents the frequencies of vibration of the system with the absence of external forces, and A_1 and A_2 are the respective amplitudes of vibrations of mass M_1 and M_2.

The substitution of these harmonic displacement functions into equations (2.34) and (2.35) yields the following simultaneous equations:

$$(-M_1 \omega^2 + k_1 + k)A_1 - kA_2 = 0 \tag{2.36}$$

$$-kA_1 + (-M_2\omega^2 + k_2 + k)A_2 = 0 \tag{2.37}$$

or in a matrix form:

$$\begin{bmatrix} -M_1\omega^2 + k_1 + k & -k \\ -k & -M_2\omega^2 + k_2 + k \end{bmatrix} \begin{Bmatrix} A_1 \\ A_2 \end{Bmatrix} = \begin{Bmatrix} 0 \\ 0 \end{Bmatrix} \tag{2.38}$$

By comparing the above equation and equation (2.32), one may recognize that equations (2.36) and (2.37), or equation (2.38) matches the general formulation of an eigenvalue problem. One will further realize that unique solutions for A_1 and A_2 (eigenvectors) are possible only if the values (eigenvalues) of ω are known.

The solution of eigenvalue ω can be obtained by using equation (2.33); that is, the determinant of the coefficient matrix in equation (2.38) be set to equal zero:

$$\begin{vmatrix} -M_1\omega^2 + k_1 + K & -k \\ -k & -M_2\omega^2 + k_2 + k \end{vmatrix} = 0 \tag{2.39}$$

The above relation will lead to a fourth-order polynomial equation for the eigenvalue ω:

$$M_1 M_2 \omega^4 - [(k_1 + k)M_2 + (k_2 + k)M_1] \\ \times \omega^2 + k(k_1 + k_2) + k_1 k_2 = 0 \tag{2.40}$$

The roots of equation (2.40) (i.e., ω values) will be the natural frequencies of the system in Figure 2.1. The substitution of each ω value into equation (2.38) will resolve the eigenvector of A_1 and A_2, which characterize the mode of vibrations for mass M_1 and M_2. ❑

EXAMPLE ■ 2.3

Find the natural frequencies of a structural member in longitudinal vibration.

Some long and slender machines or structural members are expected to sustain dynamic loads in the longitudinal direction. It is desirable to determine the natural frequencies for a particular structural member so design loading conditions can be properly selected to avoid resonant vibration of the member.

In the last example, we presented a case where a mass can vibrate from its equilibrium position as a result of a small disturbance when it is attached to an elastic spring. In reality, one may imagine that all solid structures, regardless of their geometries, are made up of an infinite number of tiny particles of finite mass interconnected by elastic links. The situation is very much like the congregation of an infinite number of mass-spring systems such as the one illustrated in Figure 2.1. One can thus conclude that structures and machine components can, in fact, vibrate from their equilibrium state with *small* external disturbances.

We will illustrate the phenomenon of free vibration of a structural member by a simple case involving a rod of finite length, as shown in Figure 2.2. The rod is made of a material with density ρ and Young's modulus E. Since the rod is assumed to deform in the longitudinal (i.e., x) direction, we may assume a harmonic function for the displacement at various points of the rod in the x direction.

Mathematically, the displacement function can be expressed as

$$\{u(x, t)\} = \{u_0(x)\} \sin \omega t \tag{2.41}$$

where $\{u_0(x)\}$ are the amplitudes of vibration at various locations along the rod and ω is the natural frequency of the rod.

The substitution of equation (2.41) into an equation of motion for a continuum solid with the absence of force function can lead to the following characteristic equation:

$$([K] - \omega^2[M])\{u_0(x)\} = 0 \tag{2.42}$$

where $[K]$ and $[M]$ are the respective stiffness and mass matrices of the rod.

The presence of these stiffness and mass matrices in equation (2.42), instead of single values, is due to the fact that the stiffness and mass are distributed throughout the material. The stiffness and mass matrices for a bar with free ends, as shown in Figure 2.2, will have

FIGURE 2.2
A longitudinally vibrating rod

the forms

$$[K] = \frac{AE}{L}\begin{bmatrix} 1 & -1 \\ -1 & 1 \end{bmatrix} \tag{2.43}$$

and

$$[M] = \frac{\rho AL}{6}\begin{bmatrix} 2 & 1 \\ 1 & 2 \end{bmatrix} \tag{2.44}$$

Now, we decompose the $[K]$ matrix in equation (2.42) into the lower triangular matrix according to Cholesky's decomposition method, as expressed in equation (2.25) or in the form

$$[K] = [L][L]^T \tag{2.45}$$

where $[L]$ is the lower triangular matrix and $[L]^T$ is its transposed form. Entries to the $[L]$ matrix can be determined by using equations (2.26) and (2.27).

By replacing the matrix $[K]$ in equation (2.42) using the above decomposed form, the following equation is obtained:

$$([L][L]^T - \omega^2[M])\{u_0(x)\} = 0$$

If we multiply the above equation by the inverse of $[L]$ matrix (i.e., $[L]^{-1}$) and use the relationship $([L]^T)^{-1} = ([L]^{-1})^T$, after rearranging terms, the following form emerges:

$$([I] - \omega^2[L]^{-1}[M]([L]^{-1})^T)([L]^T\{u_0(x)\}) = 0$$

or

$$\left([L]^{-1}[M]([L]^{-1})^T - \frac{1}{\omega^2}[I]\right)([L]^T\{u_0(x)\}) = 0 \tag{2.46}$$

A comparison of equations (2.46) and (2.32) will show that the former equation does match the general formulation of an eigenvalue problem with the following relations:

$$[A] = [L]^{-1}[M]([L]^{-1})^T \tag{2.47}$$

and the eigenvalue

$$\lambda = 1/\omega^2 \tag{2.48}$$

The eigenvector in this case is

$$\{x\} = [L]^T\{u_0(x)\} \tag{2.49}$$

The natural frequencies, $\omega_n (n = 1, 2)$, can be solved by the following characteristic equation:

$$\left|\left([A] - \frac{1}{\omega^2}[I]\right)\right| = 0 \tag{2.50}$$

in which $[A]$ is expressed in equation (2.47).

Another way to solve for the eigenvalues in the present case is to decompose the mass matrix as

$$[M] = [L][L]^T \tag{2.51}$$

where $[L]$ now stands for the lower triangular matrix of $[M]$.

The solution of eigenvalues in this case becomes

$$[A] = [L]^{-1}[K]([L]^{-1})^T \tag{2.52}$$

and $\omega_n (n = 1, 2, \ldots)$ are the roots of the following equation:

$$|([A] - \omega^2[I])| = 0 \tag{2.53}$$

❑

The following is a numerical illustration for the determination of natural frequencies for a rod with one end fixed and the other end free (Figure 2.3).

In addition to the two nodes at the end points, another node point (node 2) located midspan (i.e., $x = 10$ cm) is selected. The selection of the third node point is necessary in this case in order to maintain the minimum required number of two degrees of freedom (with node 1 at $x = 0$ being fixed).

The amplitudes of the displacement functions at the three nodes are

$$\{u_0(x)\}^T = \{u_0^1(0) = 0 \quad u_0^2(10) \quad u_0^3(20)\}$$

where u_0^2 and u_0^3 are the respective longitudinal amplitudes of displacements at nodes 2 and 3.

The arrangement in Figure 2.3 indicates a division of the rod into two sections (or elements). Each element has a length of 10 cm.

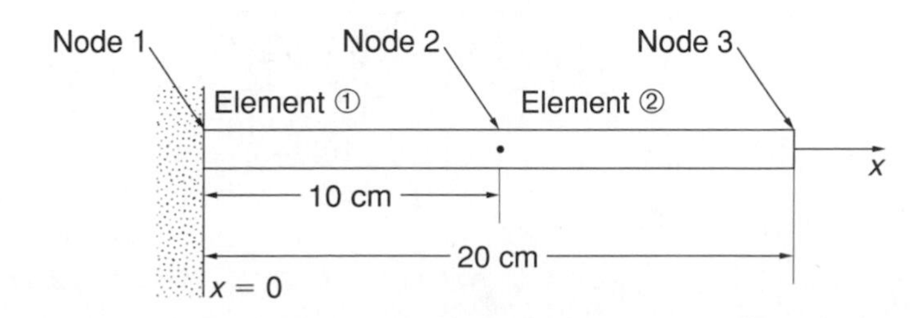

FIGURE 2.3 Natural frequencies in a rod vibrating in longitudinal direction

Since the geometry and material properties of both elements are identical, the element stiffness and mass matrices are given in equations (2.43) and (2.44). The following numerical values for the physical quantities are used in these equations:

$$\text{Length of the element, } L = 100 \text{ mm}$$

$$\text{Cross-sectional area, } A = 10 \text{ mm}^2$$

$$\text{Young's modulus, } E = 1000 \text{ MPa}$$

$$\text{Mass density, } \rho = 10^{-7} \text{ N-sec}^2/\text{mm}^4$$

$$[K]^1 = [K]^2 = \frac{AE}{L}\begin{bmatrix} 1 & -1 \\ -1 & 1 \end{bmatrix} = \begin{bmatrix} 100 & -100 \\ -100 & 100 \end{bmatrix} \text{N/mm}$$

$$[M]^1 = [M]^2 = \frac{\rho AL}{6}\begin{bmatrix} 2 & 1 \\ 1 & 2 \end{bmatrix} = \frac{10^{-4}}{6}\begin{bmatrix} 2 & 1 \\ 1 & 2 \end{bmatrix} \text{N-sec}^2/\text{mm}$$

The overall stiffness and mass matrices can be constructed by summing up the entries associated with their respective nodes on the rod and filling the empty spaces in the resultant matrices with zeros. In the present case, one realizes that node 2 is common to both element 1 and element 2. The summation of the element stiffness matrices follows the format

$$[K] = \begin{bmatrix} [K]^1 & 0 \\ 0 & [K]^2 \end{bmatrix} = \begin{bmatrix} 100 & -100 & 0 \\ -100 & (100+100) & -100 \\ 0 & -100 & 100 \end{bmatrix}$$

Likewise, the overall mass matrix can be shown to have the form

$$[M] = \frac{10^{-4}}{6}\begin{bmatrix} 2 & 1 & 0 \\ 1 & 4 & 1 \\ 0 & 1 & 2 \end{bmatrix}$$

The characteristic equation for the rod is obtained by substituting the overall stiffness and mass matrices into equation (2.42), as shown below.

$$\begin{bmatrix} 100 - \frac{10^{-4}}{3}\omega^2 & -\left(100 + \frac{10^{-4}}{6}\omega^2\right) & 0 \\ & 200 - \frac{2}{3} \times 10^{-4}\omega^2 & -\left(100 + \frac{10^{-4}}{6}\omega^2\right) \\ \text{SYM} & & 100 - \frac{10^{-4}}{3}\omega^2 \end{bmatrix} \times \begin{Bmatrix} u_0^1 \\ u_0^2 \\ u_0^3 \end{Bmatrix} = \begin{Bmatrix} 0 \\ 0 \\ 0 \end{Bmatrix} \qquad (2.54)$$

One may solve for ω from the above equations by setting the determinant of the entries of the coefficient matrix enclosed in the dotted line envelope to zero or by using Cholesky's decomposition method.

Let's separate the mass matrix $[M]$ related to nonzero nodal values u_0^2 and u_0^3 from the coefficient matrix in the above equations.

$$[M] = \begin{bmatrix} \frac{2}{3} & \frac{1}{6} \\ \frac{1}{6} & \frac{1}{3} \end{bmatrix} \times 10^{-4}$$

The lower triangular matrix for $[M]$ in equation (2.51) can be constructed by following the recurrence relations in equations (2.26) and (2.27) as

$$[L] = \begin{bmatrix} \sqrt{\frac{2}{3}} & 0 \\ \frac{1}{2\sqrt{6}} & \frac{\sqrt{7}}{2\sqrt{6}} \end{bmatrix} \times 10^{-2}$$

and its inverse is

$$[L]^{-1} = \begin{bmatrix} \sqrt{\frac{3}{2}} & 0 \\ -\frac{1}{2}\sqrt{\frac{6}{7}} & 2\sqrt{\frac{6}{7}} \end{bmatrix} \times 10^{2}$$

The $[A]$ matrix in equation (2.52) becomes

$$[A] = [L]^{-1}[K]([L]^{-1})^T = \begin{bmatrix} 300 & -\frac{900}{\sqrt{7}} \\ -\frac{900}{\sqrt{7}} & \frac{3900}{7} \end{bmatrix} \times 10^{4}$$

The determinant in equation (2.53) is shown to take the form

$$\begin{vmatrix} 3 \times 10^6 - \omega^2 & -\frac{9}{\sqrt{7}} \times 10^6 \\ -\frac{9}{\sqrt{7}} \times 10^6 & \frac{39}{7} \times 10^6 - \omega^2 \end{vmatrix} = 0$$

A fourth-order equation can be derived from the above relation as

$$7\omega^4 - (60 \times 10^6\omega^2) + (36 \times 10^{12}) = 0$$

from which we can solve for ω^2 using the standard solution procedure for quadratic equations. The numerical values for ω^2 were found to be

$$\omega^2 = 0.6492 \times 10^6 \quad \text{and} \quad 7.9223 \times 10^6$$

The four values of ω can be readily computed to be

$$\omega_1 = 0.8057 \times 10^3 \qquad \omega_2 = 2.8147 \times 10^3$$

$$\omega_3 = -0.8057 \times 10^3 \qquad \omega_4 = -2.8147 \times 10^3 \text{ cycles/sec}$$

The negative values (i.e., ω_3 and ω_4) obviously have no physical significance. Hence, the two natural frequencies of the rod are $\omega_1 = 0.8057 \times 10^3$ cycles/sec for mode 1 vibration and $\omega_2 = 2.8147 \times 10^3$ cycles/sec for the mode 2 vibration.

The eigenvector (i.e., $\{u\}^T = \{u_0^1 \quad u_0^2 \quad u_0^3\}$) for both mode 1 and mode 2 vibration can be computed by equation (2.54), with the substitution of the respective values of ω_1 and ω_2.

2.5 ▪ Numerical Differentiation

As explained in Section 2.1, the power of computers lies in their enormous memory capacity and the speed with which they perform arithmetic operations. Computers are not designed to perform calculus such as differentiation and integration. In this secton, we shall develop the concept of numerical differentiation and obtain the numerical analog of differential operators such that differential equations can be converted into simultaneous linear algebraic equations. Basic solution techniques for these equations have already been presented earlier in this chapter.

Let us take a look at Figure 2.4, where the variation in a function $y(x)$ is shown. The first derivative of $y(x)$ with respect to x at a point A (x_i, y_i) would be given by the slope of the tangent to the curve at that point.

The inset to Figure 2.4 shows a magnified view of the area around point A. In the case of a smooth curve, it is always possible to find a small segment

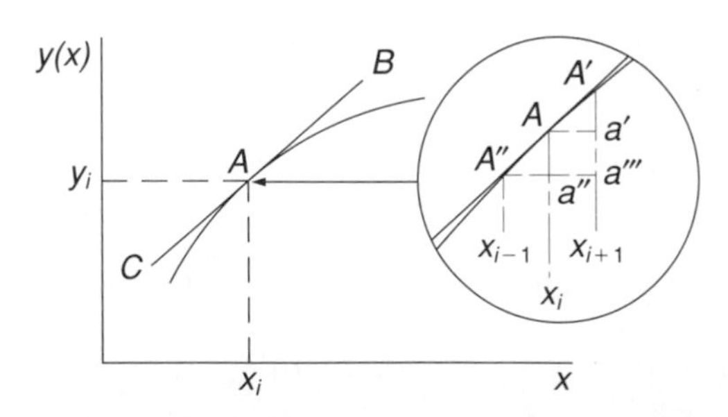

FIGURE 2.4
Graphical representation of the variation of a function

of the curve that would almost coincide with the tangent. In Figure 2.4, it appears that the tangent and the curve start to separate significantly at point x_{i+1} in the forward direction and at point x_{i-1} in the backward direction.

The slope of the tangent, $(\Delta y/\Delta x)$ for finite increments of Δx and Δy can be approximated by the derivative dy/dx, such that

$$\frac{dy}{dx} = \lim_{\Delta x \to 0} \left(\frac{\Delta y}{\Delta x}\right) \doteq \frac{\Delta y}{\Delta x}$$

We can thus express the derivative of (dy/dx) at $x = x_i$ by the slope of the tangent at that point in the following ways:

1. From triangle $A'Aa'$:

$$\left(\frac{dy}{dx}\right)\bigg|_{x=x_i} = \frac{y_{i+1} - y_i}{x_{i+1} - x_i} + 0\,(\text{error}) \tag{2.55}$$

where y_{i+1} and y_i represent the respective values of the function $y(x)$ at $x = x_{i+1}$ and $x = x_i$. Equation (2.55) describes what is known as the *forward difference method* of differentiation.

2. From triangle $AA''a''$:

$$\left(\frac{dy}{dx}\right)\bigg|_{x=x_i} = \frac{y_i - y_{i-1}}{x_i - x_{i-1}} + 0\,(\text{error}) \tag{2.56}$$

The above is the *backward difference method* of differentiation.

3. Another method of obtaining the slope is

$$\left(\frac{dy}{dx}\right)\bigg|_{x=x_i} = \frac{y_{i+1} - y_{i-1}}{x_{i+1} - x_{i-1}} + 0\,(\text{error}) \tag{2.57}$$

from the triangle $A'A''a'''$; alternatively,

$$\left(\frac{dy}{dx}\right)\bigg|_{x=x_i} = \frac{y_{i+1/2} - y_{i-1/2}}{x_{i+1/2} - x_{i-1/2}} + 0\,(\text{error}) \tag{2.58}$$

where $x_{i+1/2}$ is a point halfway between x_i and x_{i+1}, and $x_{i-1/2}$ is situated halfway between x_i and x_{i-1}. The term 0 (error) in the above expressions indicates the order of error induced by approximating the derivatives by differences of the function values. Equations (2.57) and (2.58) define two ways in which a differential can be obtained by the *central difference method.*

It should be noted that the three methods of obtaining the first derivative given above would yield different results in all cases except where the curve is a straight line, and none would be equal to the actual value. The numerical result would approach the actual value as the distance between points x_i, x_{i+1}, etc., is decreased.

The finite difference formulas for the second- and higher-order derivatives can be readily deduced by using equations (2.55) through (2.58). Following is the derivation of the second derivative of y with respect to x using the forward difference scheme:

$$\left.\frac{d^2y}{dx^2}\right|_{x=x_i} = \frac{\left.\frac{dy}{dx}\right|_{x=x_{i+1}} - \left.\frac{dy}{dx}\right|_{x=x_i}}{x_{i+1} - x_i}$$

$$= \frac{(y_{i+2} - y_{i+1})/(x_{i+2} - x_{i+1}) - (y_{i+1} - y_i)/(x_{i+1} - x_i)}{x_{i+1} - x_i} + 0\,(\text{error})$$

with $x_{i+2} - x_{i+1} = x_{i+1} - x_i = \Delta x$ for the forward difference scheme. Derivation of second derivatives by the other two difference schemes can be carried out by following a similar procedure. Expressions for the third and fourth derivatives are presented in Table 2.1.

Errors associated with the approximation of derivatives by finite difference schemes, as shown in equations (2.56) through (2.58) and given in Table 2.1, are related to the order of incremental step size, Δx. For these simple schemes, these errors can be estimated as $0(\Delta x)$ (i.e., in the order of

TABLE 2.1 The derivative in difference form

	Forward Difference	**Backward Difference**	**Central Difference**
$\frac{dy}{dx}$	$\frac{1}{\Delta x}(y_{i+1} - y_i)$	$\frac{1}{\Delta x}(y_i - y_{i-1})$	$\frac{1}{\Delta x}(y_{i+1/2} - y_{i-1/2})$, or $\frac{1}{2\Delta x}(y_{i+1} - y_{i-1})$
$\frac{d^2y}{dx^2}$	$\frac{1}{(\Delta x)^2}(y_{i+2} - 2y_{i+1} + y_i)$	$\frac{1}{(\Delta x)^2}(y_i - 2y_{i-1} + y_{i-2})$	$\frac{1}{(\Delta x)^2}(y_{i+1} - 2y_i + y_{i-1})$
$\frac{d^3y}{dx^2}$	$\frac{1}{(\Delta x)^3}(y_{i+3} - 3y_{i+2} + 3y_{i+1} - y_i)$	$\frac{1}{(\Delta x)^3}(y_i - 3y_{i-1} + 3y_{i-2} - y_{i-3})$	$\frac{1}{2(\Delta x)^3}(y_{i+2} - 2y_{i+1} + 2y_{i-1} - y_{i-2})$
$\frac{d^4y}{dx^4}$	$\frac{1}{(\Delta x)^4}(y_{i+4} - 4y_{i+3} + 6y_{i+2} - 4y_{i+1} + y_i)$	$\frac{1}{(\Delta x)^4}(y_i - 4y_{i-1} + 6y_{i-2} - 4y_{i-3} + y_{i-4})$	$\frac{1}{(\Delta x)^4}(y_{i+2} - 4y_{i+1} + 6y_i - 4y_{i-1} + y_{i-2})$

Note One may replace $y_{i\pm1}$, $y_{i\pm2}$, $y_{i\pm3}$, etc., respectively, by $y(x \pm \Delta x)$, $y(x \pm 2\Delta x)$, $y(x \pm 3\Delta x)$, etc., in the table. The increment Δx is $x_{i+2} - x_{i+1}$ for the forward difference scheme; $x_i - x_{i-1}$ or $x_{i-1} - x_{i-2}$ for the backward difference scheme; and $x_{i+1/2} - x_{i-1/2}$ for the central difference scheme.

Δx) for first-order derivatives, $0(\Delta x^2)$ for second-order derivatives, etc. In theory, a 50 percent reduction of the incremental step size can improve the accuracy of approximated first-order derivatives by 100 percent, and of approximated second-order derivatives by 200 percent. In reality, however, smaller increments do not always result in more accurate results. The reason is that smaller increments mean larger numbers of arithmetic operations, which can introduce significant roundoff errors, as indicated in section 2.3. Excessive roundoff errors can offset the improved accuracy of the results generated when incremental step sizes are reduced. An optimum incremental step size thus exists in numerical differentiation at which the net improvement of accuracy reaches its maximum value.

Several other numerical differentiation schemes are widely used in sophisticated numerical analyses. Formulations of some well-known schemes are available for numerical differentiations, such as the Runge-Kutta method [4] for greater accuracy and the Crank-Nicholson scheme [6] for initial value problems.

All of these methods are approximate, in that we approximate segments of the curve representing a continuous function with straight lines. Obviously, the smaller the increments, the smaller the deviation between the actual behavior of the variable and its approximation. It is thus conceivable that in the case of sharp changes in the variation of a dependent variable, smaller increments in the independent variable are generally required. The increment size may also depend on the type of differentiation scheme used. The forward and backward difference schemes are less accurate than the central difference scheme and therefore require smaller increments. The Crank-Nicholson scheme is more accurate than all the rest and therefore yields accurate results even for large increments. More importantly, this latter method usually generates stable solutions, while the forward difference scheme often produces results that oscillate about the true solutions.

EXAMPLE ■ 2.4

Use the finite difference method to solve the following differential equation:

$$\frac{d^2x(t)}{dt^2} + x(t) = 0 \tag{2.59}$$

with these given initial conditions:

$$x(0) = 1 \quad \text{and} \quad \dot{x}(0) = \left.\frac{dx}{dt}\right|_{t=0} = 0 \tag{2.60}$$

It can be easily shown that the exact solution of equation (2.59) is $x(t) = \cos(t)$, which is plotted in Figure 2.5 for the interval $0 \leq t \leq \pi/2$.

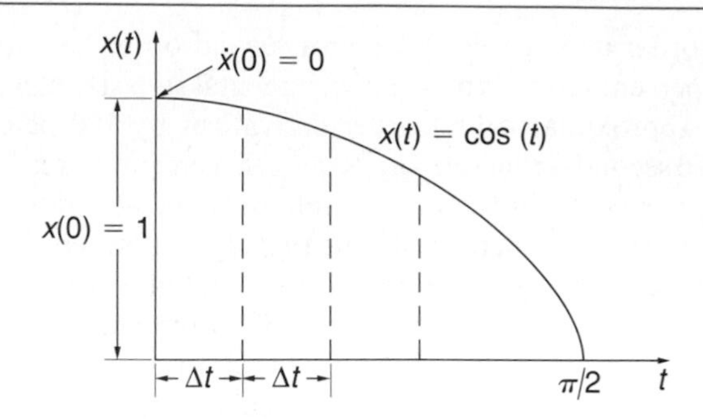

FIGURE 2.5
Solution of a finite difference equation

Suppose that we wish to use the forward difference scheme for the solution of equation (2.59). From Table 2.1, we find

$$\frac{dx}{dt} = \dot{x} = [x(t + \Delta t) - x(t)]/\Delta t$$

or

$$x(t + \Delta t) = \dot{x}(t)\,\Delta t + x(t) \tag{2.61}$$

The second derivative can be written as

$$\frac{d^2x}{dt^2} = [\dot{x}(t + \Delta t) - \dot{x}(t)]/\Delta t$$

or

$$\dot{x}(t + \Delta t) = \dot{x}(t) + \ddot{x}(t)\,\Delta t \tag{2.62}$$

We now proceed to solve the differential equation at $t = 0, 0.05$, and 0.1 with a selected increment of $\Delta t = 0.05$. The reader is advised to repeat the same exercise for at least two more values of Δt—say, $\Delta t = 0.25$ and $\Delta t = 0.5$.

Step 1 $t = 0; \Delta t = 0.05$:

Eq. (2.59): $\ddot{x}(0) = -x(0) = -1$ (given initial condition)

Eq. (2.62): $\dot{x}(0 + 0.05) = \dot{x}(0) + \ddot{x}(0)(0.05)$

$$\therefore\ \dot{x}(0.05) = -0.05$$

Eq. (2.61): $x(0 + 0.05) = \dot{x}(0)(0.05) + x(0)$

$$x(0.05) = 0 \times 0.05 + 1 = 1$$

The solution at $t = 0$ is therefore equal to

$$x(0.05) = 1 \tag{2.63}$$

Step 2 $t = 0.05, \Delta t = 0.05$:

$$\ddot{x}(0.05) = -x(0.05) = -1 \quad \text{[from (2.59) and (2.63)]}$$

Eq. (2.62): $\dot{x}(0.05 + 0.05) = \dot{x}(0.05) + (0.05)\ddot{x}(0.05)$

$$\dot{x}(0.1) = -0.05 + (-1) \times 0.05 = -0.1$$

Eq. (2.61): $\dot{x}(0.05 + 0.05) = (0.05) + x(0.05)\ddot{x}(0.05)$

$$x(0.1) = (-0.05) \times 0.05 + 1 = 0.9975$$

The solution at $t = 0.1$ is therefore equal to

$$x(0.1) = 0.9975 \tag{2.64}$$

Step 3 $t = 0.1; \Delta t = 0.05$:

$$\ddot{x}(0.1) = -x(0.1) = -0.9975 \quad \text{[from (2.59) and (2.64)]}$$

Eq. (2.62): $\dot{x}(0.1 + 0.05) = \dot{x}(0.1) + \ddot{x}(0.1)(0.05)$

$$\dot{x}(0.15) = -0.1 + (-0.9975)0.05 = -0.149875$$

Eq. (2.61): $x(0.1 + 0.05) = \dot{x}(0.1)(0.05) + x(0.1)$

$$x(0.15) = (-0.1) \times 0.05 + 0.9975 = 0.9925$$

Thus, the solution of equation (2.59) with $\Delta t = 0.05$ is

t	$x(t)_{\text{finite diff.}}$	$x(t)_{\text{exact}} = \cos(t)$
0	1.0	1.0
0.05	1.0	0.9999996
0.10	0.9975	0.9950041
0.15	0.9925	0.9887711
⋮	⋮	⋮

❑

The reader is reminded that the above technique works well for problems involving specified *initial* values. For any nth-order differential equation, all lower-order derivatives and the function itself must be specified at a given value of the independent variable. Thus, for a differential equation of the form

$$\frac{d^2y(x)}{dx^2} + \alpha\frac{dy(x)}{dx} + \beta y(x) = \gamma$$

both conditions

$$\left.\frac{dy}{dx}\right|_{x=0}$$

and $y(0)$ must be specified, as will be demonstrated in the following example.

EXAMPLE ■ 2.5

Use the finite difference method described above to calculate the deflection of a cantilever beam subjected to a concentrated load, P, acting at its free end, as illustrated in Figure 2.6.

The differential equation for the deflection of the beam, $y(x)$, is given as

$$\frac{d^2y(x)}{dx^2} = -\frac{M(x)}{EI}$$

where $M(x)$ is the bending moment distribution, given by

$$\begin{aligned} M(x) &= -P(L-x) \\ &= -3600 + 100x \\ E &= \text{modulus of elasticity} \\ &= 10^7 \text{ psi, or } 68965 \text{ MPa} \\ I &= \text{section moment of inertia} \\ &= 8 \text{ in}^4\text{, or } 333 \text{ cm}^4 \end{aligned}$$

Upon substituting the above numerical values in English units, the differential equation becomes

$$\frac{d^2y(x)}{dx^2} = (0.45 \times 10^{-4}) - (1.25 \times 10^{-6}x) \tag{2.65}$$

with

$$y(0) = \dot{y}(0) = 0 \tag{2.66}$$

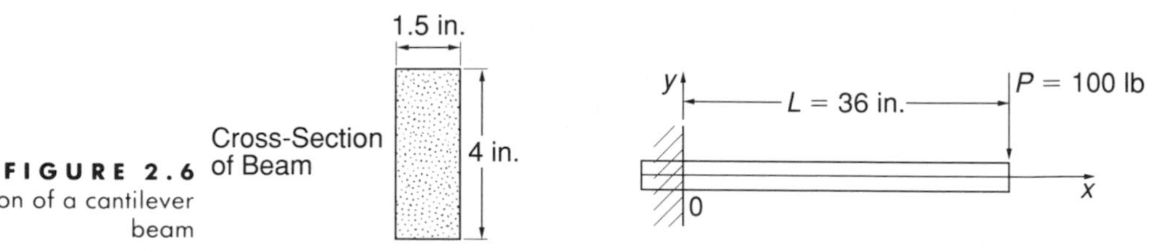

FIGURE 2.6 Deflection of a cantilever beam

```
Output Variables:      Y(I,J) = Deflection
                      Y1(I,J) = First derivative of Y(I,J), the slope
                      Y2(I,J) = Second derivative of Y(I,J),for bending moment

C      PARTICULAR DIFFERENTIAL EQUATION TO BE SOLVED IS:  D2Y+0.45F-04
C        -1.25E-06*X
C      BOUNDARY CONDITIONS ARE:     Y(0)=Y0=0,  Y'(0)=Y10=0
       DIMENSION DELTAX(5), X(400),Y2(5,400),Y1(5,400),Y(5,400)
       READ(5,1)  (DELTAX(I),I=1,5)
     1 FORMAT(5F5.1)
       DO 10 I=1,5
       WRITE(6,2) DELTAX(I)
     2 FORMAT(1H1,10X,'INCREMENT IN X = ',F5.2,//)
       WRITE (6,3)
     3 FORMAT(1H0,11X,'X',11X,'Y(X)',13X,'DYDX',13X,'D2YDX2',//)
C      L=LENGTH OF THE BEAM
       L=36
       K=L/DELTAX(I)+1
       Y1(I,1)=0.
       Y(I,1)=0.
       DO 20 J=1,K
       X(J)=(J-1)*DELTAX(I)
       Y2(I,J)=0.45E-04-1.25E-06*X(J)
       Y1(I,J+1)=Y1(I,J)+Y2(I,J)*DELTAX(I)
       Y(I,J+1)=Y(I,J)+Y1(I,J)*DELTAX(I)
    20 WRITE(6,4) X(J),Y(I,J),Y1(I,J),Y2(I,J)
     4 FORMAT(1H0,8X,F5.2,5X,E13.6,4X,E13.6,5X,E13.6)
    10 CONTINUE
       STOP
       END
```

FIGURE 2.7
A computer program for the solution of a differential equation for the deflection of a cantilever beam

By using the forward difference recurrence relations given in Table 2.1 for the second derivative in equation (2.65), together with the conditions specified in equation (2.66), and by following the procedures described in Example 2.3, the deflection of the beam at specified increments of Δx can be calculated.

The FORTRAN computer program shown in Figure 2.7 can perform the above computations at various assigned values of $\Delta x = 0.1$, 0.5, 2.0, 4.0, and 6.0 inches.

Graphic representations of the beam deflection for a few assigned values of Δx are illustrated in Figure 2.8. As expected, the accuracy of the finite difference method depends on the size of the assigned increments, Δx. No deviation from the analytical solution can be detected in the results with a small value of Δx, such as $\Delta x = 0.10$, whereas an error of 29.5 percent is observed in the case of $\Delta x = 6.0$. ❑

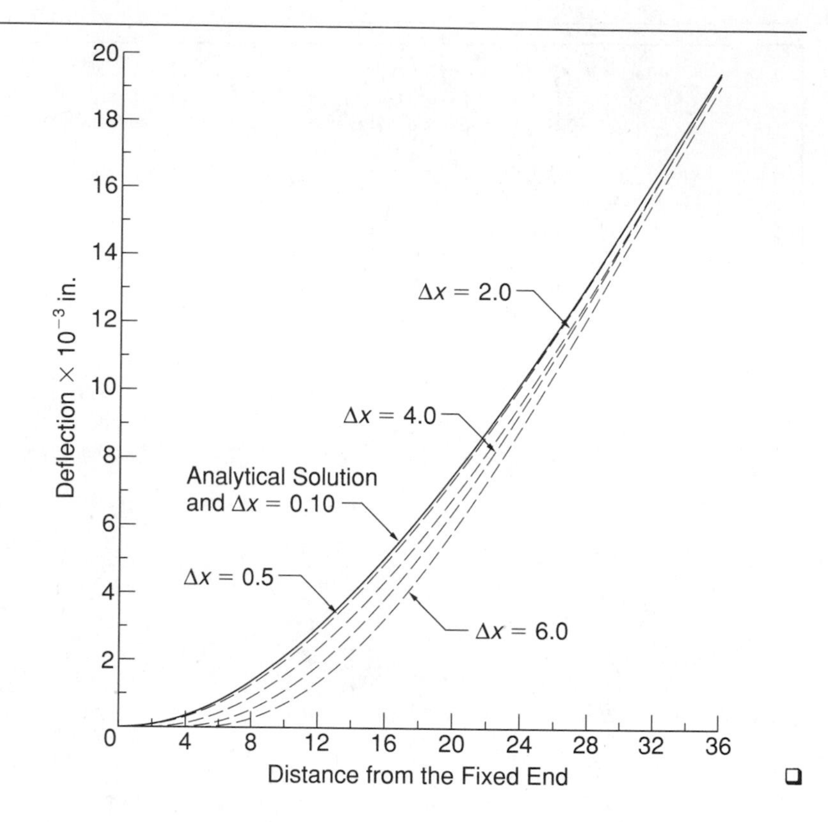

FIGURE 2.8 Deflection of a cantilever beam determined by the finite difference method with different increments

2.6 ▪ Numerical Integration

❏ GENERAL

Integration of a function over a specified interval is commonly required in engineering analysis. Practical applications—such as the computations of the total work produced by an engine, the strain energy stored in a deformed solid, and the heat stored in a thermal system—all require the evaluation of integrals. Exact evaluation of many definite integrals can be carried out by means of various techniques described in calculus. However, in many engineering applications, such as those mentioned above, and also in finite element analysis, which is presented in Chapters 6 and 7, the functions that describe the variations of physical quantities in integrals are either too complicated or are specified only at discrete points so that exact integration is not possible. Here, we will review three numerical integration methods. These are the trapezoidal rule, Simpson's one-third rule, and the Gaussian integration scheme. Detailed derivations of these and some other

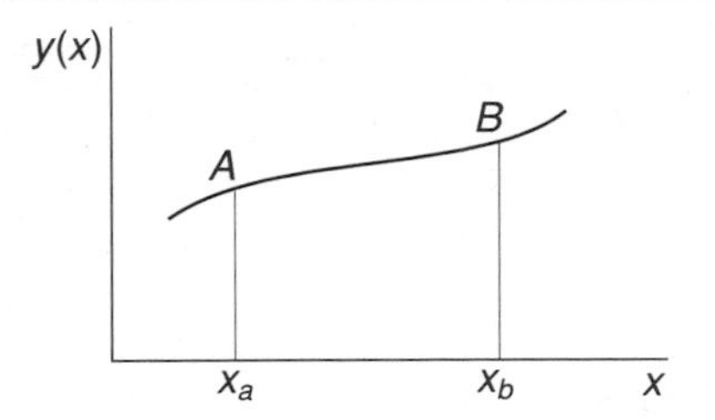

FIGURE 2.9
Numerical integration of a function $y(x)$ between two specified intervals

methods are available in [1, 5]. The first two of these methods are not commonly used in the solution of complex problems involving highly nonlinear functions. They are presented here to illustrate the principles of numerical integration. The third method, the Gaussian integration scheme, however, is extensively used in the numerical analysis of complex engineering problems.

Consider a portion of the curve AB representing a function $y(x)$, as illustrated in Figure 2.9. The area under the curve is bounded by the interval x_a to x_b and can be evaluated by the definite integral:

$$I = \int_{x_a}^{x_b} y(x)\,dx \tag{2.67}$$

There are several ways of finding an approximate estimate of this area, as will be described in the following sections.

❑ TRAPEZOIDAL RULE

This is the simplest of all numerical integration methods. The distance $x_b - x_a$ is generally divided into n segments of equal length—say, h—and the curve AB is approximated by a number of straight-line segments. Consequently, we have n trapezoids, and the area under the curve is approximately equal to the sum of areas of the individual trapezoids, as illustrated in Figure 2.10. This sum can be shown to be

$$\begin{aligned} I &= \frac{1}{2}h[(y_0 + y_1) + (y_1 + y_2) + \cdots + (y_{n-1} + y_n)] + 0\,(\text{error}) \\ &= \frac{h}{2}[y_0 + 2y_1 + 2y_2 + \cdots + 2y_{n-1} + y_n] + 0\,(\text{error}) \\ &= \frac{h}{2}\left(y_0 + y_n + 2\sum_{i=1}^{n-1} y_i\right) + 0\,(\text{error}) \end{aligned} \tag{2.68}$$

The accuracy of the above expression depends on how close the curve AB in Figure 2.10 can be approximated by the series of straight-line segments. The exact area of each individual trapezoid in Figure 2.10 is

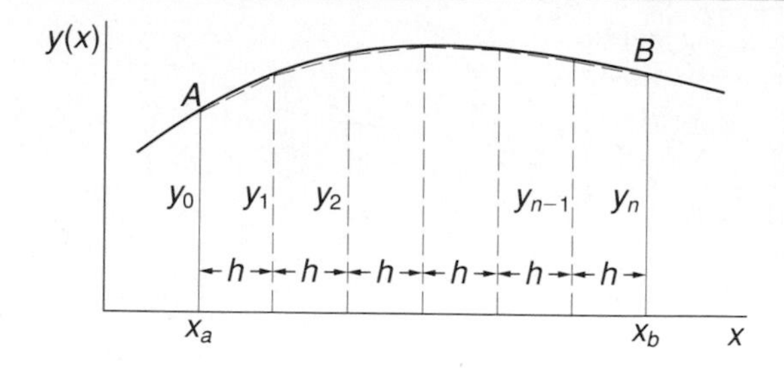

FIGURE 2.10 Graphical representation of numerical integration using trapezoidal rule

equal to

$$a_i = \left(\frac{y_i + y_{i+1}}{2}\right)h - \frac{1}{12}y_i''(x)h^3$$

The reader will notice that only the first term of the above expression is used in establishing equation (2.68). The second term, which is neglected due to mathematical complexity, is referred to as *truncation error.* This error may become significant if large interval size, h, is used. A smaller interval h can reduce the truncation error, provided that the inherent roundoff error does not become dominant.

❑ SIMPSON'S ONE-THIRD RULE

Intuitively we may say that if we replace the curve AB by a series of curves rather than by straight-line segments, as was done in the trapezoidal rule, we should get a better approximation. In Simpson's one-third rule, the curves between three points—such as points x_a, 1, and 2, or points 2, 3, and 4, in Figure 2.11—are replaced by quadratic curves (parabola) of the form

$$y(x) = a + bx + cx^2 \tag{2.69}$$

where a, b, and c are constants. The areas of all sectors, such as $A - x_a - 1 - A'$ in Figure 2.11, are individually found and added together to give

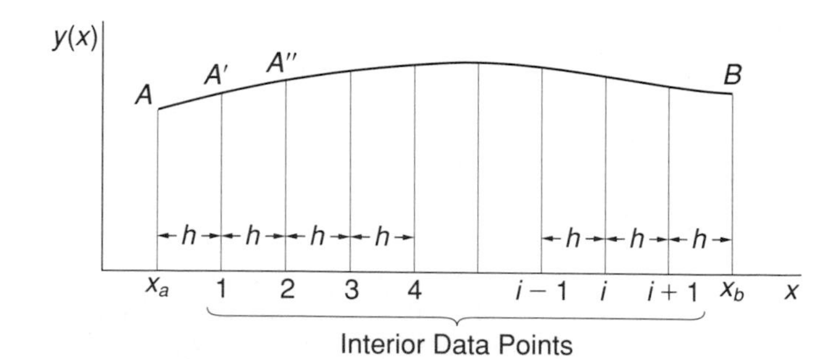

FIGURE 2.11 Graphical representation of numerical integration by Simpson's rule

the estimate of the total area under the curve. Note that the total number of sampling points must be an odd number.

Let us now consider the area between x_{i-1} and x_{i+1}. We assume that the origin of x has been moved to x_i. Equation (2.69) should satisfy the following equations if it is to pass through the points of intersection of the curve AB and the ordinates through x_{i-1}, x_i and x_{i+1}:

$$y_{i-1} = a - bh + ch^2 \tag{2.70}$$

$$y_i = a \tag{2.71}$$

$$y_{i+1} = a + bh + ch^2 \tag{2.72}$$

from which we get

$$a = y_i \tag{2.73}$$

$$b = \frac{y_{i+1} - y_{i-1}}{2h} \tag{2.74}$$

$$c = \frac{y_{i+1} - 2y_i + y_{i-1}}{2h^2} \tag{2.75}$$

The area under the assumed parabola is given as

$$\begin{aligned} I' &= \int_{-h}^{h} (a + bx + cx^2)\,dx \\ &= 2ah + 2c\frac{h^3}{3} \\ &= 2hy_i + h\frac{y_{i+1} - 2y_i + y_{i-1}}{3} \\ &= \frac{h}{3}(y_{i+1} + 4y_i + y_{i-1}) \end{aligned} \tag{2.76}$$

Adding all such areas, we get

$$\begin{aligned} I &= \frac{h}{3}[(y_0 + 4y_1 + y_2) + (y_2 + 4y_3 + y_4) + \cdots \\ &\qquad + (y_{n-2} + 4y_{n-1} + y_n)] + 0\,(\text{error}) \\ &= \frac{h}{3}(y_0 + 4y_1 + 2y_2 + 4y_3 + 2y_4 + \cdots + 2y_{n-2} \\ &\qquad + 4y_{n-1} + y_n) + 0\,(\text{error}) \end{aligned} \tag{2.77}$$

Like many other numerical integration techniques, the truncation error associated with this method is $y_i^{\text{iv}}(x)\, h^5/90$ + (higher-order terms). More accurate results are achievable by using a smaller interval size, h. However, one must not overlook the fact that smaller h values mean more arithmetic operations, which can result in significant roundoff errors.

One may also be tempted to assume that a more accurate result can be obtained by increasing the order of the approximating curve. This, unfortunately, does not happen. If a cubic curve is used instead of the parabolic curve described earlier, little improvement in the accuracy of the computation will be achieved.

❑ GAUSSIAN INTEGRATION

In the trapezoidal rule and in Simpson's one-third rule, the interval $x_b - x_a$ was divided into equal parts. Thus, we selected the sampling points and evaluated the integral in terms of the discrete values of the function at these points. These methods do not yield a good approximation because the choice of these sampling points is left to the user, who may not have good criteria that can be used to choose the optimum locations for the sample points for accurate approximation. The reader can easily verify this argument by looking at any curve and thus conclude that a far better approximation can be obtained even from the trapezoidal rule if the sampling points are located at more strategic places. Thus, it is desirable that we use some criteria for optimum accuracy in locating the sampling points.

The Gaussian integration scheme is established on the basis of strategically selected sampling points. We will find that this scheme is widely used in the finite element analysis described in Chapter 7.

The normal form of a Gaussian integral can be expressed as

$$I = \int_{-1}^{1} F(\xi)\, d\xi = \sum_{i=1}^{n} H_i F(a_i) \tag{2.78}$$

in which n is the total number of sampling points. The weighting coefficients, H_i, corresponding to sampling points located at $\xi = \pm a_i$ can be obtained from Table 2.2, which can be found in mathematical handbooks [7].

It is obvious that the form of the Gaussian integral shown in equation (2.78) is rarely seen in practice. A transformation of coordinates is required to convert the general form of integral shown in equation (2.67) into the Gaussian integral shown in equation (2.78).

Referring to the coordinates shown in Figure 2.12, if one lets

$$x = \frac{1}{2}(x_b - x_a)\xi + \frac{1}{2}(x_b + x_a) \tag{2.79}$$

the following conversion is obtained:

$$\int_{x_a}^{x_b} f(x)\, dx = \frac{x_b - x_a}{2} \int_{-1}^{1} F(\xi)\, d\xi \tag{2.80}$$

TABLE 2.2
Abscissae and weight coefficients of the Gaussian quadrature formula in equation (2.77)

n	$\pm a_i$	H_i
2	0.57735	1.00000
3	0.77459	0.55555
	0	0.88888
4	0.86113	0.34785
	0.33998	0.65214
5	0.90617	0.23692
	0.53846	0.47862
	0	0.56888
6	0.93246	0.17132
	0.66120	0.36076
	0.23861	0.46791
7	0.94910	0.12948
	0.74153	0.27970
	0.40584	0.38183
	0	0.41795
8	0.96028	0.10122
	0.79666	0.22238
	0.52553	0.31370
	0.18343	0.36268

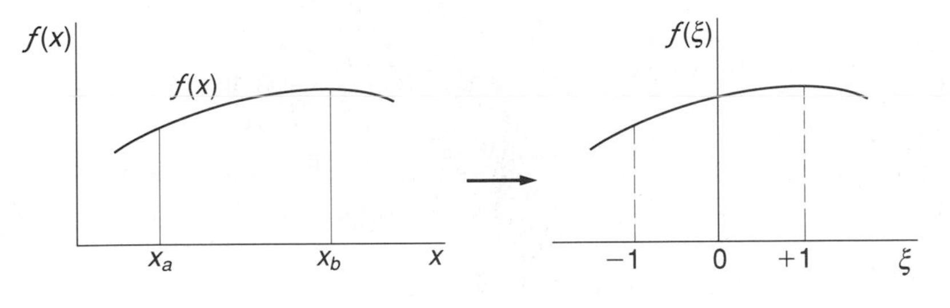

FIGURE 2.12
Transformation of coordinates for Gaussian integration

By using the relationship given in equation (2.78), the integral in equation (2.67) can be evaluated as

$$I = \int_{x_a}^{x_b} f(x)\,dx = \frac{x_b - x_a}{2} \sum_{i=1}^{n} H_i F(a_i) \tag{2.81}$$

EXAMPLE ■ 2.6

Evaluate the following integral by using the Gaussian integration scheme:

$$I = \int_0^{\pi} \sin x\,dx$$

The limits of the above integral are $x_a = 0$ and $x_b = \pi$. Thus, the transformation of coordinates follows the relationship given in equation (2.79), or

$$x = \frac{\pi}{2}\xi + \frac{\pi}{2}$$

from which one obtains $dx = \frac{\pi}{2} d\xi$.

Upon substituting the above relations into equation (2.80) and equation (2.81), the following conversion is obtained:

$$I = \int_0^{\pi} \sin x\, dx = \frac{\pi}{2}\int_{-1}^{1} \cos\frac{\pi}{2}\xi\, d\xi = \frac{\pi}{2}\sum_{i=1}^{n} H_i \cos\frac{\pi}{2} a_i \tag{2.82}$$

Let us take, for example, $n = 3$. From Table 2.2, we obtain the following sampling points, a_i, and the corresponding weighting coefficients, H_i to the fourth decimal point as

$$a_1 = 0 \qquad a_2 = +0.7746 \qquad a_3 = -0.7746$$

$$H_1 = 0.8888 \qquad H_2 = 0.5555 \qquad H_3 = 0.5555$$

Substituting the above values into (2.82):

$$I = \frac{\pi}{2}\left[0.8888 \cos(0) + 0.5555 \cos\left(\frac{\pi}{2} \times 0.7746\right) + 0.5555 \cos\left(-\frac{\pi}{2} \times 0.7746\right)\right]$$

$$= 2.0014$$

The reader may easily verify that the exact value of the integral is 2.0. ❑

EXAMPLE ■ 2.7

In the course of establishing the stiffness matrix for a tapered rod, as illustrated in Figure 2.13, for a finite element analysis, it is necessary to evaluate the following integral:

$$K_{11} = EA_0 \int_0^b \left(-\frac{3}{b} + \frac{4y}{b^2}\right)^2 \left(1 - \frac{y}{b}\right)^2 dy$$

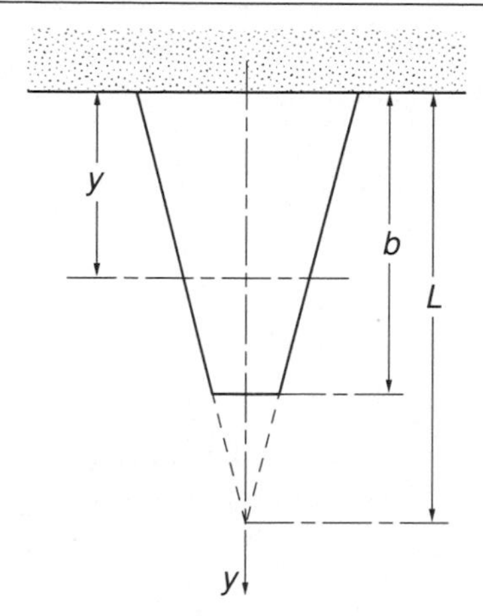

FIGURE 2.13
A hanging tapered rod

in which E is the Young's modulus of the material and A_0 is the cross-sectional area of the rod at the base. The lengths b and L are defined in Figure 2.13.

The following numerical values are used in this example:

$$A_0 = 0.0127 \text{ m}^2$$
$$L = 0.508 \text{ m}$$
$$b = 0.483 \text{ m}$$
$$E = 6.894 \times 10^{10} \text{ Pa}$$

The integral thus becomes

$$K_{11} = 8.7554 \times 10^8 \times \int_0^{0.483} (-6.2112 + 17.146y)^2(1 - 1.9685y)^2 \, dy \tag{2.83}$$

The conversion of the above integral into the Gaussian integral in equation (2.78) must be carried out first. This can be achieved by using equation (2.79) and transforming the integrand from the y domain into the ξ domain by letting

$$y = \frac{1}{2}(0.483 - 0)\xi + \frac{1}{2}(0.483 + 0) = 0.2415\xi + 0.2415 \tag{2.84}$$

It is readily seen that $dy = 0.2415\, d\xi$ from equation (2.84).

By substituting dy in the above expression and y in equation (2.84) into equation (2.83), one obtains

$$K_{11} = 2.1144 \times 10^8 \times \int_{-1}^{1} (4.1408\xi - 2.0704)^2(0.5246 - 0.4754\xi)^2 \, d\xi$$

$$= \sum_{1}^{n} H_i F(a_i) = H_1 F(a_1) + H_2 F(a_2) + H_3 F(a_3)$$

in which a_i and H_i ($i = 1, 2, 3$) can be found in Table 2.2. The reader will find that the integral in equation (2.83) has a numerical value of $K_{11} = 0.2833 \times 10^{10}$. ❑

2.7 ▪ Summary

Complicated mathematical functions and equations are often used in performing design analysis and producing geometric models in CAD. Numerical techniques must be used to convert these functions and equations into simple arithmetical operations that can be handled by digital computers. The purpose of this chapter is to provide the reader with a brief review of some key elements that are relevant to CAD and analysis.

The Gaussian elimination method was the first such element to be reviewed. It involves the elimination of terms in the sets of simultaneous equations by reducing the coefficient matrix of these equations to an upper triangular matrix. Solutions to unknown variables can then be found by the back-substitution process. Another technique, called the *Cholesky decomposition method*, was also presented. This technique can lead to the triangularization of the coefficient matrix and is often used to solve eigenvalue problems. The principle of converting derivatives into finite difference forms was presented, and the application of finite difference schemes for the solution of differential equations was illustrated. Three numerical techniques for handling definite integrals were reviewed. The first two techniques, the trapezoidal rule and Simpson's rule, illustrated numerical integration techniques; both require fixed increments for dependent variables. The third technique, the Gaussian integration scheme, automatically sets optimal sample (or integration) points. This technique gives far more accurate results than do other methods and is commonly used in finite element analysis.

Numerical techniques are necessary to convert complicated mathematical manipulations into simple arithmetical operations that are manageable by digital computers. Readers, however, must be aware of the inherent errors in such practice. Recognition of these errors and intelligent use of various numerical analysis schemes are critical to the accuracy of the results.

References

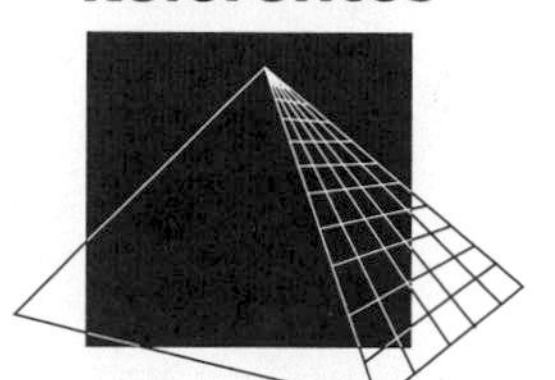

1. Ralston, A. *A First Course in Numerical Analysis.* New York: McGraw-Hill, 1965.
2. McCracken, D.D., and W.S. Dorn. *Numerical Methods and Fortran Programming.* New York: John Wiley, 1964.
3. Chapra, S.C., and R.P. Canale. *Numerical Methods for Engineers.* New York: McGraw-Hill, 1984.
4. Shoup, T.E. *Applied Numerical Methods for the Microcomputer.* Englewood Cliffs, N.J.: Prentice-Hall, 1984.
5. James, M.L., G.M. Smith, and J.C. Wolford. *Applied Numerical Methods for Digital Computation.* 3d ed. New York: Harper & Row, 1985. pp. 178–85.
6. Hildebrand, F.B. *Finite-Difference Equations and Simulations.* Englewood Cliffs, N.J.: Prentice-Hall, 1968.
7. Abramowitz M., and I.A. Stegun, eds. *Handbook of Mathematical Functions.* New York: Dover, 1965. pp. 916–20.

Problems

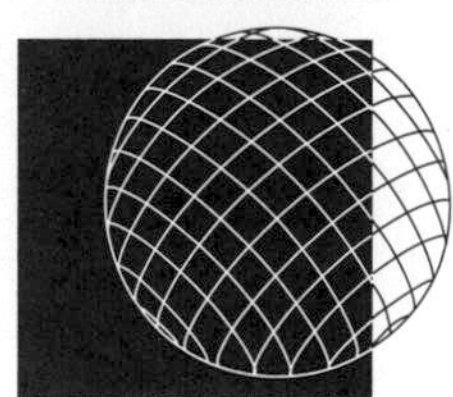

2.1 If

$$[A] = \begin{bmatrix} 3 & 0 & 1 \\ 2 & -2 & 1 \\ 0 & 2 & 0 \end{bmatrix} \quad [B] = \begin{bmatrix} 2 & -1 & 1 \\ 1 & 0 & 0 \\ 2 & 0 & 1 \end{bmatrix} \quad \text{and}$$

$$[C] = \begin{bmatrix} 1 & 2 & -1 \\ 0 & 1 & 2 \\ 1 & 1 & 1 \end{bmatrix}$$

show the following properties of matrix multiplication:

(a) $(k[A])[B] = k([A][B]) = [A](k[B])$, where k is a real number
(b) $[A]([B][C]) = ([A][B])[C]$
(c) $([A] + [B])[C] = [A][C] + [B][C]$
(d) $[A]([B] + [C]) = [A][B] + [A][C]$
(e) $([A][B])^T = [B]^T[A]^T$

2.2 Invert the matrices

$$[A] = \begin{bmatrix} \cos\theta & -\sin\theta \\ \sin\theta & \cos\theta \end{bmatrix} \quad \text{and} \quad [B] = \begin{bmatrix} 1 & 2 \\ 3 & -1 \end{bmatrix}$$

and verify your solutions.

2.3 If matrices

$$[A] = \begin{bmatrix} 0 & 1 \\ 1 & 0 \end{bmatrix} \quad \text{and} \quad [B] = \begin{bmatrix} -1 & 0 \\ 0 & 1 \end{bmatrix}$$

show that $[A][B] \neq [B][A]$.

2.4 Multiply the following matrices:

$$\begin{bmatrix} 1 & 2 & 3 \\ 3 & 1 & 2 \\ 1 & 3 & 2 \end{bmatrix} \begin{bmatrix} 2 & 0 & 0 \\ 1 & 0 & 0 \\ 3 & 0 & 0 \end{bmatrix}$$

2.5 The integration and differentiation of matrices with variable elements can be performed by integrating or differentiating the individual elements of the matrices; for example,

$$\int_v [c]\, dv = \left[\int_v c_{ij}\, dv \right], \text{ etc.}$$

Find:

(a) $\int_0^5 [T]\, dt$
(b) The inverse of the matrix $[T]$; that is, $[T]^{-1}$
(c) $d[T]^{-1}/dT$
where the elements of the matrix $[T]$ are

$$T_{11} = T^2 + 1 \qquad T_{12} = T^2$$

$$T_{21} = T^2 \qquad T_{22} = T^2 - 1$$

2.6 Referring to the following figure, if the stress components in a solid element can be expressed in matrix form as

$$\{\sigma\} = \{\sigma_1 \quad \sigma_2 \quad \sigma_3 \quad \sigma_4 \quad \sigma_5 \quad \sigma_6\}^T$$

and the strain matrix is

$$\{\varepsilon\} = \{\varepsilon_1 \quad \varepsilon_2 \quad \varepsilon_3 \quad \varepsilon_4 \quad \varepsilon_5 \quad \varepsilon_6\}^T$$

and the strain energy in the element is

$$U = 0.5(\sigma_1\varepsilon_1 + \sigma_2\varepsilon_2 + \sigma_3\varepsilon_3 + \sigma_4\varepsilon_4 + \sigma_5\varepsilon_5 + \sigma_6\varepsilon_6)$$

express U in a matrix form.

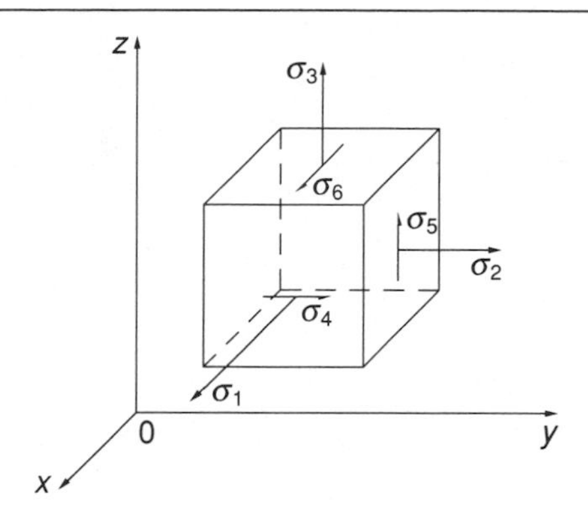

Figure for Problem 2.6

2.7 The generalized Hooke's law can be shown to be

$$\varepsilon_1 = \frac{1}{E}[\sigma_1 - \nu(\sigma_2 + \sigma_3)] \qquad \varepsilon_2 = \frac{1}{E}[\sigma_2 - \nu(\sigma_1 + \sigma_3)]$$

$$\varepsilon_3 = \frac{1}{E}[\sigma_3 - \nu(\sigma_1 + \sigma_2)] \qquad \varepsilon_4 = \frac{1}{G}\sigma_4 \qquad \varepsilon_5 = \frac{1}{G}\sigma_5 \qquad \varepsilon_6 = \frac{1}{G}\sigma_6$$

where ε_1 to ε_3 and σ_1 to σ_3 are the corresponding normal strains and stress components along the x, y, and z coordinates, as shown in the figure illustrating problem 2.6. ε_4 to ε_6 and σ_4 to σ_6 are the corresponding shearing strain and stress components on the xy, yz, and xz planes. E is the modulus of elasticity, G is the shear modulus of elasticity, and ν is Poisson's ratio.

(a) Express the normal strains and stresses in the above equations in matrix form.

(b) Express all strain and stress components in the above equations in matrix form.

(c) Express the normal stresses in terms of normal strains in the matrix form by matrix inversion.

2.8 Use the computer programs provided in Appendix I to obtain

(a) $\begin{bmatrix} 5 & 3 & 4 \\ 7 & 6 & -3 \end{bmatrix} \begin{bmatrix} 2 & 1 \\ 5 & 4 \\ 7 & 9 \end{bmatrix}$ (b) Inverse of $\begin{bmatrix} 7 & 4 & 6 \\ 1 & 8 & 5 \\ 3 & 2 & 9 \end{bmatrix}$

2.9 Use (a) the Gaussian elimination method and (b) the Cholesky decomposition method to solve the following set of simultaneous equations:

$$\begin{aligned}
2x_1 + 5x_2 - 7x_3 + 4x_4 &= 7 \\
5x_1 - 4x_2 + 3x_3 - 6x_4 &= -18 \\
-7x_1 + 3x_2 + 5x_3 + 4x_4 &= 30 \\
4x_1 - 6x_2 + 4x_3 + 8x_4 &= 36
\end{aligned}$$

2.10 Show that if $[A] = [L][L]^T$ and $[L]$ is a lower triangular matrix of the same order as $[A]$, then $[A]$ must be a symmetric square matrix.

2.11 Find the natural frequencies of a two-degrees-of-freedom mass-spring system (see Figure 2.1) with $M_1 = 2M_2$ and $k_1 = k_2 = 2k$. Also calculate the modes of vibration of mass M_1 and M_2.

2.12 Find the natural frequencies and modes of vibration of a free-rolling freight train after it hits a stopper, as illustrated below.

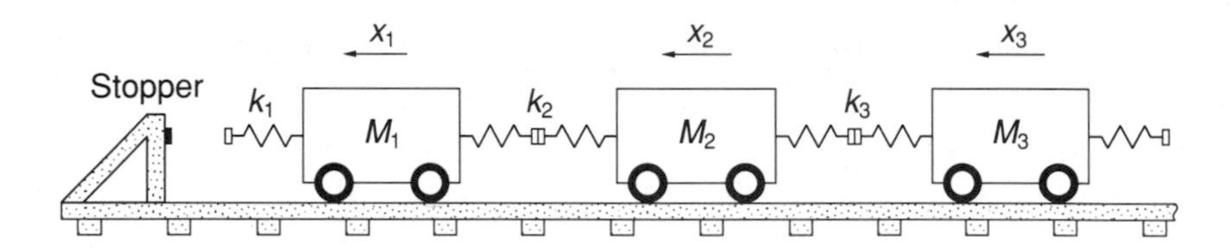

Assume the impact velocity and the friction between the wheels and the tracks can be neglected.

2.13 Solve Example 2.3 by decomposing the stiffness matrix $[K]$ and subdivide the rod into three sections with interior nodes situated at $x = 6$ cm for node 2 and $x = 12$ cm for node 3. Tabulate the modes of vibration for nodes 1, 2, 3, and 4.

2.14 Solve the numerical illustration of Example 2.3 after adding another 10-cm-long section.

2.15 Derive formulas for the third and fourth derivatives, using the three different schemes shown in Table 2.1.

2.16 Write a FORTRAN computer program for the finite difference method and use it to find the solution of the following differential equation:

$$t\frac{d^2\phi(t)}{dt^2} - 6\frac{d\phi(t)}{dt} - 15\phi(t) = 6.25te^{-2.5t} - (144t^4 + 15t^3 + 12t^2)\cos 12t$$

$$0 \le t \le 100$$

with

$$\phi(0) = 1$$
$$\phi'(0) = -2.5$$

Use $\Delta t = 0.1$ and $\Delta t = 10$. Compare your finite difference solutions with the analytical solutions, which are $\phi(t) = e^{-2.5t} + t^3 \cos 12t$.

2.17 Find the finite difference equation for each of the following differential equations by the forward difference scheme:

(a) $\frac{d}{dx}\left[(6 - x)\frac{dy(x)}{dx}\right] = 10$

(b) $\frac{\partial^2 \phi(x, y)}{\partial x^2} + \frac{\partial^2 \phi(x, y)}{\partial y^2} = 0$

(c) $\frac{\partial^2 \psi(x, y)}{\partial x^2} + \alpha \frac{\partial^2 \psi(x, y)}{\partial y^2} = 0$

2.18 Evaluate the following integral:

$$I = \int_1^6 (3 + x)^{1/2}\, dx$$

Use

(a) The trapezoidal rule with increment $h = 1$;
(b) The Gaussian integration method with three integration points ($n = 3$)

2.19 Use the three methods described in Section 2.6 to determine the following three integrals:

$f_1(x) = \exp(-x^2)$	interval x: (0, 1)
$f_2(x) = \cos 5x$	x: (0, 1)
$f_3(t) = t^3(t - 1)^{1/2} \cos 6t$	t: (1, 4)

Which method, in your view, is best in terms of both the accuracy of the results and the required computational effort?

2.20 The strain energy of a solid induced by stress components, $\{\sigma\}$, can be expressed as

$$U = \frac{1}{2}\int_v \{\varepsilon\}^T\{\sigma\}\, dv$$

where $\{\varepsilon\}$ is the strain components corresponding to the stress components and v is the volume.

In a one-dimensional case, assume that the variations of the stress and strain along a rod with a length of 10 cm and a cross-sectional area of 3 cm^2 follow these expressions:

$$\sigma(x) = 5e^{-x} + 4x^3$$

and

$$\varepsilon(x) = 10e^{-2x} + 7x$$

Use the Gaussian integration method to determine the total strain energy in the rod.

2.21 Use (a) the trapezoidal rule with $h = 0.25$; (b) Simpson's rule with $h = 0.25$; and (c) the Gaussian integration method with $n = 3$ to evaluate the following integral:

$$\int_{2.0}^{4.5} x^2 \log x \, dx$$

CHAPTER 3

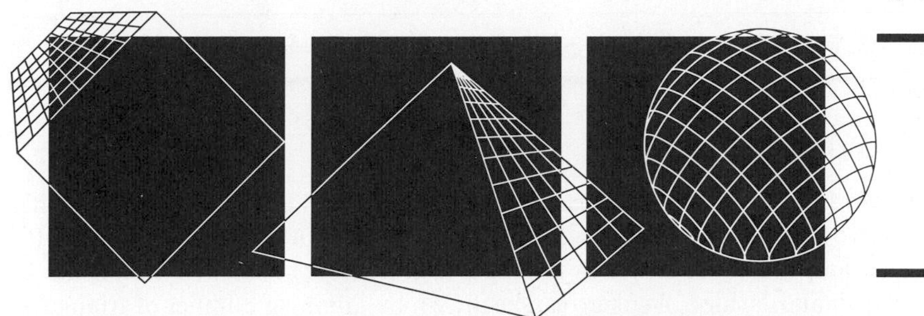

Principles of Computer Graphics

3.1 Introduction

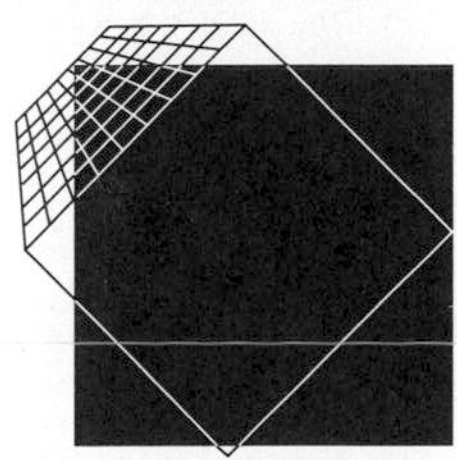

Engineers often use computer graphics for plotting graphs of two variables, pie charts, and histograms and for preparing floor plans, factory layouts, and orthographic projections. These are examples of two-dimensional graphics in which the image is two-dimensional—as on the computer monitor screen. These are somewhat simpler and more mundane uses of computer graphics. The real power of the computer is in three-dimensional graphics. It has the potential for creating realistic-looking images of solid objects (such as a block of metal, a table, a chair, or an aircraft). The mathematical formulation necessary for three-dimensional graphics, however, is far more involved than that for two-dimensional graphics. This is because significantly more data are needed to represent the object. Hidden line/surface removal and shading algorithms are also necessary to create the appearance of depth.

To fully explain the principles involved in present-day computer graphics, several volumes would be required—and quite a few of these are available. In view of the limited space available here and the needs of the target audience, only selected basics of computer graphics are discussed below. The topics have been chosen with a view toward providing insight into the most commonly used computer graphics operations.

3.2 ▪ Mathematical Formulations for Graphics

❑ LINES

Straight lines form the basis for the display of all types of shapes, whether two-dimensional or three-dimensional, in computer graphics. A circle, for example, may be drawn by joining a large number of small line segments. In a similar fashion, a sphere can be drawn by joining a number of triangular shapes, which in turn may be created by joining three straight lines.

Suppose that a straight line is to be drawn from point $P_1(x_1, y_1)$ to point $P_2(x_2, y_2)$. It is shown later that by means of window and viewport transformations, it is possible to assign these points to specific locations on a computer monitor screen. On a raster display, the location of these points is defined in terms of the sequence number (address) of a picture element (pixel or pel) and the row number to which the pixel belongs. In order to draw a continuous-looking line, the computer must be able to pick up a number of other pixels that should also be lighted in addition to the two end pixels. Several methods are available to identify the intervening pixels. A popular method uses an algorithm known as the *symmetric digital differential analyzer* (DDA). In DDA the equation of a line is expressed as a pair of parametric equations:

$$x = x_1 + (x_2 - x_1)*s \tag{3.1}$$

$$y = y_1 + (y_2 - y_1)*s \tag{3.2}$$

where the parameter s varies from 0 to 1. If s is incremented by, say, 0.1, then the DDA will generate 9 points in between the two ends of the line. For each value of s, real numbers are found for the x and the y coordinates of the intervening pixels from equations (3.1) and (3.2). A pixel, however, can have only integer coordinates. Therefore, a hardware device is usually provided that converts each fractional coordinate into the nearest integer. Thus, 9.6 is rounded up to 10, and 9.1 is rounded down to 9.

Once the addresses of all the pixels are determined, these are stored in a frame buffer. The display driver reads the array of addresses and illuminates the corresponding pixels. In order for a line to look continuous, it is necessary to increment the parameter s in a manner consistent with the resolution of the display device. If too large an increment is selected, the line may appear to be discontinuous. If too small an increment is selected, a large number of pixels near each other are illuminated, and the line appears thicker and brighter, but the drawing process becomes slower.

❑ CIRCLES

As mentioned before, a circle may be treated as a polygon. Typically, a circle is represented as a polygon inscribed within the desired circle. An efficient

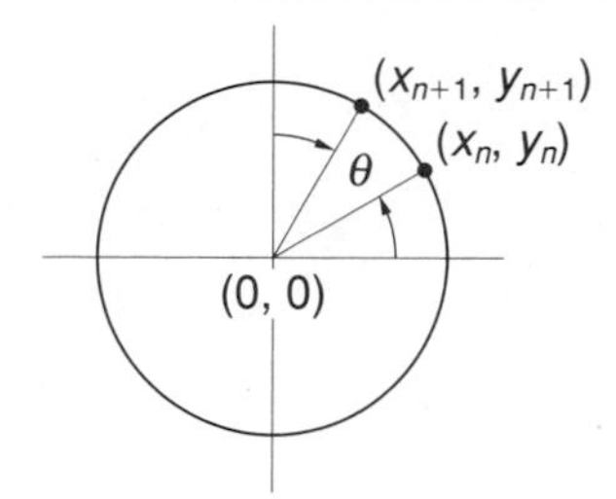

FIGURE 3.1
Terms used in equations (3.3) and (3.4)

way of generating the coordinates of the $n + 1$th vertex of the inscribing polygon is given by the following equations:

$$x_{n+1} = x_n \cos\theta + y_n \sin\theta \tag{3.3}$$

$$y_{n+1} = y_n \cos\theta - x_n \sin\theta \tag{3.4}$$

In equations (3.3) and (3.4), θ is the angle subtended by each side of the polygon at the center of the circle. (x_n, y_n) are the coordinates of the nth vertex, and (x_{n+1}, y_{n+1}) are the coordinates of the $n + 1$th vertex, as shown in Figure 3.1. The advantage of using parametric equations, such as those above, is that the points for the circle may be generated recursively and, therefore, rapidly. Values for the trigonometric functions are determined only once.

❑ ELLIPSES

Parametric equations are also used for ellipses. Equations (3.5) and (3.6) are used for updating the trigonometric functions, and equations (3.7) and (3.8) for generating the coordinates [1]:

$$\cos\gamma_{n+1} = \cos\theta\cos\gamma_n - \sin\theta\sin\gamma_n \tag{3.5}$$

$$\sin\gamma_{n+1} = \sin\theta\cos\gamma_n + \cos\theta\sin\gamma_n \tag{3.6}$$

$$x_{n+1} = x_0 + a\cos\phi\cos\gamma_{n+1} - b\sin\phi\sin\gamma_{n+1} \tag{3.7}$$

$$y_{n+1} = y_0 + b\sin\phi\cos\gamma_{n+1} + b\cos\phi\sin\gamma_{n+1} \tag{3.8}$$

where ϕ is the angle of inclination of the major axis, a and b are the lengths of the semimajor and semiminor axes, (x_0, y_0) are the coordinates of the center of the ellipse, and θ is an incremental angle. (See Figure 3.2.) The parametric angle γ varies from 0 to 2π.

❑ CURVES

Bézier Curves ▪ In many computer graphics applications, especially those related to the interactive design of shapes, it is necessary to generate

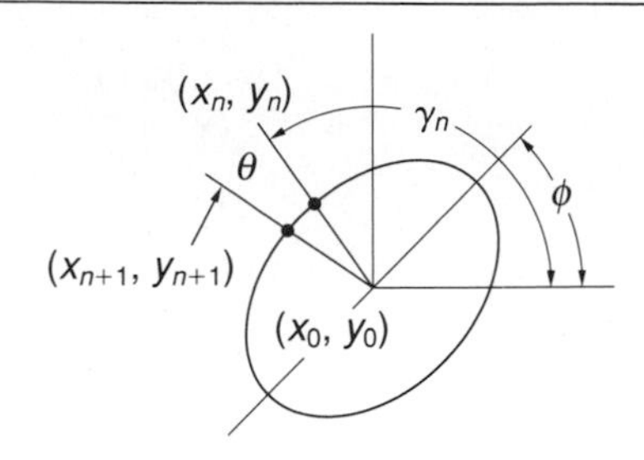

FIGURE 3.2 An ellipse with inclined axes

free-form curves and surfaces. The curves are often required to pass through or to be controlled by several key points. Many algorithms are available for drawing such curves, in either two or three dimensions. One simple and very popular method was developed by Bézier [2].

A Bézier curve, of order n, is defined in parametric form by the following equation:

$$P(s) = \sum_{i=0}^{n} p_i B(s)_{i,n} \tag{3.9}$$

where there are $n + 1$ control points, s is a single-valued parameter varying from 0 to 1, and p_i is a vector of coordinates of the ith control point (i.e., $p_i = \{x_i, y_i\}$). $B(s)_{i,n}$ is the blending function

$$B(s)_{i,n} = \frac{n!}{i!n - i!} s^i (1 - s)^{n-i} \tag{3.10}$$

$P(s) = \{X(s), Y(s)\}$ is the vector of coordinates of points on the Bézier curve. More explicitly, equation (3.9) is written in the following form for two-dimensional graphics:

$$X(s) = \sum_{i=0}^{n} x_i B(s)_{i,n} \tag{3.11}$$

$$Y(s) = \sum_{i=0}^{n} y_i B(s)_{i,n} \tag{3.12}$$

EXAMPLE ■ 3.1

Draw a Bézier curve with control points; (0, 0), (12, 3), and (8, 5.5). Suppose that s is incremented by 0.2. Then the following points are obtained for different values of s, using equations (3.10) through (3.12):

s	$X(s)$	$Y(s)$	s	$X(s)$	$Y(s)$
0	0	0	0.6	8.64	3.42
0.2	4.16	1.18	0.8	8.96	4.48
0.4	7.04	2.32	1.0	8.0	5.5

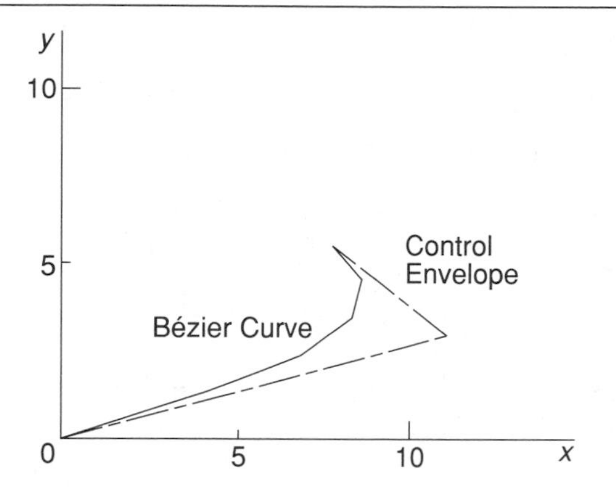

FIGURE 3.3
A Bézier curve and the control points

Typically, the values shown above are found thus:

$$X(0.2) = x_0 B(0.2)_{0.2} + x_1 B(0.2)_{1.2} + x_2 B(0.2)_{2.2}$$
$$= (0)\frac{2!}{0!2!}(0.2)^0(0.8)^2 + (12)\frac{2!}{1!1!}(0.2)^1(0.8)^1 + (8)\frac{2!}{0!2!}(0.2)^2(0.8)^0$$
$$= 4.16$$

In Figure 3.3, the Bézier curve is drawn through these points. The control points are joined by broken lines to form the control envelope. ❑

As is seen from equations (3.11) and (3.12), the blending functions act as links between the coordinates of the control points and the points on the Bézier curve. If there are $n + 1$ control points, there will be exactly the same number of separate blending functions for each value of s. At $s = 0$ (i.e., at the 0th control point), the blending function $B_{0,1} = 1$ and all other blending functions are zero. As a result, the first point on the Bézier curve coincides with the 0th control point. Similarly, the last point on the Bézier curve coincides with the $n + 1$th control point . In general, a Bézier curves does not pass through any other control points, as shown in Example 3.1.

Although simple to use and apply, the Bézier method has certain disadvantages. It is generally difficult to predict the shape of the curve since it does not pass through most of the control points. Thus, problems arise in modeling complex profiles. Usually, in such situations, multiple Bézier curves are required—joined end to end—to produce the desired shape.

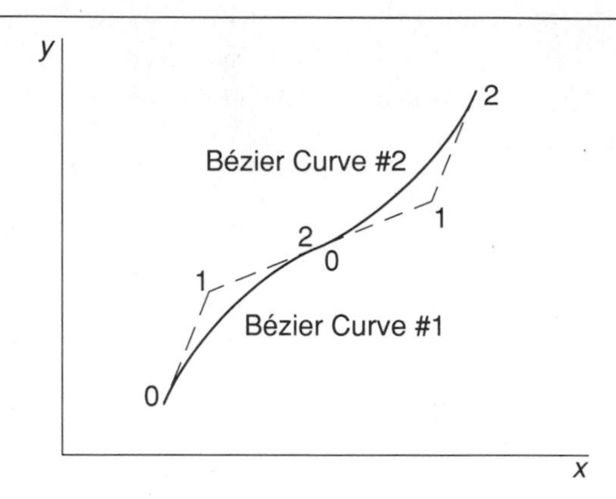

FIGURE 3.4
Two Bézier curves with continuity of first order at the junction

In joining two curves, certain continuity conditions must be met. One obvious requirement is that the two curves do indeed meet at the junction point. Mathematically, this is termed a continuity of *order zero.* A 0th order continuity on Bézier curves is obtained by making the last control point of the first curve the first control point of the second curve. Continuity of *first order* (i.e., continuity of slope) at the junction may be guaranteed by placing the last two control points of the first curve and first two control points of the second curve on a straight line. The first control point of the second curve and the last control point of the first curve must still be coincident. See Figure 3.4.

The requirements for the above types of continuity become quite restrictive in most cases, causing great difficulty in creating arbitrary (free-form) curves. If continuity of *second order* (i.e., continuity of curvature) is also sought at the junctions, the Bézier method becomes extremely cumbersome to use.

Piecewise Polynomial Functions ▪ A more flexible method for producing arbitrary curves uses piecewise polynomial (pp) functions. A convenient method of generating pp functions is that of polynomial B-splines. Another form of B-spline is the *beta spline* [3], in which *bias* and *tension* parameters are used to produce additional flexibility. Rational B-spline curves, in which the scale factors (Section 3.6) in the homogeneous coordinate system are different from 1, are also powerful in generating free-form curves, conic sections, circles, etc. Some discussion on nonuniform rational B-splines, known as *NURBS,* follows the discussion on the polynomial B-splines. It is, however, advised that students familiarize themselves with the concept of the homogeneous coordinate system before reading about NURBS.

Polynomial B-splines. A piecewise polynomial function consists of a number of different polynomial functions joined together to form a single curve. The points at which segments of the polynomial curves are joined are known as *knots* or *joints.* There are several types of piecewise polynomials possible (e.g., the piecewise cubic polynomial, such as that used in Section 3.3 for curve fitting). Piecewise cubic Hermite polynomials (splines) are also used for curve fitting.

A polynomial spline $P(s)$ of order k (degree $k - 1$) is a pp function comprised of polynomial segments of degree less than k. Generally, it is assumed that the spline has continuity of order $k - 1$ at all *interior knots* (i.e., at all knots away from the ends of the curve). It is possible nonetheless, and often advantageous, to maintain a higher or lower order of continuity at one or all of the interior knots.

A spline $P(s)$ may be represented as a linear combination of basis or B-splines in the following manner:

$$P(s) = \sum_{i=1}^{n} p_i B(s)_{i,k} \tag{3.13}$$

In the above expression, index i represents the designation of the elements of the knot sequence vector [4]:

$$t = \{t_i\}_1^{k+n} \tag{3.14}$$

where $k + n$ is the total number of elements in the vector. As the name implies, the knot sequence vector contains the identification numbers of all or some of the knots. Depending on which knots have been included and how many times in the vector t, the nature of the spline curve varies a good deal, as is shown in Example 3.2. The formulation of a knot sequence is thus critical. More information on the techniques of forming a knot sequence vector is given in the following paragraphs.

The knot sequence in the vector t is nondecreasing (i.e., knot $m + 1$ follows knot m, etc.). If a knot is included once, the order of continuity of the spline at that knot is $k - 2$. Such a knot is known as a *simple knot*. If the knot appears twice, it is known as a *double knot* [5]. The order of continuity at a double knot is $k - 3$. If a knot is repeated r times, the order of continuity at that knot is $k - r - 1$. Thus, if a knot is repeated $k - 1$ times, the order of continuity of the spline at that knot is zero.

The B-splines given in equation (3.13) may be calculated from the following recurrence formula [4]:

$$B_{i,1} = 1, \quad \text{when } t_i < s < t_{i+1}$$

otherwise,

$$B_{i,1} = 0$$

$$B(s)_{i,k} = \frac{s - t_i}{t_{i+k-1} - t_i} B(s)_{i,k-1} + \frac{t_{i+k} - s}{t_{i+k} - t_{i+1}} B(s)_{i+1,k-1} \tag{3.15}$$

In order to define the complete set of B-splines in equation (3.13), it is necessary to augment the set of interior knots—1, 2, . . . , $m-1$, m—by k additional knots preceding the first knot and by k knots following the mth knot. This is generally done by repeating the 0th and the $m + 1$th (exterior) knots several times in the vector t. Thus, a knot sequence for a spline of order k typically contains the following:

$$t = \{\underbrace{0, 0, \ldots, 0}_{k \text{ terms}}, 1, 2, \ldots, m-1, m, \underbrace{m+1, m+1, \ldots, m+1}_{k \text{ terms}}\} \tag{3.16}$$

The knot sequence in equation (3.16) thus contains $2k + m$ elements. By definition, therefore, in equation (3.13)

$$n = k + m \tag{3.17}$$

If a knot—say, P—is repeated p times and a knot R is repeated r times,

$$n = k + m + (p-1) + (r-1) + \cdots \tag{3.18}$$

B-splines exhibit local support (i.e., only a few knots affect the nature of the curves in a given region). At any value of s a maximum of k B-splines have nonzero values. It can be shown that the sum total of numerical values of all nonzero B-splines at any point is equal to unity [4].

The numerical value of parameter s in equation (3.13) varies from t_k (i.e., the sequence number of the knot in the kth position in vector t) to t_{n+1}, or from 0 to $m + 1$, assuming that the labeling starts at knot 0 and the last knot is labeled $m + 1$. B-spline coefficients, p_i, are determined in several ways. One convenient way [4] is to have

$$p_i = g(t_i^*) \tag{3.19}$$

where $g(t)$ is a function describing data points and

$$t_i^* = \frac{t_{i+1} + t_{i+2} + \cdots + t_{i+k-1}}{k-1} \tag{3.20}$$

In most engineering applications, one does not have a function, but rather a set of data points that control the shape of the piecewise polynomial curves. The function $g(t)$, in such cases, is determined by some appropriate interpolation method. Suppose that $t_i^* = 4.5$ is predicted by equation (3.20) for some i; then the coefficient p_i may be determined from the data at the fourth and fifth knots.

EXAMPLE ■ 3.2

Draw a pp of order 4 using B-splines where the knots are located as given below.

(1, 1), (2, 3), (3, 3), (4, 4), (5, 4), (6, 3), (6, 1)

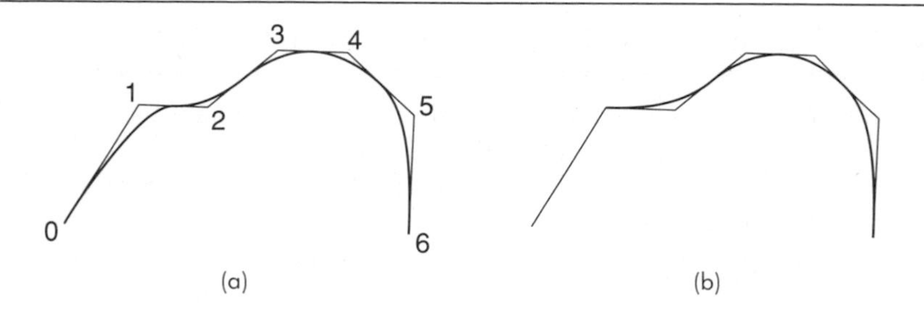

FIGURE 3.5
Spline of Example 3.2: (a) simple interior knots; (b) continuity of order 0 at knot 1

A total of seven knots is defined above. The knots at (1, 1) and (6, 1) are external knots (i.e., knot number 0 and knot number $m + 1$, respectively, as defined in equation (3.16). Let's say that the vector t is formed as follows:

$$t = \{0, 0, 0, 0, 1, 2, 3, 4, 5, 6, 6, 6, 6\}$$

As can be seen, the two external knots have been repeated four times each so that the necessary B-splines can be calculated. The internal knots occur only once, which means that the order of continuity of the pp is everywhere 2. The resulting 4th order curve is shown in Figure 3.5(a).

Suppose that the vector t is reformulated as shown below.

$$t = \{0, 0, 0, 0, 1, 1, 1, 2, 3, 4, 5, 6, 6, 6, 6\}$$

As explained before, with the above knot sequence vector, continuity of the pp curve produced at knot 1 will be of order 0. That fact is obvious from Figure 3.5(b). ❑

Nonuniform rational B-splines. The polynomial B-splines discussed above are often not flexible enough to produce perfect circles, ellipses, and other common types of conics. NURBS, however, have been found to generate both free-form curves and conics quite accurately [6].

The rational B-spline curve is defined by the following equation:

$$P(s) = \frac{\sum_{i=1}^{n} w_i p_i B(s)_{i,k}}{\sum_{i=1}^{n} p_i B(s)_{i,k}} \tag{3.21}$$

where w_i is the scale factor at the control knot i, $P(s) = \{X(s) \quad Y(s) \quad w(s)\}$ is the vector of homogeneous coordinates of the points on the curve, and $p_i = \{x_i w_i \quad y_i w_i \quad w_i\}$ is the vector of homogeneous coordinates of the ith knot. The knot vector is formed in exactly the same manner as in the case of the polynomial B-splines discussed above. As before, the continuity of the curve is governed by the composition of the knot vector. The term $B(s)_{i,k}$ is the B-spline basis function defined in equation (3.15).

3.3 ▪ Basic Curve-Fitting Techniques

In many engineering applications, physical quantities are specified at a few discrete points, but not in the entire domain. The reader will find many such examples in practice. For instance, the data on strains in a structure are available at only a few selected points as measured by strain gauges. Similarly, the temperatures of only a few points are determined experimentally by thermocouples attached at strategically important points. Interpolation or extrapolation procedures are used to determine values (such as strains or temperatures in the two examples mentioned above) at points other than those specified or measured at the discrete points. Curve fitting is one of the techniques that can be used to achieve this purpose.

Curve-fitting techniques have become increasingly important with the advent of CAD and drafting. As described in Chapter 1, geometric modeling of solid components is an essential part of CAD. The geometry of a solid can be described by the topography of its surfaces, which consist of an infinite number of points. Each of these points can be uniquely described by a set of coordinates in the space. The description of the geometry of a solid surface and the storage and retrieval of this information thus, in theory, require the specification and storage of an infinite number of sets of coordinates. This, of course, is not possible in reality. In such instances, curve-fitting techniques become extremely useful. Mathematical functions that satisfy the coordinates of a finite number of specified points on the surface can be readily derived, and these can then be used to provide interpolated coordinates of all intermediate points. Complex surface information of a solid can thus be described (approximately) by mathematical functions.

The essence of most curve-fitting techniques is to derive continuous functions that satisfy the values of certain variables specified at given points. The functions so derived yield approximate values of the variables at all other points. In a practical sense, it means that mathematical functions derived by curve-fitting techniques represent approximate variation of the physical quantity over a specified domain.

A number of curve-fitting techniques are described in the literature [7, 8]. Here, we will introduce the three most basic techniques: (1) polynomial fitting, (2) polynomial regression with least square fit, and (3) spline interpolation. These techniques are not necessarily the most efficient ones for CAD applications. They are introduced primarily to illustrate the principles of curve-fitting by mathematical functions.

❑ POLYNOMIAL CURVE-FITTING

The mathematical function used to represent the point values of a dependent variable Y_i at discrete points X_i ($i = 1, 2, \ldots, n + 1$) may be assumed to be a polynomial function such as

$$Y(X) = A_0 + A_1X + A_2X^2 + \cdots + A_nX^n \tag{3.22}$$

where $A_0, A_1, \ldots, A_n$ are constants to be determined by solving a set of simultaneous equations that are obtained by substituting pairs of specified values (X_i, Y_i) for the variables X and Y in equation (3.22).

The polynomial technique is simple to use. However, its accuracy depends on the number and the choice of data points. Equation (3.22) indicates that a quadratic function is the highest order of polynomial function that one may use for the case of three data points. Likewise, a cubic function would require four data points. Thus, if N is the number of available data points, the order of the polynomial function $n = N - 1$.

EXAMPLE ■ 3.3

A curve is known to pass through the following three data points: (1, 1.943), (2.75, 7.886), and (5, 1.738), as shown in Figure 3.6. Find a polynomial function to describe the curve using these three data points.

Since the total number (N) of points to be passed through by the curve is 3, the highest order (n) of the polynomial function in Figure 3.6 is $3 - 1 = 2$. The polynomial function is thus in the form of a quadratic function:

$$Y(X) = A_0 + A_1X + A_2X^2$$

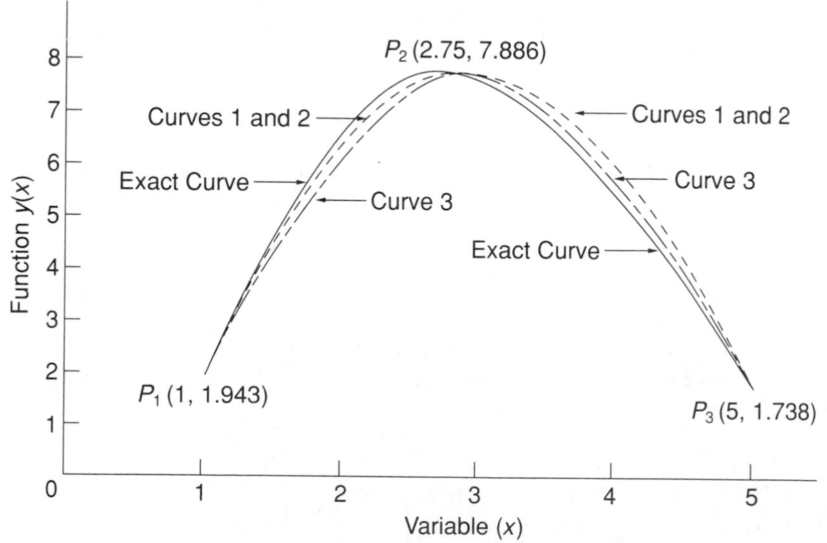

FIGURE 3.6
Polynomial curve-fitting with three points

By substituting for the three point values in the above expression, a set of three simultaneous equations is obtained for the three unknown coefficients:

$$A_2 + A_1 + A_0 = 1.943$$
$$7.5625A_2 + 2.75A_1 + A_0 = 7.886$$
$$25A_2 + 5A_1 + A_0 = 1.738$$

From these, one may solve for the coefficients as follows:
$A_0 = -5.6663$, $A_1 = 9.1414$, and $A_2 = -1.5321$.

The function that passes through these three data points thus takes on the following form:

$$Y(X) = -1.5321X^2 + 9.1414X - 5.6663$$

Graphical representation of this function is shown as curve 1 in Figure 3.6. Correlation with the prescribed curve is good between points P_1 and P_2, but is rather poor for the second half between points P_2 and P_3. ❑

EXAMPLE ■ 3.4

Derive a polynomial function to fit part of the profile of a gear tooth between points A and D in Figure 3.7.

Two additional data points, B and C, are chosen. Coordinates of these points, as read from Figure 3.7, are listed below.

Data point	X_i	Y_i
A	0	3.0
B	1.6	2.6
C	3.6	5.6
D	5.0	9.0

The order, n, of the polynomial function with four data points ($N = 4$) is 3. A cubic function is thus required and may be expressed as follows:

$$Y(X) = A_0 + A_1X + A_1X^2 + A_3X^3 \qquad \text{(a)}$$

On substituting into equation (a) the coordinates of all four data points given in the table above, the following simultaneous equations are obtained:

$$\begin{bmatrix} 1 & 0 & 0 & 0 \\ 1 & 1.6 & 2.56 & 4.096 \\ 1 & 3.6 & 12.96 & 46.66 \\ 1 & 5 & 25 & 125 \end{bmatrix} \begin{Bmatrix} A_0 \\ A_1 \\ A_2 \\ A_3 \end{Bmatrix} = \begin{Bmatrix} 3 \\ 2.6 \\ 5.6 \\ 9 \end{Bmatrix}$$

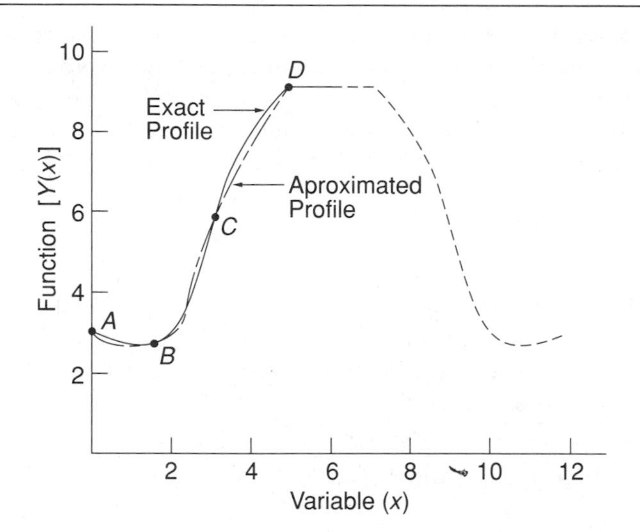

FIGURE 3.7
Approximation of gear tooth profile

The solution to the above equations can be obtained either by matrix inversion (described in Appendix I) or by the methods presented in Chapter 2. The solution is as follows: $A_0 = 3$, $A_1 = -1.273$, $A_2 = 0.7077$, and $A_3 = -0.043$.

The function that fits the four data points from points A to D is thus given as follows:

$$Y(X) = 3 - 1.273X + 0.7077X^2 - 0.043X^3$$

The extent to which the above function fits the gear profile is shown in Figure 3.7. ❑

❑ POLYNOMIAL REGRESSION WITH LEAST SQUARES FIT

The objective of the curve-fitting techniques is to represent a set of data points (e.g, (X_i, Y_i) for $i = 1, 2, \ldots, N$) with sufficient accuracy by a functional relationship. An appropriate question to ask then is how to find a function and the corresponding curve with a minimum amount of deviation from the original curve. The deviation at a data point is defined as the difference between the given Y value (Y_i) and the corresponding Y value computed from the functional relationship $Y(X_i)$. It is logical to expect that in a well-formulated functional relationship the sum of the deviations should be as small as possible. Mathematically, we seek the minimization of

$$D = \sum_{i=1}^{N} [Y_i - Y(X_i)] \tag{3.23}$$

in which N is the number of data points.

Two major deficiencies exist in equation (3.23): (1) it does not provide for the computation of absolute deviations, and (2) the expression in equation (3.23), being a function of the first order, cannot be differentiated to solve for the minimum value. An expression for the deviation that is free of these deficiencies takes the following form:

$$D = \sum_{i=1}^{N} [Y_i - Y(X_i)]^2 \tag{3.24}$$

The function $Y(X)$ in equation (3.24) can be expressed in many different forms. A common choice is a polynomial similar to that shown in equation (3.22). On substituting $Y(X)$ from equation (3.22) into equation (3.24), the following expression for the error function is obtained:

$$D = \sum_{i=1}^{N} [Y_i - (A_0 + A_1 X_i + A_2 X_1^2 + \cdots + A_n X_i^n)]^2 \tag{3.25}$$

Since equation (3.25) is a linear equation, the minimization of D requires that

$$\frac{\partial D}{\partial A_0} = \frac{\partial D}{\partial A_1} = \frac{\partial D}{\partial A_2} = \cdots = \frac{\partial D}{\partial A_n} = 0$$

from which one may derive n linear algebraic equations to solve for the coefficients $A_0, A_1, \ldots, A_n$.

The procedure described above involves the minimization of the sum of the squares of the deviations. It is thus usually referred to as the least squares method. The least squares method gives better results than does the polynomial fitting method when $N > n + 1$.

EXAMPLE ■ 3.5

Use the least squares method to find a quadratic polynomial function to fit data points (X_i, Y_i), $i = 1, 2, \ldots, N$.

Let the required polynomial be of the following form:

$$Y(X) = A_0 + A_1 X + A_2 X^2 \tag{a}$$

The square of the deviations, as per equation (3.25), is given as follows:

$$D = \sum_{i=1}^{N} [Y_i - (A_0 + A_1 X_i + A_2 X_i^2)]^2$$

Minimization of D in the above expression results in equation (b):

$$\frac{\partial D}{\partial A_0} = -2 \sum_{i=1}^{N} (Y_i - A_0 - A_1 X_i - A_2 X_i^2) = 0$$

or

$$NA_0 + (\Sigma X_i)A_1 + (\Sigma X_i^2)A_2 = \Sigma Y_i \quad \text{(b)}$$

All summation signs in equation (b) and subsequent equations imply $i = 1, 2, \ldots, N$.

Applying the other two conditions of $\partial D/\partial A_1 = \partial D/\partial A_2 = 0$ results in the following equations:

$$(\Sigma X_i\ )A_0 + (\Sigma X_i^2)A_1 + (\Sigma X_i^3)A_2 = \Sigma(X_i Y_i) \quad \text{(c)}$$

and

$$(\Sigma X_i^2)A_0 + (\Sigma X_i^3)A_1 + (\Sigma X_i^4)A_2 = \Sigma(X_i^2 Y_i) \quad \text{(d)}$$

The coefficients A_0, A_1, and A_2 may be evaluated from equations (a) through (d) and the quadratic polynomial found. ❑

EXAMPLE ■ 3.6

Use the least squares method to find a quadratic polynomial function that fits the data points given in Example 3.3.

On substituting the coordinates of the data points into equations (a) through (d) from Example 3.5, the following simultaneous equations are obtained:

$$3.3000A_0 + 8.7500A_1 + 33.5625A_2 = 11.5670 \quad \text{(a)}$$

$$8.7500A_0 + 33.5625A_1 + 146.7968A_2 = 32.3195 \quad \text{(b)}$$

$$33.5625A_0 + 146.7968A_1 + 683.1914A_2 = 105.0308 \quad \text{(c)}$$

The coefficients are evaluated as follows: $A_0 = -5.6654$, $A_1 = 9.1406$, and $A_2 = 1.532$. The polynomial function that fits the three specified data points is given as

$$Y(X) = +1.532X^2 + 9.1406X - 5.6654 \quad \text{(d)}$$

Notice that the solution obtained here, shown as curve 2 in Figure 3.6, is the same as that obtained in Example 3.3. In other words, there is no improvement over the polynomial fit. This is so because in the example $N = n + 1$. If a fourth data point is added to the list of three considered above, a closer fit will be obtained. ❑

❑ SPLINE INTERPOLATION

Splines are used for curve-fitting, in addition to generating free-form curves as explained previously. In the commonly used method of spline interpolation, a separate function is used between two adjacent data points,

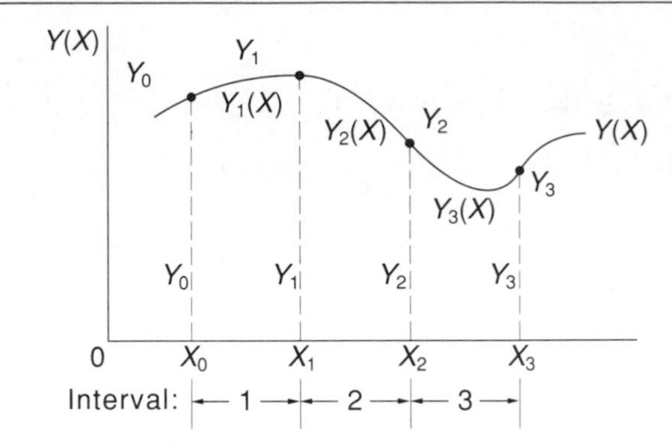

FIGURE 3.8 Curve-fitting by spline interpolation

guaranteeing a smooth transition between the spline functions at the data points. Cubic splines are employed by engineers in most practical situations. Given below are the considerations that are taken into account in determining cubic spline functions.

Take, for example, the situation depicted in Figure 3.8. Four data points are specified at $X = X_0$, $X = X_1$, $X = X_2$, and $X = X_3$. Assume that the spline functions are $Y_1(X)$, $Y_2(X)$, and $Y_3(X)$ in the intervals $X_1 - X_0$, $X_2 - X_1$, and $X_3 - X_2$, respectively.

In order that the splines pass through the terminal (end) data points, the following conditions should be satisfied:

$$Y_1(X_0) = Y_0$$
$$Y_3(X_3) = Y_3$$

In order that continuity of the curve is maintained at the interior data points, the following conditions should be satisfied:

$$Y_1(X_1) = Y_2(X_1) = Y_1$$
$$Y_2(X_2) = Y_3(X_2) = Y_2$$

Also, in order that the first derivative of the connecting curves is continuous at the interior points, the following must hold true:

$$Y'_1(X_1) = Y'_2(X_1)$$
$$Y'_2(X_2) = Y'_3(X_2)$$

where $Y'_i(X_i)$ denotes the first derivative of function Y_i with respect to X. The second derivative of the functions is set to zero at the terminal points. At the interior points, the second derivatives of the connecting curves must be equal, therefore, the following conditions should be satisfied:

$$Y''_1(X_1) = Y''_2(X_1)$$
$$Y''_2(X_2) = Y''_3(X_2)$$

As in the polynomial regression technique, the accuracy of the spline interpolation depends on the choice of the spline functions. However, as mentioned above, cubic splines offer the most practical tool. Assume that the following polynomial function is employed for spline interpolation:

$$Y(X) = AX^3 + BX^2 + CX + D \tag{3.26}$$

On evaluating and substituting the coefficients in equation (3.26), the following equation is obtained for the spline functions [7]:

$$\begin{aligned} Y_i(X) = &\frac{Y''(X_{i-1})}{6(X_i - X_{i-1})}(X_i - X)^3 \\ &+ \frac{Y''(X_i)}{6(X_i - X_{i-1})}(X - X_{i-1})^3 \\ &+ \left[\frac{Y(X_{i-1})}{X_i - X_{i-1}} - \frac{Y''(X_{i-1})(X_i - X_{i-1})}{6}\right](X_i - X) \\ &+ \left[\frac{Y(X_i)}{X_i - X_{i-1}} - \frac{Y''(X_i)(X_i - X_{i-1})}{6}\right](X - X_{i-1}) \end{aligned} \tag{3.27}$$

The second derivatives in equation (3.27) can be evaluated from the following equations [7]:

$$\begin{aligned} &(X_i - X_{i-1})Y''(X_{i-1}) + 2(X_{i+1} - X_{i-1})Y''(X_i) \\ &\quad + (X_{i+1} - X_i)Y''(X_{i+1}) \\ &= \frac{6}{(X_{i+1} - X_i)}[Y(X_{i+1}) - Y(X_i)] \\ &\quad + \frac{6}{(X_i - X_{i-1})}[Y(X_{i-1}) - Y(X_i)] \end{aligned} \tag{3.28}$$

EXAMPLE ■ 3.7

Use a cubic spline function to fit the three data points specified in Example 3.3.

Since three data points are involved in this problem, two intervals and thus two cubic spline functions have to be found. Assume that these functions are

$$Y_1(X) = A_1X^3 + B_1X^2 + C_1X + D_1 \tag{a}$$

in the interval $1 \le X \le 2.75$, and

$$Y_2(X) = A_2X^3 + B_2X^2 + C_2X + D_2 \tag{b}$$

in the interval $2.75 \le X \le 5$.

For the first interval (i.e., $i = 1$), we obtain the following relationship from equation (3.28):

$$1.75Y''(X_0) + 8Y''(X_1) + 2.25Y''(X_2) = -37.3227$$

Since the second derivatives should be zero at the terminal points, the equation above reduces to the following form:

$$Y''(X_1) = -4.66533 \quad \text{(c)}$$

One may readily obtain the spline function in the first interval by substituting $Y''(X_1)$ from equation (c) above and $Y''(X_0) = 0$ into equation (3.27). The spline function reads thus:

$$Y_1(X) = -0.4443X^3 + 1.3329X^2 + 3.4238X - 2.3694 \quad \text{(d)}$$

The spline function in the second interval (i.e., $i = 2$) may be obtained by substituting $Y''(X_1)$ from equation (c) above and $Y''(X_2) = 0$ into equation (3.27). The spline function reads thus:

$$Y_2(X) = 0.3456X^3 - 5.1837X^2 + 21.4366X - 19.0498 \quad \text{(e)}$$

The functions $Y_1(X)$ and $Y_2(X)$ are plotted as curve 3 in Figure 3.6. It is evident that a better fit to the exact curve is obtained by the cubic spline interpolation. ❑

EXAMPLE ■ 3.8

Use cubic spline functions to fit the profile of the gear tooth described in Example 3.4.

Since there are four data points chosen for fitting, a total of $4 - 1 = 3$ functions need to be determined. By following the procedure described in Example 3.7, the spline functions are found to be

$$Y_1(X) = 0.1396X^3 - 0.6074X + 3$$
(between points A and B)
$$Y_2(X) = -0.0763X^3 + 1.0362X^2 - 2.2651X + 3.8842$$
(between points B and C)
$$Y_3(X) = -0.0506X^3 + 0.7591X^2 - 1.2677X + 2.6871$$
(between points C and D)

Fitting of the functions given above to the gear profile is illustrated in Figure 3.9.

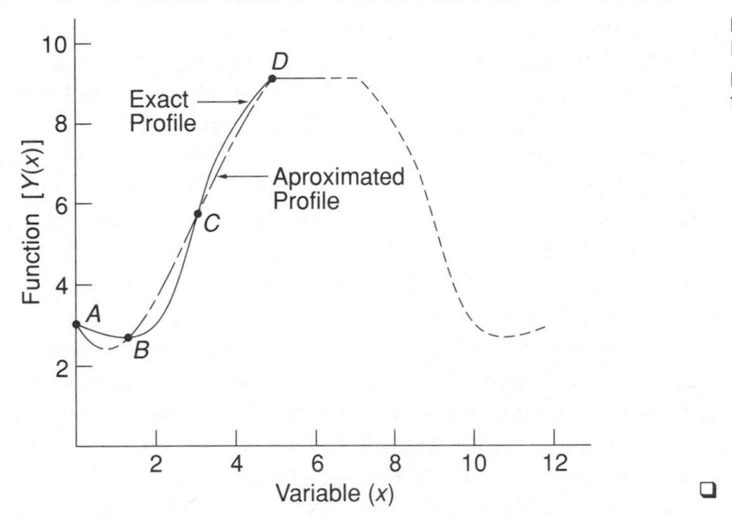

FIGURE 3.9
Fitting a gear tooth profile with cubic spline functions

❑

3.4 ▪ Algorithms for Raster-Scan Graphics

At the present time, the *raster-scan technique* is the most commonly used image display technique in computer graphics. In a raster-scan display, the intensity of each phosphor dot (picture element, pel, or pixel) is controlled separately, whether or not it contains any details of the graphics image. The image is generated on the display screen by scanning one line (of pixels) at a time, and each horizontal line of pixels is referred to as a *raster line.*

Most of the raster graphics display devices are operated through a frame buffer. A frame buffer consists of a large memory bank. In the simple monochrome display, the buffer contains one memory bit for each picture element in the raster display device. Thus, if there are 640 × 400 pixels on the display device (screen), the buffer contains the same number (i.e., 256,000) of memory bits in a single bit-plane. The memory bits have binary values of either a 0 or a 1. A zero value in the memory corresponds to an unlit pixel, and 1 corresponds to a lighted pixel. If there are n bit-planes in the buffer (each having the same size memory), the binary value corresponding to each pixel from each plane is loaded into a register and the values summed up. These values are then interpreted as intensity values. Obviously, in an n bit-plane buffer, each pixel may have 2^n levels of intensity. By varying the values of intensity within an image, a shading effect (called a *gray scale*) is produced in a monochrome composition.

For color graphics displays, there are at least three bit-planes (corresponding to the three basic colors—red, green, and blue) in the frame buffer. Each bit-plane drives a different electron color gun. By turning on one, two, or all three guns ($2^3 =$), eight different colors may be produced at each pixel. By providing more than one bit-plane corresponding to each color, display intensity of the primary colors is individually controlled. Thus, if there are eight bit-planes for each color, 2^8 shades may be produced. As explained in Chapter 1, by mixing these colors together, a total of $2^{8 \times 3} = 16{,}777{,}216$ different colors can be produced.

From the above discussion, it is clear that for creating the image of any vector, character, or solid area, a dot pattern (mask) must be determined. The process of obtaining the dot pattern is known as *scan conversion*. In 3D graphics, where several solid areas may be involved in one image, the *priority* of each area should be determined. The priorities of the areas are used in ascertaining which areas obscure other areas. Thus, the concept of priority is useful in hidden surface removal algorithms, described in the next chapter. For shaded images, a shading rule is required. By means of appropriate shading rules, it is possible to recreate surface texture and convey a sense of depth. In the following two sections, algorithms for scan conversion are discussed. Shading techniques such as those developed by Phong [9] and Gouraud [10] have not been included due to space limitation.

3.5 ▪ Algorithms for Scan Conversion

❑ STRAIGHT LINES

Equations (3.1) and (3.2) are parametric equations for straight lines. A line can thus be scan converted by finding the address of each pixel that falls on or near the line. The drawback of this approach, however, is that the calculations involve floating point (or binary fractions) calculations. Frequent rounding is thus required. An algorithm suggested by Bresenham [11] uses integer arithmetic. This algorithm is efficient and fast, and is therefore quite popular.

❑ RECTANGLES

A rectangle can be scan converted quite easily if its sides run along the scan lines. Let's assume that the coordinates of the lower left corner of a rectangle are (1, 2) and that the rectangle is four pixels in height and five pixels in width. See Figure 3.10. It is obvious that the pixels in columns 1 through 5 belonging to rows 2 through 5 must be illuminated for a solid area representation of the rectangle. The scan conversion algorithm should therefore allocate the intensity of the rectangle to these pixels and set the intensity everywhere else on the display equal to the background value.

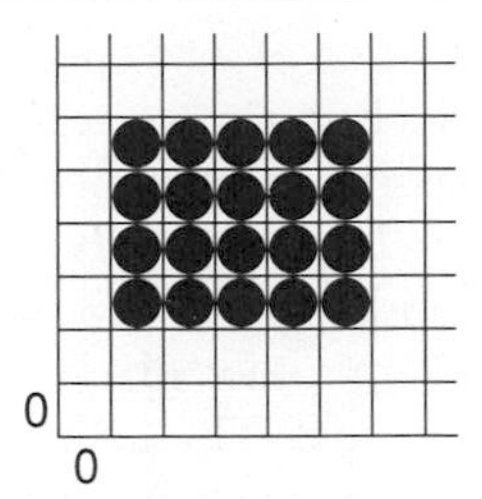

FIGURE 3.10
Scan conversion of a rectangle

❑ POLYGONS

The examples of scan conversion discussed above can now be extended to a polygon. Consider, for example, a polygon whose vertices have coordinates (1, 4), (1, 7), (8, 6), (8, 2), (5, 1), and (5, 9). See Figure 3.11. Now consider a point—say, *A* (5, 5)—in Figure 3.11. If a horizontal line is drawn from this point toward the left, it intersects the edges of the polygon only once. A horizontal line drawn from point *B* (10, 3) toward the left, on the other hand, intersects the edges twice. As can be seen, point *A* is situated inside the polygon, and point *B* is outside. From these observations, a generalized result can be obtained showing that the horizontals drawn toward the left or the right from every point within a polygon of any shape intersect the edges of the polygon an odd number of times, while the horizontals from every point outside the polygon intersect the polygon an even (or zero) number of times. The above test is often referred to as the *inside test*. Every pixel inside the polygon is given the intensity desired for the polygon and those outside are allocated the intensity of the background. This information is then stored in the frame buffer for scanning of the image.

The disadvantage of scan conversion based on the inside test is obvious: a large number of tests need to be performed. To reduce the number of tests, the *scan-line coherence* property of solid areas is utilized. Scan-line coherence implies that there are neighboring pixels on a scan-line that are either inside or outside a polygon, just as the pixel tested by the inside test.

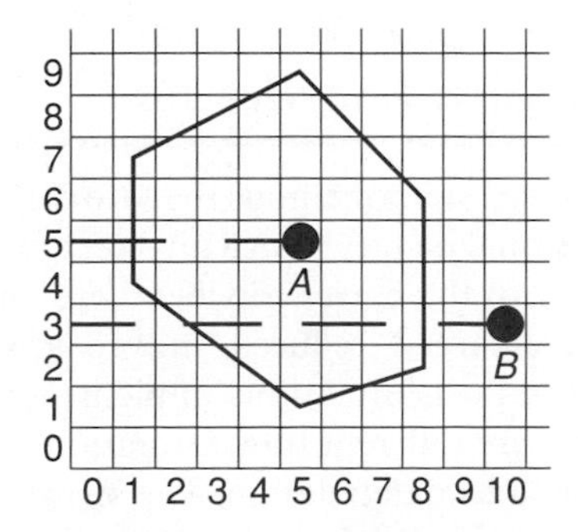

FIGURE 3.11
A polygon

Thus, by determining the position of a few pixels vis-à-vis the polygon, the positions of several other pixels in a row are also ascertained.

Consider the scan-line passing through point B in Figure 3.11. The line intersects the polygon at $x = 2$ and $x = 8$. It is obvious, therefore, that all pixels lying between $x = 2$ and $x = 8$ at $y = 3$ are inside the polygon. No further tests are needed for any other pixel on this scan-line. Pixels not inside the polygon are outside. Thus, by considering the scan-line coherence, the positions of all pixels on a horizontal line with respect to a given polygon are determined by a single test. Using the inside test scheme, many more tests (say, 640 or 720 depending on the resolution of the screen) would have been necessary, resulting in a much longer computation time. A popular algorithm that uses the scan-line coherence property is the *YX Algorithm* [12].

Another form of coherence is utilized for reducing the effort required for scan conversion still further. The method, as shown below, obviates the need for determining the points of intersection of an edge with all horizontals. Also, the points of intersection are determined recursively. This other type of coherence is called *edge coherence*. In edge coherence it is assumed that if the ith scan-line intersects an edge, it is probable that the $i + 1$th scan-line also intersects the same edge. If the point of intersection of the ith scan-line has coordinates (x_i, y_i), then the point of intersection, if there is one, with the $i + 1$th scan-line will have coordinates

$$(x_{i+1}, y_{i+1}) = (x_i + \Delta x, y_i + 1) \tag{3.29}$$

where Δx is the change in x for any unit change in the y of the edge. Δx may be readily determined from the slope of the edge. From the values obtained from equation (3.29) and the coordinates of the end points of the particular edge, it is easy to check whether an intersection is possible. If no intersection is indicated, no further test is performed on the edge. A popular algorithm based on edge coherence is the *Y–X Algorithm* [12].

3.6 ▪ Two-Dimensional Transformations: Homogeneous Coordinates

Thus far, we have discussed some methods that are commonly employed for producing two-dimensional images from a given set of data. The computer, however, provides us with the power to view these images in a variety of ways. Images may be enlarged, reduced, moved around (translated), or rotated using geometric transformations implemented through transformation matrices. There are different transformation matrices for performing different types of image manipulation. The structure of these matrices is described later. It is important, however, to realize that if the transformation matrices are formed using Cartesian coordinates, matrix addition or subtraction is required to implement translation, while matrix multiplication is required for all other types of image manipulation (i.e., enlargement/reduction, rotation, etc.) This causes difficulty during programming. If, however, the transformation matrices are formed with homogeneous coordinates (described in the next paragraph), the most common types of image manipulation can be carried out with only one type of matrix operation—namely, multiplication.

In a homogeneous coordinate system, a scale factor is introduced along with the Cartesian coordinates. Thus, a point (x, y, z) is represented with the set (x', y', z', w), where $w\ (\neq 0)$ is the scale factor. The relationships between the Cartesian coordinates and the homogeneous coordinates of a point are given by

$$x = x'/w$$
$$y = y'/w$$
$$z = z'/w \qquad (3.32)$$

Thus, the homogeneous coordinates of a point P_1 with Cartesian coordinates $(1, 2, 1)$ may be expressed as $\{2, 4, 2, 2\}$ or $\{3, 6, 3, 3\}$ or $\{1, 2, 1, 1\}$. In computer graphics applications, $w = 1$ is often used.

3.7 ▪ Transformations of a Point

❑ ROTATION

Let's assume that the computer monitor screen is designated as the xy plane, with the z axis standing out toward the viewer. If it is desired to rotate a point $P\ \{x, y, z, 1\}$ through a positive (counterclockwise) angle α about the origin, then the following transformation matrix is used:

$$R(\alpha) = \begin{bmatrix} \cos\alpha & \sin\alpha & 0 \\ -\sin\alpha & \cos\alpha & 0 \\ 0 & 0 & 1 \end{bmatrix} \qquad (3.33)$$

The following matrix multiplication transforms point P into point P'. See Figure 3.12.

$$P'\{x' \quad y' \quad 1\} = P\{x \quad y \quad 1\}R(\alpha) \qquad (3.34)$$

Thus, $x' = x\cos\alpha - y\sin\alpha$ and $y' = x\sin\alpha + y\cos\alpha$. If $\alpha_1, \alpha_2, \alpha_3, \ldots$ rotations are applied, in that order, then the transformation is obtained by

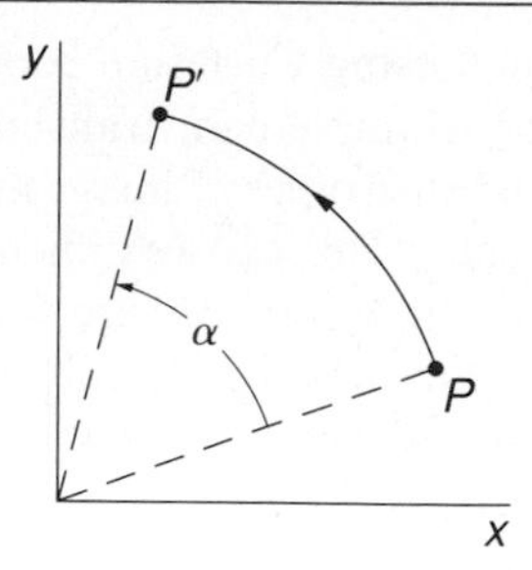

FIGURE 3.12 Rotation of point P through an angle α

chain multiplication as follows:

$$P'' = PR(\alpha_1)R(\alpha_2)R(\alpha_3), \ldots$$

or

$$P'' = PR'$$

where

$$R' = \begin{bmatrix} \cos(\alpha_1 + \alpha_2 + \alpha_3 + \cdots) & \sin(\alpha_1 + \alpha_2 + \alpha_3 + \cdots) & 0 \\ -\sin(\alpha_1 + \alpha_2 + \alpha_3 + \cdots) & \cos(\alpha_1 + \alpha_2 + \alpha_3 + \cdots) & 0 \\ 0 & 0 & 1 \end{bmatrix}$$

EXAMPLE ■ 3.9

A point P lies originally at the position $(\sqrt{2}, 0)$. Find its coordinates if it is rotated 45° counterclockwise about the origin. If it is given a subsequent rotation of 45°, what will its coordinates be?

The three positions of the point are shown in Figure 3.13(a). As can be seen, point P $(\sqrt{2}, 0)$ maps into point P' (1, 1) as a result of the first rotation. After the second rotation, it maps into point P'' $(0, \sqrt{2})$. Let's

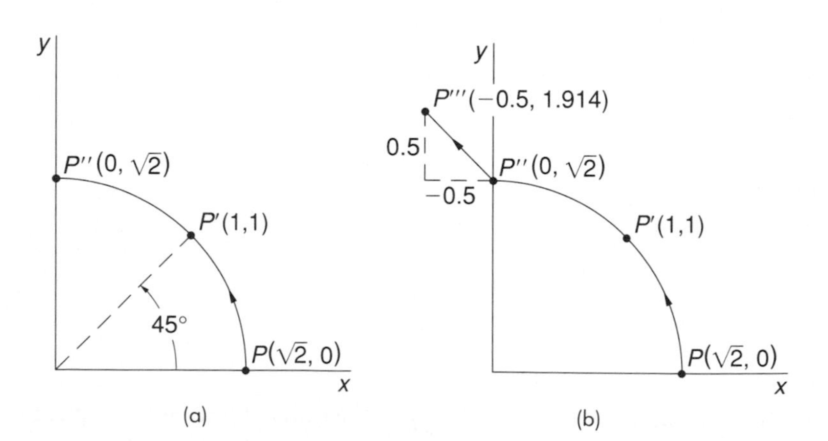

FIGURE 3.13 Rotation and translation of a point

now check whether the same results are predicted by the methods described above.

The rotation matrix for the first 45° rotation is given, from equation (3.33), as follows:

$$R' = \begin{bmatrix} \cos 45^\circ & \sin 45^\circ & 0 \\ -\sin 45^\circ & \cos 45^\circ & 0 \\ 0 & 0 & 1 \end{bmatrix} = \begin{bmatrix} \frac{1}{\sqrt{2}} & \frac{1}{\sqrt{2}} & 0 \\ \frac{-1}{\sqrt{2}} & \frac{1}{\sqrt{2}} & 0 \\ 0 & 0 & 1 \end{bmatrix}$$

The homogeneous coordinates of point P' (x', y') are then given as

$$\{x' \quad y' \quad 1\} = \{\sqrt{2} \quad 0 \quad 1\} \begin{bmatrix} \frac{1}{\sqrt{2}} & \frac{1}{\sqrt{2}} & 0 \\ \frac{-1}{\sqrt{2}} & \frac{1}{\sqrt{2}} & 0 \\ 0 & 0 & 1 \end{bmatrix} = \{1 \quad 1 \quad 1\}$$

Thus, $x' = 1$ and $y' = 1$ as expected.

The rotation matrix for the full 90° counterclockwise rotation is given as follows:

$$R'' = \begin{bmatrix} \cos 90^\circ & \sin 90^\circ & 0 \\ -\sin 90^\circ & \cos 90^\circ & 0 \\ 0 & 0 & 1 \end{bmatrix} = \begin{bmatrix} 0 & 1 & 0 \\ -1 & 0 & 0 \\ 0 & 0 & 1 \end{bmatrix}$$

The homogeneous coordinates of point P'' (x'', y'') are given as

$$\{x'' \quad y'' \quad 1\} = \{\sqrt{2} \quad 0 \quad 1\} \begin{bmatrix} 0 & 1 & 0 \\ -1 & 0 & 0 \\ 0 & 0 & 1 \end{bmatrix} = \{0 \quad \sqrt{2} \quad 1)$$

Thus, $x'' = 0$ and $y'' = \sqrt{2}$, as shown in Figure 3.13(a). ❑

❑ TRANSLATION

In case point P is to be translated by a distance Δx in x direction and Δy in y direction, then the following transformation would be used:

$$T(\Delta x, \Delta y) = \begin{bmatrix} 1 & 0 & 0 \\ 0 & 1 & 0 \\ \Delta x & \Delta y & 1 \end{bmatrix} \tag{3.35}$$

The translation is carried out using the following matrix multiplication:

$$P'\{x' \quad y' \quad 1\} = P\{x \quad y \quad 1\} T(\Delta x, \Delta y)$$

If the overall translation takes place in several steps of $(\Delta x_1, \Delta y_1)$, $(\Delta x_2, \Delta y_2), \ldots,$ the transformation matrix is constructed as follows:

$$T' = \begin{bmatrix} 1 & 0 & 0 \\ 0 & 1 & 0 \\ \Delta x_1 + \Delta x_2 + \cdots & \Delta y_1 + \Delta y_2 + \cdots & 1 \end{bmatrix}$$

EXAMPLE ▪ 3.10

Assuming that point P'' in Figure 3.13(a) is translated by distances $(-0.5, 0.5)$, find its new position, P'''.

Without explicitly showing the matrix multiplication, one can easily see that the coordinates of P''' are $(-0.5, 1.914)$. Let's now verify the above observation with the help of the translation matrix. The translation matrix for the given displacements, from equation (3.35), is given as follows:

$$T = \begin{bmatrix} 1 & 0 & 0 \\ 0 & 1 & 0 \\ -0.5 & 0.5 & 1 \end{bmatrix}$$

The homogeneous coordinates of P''' are

$$\{x''' \quad y''' \quad 1\} = \{0 \quad \sqrt{2} \quad 1\} \begin{bmatrix} 1 & 0 & 0 \\ 0 & 1 & 0 \\ -0.5 & 0.5 & 1 \end{bmatrix}$$

$$= \{-0.5 \quad 1.914 \quad 1\}$$

Thus, $x''' = -0.5$ and $y''' = 1.914$, as shown in Figure 3.13(b). ❑

❑ SCALING

This type of transformation is used in enlarging or reducing images. The exact implication of the scaling transformation is demonstrated later. Let's assume for the time being that it is required to scale the x coordinate of a point by a factor of j and its y coordinate by a factor of k. Then the following matrix multiplication should be applied:

$$S(j, k) = \begin{bmatrix} j & 0 & 0 \\ 0 & k & 0 \\ 0 & 0 & 1 \end{bmatrix} \tag{3.36}$$

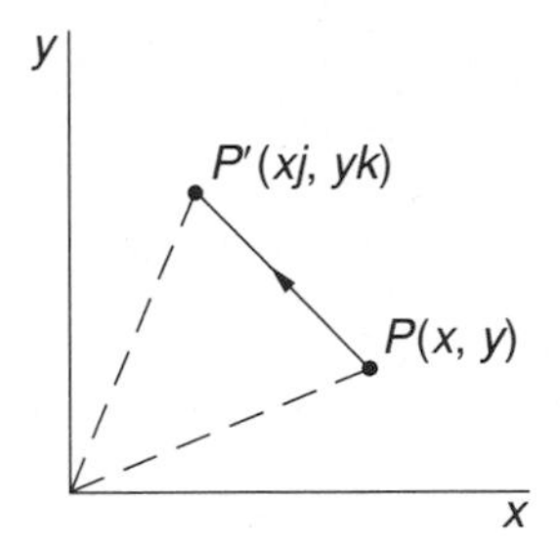

FIGURE 3.14
Effect of scaling on the position of a point

And, as a result of the following transformation, point P is mapped into point P', as shown in Figure 3.14:

$$P'\{x' \quad y' \quad 1\} = P\{x \quad y \quad 1\}S(j, k)$$

where

$$x' = xj \tag{3.37}$$

$$y' = yk \tag{3.38}$$

If successive scaling of (j_1, k_1), $(j_2, k_2), \ldots,$ is applied, the combined scaling matrix is given as follows:

$$S' = \begin{bmatrix} j_1 j_2 & \cdots & 0 & & 0 \\ 0 & & k_1 k_2 & \cdots & 0 \\ 0 & & 0 & & 1 \end{bmatrix}$$

❑ REFLECTION

Postmultiplication of the homogeneous coordinates of a point with a matrix of the form shown in equation (3.39) results in the reflection of the point about the x axis.

$$M(x) = \begin{bmatrix} 1 & 0 & 0 \\ 0 & -1 & 0 \\ 0 & 0 & 1 \end{bmatrix} \tag{3.39}$$

A reflection transformation is shown in Figure 3.15(a). If, in equation (3.39), the diagonal element in the first column, rather than the element in the second column, is negative, reflection about the y axis is obtained. If the diagonal elements in the first and second columns are both negative, reflection is obtained about the origin, as shown in Figure 3.15(b).

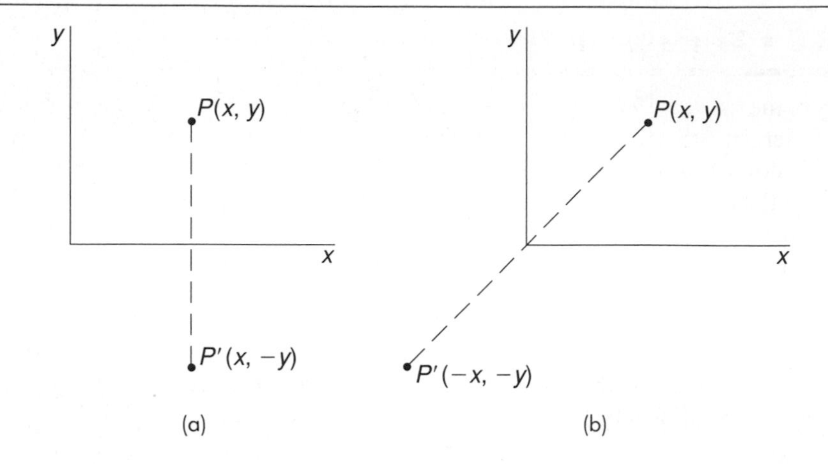

FIGURE 3.15 Reflection of a point P: (a) about the x axis; (b) about the origin

❑ CONCATENATION OF TRANSFORMATIONS

Often more than one transformation is applied to an existing image to obtain a new image. In such cases, chain multiplication of the position vectors of the points is carried out exactly in the order the transformations are required. Suppose that a point is first translated by amounts $(\Delta x, \Delta y)$, that it is rotated through an angle (α) about the z axis, and that its reflection about the x axis is then obtained. The following matrix operations are necessary to obtain a concatenation of the above transformations:

$$P' = PT(\Delta x, \Delta y)R(\alpha)M(x)$$

EXAMPLE ■ 3.11

Find the new coordinates of point P $(\sqrt{2}, 0)$ if it is first rotated 90° counterclockwise about the z axis and then translated by distances $(-0.5, 0.5)$.

Notice that the transformations applied to the above point are the same as those applied in Examples 3.9 and 3.10. Here, however, only the final coordinates are required. It is advisable in the interest of expediency that a concatenated transformation be applied. Using the matrices from Examples 3.9 and 3.10, the coordinates are found as follows:

$$\{x''' \quad y''' \quad 1\} = \{\sqrt{2} \quad 0 \quad 1\}\begin{bmatrix} 0 & 1 & 0 \\ -1 & 0 & 0 \\ 0 & 0 & 1 \end{bmatrix}\begin{bmatrix} 1 & 0 & 0 \\ 0 & 1 & 0 \\ -0.5 & 0.5 & 1 \end{bmatrix}$$

$$= \{-0.5 \quad 1.914 \quad 1\}$$

The result is exactly the same as that obtained in Example 3.10. ❑

3.8 ▪ Transformation of Plane Objects

A plane figure is built up by joining a number of straight lines, as mentioned earlier. In order to apply any transformation to a plane object, the same transformations are applied to every individual vertex of the object, and then the vertices are connected together in their new positions in the same sequence as before. Consider the polygon shown in Figure 3.16(a), which is drawn by joining vertex 1 to vertex 2, vertex 2 to vertex 3, etc. Let's say it is desired to rotate the polygon through a 30° angle in a clockwise direction about the z axis (or the origin). The transformation matrix for the rotation is created by using equation (3.33) as follows:

$$R(-30^\circ) = \begin{bmatrix} \cos(-30^\circ) & \sin(-30^\circ) & 0 \\ -\sin(-30^\circ) & \cos(-30^\circ) & 0 \\ 0 & 0 & 1 \end{bmatrix} = \begin{bmatrix} 0.866 & -0.5 & 0 \\ 0.5 & 0.866 & 0 \\ 0 & 0 & 1 \end{bmatrix}$$

The position vector, $\{x_k \quad y_k \quad 1\}$, of each of the vertices may then be operated on by the above matrix to obtain the new positions:

$$P'_k = P_k R(-30^\circ)$$

Alternatively, a matrix of homogeneous coordinates of all the vertices is formed, and the complete transformation is performed in one step, as

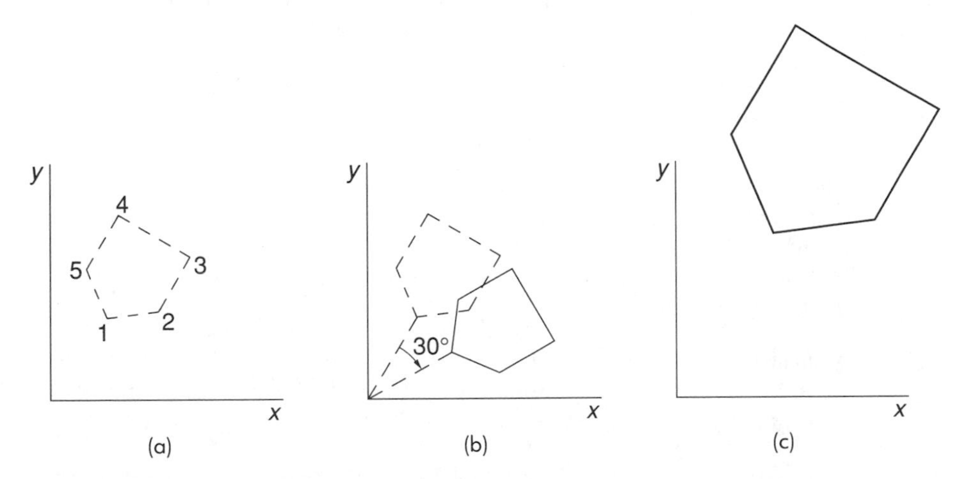

FIGURE 3.16 Transformation of a polygon: (a) in its original position; (b) rotated 30°; (c) scaled

follows:

$$\begin{bmatrix} x'_1 & y'_1 & 1 \\ x'_2 & y'_2 & 1 \\ \vdots & \vdots & \vdots \\ x'_n & y'_n & 1 \end{bmatrix} = \begin{bmatrix} x_1 & y_1 & 1 \\ x_2 & y_2 & 1 \\ \vdots & \vdots & \vdots \\ x_n & y_n & 1 \end{bmatrix} R(-30^\circ)$$

Once the transformed position of each of the vertices is obtained, the polygon is redrawn in its new position, following the sequence 1–2–· · ·–5–1, as shown in Figure 3.16(b).

Suppose that the polygon shown in Figure 3.16(a) must be enlarged to twice its size. The scaling transformation matrix for the above transformation is

$$S(2, 2) = \begin{bmatrix} 2 & 0 & 0 \\ 0 & 2 & 0 \\ 0 & 0 & 1 \end{bmatrix}$$

Transformation is applied to each vertex of the polygon, and its new shape and position are obtained by joining the vertices in the appropriate sequence, as shown in Figure 3.16(c). Notice that as a result of applying the above scaling transformation, the polygon has moved from its previous position. If the polygon must remain in a certain area on the screen when scaled, this is achieved by the technique explained below.

Suppose that the second vertex of the enlarged polygon in Figure 3.16(c) must remain at its original position. With this latter constraint, it is necessary to apply a translation transformation in addition to the scaling shown above. The following translation matrix may be used to postmultiply the coordinates and thus move the second vertex to its original position:

$$T = \begin{bmatrix} 1 & 0 & 0 \\ 0 & 1 & 0 \\ x_2 - x'_2 & y_2 - y'_2 & 1 \end{bmatrix}$$

where (x_2, y_2) are the original and (x'_2, y'_2) are the transformed coordinates of the second vertex. The desired transformation is then obtained by the following concatenation:

$$P' = PS(2, 2)T$$

Alternatively, the origin of the coordinate system may be moved to the second vertex and scaling applied subsequently. The translation matrix for shifting the origin is given as follows:

$$T = \begin{bmatrix} 1 & 0 & 0 \\ 0 & 1 & 0 \\ -x_2 & -y_2 & 1 \end{bmatrix}$$

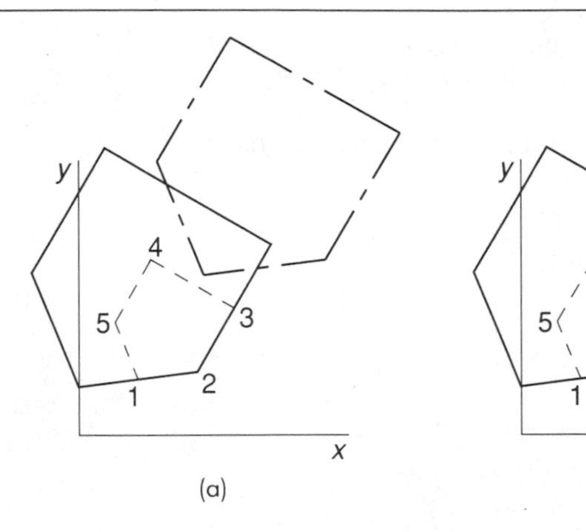

FIGURE 3.17 Methods for controlling the position of a scaled object: (a) scale and translate; (b) shift the origin and scale

The concatenation of the transformations is obtained as follows:

$$P' = PTS(2, 2)$$

Parts (a) and (b) of Figure 3.17 show the difference between the two transformations discussed above.

3.9 ▪ Image Manipulation About Arbitrary Axes

❑ ROTATION ABOUT AN ARBITRARY AXIS

Suppose that a point must be rotated not about the z axis, but about an axis parallel to the z axis, and that the axis of rotation passes through point $C(x_c, y_c)$. See Figure 3.18(a). The transformation matrix for this operation

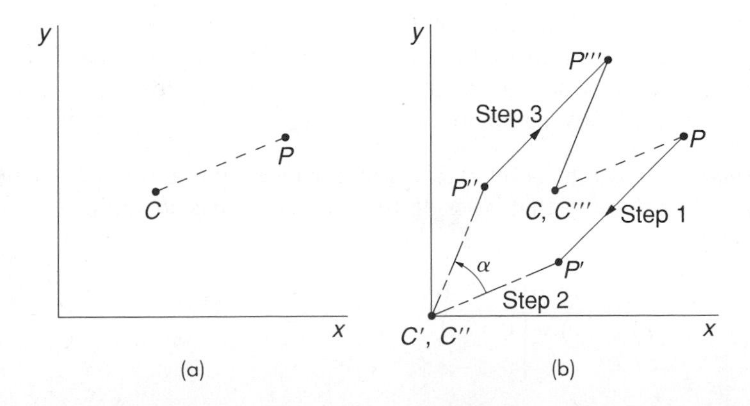

FIGURE 3.18 (a) a point in its original position; (b) the steps involved in rotation about a point

is easily derived using the matrices previously described. The derivation takes place in three steps, as outlined below [13]:

Step 1 Move point C to the origin using a translation matrix:

$$T(-x_c, -y_c) = \begin{bmatrix} 1 & 0 & 0 \\ 0 & 1 & 0 \\ -x_c & -y_c & 1 \end{bmatrix}$$

Step 2 Rotate about the z axis through the angle specified—say, α:

$$R(\alpha) = \begin{bmatrix} \cos\alpha & \sin\alpha & 0 \\ -\sin\alpha & \cos\alpha & 0 \\ 0 & 0 & 1 \end{bmatrix}$$

Step 3 Move point C back to its original position:

$$T(x_c, y_c) = \begin{bmatrix} 1 & 0 & 0 \\ 0 & 1 & 0 \\ x_c & y_c & 1 \end{bmatrix}$$

Thus, the overall rotation matrix is given by the following concatenation of matrices:

$$T = T(-x_c, -y_c)R(\alpha)T(x_c, y_c)$$

$$= \begin{bmatrix} \cos\alpha & \sin\alpha & 0 \\ -\sin\alpha & \cos\alpha & 0 \\ x_c(1-\cos\alpha) + y_c \sin\alpha & -x_c \sin\alpha + y_c(1-\cos\alpha) & 1 \end{bmatrix}$$

Figure 3.18(b) illustrates the steps in the above transformations.

❑ REFLECTION ABOUT AN ARBITRARY AXIS

Assume that the reflection of point P is to be obtained about a given axis ab. The intercept of axis ab on the y axis is equal to c. See Figure 3.19(a). The necessary transformation matrix for the above transformation may be derived in the following manner [13]:

Step 1 Move the point at $x = 0$ on the given axis a distance $-c$ in the y direction so that it lies on the origin and axis ab occupies position $a'b'$:

$$T(0, -c) = \begin{bmatrix} 1 & 0 & 0 \\ 0 & 1 & 0 \\ 0 & -c & 1 \end{bmatrix}$$

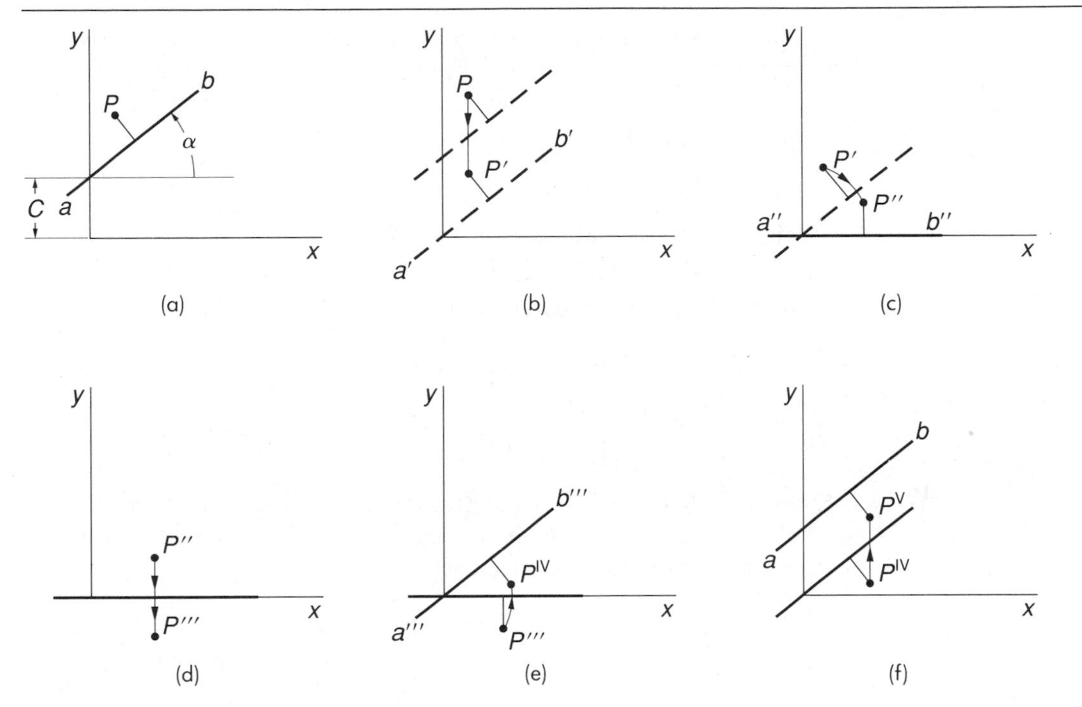

FIGURE 3.19 (a) point in relation to an arbitrary axis; (b)–(f) derivation of the reflection matrix

The transformation is shown in Figure 3.19(b). Point P now occupies position P'.

Step 2 Rotate the axis through an angle $-\alpha$ about the origin so that it coincides with the x axis:

$$R(-\alpha) = \begin{bmatrix} \cos\alpha & -\sin\alpha & 0 \\ \sin\alpha & \cos\alpha & 0 \\ 0 & 0 & 1 \end{bmatrix}$$

Axis ab occupies position $a''b''$, and point P occupies position P'', [see Figure 3.19(c)] as a result of the transformations implemented so far.

Step 3 Obtain the reflection about the x axis:

$$M(x) = \begin{bmatrix} 1 & 0 & 0 \\ 0 & -1 & 0 \\ 0 & 0 & 1 \end{bmatrix}$$

The current position of point P is given by point P'''. See Figure 3.19(d).

Step 4 Rotate the axis back to its original slope:

$$R(\alpha) = \begin{bmatrix} \cos\alpha & \sin\alpha & 0 \\ -\sin\alpha & \cos\alpha & 0 \\ 0 & 0 & 1 \end{bmatrix}$$

Point P now occupies position P^{iv}, and axis ab has the attitude depicted by line $a'''b'''$ in Figure 3.19(e).

Step 5 Move the origin point up by c:

$$T(0, c) = \begin{bmatrix} 1 & 0 & 0 \\ 0 & 1 & 0 \\ 0 & c & 1 \end{bmatrix}$$

Axis ab is back to its original position and attitude, and point P^{v} is the reflection of point P about the axis ab. The concatenated transformation matrix is thus given as

$$T = T(0, -c)R(-\alpha)M(x)R(\alpha)T(0, c)$$

$$= \begin{bmatrix} \cos 2\alpha & \sin 2\alpha & 0 \\ \sin 2\alpha & -\cos 2\alpha & 0 \\ -c\sin 2\alpha & c(\cos 2\alpha + 1) & 1 \end{bmatrix}$$

3.10 ▪ Graphics Packages and Graphics Standards

For application programs in a high-level language, a graphics package is required for producing (rendering) images of objects. A typical graphics package contains the following features:

1. Graphics primitive subroutines for points, lines, circles, symbols, and characters
2. Window, viewport, and clipping subroutines
3. Miscellaneous subroutines for entering and leaving the graphics mode, selecting colors, hard copying, etc.

Graphics packages were originally written in low-level machine language by computer hardware manufacturers and worked only on the specific computers in question. An application program that used the graphics package could thus be run only on the specific machine. This situation was unsatisfactory for developing portable programs (i.e., programs that could work on all computers and display devices). It was therefore felt that some method should be found for obviating machine dependence of graphics programs. The route taken toward this goal was the use of graphics standards.

As you are no doubt aware, English has been standardized as the language of international aviation. A pilot, therefore, flying over any territory in the world can communicate with the nearest control tower to seek help and directions. In computer programming, a similar approach was taken by standardizing FORTRAN, COBOL, Pascal, etc. Introduction of these high-level languages caused an explosion in computer utilization. What has worked in other aspects of life should work in computer graphics as well. Driven by such conviction, computer scientists in industry and academia began to develop graphics standards.

In 1977 a set of specifications for graphical programming was established by the Special Interest Group on Graphics (SIGGRAPH) of the Association for Computer Machinery (ACM). These specifications were given the name CORE Graphics System (or Core). Another set of graphics standards, developed in Europe, is known as the Graphical Kernel System (GKS). The International Standard Organization (ISO) and the American National Standard Institute (ANSI) have accepted GKS as their official standard [14, 15]. GKS and Core consist of a number of subroutines, which application programmers (i.e., the people who write programs as opposed to those who use somebody else's programs) may incorporate within a program for graphical images. GKS defines about 200 different subroutines. These subroutines are independent of any programming language. Some *language bindings* are also standardized, and work is continuing on more. Language bindings allow the programmer to call the GKS subroutines from various high-level programming languages. Once all language bindings have been standardized, it will be possible to call any GKS graphics subroutine from any high-level language.

Core is a full three-dimensional graphics system. GKS is at the moment a two-dimensional system. A draft standard—namely, GKS–3D [16]—was released in 1987 for applications involving three-dimensional graphics. A brief discussion of GKS–3D and PHIGS, yet another graphics standard, is given in Section 3.14.

3.11 ▪ GKS

As might be expected, GKS is language and device independent. This means that the system can be adopted for use in almost any programming language and on almost any graphics input/output device.

Device independence in GKS is achieved through the use of virtual devices. A *virtual device* is an idealized device with defined characteristics. In general, any commercial graphics device does not have the exact characteristics of the virtual device. The characteristics of the virtual devices, however, may be selected so as to simulate many commercial devices.

3.12 ▪ Graphics Primitives in GKS

Four main types of graphics primitives are provided in GKS: (1) polyline, (2) polymarker, (3) fill area, and (4) text. In the paragraphs below, each of these primitives is described in brief; more details may be found in specialized references [17, 18].

❑ POLYLINE

Polyline is a primitive for drawing a sequence of straight lines connected end to end. The GKS function for the polyline primitive is POLYLINE (N,X,Y). It may be called from a FORTRAN program using the statement CALL GPL (N,X,Y), which is the FORTRAN binding for the function. In the above function, N defines the size of the coordinate arrays [$X(N)$, $Y(N)$] for the points that must be joined to form the lines. The line drawing begins at point [$X(1)$, $Y(1)$] and ends at [$X(N)$, $Y(N)$]. If a FORTRAN program contains the following statements:

```
CALL GPL(N,X,Y)
CALL GPL(M,X1,Y1)
```

then two independent sequences of lines not connected to each other are drawn, as shown in Figure 3.20.

It is possible to specify such attributes as width, color, and line type (i.e., dashed, continuous, dotted, etc.) for the polylines. All these attributes can be bundled together and given an index number. It is possible to have as many

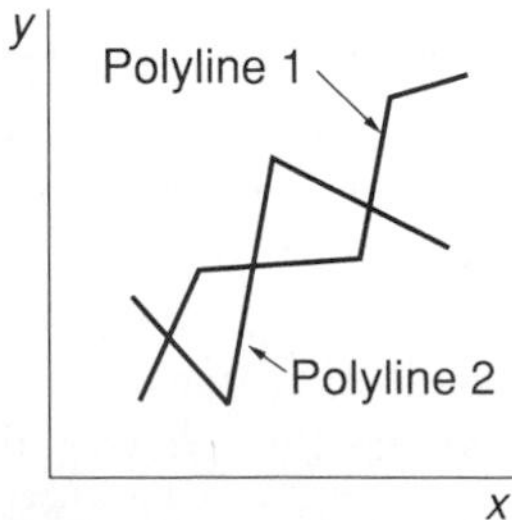

FIGURE 3.20
Polylines

attribute bundles as desired. An attribute bundle can be invoked by the use of the function SET POLYLINE INDEX (J). The FORTRAN binding for the function is CALL GSPLI(J). Once the attributes have been selected in an application program, they apply to all polylines drawn subsequently. One can, of course, alter the attributes as many times as desired within a single application program.

❑ POLYMARKER

At times one may wish not to connect data points, but rather to plot them as discrete points. For such applications, function POLYMARKER (N,X,Y) has been provided. The FORTRAN 77 binding for the polymarker is CALL GPM(N,X,Y), where N is the number of data points and $[X(N), Y(N)]$ are the coordinates of those points. Attributes for the data points are selected through the function SET POLYMARKER INDEX (N). The FORTRAN 77 binding for the above function is CALL GSPMI(N). As with the polylines, the index number is associated with a number of attributes that the user may select. Depending on the attributes selected, different types of markers are drawn.

❑ FILL AREA

This primitive is used to fill a predefined area with different types of hatching or patterns. The name given to the fill area function is FILL AREA (N,X,Y), and its FORTRAN 77 binding is CALL GFA (N,X,Y). The area in question in the above case is bounded by straight lines joining points $[X(1), Y(1)], \ldots, [X(N), Y(N)], [X(1), Y(1)]$. An attribute statement should precede the call to the fill area function. The attributes are set by the function SET FILL AREA INDEX (K), whose FORTRAN 77 binding is CALL GSFAI(K). Depending on the value of K, the enclosed area is either filled or not. If one specifies a hollow fill, only the boundary of the area is drawn, and the area within the boundary is left empty, as shown in Figure 3.21(a). If a solid fill is prescribed, the area is completely filled with a single color, as shown in Figure 3.21(b). For other types of fill (e.g., pattern and

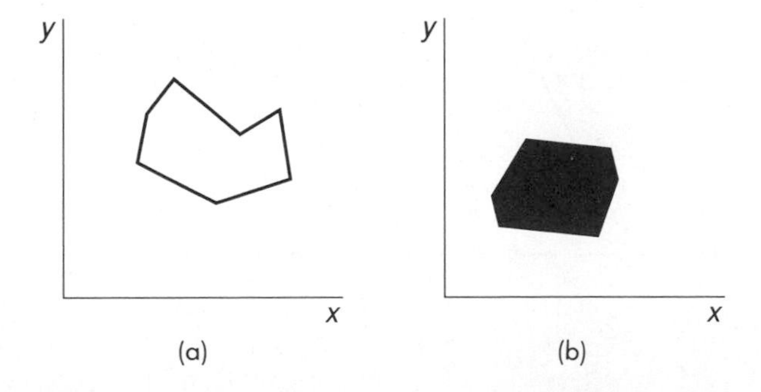

FIGURE 3.21
Area fills: (a) hollow; (b) solid

hatch), no boundary is drawn, but the area is either hatched or filled with some user-selected pattern. In Appendix II the functions used to specify the remaining attributes are discussed.

❑ TEXT

A text primitive is used to place text strings on a graphics image. The text function is TEXT(X,Y,'text'). The function SET CHARACTER HEIGHT (H) is used to fix the height of the characters, and the function SET CHARACTER UP VECTOR (X,Y) is used to define a vector that orients the characters. The quantities X and Y specify offsets from the character up vector in the x and y directions. If X and Y are both positive, then the character up vector is inclined as shown in Figure 3.22(a). If, say, X is negative and Y is positive, then the character up vector is inclined as shown in Figure 3.22(b). A text path attribute is defined by the function SET TEXT PATH (XYZ) in which

XYZ = RIGHT causes the characters to be placed from left to right in a direction perpendicular to the character up vector. This is also the default attribute.

XYZ = LEFT causes the characters to be placed from right to left in a direction perpendicular to the character up vector.

XYZ = DOWN causes the characters to be placed one below the other in a direction opposite to that of the character up vector.

XYZ = UP causes the characters to be placed one above the other in the direction of the character up vector.

The coordinates (X, Y) in the function TEXT act as a reference point for the character string. The function SET TEXT ALIGNMENT (H,V) is used

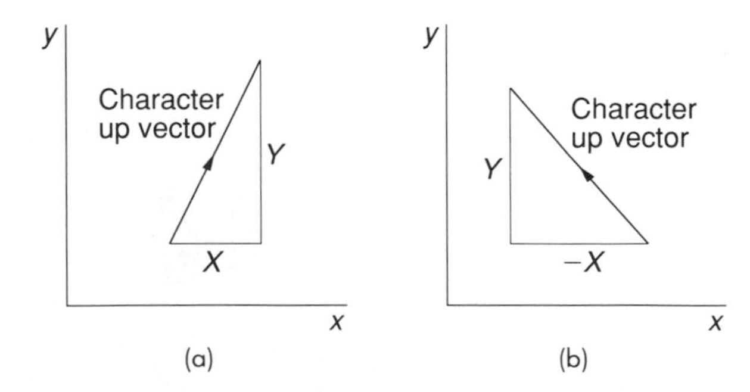

FIGURE 3.22
Character up vector with (a) positive X-offset and (b) negative X-offset

to align the text with reference to the reference point. Parameters *H* and *V*, respectively, determine the horizontal and vertical positions of the strings. By assigning

H = LEFT, the text is placed to the right of the reference point (i.e., the reference point stays to the left of the string on a horizontal line).

H = RIGHT, the text is placed such that the reference point is to the right of the character string.

H = CENTER, the text is spread out equally on either side of the reference point along a horizontal line.

V = TOP, the top of the character string is aligned with the reference point.

V = BOTTOM, the bottom of the character string is aligned with the reference point.

V = BASE, CAP, or HALF, the base, the top or the middle font line of the characters is aligned with the reference point.

One can also assign H = NORMAL, which is the same as H = LEFT, and V = NORMAL, which is the same as V = BASE. If alignment attributes are omitted altogether, the normal values for *H* and *V* are assumed by default. In the above discussions, vertical direction is in the direction of the character up vector, and horizontal direction is perpendicular to the character up vector.

Several other attributes of the text primitives are defined by a text index. The GKS function for selecting a predefined text index is the function SET TEXT INDEX (N). In Appendix II, the function used to assign the remaining attributes is described.

FORTRAN 77 bindings for the text functions are given below.

Function Name	FORTRAN 77 binding
TEXT	CALL GTXS (X,Y,'text')
SET CHARACTER HEIGHT	CALL GSCHH (H)
SET CHARACTER UP VECTOR	CALL GSCHUP (X,Y)
SET TEXT PATH	CALL GSTXP (XYZ)
SET TEXT ALIGNMENT	CALL GSTXAL (H,V)
SET TEXT INDEX	CALL GSTXI (N)

EXAMPLE ■ 3.12

Write a graphics program, using GKS functions, to produce a title page for this book, as shown in Figure 3.23.

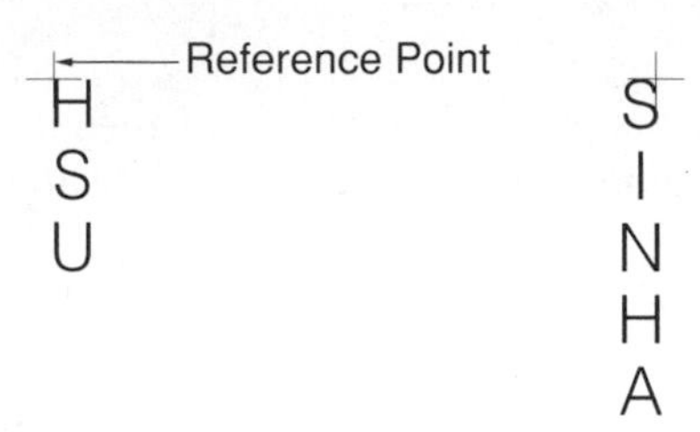

FIGURE 3.23
A proposed title page

```
01   SET CHARACTER UP VECTOR (0,1)
     SET TEXT PATH (RIGHT)
     SET TEXT ALIGNMENT (LEFT,BASE)
     TEXT (5,8,'COMPUTER AIDED DESIGN')
05   SET TEXT PATH (DOWN)
     SET TEXT ALIGNMENT (LEFT,TOP)
     TEXT (6,6,'HSU')
     SET TEXT ALIGNMENT (RIGHT,TOP)
     TEXT (10,6,'SINHA')
```

In line 01, the character up vector is specified as a vertical line. In lines 02–04, it is determined that the string COMPUTER AIDED DESIGN will be placed horizontally from left to right and in such a way that the reference point at (5, 8) is to the left of the characters and aligned with the base of the characters. The implications of the other statements are obvious.

3.13 ▪ Other Graphics Standards

GKS and Core have already been discussed. These establish a programmer-level interface between application programs and graphics utility programs. As will be obvious from the discussions below, certain other types of standards are needed as well. At the present time, work is in progress on the following.

❑ VIRTUAL DEVICE INTERFACE

Virtual Device Interface (VDI) refers to the standards used for device drivers. *Device drivers* are programs that control the operation of the graphics peripherals. Thus, the device-independent application programs will be able to use the VDI to pass on data and control commands to real devices for generating images. The VDI is also often referred to as Computer Graphics Interface (CGI) or CG-VDI. At the present time, an ISO draft proposal on VDI [19] is being considered for adoption by member countries including the United States.

❑ COMPUTER GRAPHICS METAFILE

Computer Graphics Metafile (CGM) provides a mechanism for storing and transmitting graphical data in a device-independent way so that metafiles are transportable. An obvious implication of CGM is that it will make it possible for metafiles generated by one system to be used by any other system. ISO and ANSI standards on CGM have been developed [20, 21].

❑ NORTH AMERICAN PRESENTATION LEVEL PROTOCOL

North American Presentation Level Protocol (NAPLP), developed by AT&T, is a standard for transmitting text and graphics over telecommunication lines.

The following standards, similar in nature to GKS and Core, also deserve mention in the present discussion.

❑ PHIGS

PHIGS is the acronym for Programmer's Hierarchical Interactive Graphics System [22, 23]. PHIGS is actually a modern development, an offshoot of GKS and GKS–3D. Most of the primitives and the attributes are identical in all the above standards. PHIGS differs from GKS primarily in its organization and manipulation of data. It allows rapid updating of graphics models. Also, attributes may be changed dynamically. In the offing is PHIGS+, an extended version of the standard, which has specifications for surface-rendering techniques (i.e., shading and shadowing).

❑ GKS–3D

As mentioned earlier, GKS–3D includes functions for three-dimensional graphics applications. It allows three-dimensional coordinates to be entered from three-dimensional input devices and permits work-station-dependent hidden line/hidden surface removal. The viewing of solid objects from different angles is also permitted.

3.14 ▪ Principles of Three-Dimensional Graphics

Creating three-dimensional graphics involves techniques for presenting images of three-dimensional objects on a two-dimensional display device. As mentioned earlier, owing to the mismatch between the dimensions of the object and the display device, drawing procedures for three-dimensional graphics can be quite complex. One has to devise ways of representing depth on a plane surface. There are several ways of doing this. One most obvious method is to present images of the same type as are perceived by the human eye or by a camera. Such images, though realistic, are somewhat difficult to generate. Over the years, engineers have devised some simpler, though not so realistic, ways of conveying depth information of three-dimensional objects. In the next chapter, some aspects of three-dimensional graphics are covered. In the remainder of this chapter, the mathematical techniques for three-dimensional image manipulation are presented, which may be seen as an extension of the theory already discussed.

3.15 ▪ Three-Dimensional Transformations: Simple Cases

Translation, rotation, scaling, reflection, and the like, of three-dimensional objects are carried out using matrices similar to those given in Sections 3.8 through 3.10.

❑ TRANSLATION

A point P (x, y, z) is translated through the distances Δx, Δy, and Δz in the Cartesian coordinate system by the following multiplication:

$$\{x' \quad y' \quad z' \quad 1\} = \{x \quad y \quad z \quad 1\} T(\Delta x, \Delta y, \Delta z) \qquad (3.40)$$

where T is the transformation matrix expressed in terms of homogeneous coordinates:

$$T(\Delta x, \Delta y, \Delta z) = \begin{bmatrix} 1 & 0 & 0 & 0 \\ 0 & 1 & 0 & 0 \\ 0 & 0 & 1 & 0 \\ \Delta x & \Delta y & \Delta z & 1 \end{bmatrix} \qquad (3.41)$$

❑ ROTATION

The matrix for rotation through an angle α about the x axis is given below.

$$R(x, \alpha) = \begin{bmatrix} 1 & 0 & 0 & 0 \\ 0 & \cos\alpha & \sin\alpha & 0 \\ 0 & -\sin\alpha & \cos\alpha & 0 \\ 0 & 0 & 0 & 1 \end{bmatrix} \tag{3.42}$$

For a rotation through an angle β about the y axis, the following matrix is used:

$$R(y, \beta) = \begin{bmatrix} \cos\beta & 0 & -\sin\beta & 0 \\ 0 & 1 & 0 & 0 \\ \sin\beta & 0 & \cos\beta & 0 \\ 0 & 0 & 0 & 1 \end{bmatrix} \tag{3.43}$$

For a rotation through an angle γ about the z axis, the transformation matrix reads

$$R(z, \gamma) = \begin{bmatrix} \cos\gamma & \sin\gamma & 0 & 0 \\ -\sin\gamma & \cos\gamma & 0 & 0 \\ 0 & 0 & 0 & 0 \\ 0 & 0 & 0 & 1 \end{bmatrix} \tag{3.44}$$

In all the above derivations, positive angles are measured in a counterclockwise manner while observing at the origin from the positive side of the corresponding axis.

❑ SCALING

The scaling matrix is as follows:

$$S(i, j, k) = \begin{bmatrix} i & 0 & 0 & 0 \\ 0 & j & 0 & 0 \\ 0 & 0 & k & 0 \\ 0 & 0 & 0 & 1 \end{bmatrix} \tag{3.45}$$

The above matrix scales the x coordinate of a point by a factor of i, the y coordinate by a factor of j, and the z coordinate by a factor of k.

❑ REFLECTION

Reflection of a point about a given plane is obtained by moving the point to the other side of the plane while keeping its distance from the plane the same. Thus, the reflection of point $P\ (x, y, z)$ about the xy plane is point P'

$(x, y, -z)$, which is obtained by the following matrix multiplication:

$$P'\{x \quad y \quad -z \quad 1\} = P\{x \quad y \quad z \quad 1\}M(xy)$$

where $M(xy)$ is the reflection matrix given by

$$M(xy) = \begin{bmatrix} 1 & 0 & 0 & 0 \\ 0 & 1 & 0 & 0 \\ 0 & 0 & -1 & 0 \\ 0 & 0 & 0 & 1 \end{bmatrix} \qquad (3.46)$$

Similarly, for obtaining reflection about the xz and yz planes respectively, the matrices used are:

$$M(xz) = \begin{bmatrix} 1 & 0 & 0 & 0 \\ 0 & -1 & 0 & 0 \\ 0 & 0 & 1 & 0 \\ 0 & 0 & 0 & 1 \end{bmatrix} \qquad (3.47)$$

and

$$M(yz) = \begin{bmatrix} -1 & 0 & 0 & 0 \\ 0 & 1 & 0 & 0 \\ 0 & 0 & 1 & 0 \\ 0 & 0 & 0 & 1 \end{bmatrix} \qquad (3.48)$$

3.16 ▪ Three-Dimensional Transformations: Complex Cases

The matrices given in equations (3.41) through (3.48) are often referred to as primitive transformation matrices. It is possible to derive matrices for other transformations from these primitive matrices. A few such cases are demonstrated below.

❑ CONCATENATED TRANSFORMATION

A series of different types of transformations can be achieved by chain multiplication of the appropriate matrices. Thus, if point P is to be first translated by distances $(\Delta x, \Delta y, \Delta z)$ and then rotated about the x axis through an angle α followed by rotations of angles β and γ about the y and z axes, the entire transformation may be carried out by the following multiplication:

$$P' = PT(\Delta x, \Delta y, \Delta z)R(x, \alpha)R(y, \beta)R(z, \gamma)$$

EXAMPLE ■ 3.13

A point $P(1, 1, 2)$ is to be translated through (0.5, 0.6, 0.4), followed by rotations of 10° about the x axis, 15° about the y axis, and 15° about the z axis. Find the transformation matrix for the concatenated transformation and the coordinates of the final position of the point.

The concatenation transformation matrix is given as

$$T = T(0.5, 0.6, 0.4)R(x, 10)R(y, 15)R(z, 15)$$

$$= \begin{bmatrix} 0.966 & 0 & -0.259 & 0 \\ 0 & 1 & 0 & 0 \\ 0.259 & 0 & 0.966 & 0 \\ 0 & 0 & 0 & 1 \end{bmatrix}$$

Thus, the coordinates of the final position of the point are given as

$$P'\{x' \quad y' \quad z' \quad 1\} = P\{1 \quad 1 \quad 2 \quad 1\}T$$
$$= \{1.76 \quad 1.67 \quad 2.16 \quad 1\}$$

The path traced by the above point is shown in Figure 3.24.

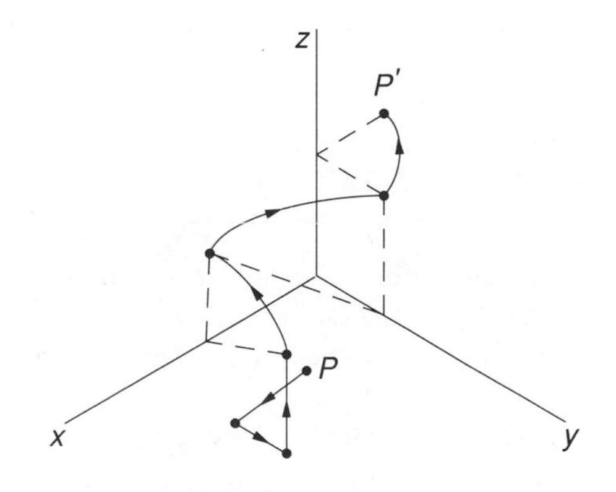

FIGURE 3.24
Path of point P

❑

❑ TRANSFORMATION OF SOLID OBJECTS

Transformation operations are applied to solid objects by multiplying coordinates of every point on the object by the same transformation matrix. Usually, solids are identified by a few vertices, and it is sufficient to apply the transformation to only a finite number of such vertices. Thus, if the unit cube shown in Figure 3.25 is to be translated through distances (2, 2, 6), the following matrix multiplication is necessary to redefine the new positions of

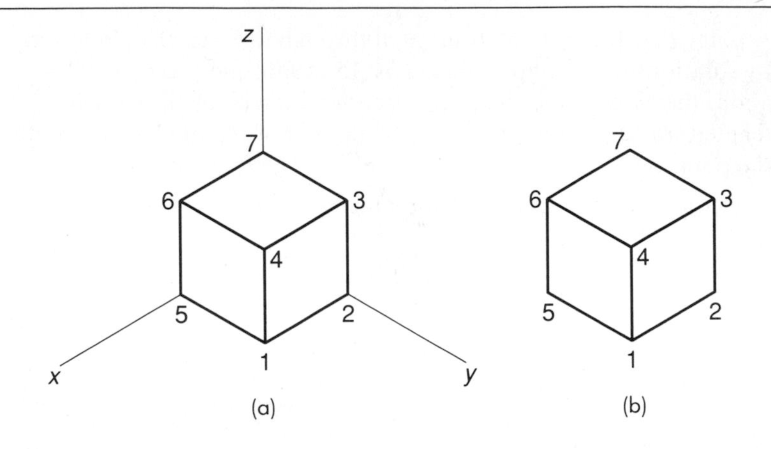

FIGURE 3.25 A cube: (a) original position; (b) transformed position

each of its vertices:

$$\begin{bmatrix} 1 & 1 & 0 & 1 \\ 0 & 1 & 0 & 1 \\ \vdots & \vdots & \vdots & \vdots \\ 1 & 0 & 1 & 1 \end{bmatrix} \begin{bmatrix} 1 & 0 & 0 & 0 \\ 0 & 1 & 0 & 0 \\ 0 & 0 & 1 & 0 \\ 2 & 2 & 6 & 1 \end{bmatrix} = \begin{bmatrix} 3 & 3 & 3 & 1 \\ 2 & 3 & 3 & 1 \\ \vdots & \vdots & \vdots & \vdots \\ 3 & 2 & 4 & 1 \end{bmatrix}$$

The cube can now be reconstructed in its new position by joining the vertices in the same order as they were joined in the original position of the cube.

❑ ROTATION ABOUT AN ARBITRARY AXIS

Let's assume that a straight line AB is given in three-dimensional space. One of the ways of identifying the straight line is through the set of parameters known as *direction cosines*. These are cosines of the angles that the straight line makes with the principal axes of the Cartesian coordinate system. The direction cosines of the line AB are $(\cos\alpha, \cos\beta, \cos\gamma)$, as shown in Figure 3.26. Let P (p_x, p_y, p_z) and Q (q_x, q_y, q_z) be two points on the straight line. It can be shown easily that

$$\cos\alpha = \frac{q_x - p_x}{L} \quad \cos\beta = \frac{q_y - p_y}{L} \quad \cos\gamma = \frac{q_z - p_z}{L}$$

where

$$L = \sqrt{(q_x - p_x)^2 + (q_y - p_y)^2 + (q_z - p_z)^2}$$

Suppose that point R (x, y, z) is to be rotated around the line AB through an angle θ in the counterclockwise direction. See Figure 3.26. The required

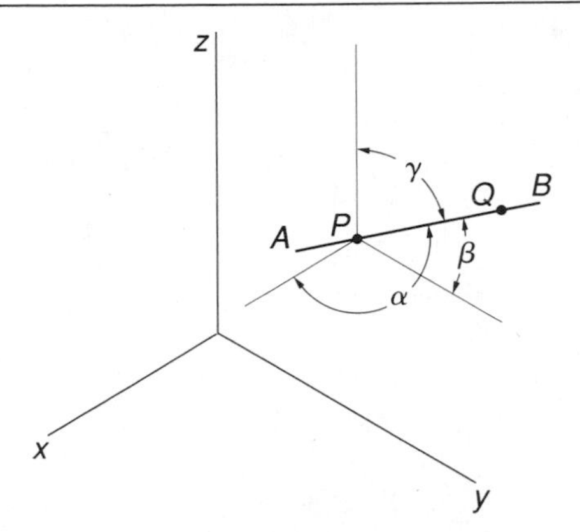

FIGURE 3.26
An arbitrary axis in the three dimensional space

rotation may be viewed as being attained not directly, but in a series of steps for which transformation matrices exist [13]:

Step 1 Move points P, Q, and R through distances $(-p_x, -p_y, -p_z)$ such that point P now coincides with the origin. The transformation matrix is

$$TR1 = \begin{bmatrix} 1 & 0 & 0 & 0 \\ 0 & 1 & 0 & 0 \\ 0 & 0 & 1 & 0 \\ -p_x & -p_y & -p_z & 1 \end{bmatrix}$$

Step 2 Rotate about the y axis in a clockwise manner through an angle ϕ_1 [see Figure 3.27(a)] such that the line AB now lies in the yz plane.

Use the following rotation matrix:

$$R(y, -\phi_1) = \begin{bmatrix} \cos\phi_1 & 0 & \sin\phi_1 & 0 \\ 0 & 1 & 0 & 0 \\ -\sin\phi_1 & 0 & \cos\phi_1 & 0 \\ 0 & 0 & 0 & 1 \end{bmatrix}$$

where

$$\sin\phi_1 = \frac{q_x - p_x}{L1}$$

$$\cos\phi_1 = \frac{q_z - p_z}{L1}$$

$$L1 \quad \sqrt{(q_x - p_x)^2 + (q_z - p_z)^2}$$

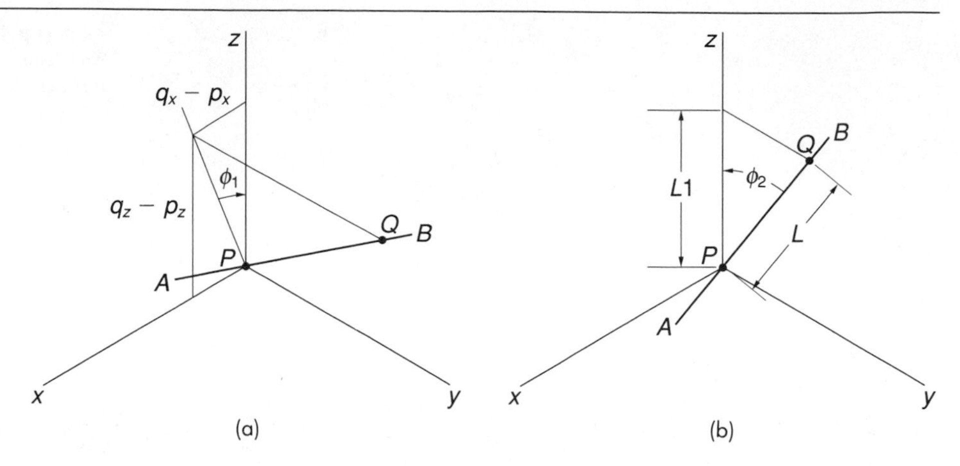

FIGURE 3.27 Line *AB*: (a) after translation; (b) after translation and rotations

Step 3 Rotate counterclockwise about the x axis through an angle ϕ_2. See Figure 3.27(b). Use the following transformation matrix:

$$R(x, \phi_2) = \begin{bmatrix} 1 & 0 & 0 & 0 \\ 0 & \cos\phi_2 & \sin\phi_2 & 0 \\ 0 & -\sin\phi_2 & \cos\phi_2 & 0 \\ 0 & 0 & 0 & 1 \end{bmatrix}$$

where

$$\sin\phi_2 = \frac{q_y - p_y}{L}$$

$$\cos\phi_2 = \frac{L1}{L}$$

This operation makes line AB coincident with the z axis.

Step 4 Rotate counterclockwise through an angle θ about the z axis.

$$R(z, \theta) = \begin{bmatrix} \cos\theta & \sin\theta & 0 & 0 \\ -\sin\theta & \cos\theta & 0 & 0 \\ 0 & 0 & 1 & 0 \\ 0 & 0 & 0 & 1 \end{bmatrix}$$

We have now affected the rotation about AB originally sought. However, AB is not where it should be. In the next three steps, line AB is restored, to its original position and attitude.

Step 5 Rotate through angle $-\phi_2$ about the x axis.

$$R(x, -\phi_2) = \begin{bmatrix} 1 & 0 & 0 & 0 \\ 0 & \cos\phi_2 & -\sin\phi_2 & 0 \\ 0 & \sin\phi_2 & \cos\phi_2 & 0 \\ 0 & 0 & 0 & 1 \end{bmatrix}$$

Step 6 Rotate through an angle ϕ_1 about the y axis.

$$R(y, \phi_1) = \begin{bmatrix} \cos\phi_1 & 0 & -\sin\phi_1 & 0 \\ 0 & 1 & 0 & 0 \\ \sin\phi_1 & 0 & \cos\phi_1 & 0 \\ 0 & 0 & 0 & 1 \end{bmatrix}$$

Step 7 Move through distances (p_x, p_y, p_z).

$$TR2 = \begin{bmatrix} 1 & 0 & 0 & 0 \\ 0 & 1 & 0 & 0 \\ 0 & 0 & 1 & 0 \\ p_x & p_y & p_z & 1 \end{bmatrix}$$

The required transformation matrix is obtained by the following multiplication:

$$T = TR1R(y, -\phi_1)R(x, \phi_2)R(z, \theta)R(x, -\phi_2)R(y, \phi_1)TR2$$

$$= \begin{bmatrix} T_{11} & T_{12} & T_{13} & 0 \\ T_{21} & T_{22} & T_{23} & 0 \\ T_{31} & T_{32} & T_{33} & 0 \\ T_{41} & T_{42} & T_{43} & 1 \end{bmatrix}$$

where

$$T_{11} = C\theta(C1^2 + S1^2S2^2) + (S1C2)^2$$

$$T_{12} = S1C2S\theta + S1S2C2(1 - C\theta)$$

$$T_{13} = S1C1C2^2(1 - C\theta) - S2S\theta$$

$$T_{21} = -S\theta C1C2^2 + S1S2C2(1 - C\theta)$$

$$T_{22} = C\theta S2^2 + C2^2$$

$$T_{23} = S1C2S\theta + C1C2S2(1 - C\theta)$$

$$T_{31} = S\theta S2 + S1C1C2^2(1 - C\theta)$$

$$T_{32} = -C2(S1S\theta - C1S2(1 - C\theta))$$

$$T_{33} = C\theta(S1^2 + C1^2S2^2) + C1^2C2^2$$

$$T_{41} = -p_x(C1C\theta + S1S2S\theta) + p_yC2S\theta - p_z(C1C2S\theta - S1C\theta)$$

$$T_{42} = -p_x(C1C2S\theta + S1S2C2(1 - C\theta)) - p_y(C2^2C\theta + S2^2) + p_z(S1C2S\theta - C1S2C2(1 - C\theta))$$

$$T_{43} = p_x(C1S2S\theta - S1S2^2C\theta - S1C2^2) - p_y(S2C2(1 - C\theta)) - p_z(S1S2S\theta + C1S1^2C\theta + C1C2^2)$$

$$S1 = \sin\phi_1,\ C1 = \cos\phi_1,\ S2 = \sin\phi_2,\ C2 = \cos\phi_2$$

$$S\theta = \sin\theta,\ C\theta = \cos\theta$$

3.17 ▪ Summary

The mathematical theory underlying some of the more commonly used two-dimensional graphics elements—namely, straight lines, circles, and ellipses—are given. An introduction is provided to Bézier curves and splines of different types, while the mathematical formulations for Bézier curves and B-splines are presented in detail. The purpose of curve fitting is explained, and the mathematical techniques for polynomial curve fitting, polynomial regression with least squares, and spline interpolation are illustrated by means of several examples. Techniques for image manipulations (such as translation, rotation, and scaling) in two- and three-dimensions are described. Also covered are homogeneous coordinates and their use in the formulation of transformation matrices.

The reader is informed about the necessity of standards in relation to computer graphics. Special features of such graphics packages as Core, GKS, GKS-3D, PHIGS, and PHIGS+ are explained. Extensive details on GKS are provided in the text and Appendix II. Brief introductions to virtual device interfaces, computer graphics metafiles, and the North American Presentation Level Protocols are provided.

Generally, sufficient information is given so that a novice in computer graphics can appreciate the technical aspects of the commercial drafting packages that he/she may use.

— References

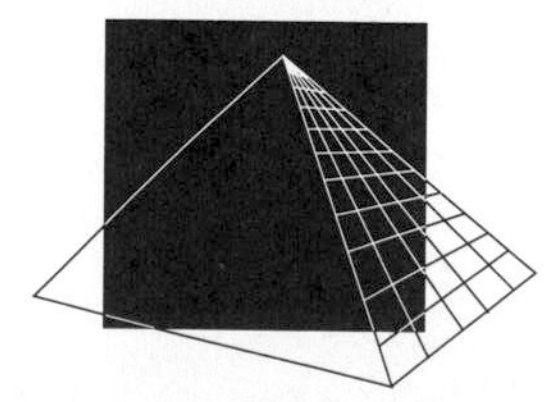

1. Rogers, D.F., and J.A. Adams. *Mathematical Elements for Computer Graphics.* New York: McGraw-Hill, 1976.
2. Bézier, P. "Mathematical and Practical Possibilities of UNISURF." In *Computer Aided Geometric Design.* Edited by R.E. Barnhill and R.F. Riesenfeld. New York: Academic Press, 1974.
3. Barsky, B.A. *Computer Graphics and Modeling Using Beta-Splines.* New York: Springer-Verlag, 1988.

4. deBoor, Carl. *A Practical Guide to Splines.* New York: Springer-Verlag, 1978.
5. Cox, M.G. "A Survey of Numerical Methods." In *The State of Art in Numerical Analysis.* Edited by D. Jacobs. London: Academic Press, 1977.
6. Tiller, W. "Rational B-Spline for Curve and Surface Representation." *IEEE Computer Graphics and Applications* 3, no. 6 (1983): 61–69.
7. Chapra, S.C., and R.P. Canale. *Numerical Methods for Engineers.* New York: McGraw-Hill, 1984.
8. Artwick, B.A. *Applied Concepts in Microcomputer Graphics.* Reading, Mass.: Addison-Wesley, 1984.
9. Phong, Bui-Tuong, "Illumination for Computer-Generated Pictures." *Communications of the ACM* 18, no. 6 (1975): 311–17.
10. Gourard, H. "Continuous Shading of Curved Surfaces." *IEEE Transactions on Computers* C-20, no. 6 (1971): 623–28.
11. Bresenham, J.E. "Algorithm for Computer Control of a Digital Plotter." *IBM Systems Journal* 4, no. 1 (1965): 25–30.
12. Newman, W.M., and R.F. Sproull. *Principles of Interactive Computer Graphics.* New York: McGraw-Hill, 1979.
13. Park, Chan S. *Interactive Microcomputer Graphics.* Reading, Mass.: Addison-Wesley, 1985.
14. ANS X3.124–1985. *Graphical Kernel System* (*GKS*). New York: American National Standards Institute, 1985.
15. ISO 7942. *Information processing systems—computer graphics—Graphical Kernel System (GKS) functional description.* International Standards Organization, 1985.
16. ISO/DIS 8805. *Information processing systems–computer graphics—Graphical Kernel System for three dimensions (GKS-3D) functional description.* International Standards Organization, 1985.
17. Hopgood, F.R.A., D.A. Duce, J.R. Gallop, and D.C. Sutcliffe. *Introduction to the Graphical Kernel System (GKS).* New York: Academic Press, 1983.
18. Enderle, G., K. Kensy, and G. Plaff. *Computer Graphics Programming: GKS—The Graphics Standard.* New York: Springer-Verlag, 1984.
19. ISO DP 9636. *Computer Graphics Interfacing Techniques for Dialogues with Graphical Devices.* International Standards Organization, 1990: Parts 1 through 6.
20. ISO 8632. *Information processing systems—computer graphics—Metafile for transfer and the storage of picture description information.* International Standards Organization, 1987: Parts 1 through 4.
21. ANS X3.122–1986. *Computer Graphics Metafile.* New York: American National Standards Institute, 1986.
22. ISO DIS 9592. *Information Processing Systems—computer graphics—Programmer's Hierarchical Interactive Graphics System*

(PHIGS). International Standards Organization, 1989: Parts 1 through 3.

23. ANS X3.144–1988. *Programmer's Hierarchical Interactive Graphics System.* New York: American National Standards Institute, 1988.

Problems

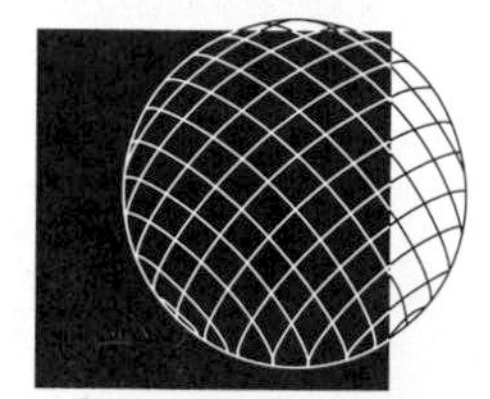

3.1 Plot points A (5, 6) and B (10, 9) on graph paper, and draw a straight line through them. Mark on the graph those pixels that will be lighted up, assuming the graph paper is a raster display screen and the addresses given above are the pixel numbers. Use symmetric DDA to identify the lighted pixels. Comment on the difference in the nature of the lines produced for $s = 0.25$ and $s = 0.05$.

3.2 On a piece of graph paper draw a Bézier curve with the following control points: (1, 1.2), (5, 7), (3, 9), and (0.5, 9.6).

3.3 Draw a second Bézier curve, to join the curve drawn in problem 3.2, where the control points are (0.5, 9.6), (−2, 10.2), (−3, 11), and (−4, 9.5). Will there be continuity of slope at the junction?

3.4 Draw a piecewise polynomial of order 3 based on the following knots: (1, 1.4), (5, 7), (3.2, 10), and (0, 9.6). Maintain continuity of slopes everywhere.

3.5 Use cubic spline functions to describe the profile of the cam shown below. It is given that the following points lie on the cam profile: (2, 0), (0, 3), and (3, 10).

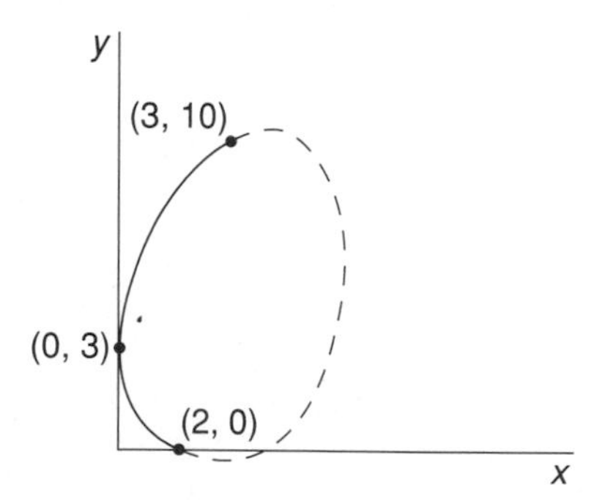

3.6 The function whose plot is shown in Figure 3.6 has the following form:

$$Y(X) = 8X^3e^{-X} - X$$

Use the three curve-fitting techniques described in Section 3.3 to fit the curve from the following data points: (1, 1.943), (1.60, 5.016), (3.4, 7.094), and (5, 1.738).

3.7 The two ends of a straight line have coordinates P_1 (0.5, 1.5) and P_2 (1, 2.5). The line must be rotated through 40° counterclockwise about the origin in the xy plane and then translated four units in the x direction. Write the necessary transformation matrix, and determine the new coordinates of the two end points.

3.8 Reflection of the line described in problem 3.5 is to be obtained about a straight line that passes through points (0, 0.2) and (2, 3.2). Write the necessary transformation matrix for the above operation, and determine the new coordinates of the end points of the line.

3.9 The line segment described in problem 3.5 is to be stretched to twice its present size so that the end at point P_1 occupies the position (0.3, 0.9) and there is no change in the slope of the line. Write the transformation matrix and the coordinates of the new position of end P_2.

3.10 A scaling factor of 2 in the y direction is to be applied to the line segment described in problem in 3.5. No scaling is applied in the x direction, and there is no change in the position of the point P_1. The line is to be rotated subsequently through 30° in a counterclockwise direction. Determine the necessary transformation matrix for the operation and the new coordinates of point P_2.

3.11 In problem 3.10, if the order of operations is reversed (i.e., if rotation is applied before scaling), determine the transformation matrix and the coordinates of point P_2.

3.12 The vectices of a triangle are situated at points (0.5, 1.0), (0.8, 1.2), and (0.2, 1.45). Find the coordinates of the vertices if the triangle is first rotated 10° counterclockwise about the origin and then scaled to twice its size. Draw the two triangles on graph paper.

3.13 In problem 3.12, if the vertex at (0.5, 1.0) must remain at its original position after the transformations, find a suitable matrix for such a transformation. Verify the correctness of your matrix by actual computations.

3.14 Assume that the triangle given in problem 3.12 is first rotated 10° counterclockwise about an axis through the vertex (0.5, 1.0) and then scaled to twice its size with the same vertex as the base point. Find the coordinates of the other two vertices.

3.15 Repeat the exercise in problem 3.14 by selecting an axis and base point at the vertex (0.8, 1.2). Plot the original triangle and the transformed triangles from problem 3.14 and this problem on graph paper.

3.16 In Figure 3.24, the path of a point for certain transformations is given. In order to prove that the 3D transformations are noncommutative, find the coordinates of the point for the following sequence:

(a) translation by (0.5, 0.6, 0.4)
(b) rotation about y axis by 15°
(c) rotation about x axis by 10°
(d) rotation about z axis by 15°

3.17 The scaling matrix given in equation (3.45) uses the origin as the base point. Derive a matrix using an arbitrary point (x_b, y_b, z_b) as the base.

3.18 A straight line passes through points (1, 1, 2) and (2, 3, 3). A point presently occupying the position (4, 5, 6) is to be rotated through +20° about the above line. Find the new coordinates of the point.

CHAPTER 4

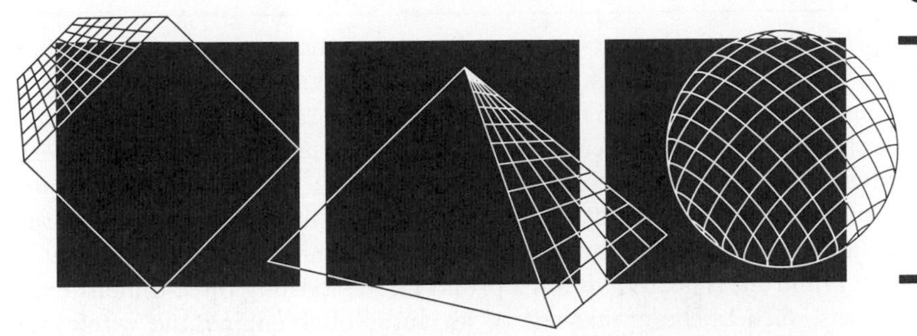

Computer Graphics and Design

4.1 Introduction

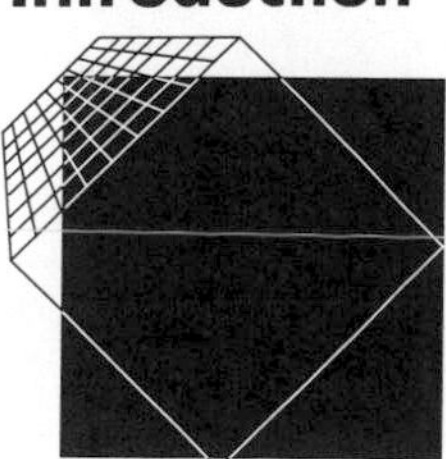

Some basic principles of computer graphics were discussed in Chapter 3. In this chapter, the role of computer graphics in mechanical design is examined. The relevant theory and the algorithms that make computer graphics applications possible in this domain are also considered.

4.2 ▪ Geometric Modeling

Geometric modeling pertains to the process of graphically representing the shape and size of physical objects. Such models are used in design to give a visual perspective of objects not yet existing. A designer can keep changing the details of a design until a satisfactory product is found. Some types of geometric models are also used for evaluating mass, volume, and moments of inertia and for predicting the response of mechanical systems to temperature and stress fields, dynamic forces, etc. From the above, it may be concluded that geometric modeling represents one of the most valuable contributions that computer graphics has made to design engineering. In the next few sections, various tools, techniques, and methods presently available for geometric modeling are investigated.

One can use the lowest-level graphics primitive (i.e., the straight line) to create wireframe models of all types of objects. For modeling two-dimensional objects, such as a circuit board, one can employ lines to draw circles, polygons, and other arbitrary shapes such as Bézier or spline curves. Most two-dimensional computer-aided drafting packages, however, do provide subroutines for drawing these shapes. Thus, the application programmer is saved the trouble of developing subroutines for circles, etc., every time one is needed within a program. In modeling three-dimensional objects, such as the crankshaft of an automobile engine, the wireframe approach is generally inadequate. Besides being confusing to look at, a wireframe model does not lend itself to any rapid method of hidden line removal. It is also impossible to obtain the weight, moment of inertia, and volume of the objects represented by wireframe models. All viable three-dimensional modeling packages, therefore, provide certain higher-level primitives, such as cylinders, spheres, and surface patches. These higher-level primitives can be utilized in a number of ways for creating three-dimensional models.

There are two distinct methods of modeling three-dimensional objects: (1) *surface modeling* and (2) *solid modeling*. The salient features of these methods are outlined in the next two sections.

4.3 ▪ Surface Modeling

Surface models are most suited for displaying the surface when interior details are of no interest. A model is generated by bounding the exterior of the object by a number of mathematically defined surfaces. These surfaces may be simple (such as planes, cylinders, and cones) or complex (such as surfaces of revolution or sculpted surfaces based on B-splines or Bézier curves).

Figure 4.1 presents an example of the use of polygons for surface modeling. The object modeled is a funnel. The graphic model does resemble the object, but the representation is not very accurate. If a somewhat larger number of polygons are employed, the accuracy of the representation would improve. Ideally, an infinite number of planar polygons are required to make the representation exact. From the above example, it is clear that a

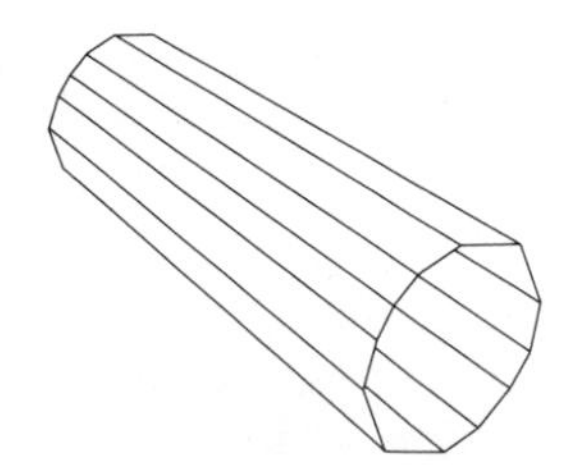

FIGURE 4.1
Polygon mesh for a funnel

polygonal surface model is an expensive method of geometric modeling insofar as computer time and memory requirements are concerned. On the other hand, the mathematics of polygonal surface models is quite easy to manipulate. Polygonal surface models are most beneficially applied in cases where an object contains simple straight edges, as in modern high-rise buildings, bridges, furniture, etc. However, if curved surface patches are used instead of polygons, data storage and handling requirements may be considerably reduced.

Polygonal models can be generated in several ways. Some simpler methods are outlined below [1].

❑ EXPLICIT POLYGON METHOD

In this method [1], every polygon is defined by an array of vertex coordinates, such as

$$VA = \{(x_1, y_1, z_1) \quad (x_2, y_2, z_2) \quad \ldots \quad (x_n, y_n, z_n)\}$$

The polygons are generated by joining the vertices as they occur in the vertex list.

In Figure 4.2, a simple polygon mesh is shown. It can be seen that most of the vertices are common to two or more polygons. Obviously, the common vertices are stored as many times as they occur in the different polygons. For example, vertex 1 will be stored in four different vertex arrays in the above case. Because some vertex coordinates need to be stored in several arrays, the explicit polygon method is inefficient in its utilization of computer memory. The method is also slow in terms of execution because several edges are often drawn more than once. In the example shown in Figure 4.2, edges 1–2, 1–4, 1–6, and 1–1 are each drawn twice.

In spite of its weaknesses, the explicit polygon method is worth considering in many instances because of its simplicity.

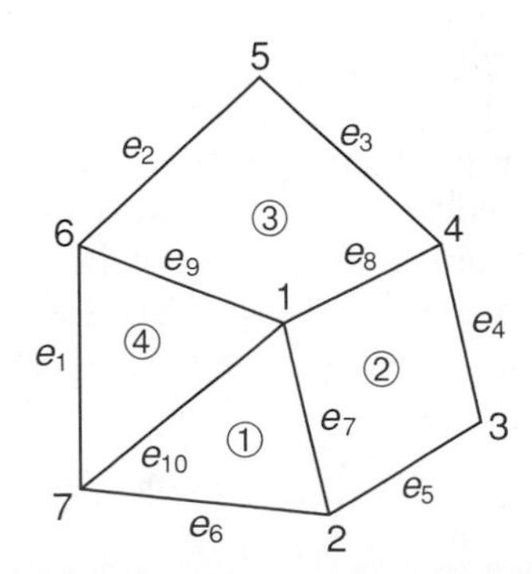

FIGURE 4.2
A polygon mesh

❑ POLYGONS BY POINTERS

In order to avoid the necessity of storing the same vertex coordinates in more than one array, the method of drawing polygons by pointers is employed [1]. The coordinates of all the vertices in the polygon mesh are stored in a single array. Then for each polygon, pointers are formed that contain, sequentially, the serial numbers of the vertices as they occur in the vertex array. The pointer for polygon number 1 in the mesh shown in Figure 4.2 may read $P = \{1 \quad 7 \quad 2 \quad 1\}$. The polygon may then be drawn by joining vertices 1, 7, 2, and 1.

The method of pointers does economize on the use of computer memory by storing the coordinates of the vertices only once. It does, however, need to store the pointers. Thus, the economy is not substantial unless the shape of the model is quite complex. The method does not address the problem of coincident edges (i.e., it draws them more than once as is done in the explicit polygon method).

❑ METHOD OF EXPLICIT EDGES

A single array of vertex coordinates is formed, as in the method of polygons by pointers [1]. The pointers, however, contain lists of edges for each polygon. Thus, the pointer for polygon 1 in Figure 4.2 reads $P = \{e_6 \quad e_7 \quad e_{10}\}$. An edge is described by an array containing the serial numbers of the vertices it joins and the polygon(s) to which it belongs. Thus, $e_6 = \{2, 7, 1, 0\}$ and $e_7 = \{1, 2, 1, 2\}$. The first two numbers identify the end points, and the last two numbers identify the polygons.

Since each edge is labeled separately, flags are provided within the program to avoid drawing any edge twice. Thus, the method of explicit edges obviates both of the problems encountered in the method of explicit polygons. However, one must realize that programming with this approach is somewhat more complex and takes additional effort in data entry.

❑ METHOD OF POLYHEDRONS

At times, simple solid objects may be treated as *polyhedrons*. See Figure 4.3. In this approach, polygons are employed to completely enclose the object (i.e., there are no openings). It is easy to generate a polygonal surface model of a polyhedron because certain distinct guidelines are available for assembly of the polygons. In the polyhedron, two polygons share a common edge. At least three edges meet at each vertex point of the polyhedron. For each polyhedron, the following Euler equation [2] is satisfied:

$$V + F - E = 2 \tag{4.1}$$

where, V, F, and E are, respectively, the number of vertices, faces, and edges.

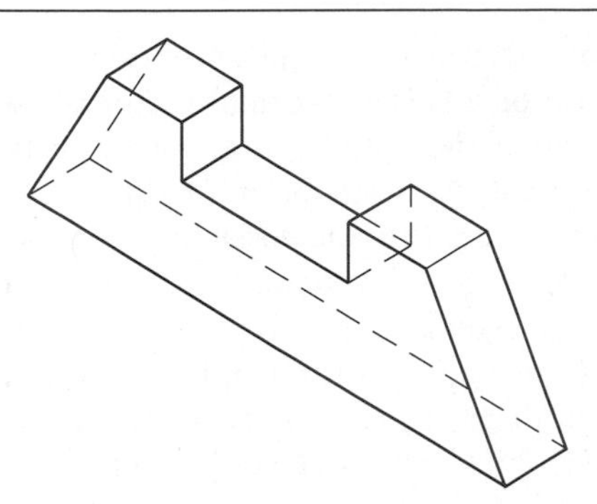

FIGURE 4.3
A polyhedron

Surface modelers are much easier to implement (compared with the solid modelers described in the next section) and, if implemented efficiently, require less computational time and memory. They are useful for generating finite element meshes, for rendering, and for describing machine tool paths. They are also useful for simulating the interaction of different objects in plant layouts and machinery designs. Information concerning geometric clearances, working space requirements, etc., can readily be evaluated. Surface models suffer from the disadvantage, however, that no information is available for distinguishing between the inside and outside regions of the solid. In fact, there are no details regarding the inside of objects available in surface models; therefore, these cannnot be used for evaluating weights, etc.

4.4 ▪ Solid Modeling

In solid modeling, an attempt is made to include more detailed geometric information about an object (i.e., internal as well as external). This makes it possible to section an object and view its internal details. As mentioned earlier, the additional information can also be utilized for finding geometric properties (volume, area, length, etc.) as well as for conducting finite element analysis.

Over the years, several techniques for solid modeling have evolved; these include boundary representation, constructive solid geometry, the spatial subdivision method, and analytic solid modeling. The first two of these techniques are described below.

❑ BOUNDARY REPRESENTATION

Boundary representation (B-Rep) is a technique similar to the surface modeling techniques described above. The difference, however, lies in the fact that the data corresponding to each panel or other type of surface also includes information regarding the interior (or exterior) of a solid object. In general, the direction of a surface normal vector is used to make such a distinction. For example, the positive direction of a surface normal vector may be used to represent the inside of an object.

The B-Rep scheme does afford the possibility of obtaining volume, mass, moment of inertia, and cut-away views since it distinguishes between the interior and exterior of solid objects. However, its use is limited to either homogeneous solids or objects with simple interior details. Some of the commercial solid modelers based on the B-Rep method are MEDUSA (Prime) and ROMULUS (Evans and Sutherland).

❑ CONSTRUCTIVE SOLID GEOMETRY

Constructive solid geometry (CSG) is often referred to as a building block approach, the building blocks in question being the higher-level graphics primitives. Three-dimensional models are created by combining the appropriate types of primitives based on the rules of Boolean operations. The operators commonly used are *union* ($\cup$), *difference* ($-$), and *intersection* ($\cap$). The union operator joins two primitives; thus, the union of the cylinder A with the plate B creates the vane in Figure 4.4(a). The difference operator subtracts one primitive from the other; thus, $A - B$ yields the fork shown in Figure 4.4(b). In other words,

$$A - B = (\text{volume of object } A) - (\text{volume common to } A \text{ and } B)$$

The intersection operator eliminates all parts of the primitives except the regions that are common to both; thus, $A \cap B$ results in the piece shown in Figure 4.4(c).

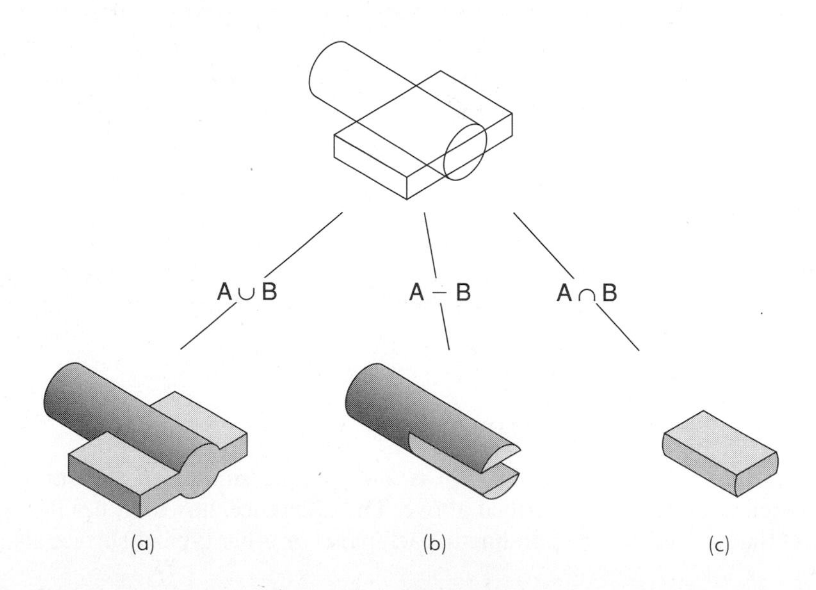

FIGURE 4.4
Boolean operations

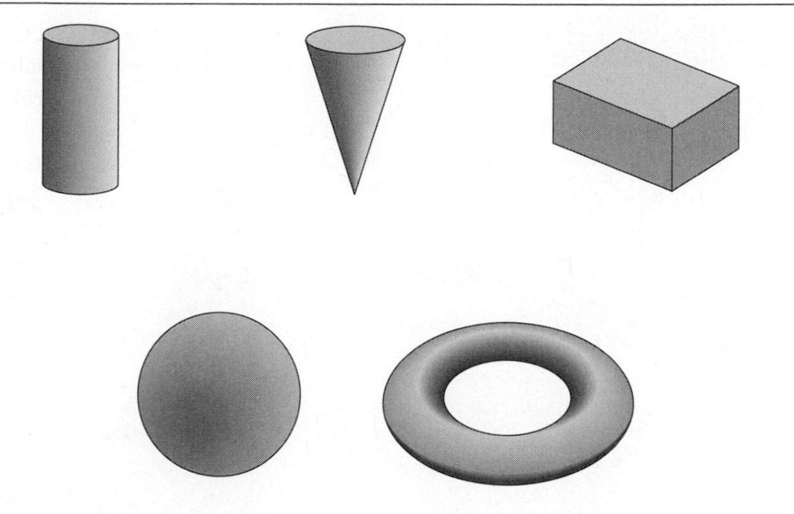

FIGURE 4.5
Typical CSG primitives

Steps Involved in CSG ■ As a first step toward model building, appropriate primitives are selected from a menu. A typical assortment of primitives is shown in Figure 4.5. The primitives are then subjected to some so-called *unary operations*, such as scaling to the right dimensions, rotating to the correct orientation, translating to the correct position, and mirroring. Two or more of the basic primitives thus created are then operated on to form a new primitive. A typical modeling operation may require several such operations, as illustrated in Example 4.1 below.

EXAMPLE ■ 4.1

Construct a solid model of the part shown in Figure 4.6, using the primitives shown in Figure 4.5.

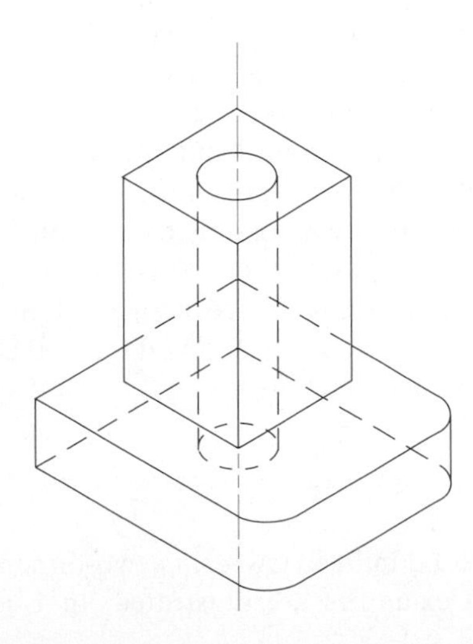

FIGURE 4.6
A stepped block

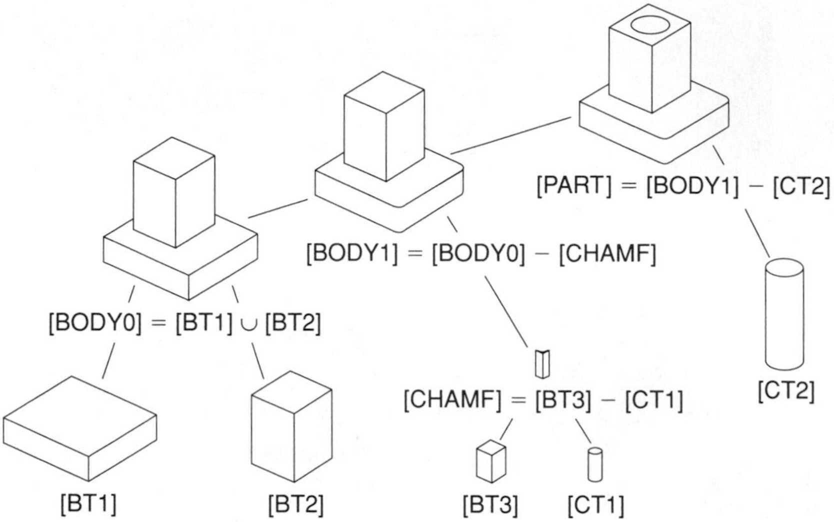

FIGURE 4.7 A binary tree representation

A *binary tree representation* of the model-building process is shown in Figure 4.7. At the bottom of the tree are shown the basic primitives [BT1], [BT2], [BT3], and [CT1], which are obtained by scaling brick and cylindrical primitives. [BT1] and [BT2] are combined to form [BODY0]. The [BT3] − [CT1] operation is used to obtain the chamfer tool [CHAMF]. The operation [BODY0] − [CHAMF] is applied twice to obtain [BODY1] with chamfered corners. A second cylindrical primitive [CT2] is then created, and the final model is obtained by the difference operation [BODY1] − [CT2]. ❑

Since CSG uses solid primitives, internal details of the object are automatically contained in the model. It goes without saying, therefore, that CSG models may be sectioned to reveal internal details and may be used for calculating mass, volume, moment of inertia, etc. The drawback of this modeling method is that the user is limited by the types of primitives available in the modeler. It is not possible to guarantee that any arbitrary shape may be modeled by this method. Most of the geometric modelers based on the CSG scheme have therefore incorporated the B-rep scheme as well. CATIA (IBM) and UNISOLIDS (McDonnell Douglas) are examples of so-called hybrid solid modelers.

❑ SWEEPING

Solid models can also be built by *sweeping* two-dimensional surfaces. In Figure 4.8, two such examples are illustrated. In Figure 4.8(a), a solid

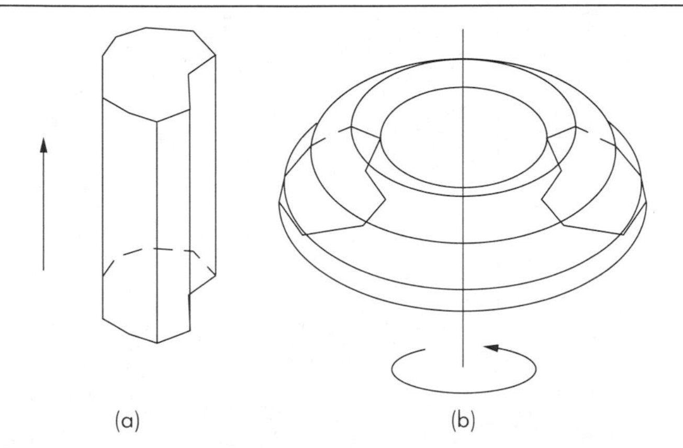

FIGURE 4.8
Solid modeling by sweeping

cylinder is extruded by sweeping an arbitrary shape in the vertical direction. An axisymmetrical ring constructed by sweeping the surface along the vertical axis is shown in Figure 4.8(b).

4.5 ▪ Viewing in Three-Dimensions

Thus far, in the present chapter, several methods for mathematically generating three-dimensional geometric models have been discussed. A geometric model is much like a piece of sculpture. A sculpture can exist only in three-dimensional space, but, unfortunately, in computer graphics, three-dimensional space does not exist for the display of objects. Computer models must therefore be viewed on a two-dimensional screen. Given below are the methods used for mapping the image of three-dimensional models onto two-dimensional surfaces.

The techniques adopted in computer graphics for creating images of three-dimensional objects are based on the principles of projection, details of which are found in engineering graphics books [3].

For obtaining projections, a center of projection (CP) is first selected. Straight-line projectors are then drawn from the CP, linking every point on the solid that requires projection. The trace of these projectors onto a specified projection plane gives the desired projection. Take, for example, the quadrilateral shown in Figure 4.9. The projection of the quadrilateral *ABCD* is another quadrilateral *abcd*, which is obtained in the manner explained above. As can be easily visualized, all the edges of the object are foreshortened in the image. In all likelihood the angles between the edges are also distorted.

The extent of foreshortening and the distortion of the angles depend on the orientation of the projection plane with respect to the projectors, the distance of the CP from the quadrilateral, and the distance of the projection

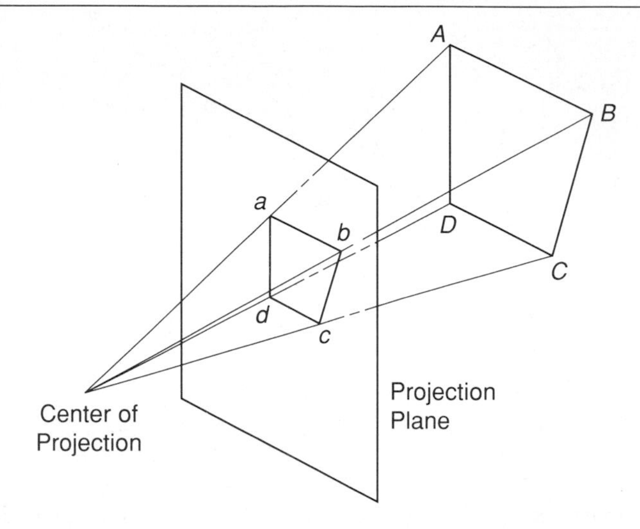

FIGURE 4.9 Projection of a quadrilateral

plane from the CP and the object. In all projection algorithms, therefore, it is necessary that these parameters be explicitly stated.

When the distance of the CP from the object is finite, the projection obtained is called a *perspective projection* because in such cases the projection tends to look somewhat like the image seen by a human observer. If the CP is moved to infinity, then the projectors become parallel to each other. The image produced in such cases is called a *parallel projection.* In parallel projection, the location of the projection plane has no bearing on the shape or the size of the image. The only relevant projection parameter is the direction of projection (i.e., the angle that the projectors make with the projection plane). In parallel projection, the lines on the object that are parallel to the projection plane maintain their true length in the image.

4.6 ▪ Principles of Projections

❑ PARALLEL PROJECTIONS

The parallel projection is most commonly used in engineering drafting. The reason for its popularity is, of course, its simplicity. Engineers obtain parallel projections using simple drafting tools, such as a T-square and triangles, and a drafting board. Parallel projections may be subdivided into three categories: orthographic projections, oblique projections, and axonometric projections.

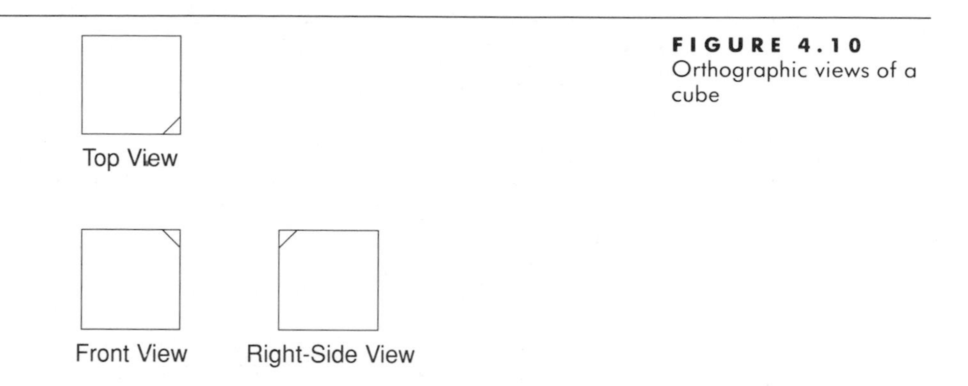

FIGURE 4.10
Orthographic views of a cube

Orthographic and Oblique Projections ■ In *orthographic projections*, the projectors are normal to the projection plane and parallel to one of the principal axes. The principal axes are those along which the height or the width or the depth of an object is measured. In *oblique projections*, the projectors are inclined to the projection plane at angles other than 90°.

Most engineering drawings contain three orthogonal projections: projections of a part on frontal, top, and right-side planes. Each projection plane is normal to one of the principal axes of the object. The projection on the frontal plane is called the *front view* (*or elevation*). The projection on one of the side planes is called the *right- or left-side view*. The projection on the top plane is called the *top view* (*or plan view*). Figure 4.10 shows three orthographic views of a cube.

Each orthographic projection contains details of only one face of the object, as can be seen from Figure 4.10. For the untrained eye, this can present considerable difficulty in visualizing the actual shape of the object. Even among seasoned engineers this is often a source of misinterpretation. The advantages of orthographic projections, however, are the simplicity of drawing and the fact that most of the lengths and angles remain unchanged. Thus, orthographic projections are ideally suited for manufacturing purposes.

Axonometric Projections ■ When the projection plane is inclined to all the principal axes, *axonometric projections* are obtained. Axonometric projections contain details of three mutually perpendicular faces of the object. There are several types of axonometric projections: isometric, dimetric, and trimetric.

Of the above three types, the isometric projection is most commonly used in engineering practice. In the isometric projection, the projection plane makes equal angles with all the principal axes of the object. An isometric projection of the cube shown in Figure 4.10 is shown in Figure 4.11.

FIGURE 4.11
Isometric projection of a cube

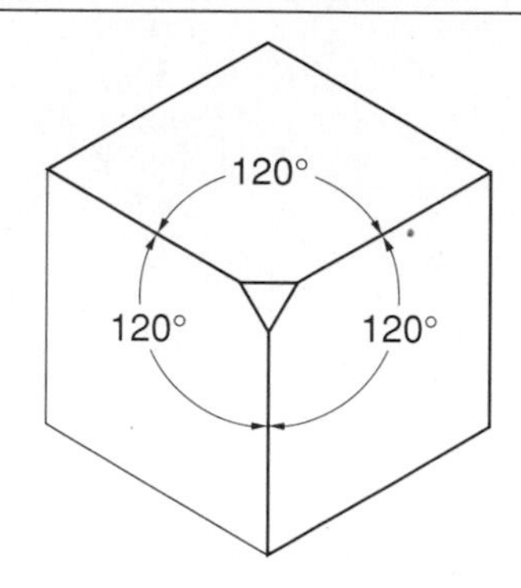

❑ PERSPECTIVE PROJECTIONS

In the perspective projection, the projectors are inclined at varying angles to the projection plane since they converge at the center of projection. Perspective projections are of three types: (1) one-point perspective, (2) two-point perspective, and (3) three-point perspective.

To understand the difference among the three perspectives, reference should be made to Figures 4.12 and 4.13. A block is shown placed on the *ground plane*. An observer stands on the ground plane and looks toward the object. The projection plane is positioned between the viewer and the

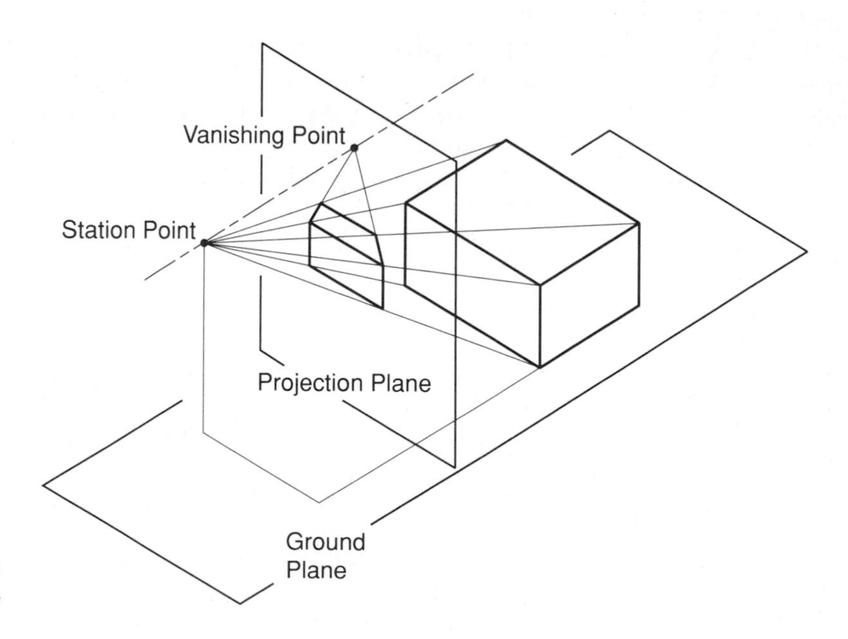

FIGURE 4.12
One-point perspective

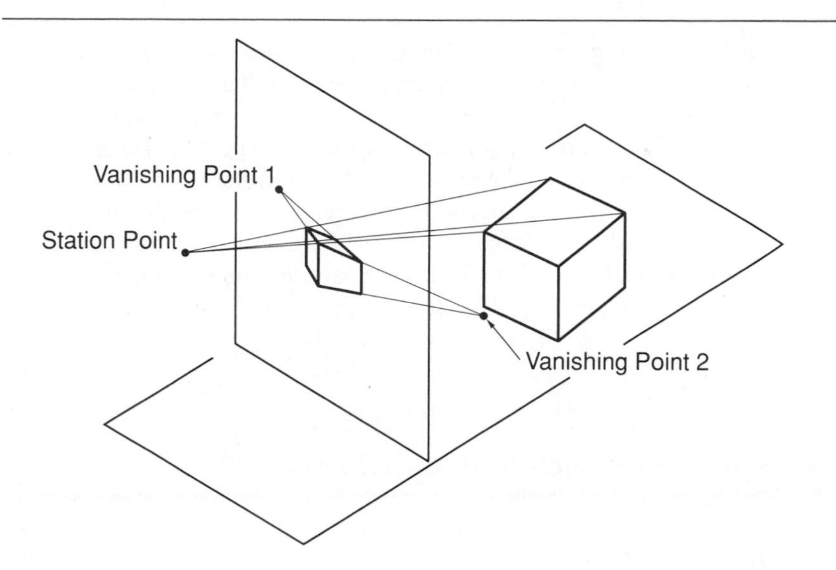

FIGURE 4.13
Two-point perspective

object. The eye of the observer is at the *viewing* (*station*) *point*. One of the principal axes of the block is perpendicular to the projection plane in Figure 4.12. Each corner of the block is joined to the station point by a straight line. The station point is thus the center of projection. The intersection of these lines (*lines of sight*) with the projection plane determines the position of the vertices on the image. The perspective projection is completed by joining the vertices on the projection plane.

As can be seen, all vertical edges of the block appear vertical in the image. Similarly, the horizontal edges remain horizontal. But the edges perpendicular to the projection plane appear as inclined lines and converge to a single point on the projection plane. The projection is therefore known as a one-point perspective. The point at which the depth lines converge is known as the *vanishing point*. The vanishing point is at the same height above the ground plane as the station point. The plane, parallel to the ground plane, containing the vanishing point and the station point is referred to as the *horizontal plane*.

Suppose now that the block shown in Figure 4.12 is rotated on the ground plane about an axis perpendicular to the plane so that none of its principal axes is perpendicular to the projection plane. The image produced in this orientation of the block is shown in Figure 4.13. This is a two-point perspective since all edges that are not vertical seem to converge at two points. The edges in the depth direction converge to one vanishing point, and the edges in the width direction converge to a second vanishing point. The vanishing points and the station point are situated at the same height above the ground plane.

If the block in Figure 4.13 is tilted toward or away from the projection plane, its vertical edges will also converge to a point like the other edges. We would then have a three-point perspective. The three-point perspective is the most sophisticated projection technique that can be used. It generates a realistic image of objects in the same manner as a camera does. Mathematical formulations for the projection, however, are somewhat tedious. In most computer-aided drafting packages, therefore, only the two-point perspective is used. It reveals probably the same amount of information as the three-point perspective, while keeping the mathematics simpler.

4.7 ▪ Mathematics of Projections

❑ ISOMETRIC PROJECTION

Figure 4.14(a) shows a cube with one of its corners situated at the origin of the coordinate system and its principal axes coincident with the coordinate axes. Assume that an isometric projection is to be generated on a projection plane normal to the *z* axis. In isometric projection, the principal axes of the object are inclined equally to the projection plane. One way of orienting the object in this attitude is to rotate it about the *y* axis through an angle of 45° and then about the *x* axis through an angle of 35.26°. The results of these rotations are shown in Figure 4.14(b) and (c).

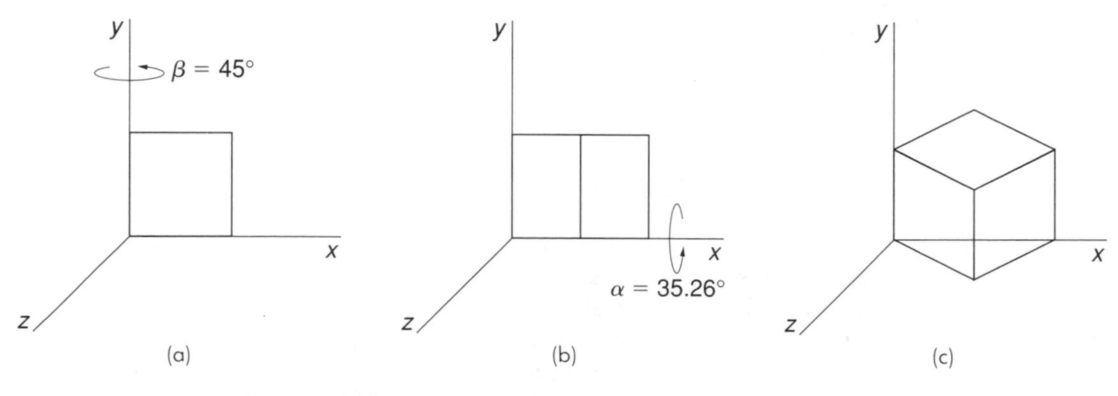

FIGURE 4.14 Rotations in the isometric projection

The three-dimensional rotation matrices have been discussed in Chapter 3. Using equations (3.42) and (3.43), the transformation matrix (R_p) for isometric projection is obtained thus:

$$R1 = \begin{bmatrix} \cos 45^\circ & 0 & -\sin 45^\circ & 0 \\ 0 & 1 & 0 & 0 \\ \sin 45^\circ & 0 & \cos 45^\circ & 0 \\ 0 & 0 & 0 & 1 \end{bmatrix}$$

$$R2 = \begin{bmatrix} 1 & 0 & 0 & 0 \\ 0 & \cos 35.36^\circ & \sin 35.26^\circ & 0 \\ 0 & -\sin 35.25^\circ & \cos 35.26^\circ & 0 \\ 0 & 0 & 0 & 1 \end{bmatrix} \tag{4.2}$$

$$R_p = R1R2$$

$$= \begin{bmatrix} 0.7071 & 0.4082 & -0.5774 & 0 \\ 0 & 0.8165 & 0.5774 & 0 \\ 0.7071 & -0.4082 & 0.5774 & 0 \\ 0 & 0 & 0 & 1 \end{bmatrix}$$

EXAMPLE ■ 4.2

(a) If the edges of the cube shown in Figure 4.14 are each of unit length, find the coordinates of the corners in an isometric projection. (b) The cube is translated in such a manner that the corner originally at the origin moves to a point (5, 6, 3). If isometric transformations are now carried out on the cube, what will be the coordinates of its vertices?

(a) A matrix of the coordinates of the vertices of the cube may be formed thus:

$$P_0 = \begin{bmatrix} 0 & 0 & 0 & 1 \\ 1 & 0 & 0 & 1 \\ 1 & 0 & 1 & 1 \\ 0 & 0 & 1 & 1 \\ 0 & 1 & 0 & 1 \\ 1 & 1 & 0 & 1 \\ 1 & 1 & 1 & 1 \\ 0 & 1 & 1 & 1 \end{bmatrix} \quad (\text{columns: } x \; y \; z \; w)$$

The transformed coordinates of the vertices are then found by the following equation:

$$P_{p0} = P_0 R_p = \begin{bmatrix} 0 & 0 & 0 & 1 \\ 0.7071 & 0.4082 & -0.5774 & 1 \\ 1.4142 & 0 & 0 & 1 \\ 0.7071 & -0.4082 & 0.5774 & 1 \\ 0 & 0.8165 & 0.5774 & 1 \\ 0.7071 & 1.2247 & 0 & 1 \\ 1.4142 & 0.8165 & 0.5774 & 1 \\ 0.7071 & 0.4082 & 1.1548 & 1 \end{bmatrix}$$

(b) The vertex coordinate matrix for the relocated cube is

$$P_1 = \begin{array}{c} \begin{matrix} x & y & z & w \end{matrix} \\ \begin{bmatrix} 5 & 6 & 3 & 1 \\ 6 & 6 & 3 & 1 \\ 6 & 6 & 4 & 1 \\ 5 & 6 & 4 & 1 \\ 5 & 7 & 3 & 1 \\ 6 & 7 & 3 & 1 \\ 6 & 7 & 4 & 1 \\ 5 & 7 & 4 & 1 \end{bmatrix} \end{array}$$

The transformed vertex coordinates are

$$P_{p1} = P_1 R_p = \begin{bmatrix} 5.6568 & 5.7148 & 2.3096 & 1 \\ 6.3639 & 6.1230 & 1.7322 & 1 \\ 7.0710 & 5.7148 & 2.3096 & 1 \\ 6.3639 & 5.3066 & 2.8870 & 1 \\ 5.6568 & 6.5312 & 2.8870 & 1 \\ 6.3639 & 6.9394 & 2.3096 & 1 \\ 7.0710 & 6.5312 & 2.8870 & 1 \\ 6.3639 & 6.1230 & 3.4644 & 1 \end{bmatrix}$$ ❑

From the results obtained above, notice that in the first case, the corner situated at the origin remains at the origin after the transformation. This is because the cube undergoes rotations only. In the second case, however, the isometric transformation causes translation in addition to rotations. As a result, none of the vertices of the cube remains at its original position. In some cases, when the object is situated at a large distance from the origin, the accompanying translation associated with the isometric transformation is quite large. It is possible that the object may move outside the viewing window, in which case only a portion of the object may be visible. To avoid excessive translation, often the object is first moved close to the origin, the isometric transformation is then applied, and the object is then moved back to a position close to the original.

❑ PERSPECTIVE PROJECTION

As explained earlier, a two-point perspective is obtained by rotating the object about its vertical axis. A good angle to use for the rotation is 30°. Assume for the sake of generality that the rotation selected is β; the rotation matrix, according to equation (3.43), is then

$$R1 = \begin{bmatrix} \cos\beta & 0 & -\sin\beta & 0 \\ 0 & 1 & 0 & 0 \\ \sin\beta & 0 & \cos\beta & 0 \\ 0 & 0 & 0 & 1 \end{bmatrix} \qquad (4.3)$$

For a three-point perspective, another rotation about the x axis follows the above. For a one-point perspective, no rotation is required.

Assume now that the center of projection is situated along the z axis, as shown in Figure 4.15. Let the projection plane (computer screen) be situated

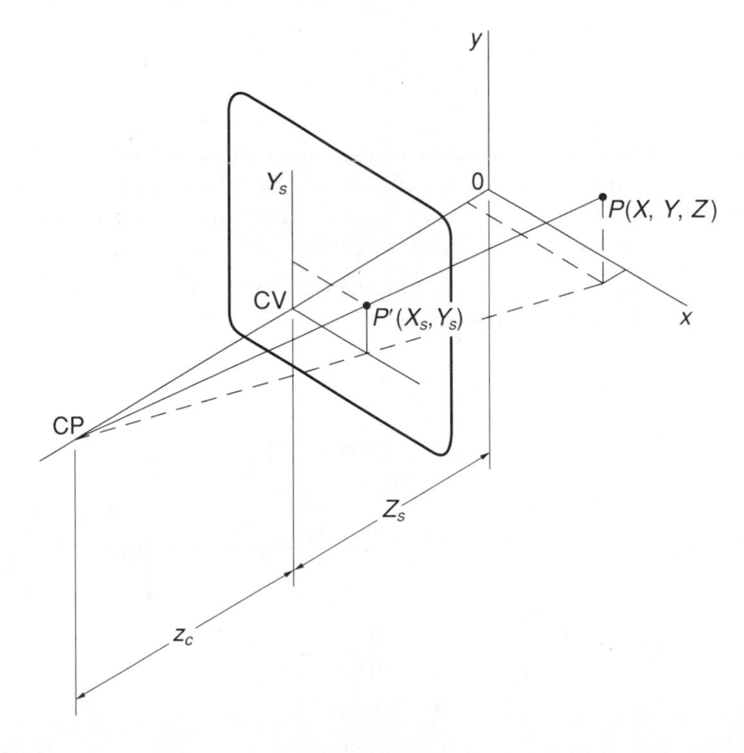

FIGURE 4.15
Parameters in the perspective projection

a distance z_c from the CP. Let the location of a point P, after the rotation described above, be (X, Y, Z). At this stage, we define another coordinate system, the *screen coordinate system*, which has its origin at the point of intersection with the z axis, known as the *center of vision* (CV). $P'(x_s, y_s)$ is the projection of point P on the computer screen. The distance of the screen from the origin O is Z_s.

By looking at the similar triangles in Figure 4.15, it is quite easy to derive the following relations for x_s and y_s:

$$x_s = X\frac{z_c}{z_c + Z_s - Z}$$

$$y_s = Y\frac{z_c}{z_c + Z_s - Z}$$

The homogeneous coordinates of point P' are obtained from the homogeneous coordinates of point P, using the following matrix multiplication:

$$\{x_s \quad y_s \quad 1\} = \{X \quad Y \quad Z \quad 1\}\frac{z_c}{z_c + Z_s - Z}\begin{bmatrix} 1 & 0 & 0 \\ 0 & 1 & 0 \\ 0 & 0 & \dfrac{-1}{z_c} \\ 0 & 0 & 1 + \dfrac{Z_s}{z_c} \end{bmatrix}$$

$$= \{X \quad Y \quad Z \quad 1\}R2 \tag{4.4}$$

EXAMPLE ■ 4.3

Obtain a two-point perspective of the cube mentioned in Example 4.2 above. Assume that one of the corners of the cube is located at the origin. The computer screen is placed two units in front of the origin, and the observer is at a distance three units in front of the screen.

Obviously, $z_c = 3$ and $Z_s = 2$. Assume a rotation of 30° about the y axis. The rotation matrix is given as

$$R1 = \begin{bmatrix} 0.8667 & 0 & -0.5 & 1 \\ 0 & 1 & 0 & 0 \\ 0.5 & 0 & 0.8667 & 0 \\ 0 & 0 & 0 & 1 \end{bmatrix}$$

The transformation matrix $R2$ as defined in equation (4.4) is

$$R2 = \frac{3}{5 - Z}\begin{bmatrix} 1 & 0 & 0 \\ 0 & 1 & 0 \\ 0 & 0 & -0.333 \\ 0 & 0 & 1.6667 \end{bmatrix}$$

The coordinate matrix for the vertices is found in two steps as follows:

1. Rotation of the object about the y axis yields

$$P_1 = P_0 R1 = \begin{bmatrix} 0 & 0 & 0 & 1 \\ 0.8667 & 0 & -0.50 & 1 \\ 1.3667 & 0 & 0.3667 & 1 \\ 0.50 & 0 & 0.8667 & 1 \\ 0 & 1 & 0 & 1 \\ 0.8667 & 1 & -0.50 & 1 \\ 1.3667 & 1 & 0.3667 & 1 \\ 0.50 & 1 & 0.8667 & 1 \end{bmatrix}$$

2. Scaling with the matrix $R2$ yields

$$P_s = P_1 R2 = \begin{bmatrix} 0 & 0 & 1 \\ 0.520 & 0 & 1 \\ 0.8849 & 0 & 1 \\ 0.3629 & 0 & 1 \\ 0 & 0.5999 & 1 \\ 0.520 & 0.5454 & 1 \\ 0.8849 & 0.6475 & 1 \\ 0.3629 & 0.7259 & 1 \end{bmatrix}$$

The perspective view is shown in Figure 4.16.

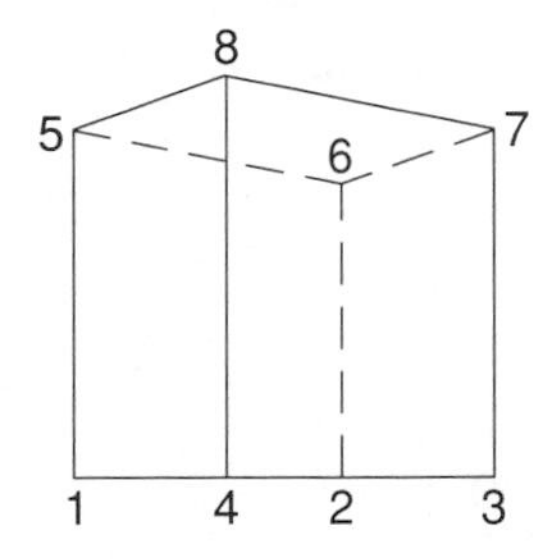

FIGURE 4.16
A two-point perspective projection of a cube

❑

The formulation given above for the perspective projection is somewhat restrictive in that it forces the user to specify a center of projection along the z axis. Such a perspective projection is known as a *central projection.* It is often better, however, to have the flexibility of selecting the CP in a more general way. Let the x and y coordinates of the CP be (x_c, y_c); a translation matrix may then be formed as shown below.

$$T_c = \begin{bmatrix} 1 & 0 & 0 & 1 \\ 0 & 1 & 0 & 1 \\ 0 & 0 & 1 & 1 \\ -x_c & -y_c & 0 & 1 \end{bmatrix} \tag{4.5}$$

A perspective projection may now be drawn using the following sequence of operations:

1. Rotate the model using matrix $R1$, equation (4.3).
2. Translate the model using matrix T_c, equation (4.5).
3. Scale the coordinates using matrix $R2$, equation (4.4).
4. Translate the projection by distances (x_c, y_c) to correspond to the designated CP, using the following matrix:

$$T_r = \begin{bmatrix} 1 & 0 & 0 \\ 0 & 1 & 0 \\ x_c & y_c & 1 \end{bmatrix} \tag{4.6}$$

EXAMPLE ▪ 4.4

Draw a two-point perspective of the cube mentioned in Example 4.2 by selecting a CP of (0, 2). Assume the distance of the screen from the origin is 2 and the distance of the viewer from the screen is 3.

As in Example 4.3 above, $z_c = 3$ and $Z_s = 2$.

(i) Assuming a rotation of 30°, as in Example 4.3, the coordinate matrix $P1$ derived earlier is used here, too.

(ii) The model is translated by distances $(-0, -2)$, using the following matrix multiplication:

$$P_2 = P_1 \begin{bmatrix} 1 & 0 & 0 & 1 \\ 0 & 1 & 0 & 1 \\ 0 & 0 & 1 & 1 \\ 0 & -2 & 0 & 1 \end{bmatrix} = \begin{bmatrix} 0 & -2 & 0 & 1 \\ 0.8667 & -2 & -0.50 & 1 \\ 1.3667 & -2 & 0.3667 & 1 \\ 0.50 & -2 & 0.8667 & 1 \\ 0 & -1 & 0 & 1 \\ 0.8667 & -1 & -0.50 & 1 \\ 1.3667 & -1 & 0.3667 & 1 \\ 0.50 & -1 & 0.8667 & 1 \end{bmatrix}$$

(iii) The vertices are then projected onto the screen, using matrix $R2$ from Example 4.3:

$$P3 = P2R2 = \begin{bmatrix} 0 & -1.2 & 1 \\ 0.4727 & -1.0909 & 1 \\ 0.8849 & -1.2950 & 1 \\ 0.3629 & -1.4517 & 1 \\ 0 & -0.6 & 1 \\ 0.4727 & -0.5454 & 1 \\ 0.8849 & -0.6479 & 1 \\ 0.3629 & -0.7258 & 1 \end{bmatrix}$$

(iv) The image is translated using matrix T_r, where

$$T_r = \begin{bmatrix} 1 & 0 & 0 \\ 0 & 1 & 0 \\ 0 & 2 & 1 \end{bmatrix}$$

$$P_s = P3T_r = \begin{bmatrix} 0 & 0.8 & 1 \\ 0.4772 & 0.9091 & 1 \\ 0.8849 & 0.7050 & 1 \\ 0.3629 & 0.5483 & 1 \\ 0 & 1.40 & 1 \\ 0.4727 & 1.4546 & 1 \\ 0.8849 & 1.3521 & 1 \\ 0.3629 & 1.2742 & 1 \end{bmatrix}$$

The resulting projection of the cube is shown in Figure 4.17.

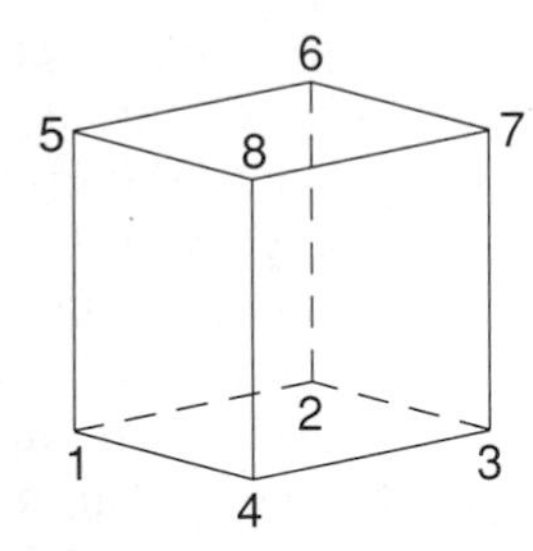

FIGURE 4.17
A two-point perspective projection from an arbitrary CP

❑

4.8 ▪ Hidden Line/Surface Removal Algorithms

Engineers are fully aware of the need to remove hidden lines/surfaces from drawings of three-dimensional parts and assemblies. These are the lines/surfaces that are not visible to an observer looking at the object. Upon eliminating these lines/surfaces, the drawing appears less confusing and more realistic. In manual drafting, the lines/surfaces are removed by observations based on certain rules. In computer graphics, mathematical formulas are needed to sort out the hidden lines/surfaces.

Mathematics for hidden line/surface removal has proved to be the most difficult and intractable of all computer graphics calculations. As a matter of fact, an all-inclusive hidden line/surface removal algorithm is not yet available. Most of the algorithms are limited to some specific types of objects. This has resulted in a plethora of hidden line/surface removal algorithms. A few popular types of hidden line/surface removal algorithms that are easy to implement in raster graphics are discussed below.

A hidden line/surface algorithm may be classified as an image space algorithm or an object space algorithm [4]. *Object space algorithms* use the coordinates that describe the actual location and orientation of the objects (world coordinates), whereas *image space algorithms* use the coordinates of the projection of the objects on the display surface (i.e., screen coordinates).

In object space algorithms, calculations are performed to the highest level of floating point accuracy, consistent with the processor available. Due to the limited resolution of the display surfaces, generally such a high level of accuracy is not required. If, however, enlargement of the image is required, then the image created by the object space algorithm remains distortion-free. Image space algorithms perform calculations that are consistent with the resolution of the display surface. Thus, for a 640×400 raster screen, image space algorithms calculate the intensities for 256,000 pixels. This amount of data may not be adequate for enlargement of the image to, say, four times its original size. It will be found that in the enlarged image, the edges will not always pass through the end points that they are supposed to.

The computation time for a typical object space algorithm generally varies as n^2, the square of the number of objects. The computation time is proportional to nN in image space algorithms, where N is the number of picture points on the display surface. Usually, $N > n$; it may appear, therefore, that object space algorithms should be much quicker. Owing, however, to the fact that coherence properties of the image are utilized in the algorithms, image space algorithms often turn out to be faster than object space algorithms. Descriptions of some hidden surface removal algorithms follow.

❑ THE *z*-BUFFER ALGORITHM

The z-buffer algorithm is an image space algorithm that was first suggested by Catmull [5]. It requires that a z value be determined for each pixel. The data on the z values are stored in a buffer similar to the frame buffer. The algorithm works as follows:

1. The depth $z_0(x, y)$ of all pixels is set to 1.0 (in the depth buffer), and the intensity of all pixels is set equal to $i_0(x, y)$, the background intensity, in the frame buffer.
2. $z_i(x, y)$, the depth value for a point belonging to the ith polygon whose image is at the screen address (x, y), is found.

If $z_i < z_0$, then $z_0(x, y) = z_i(x, y)$ in the depth buffer, and in the frame buffer $i_0(x, y)$ is replaced by $i_i(x, y)$, the intensity of the polygon. All other polygons are treated likewise in turn.

As a result of the above steps, the frame buffer contains intensities of the pixels sorted according to the visibility of the polygons. The z-buffer algorithm is the simplest image space hidden surface removal algorithm. The problem here is with the large number of calculations. One way of simplifying the calculations is to compute z values recursively from the equations of the planes containing the polygons. Let's assume that a plane containing a polygon is given by

$$Ax + By + Cz + D = 0$$

Then

$$z(x, y) = -(Ax + By + D)/C$$

and,

$$z(x + 1, y) = -(Ax + By + D)/C - A/C \tag{4.7}$$

Thus, $z(x + 1, y) = z(x, y) - A/C$ (i.e., z values for pixels on a scan-line may be calculated from single subtractions).

❑ GEOMETRIC COMPUTATIONS

The z-buffer algorithm described above conducts a depth test at each pixel for each polygon. There are other algorithms that are more discerning. In these algorithms, certain tests are performed to establish visibility of polygons. Those polygons, or parts of polygons, that are not visible are not entered into the calculations. Thus, the amount of work is often reduced to one-half when compared with the z-buffer algorithm. In this section, some of the tests that are commonly applied in the more advanced type of hidden surface algorithms are described [6].

Visibility or Back-Face Test ▪ This test affords us the option of eliminating those faces of a solid object that are not visible to the observer when located at a specific position. Such invisible faces are often called the *back-faces*. Consider the cube shown in Figure 4.18. The edge 1–2 and the diagonal 1–3 may be expressed vectorially as $l = ai + bj + ck$ and $m = di + ej + fk$, respectively, where i, j, and k are unit vectors along the x, y,

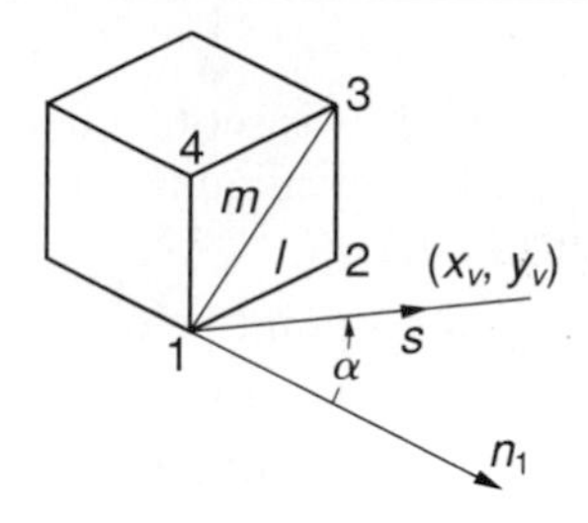

FIGURE 4.18
The visibility test

and z axes, respectively. If the coordinates of the ith vertex of the cube are (x_i, y_i, z_i), then $a = x_2 - x_1$, $b = y_2 - y_1$, $c = z_2 - z_1$, etc. The surface normal to the face 1–2–3–4 is then given by the cross-product $n_1 = l \times m = u_1 i + v_1 j + w_1 k$.

Suppose that the center of projection is located at (x_v, y_v, z_v). Then the line of sight through vertex 1 is mathematically expressed as $s = (x_v - x_1)i + (y_v - y_1)j + (z_v - z_1)k$. Dot product, $n_1 \cdot s = |n_1||s| \cos \alpha$, where α is the angle between the line of sight and the normal to the face 1–2–3–4, can be used to test the visibility of the polygon. If $\alpha > 90°$, then face 1–2–3–4 is not visible; otherwise, it is. For $\alpha > 90°$, the product $n_1 \cdot s < 0$.

A visibility test may be applied to each face of the cube or to any other solid with plane faces. In every case, where the sign of the dot product $(n_1 \cdot s)$ is negative, the corresponding face is invisible. These faces are therefore eliminated from any further calculations.

Minimax Test ▪ This test and the intersection test described below are used to determine whether two polygons overlap. Suppose that two polygons are labeled I and II (see Figure 4.19). The polygons may overlap each other if any one set of conditions from equations (4.8) through (4.11) is true.

1. From Figure 4.19(a)

$$\begin{aligned} &\max x_{\mathrm{II}} > \max x_{\mathrm{I}} \\ &\max x_{\mathrm{I}} > \min x_{\mathrm{II}} \end{aligned} \tag{4.8}$$

2. From Figure 4.19(b)

$$\begin{aligned} &\max x_{\mathrm{II}} < \max x_{\mathrm{I}} \\ &\max x_{\mathrm{II}} > \min x_{\mathrm{I}} \end{aligned} \tag{4.9}$$

3. From Figure 4.19(b)

$$\begin{aligned} &\max y_{\mathrm{II}} > \max y_{\mathrm{I}} \\ &\max y_{\mathrm{I}} > \min y_{\mathrm{II}} \end{aligned} \tag{4.10}$$

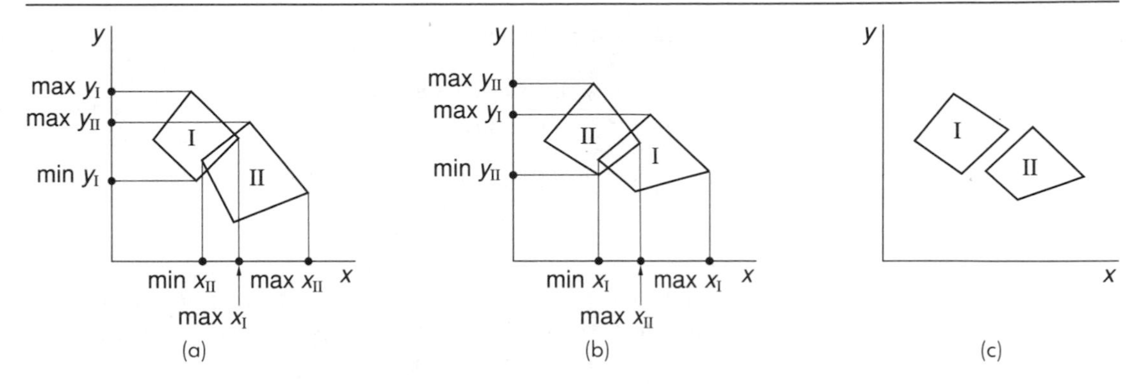

FIGURE 4.19 The minimax test

4. From Figure 4.19(a)

$$\max y_{II} < \max y_I$$
$$\max y_{II} > \min y_I \tag{4.11}$$

The minimax test indicates the possibility of overlap. There are instances where two polygons do satisfy the above conditions, yet they do not overlap each other, as is the case in Figure 4.19(c). For a definitive answer in this case, an intersection test is applied, as described below.

Intersection Test ■ When two polygons overlap, at least two types of intersections can occur between the different edges of the polygons. The points of intersection may be on two different edges or on the same edge, as shown in Figure 4.20(a) and (b). These points, if they exist, are located by first finding the equation of each edge of the two polygons and then, by permutation, checking every possible pair of edges for intersection. Equations of the edges are found from the coordinates of the end points. Assuming that the equations of the edges AB and ab in Figure 4.20 are $y = m_1x + c_1$ and $y = m_2x + c_2$, respectively, the point of intersection

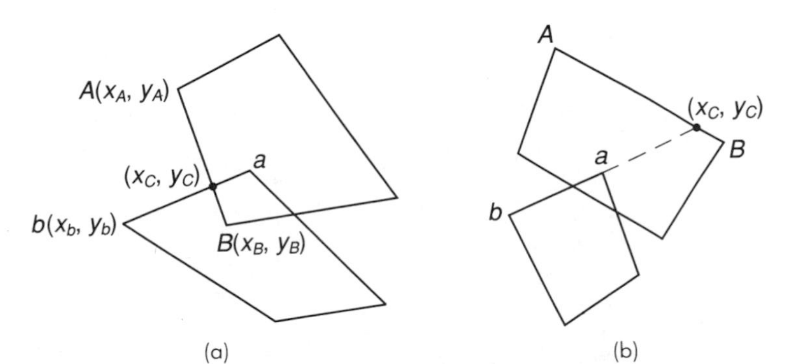

FIGURE 4.20
The intersection test

between the edges is then defined as follows:

$$x_c = \frac{(y_a - y_A) - (m_2 x_a - m_1 x_A)}{m_1 - m_2} \tag{4.12}$$

$$y_c = \frac{(y_a - m_2 x_a)m_1 - (y_A - m_1 x_A)m_2}{m_1 - m_2} \tag{4.13}$$

In the above, $m_1 = (y_A - y_B)/(x_A - x_B)$ and $m_2 = (y_a - y_b)/(x_a - x_b)$.

If it turns out that (x_c, y_c) lies on both edges AB and ab, as shown in Figure 4.20(a), then there is an intersection. Otherwise, there is no intersection, as shown in Figure 4.20(b).

Containment Test ▪ By applying this test, it can be determined whether a point is inside or outside a polygon. In Figure 4.21(a), a polygon *abcdefg* and a point P are shown. Since P is inside the polygon, angles aPb, bPc, etc., total 360°, if the angles measured in one sense are treated as positive and those measured in the opposite sense are treated as negative. Angles aPb, aPc, etc., total zero in the case shown in Figure 4.21(b) if P is outside the polygon.

❑ THE PRIORITY ALGORITHM

In the priority algorithm developed by Newell, Newell, and Sancha [7], the polygons are arranged according to their priority. The polygon farthest away is given lowest priority because the chances of its being obscured by the polygons in front of it are very high. The polygon closest to the viewer is given the highest priority because the chances of its being obscured by other polygons are minimal. Scan conversion is carried out one polygon at a time, starting with the polygon with the lowest priority. Thus, in case of an overlap, the details of the polygon with the higher priority are displayed, and those of the polygon with the lower priority are eliminated. The

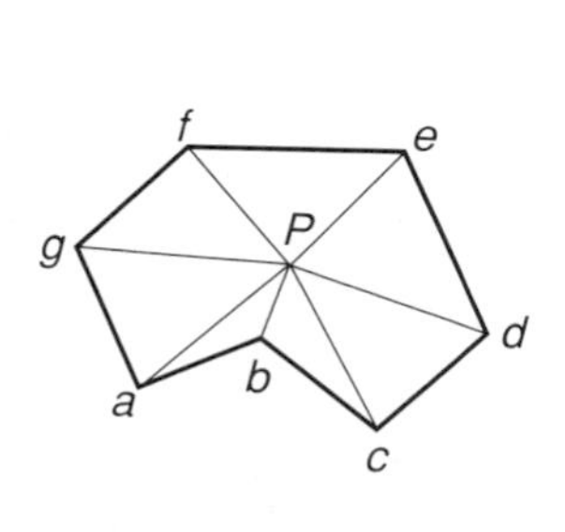

(a)

b c a d e f P

(b)

FIGURE 4.21
The containment test

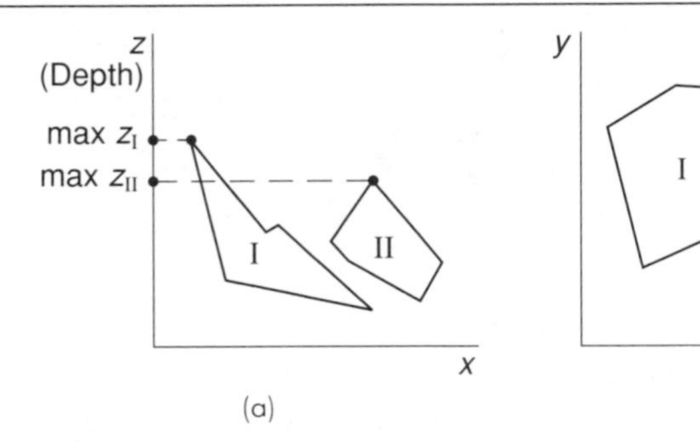

FIGURE 4.22 Projection of polygons: (a) in the *xz* plane; (b) in the *xy* plane

algorithm is a hybrid between an object space algorithm and an image space algorithm. Some of the calculations are performed in the object space, and some are performed in the image space.

The prioritized sorting of polygons according to their depth is not always a straightforward process. At times, certain ambiguities arise that must be resolved if a correct image is to be generated. Take, for example, the polygons in Figure 4.22. Assuming that the *xy* plane is the projection plane, then the *z* direction represents the depths of the polygons. Projections in the *xz* plane [see Figure 4.22(a)] show that polygon I extends beyond polygon II in the *z* direction. If depth sorting is performed on the basis of the z_{max} value of the polygons, then I has a lower priority than II—this is obviously not correct. It can easily be shown that polygon I obscures part of polygon II [see Figure 4.22(b)]. Obviously, polygon I should have a higher priority in this case. In order to safeguard against this type of pitfall, the steps suggested in the next paragraph are adopted.

Once sorting is completed on the basis of z_{max}, the following tests are conducted to check if the polygon—say, I—at the end of the list obscures any portion of another polygon—say, II. I does not obscure II if any one of the following conditions is satisfied:

1. I and II do not overlap in the depth direction, and polygon II is closer to the viewer than polygon I is. A minimax test in the *xz* plane should reveal the overlap in the *z*-direction.
2. Minimax and intersection tests indicate that I and II do not overlap in the *xy* plane.
3. With respect to the plane of polygon II, the vertices of I are on the side opposite to the viewer.
4. The vertices of II are on the same side of the plane of polygon I as the viewer is.

It may come as a surprise to the reader that the above tests are still inconclusive in some situations. Take, for example, the polygons shown in Figure 4.23. In Figure 4.23(a), polygon I obscures II, II obscures III, and III

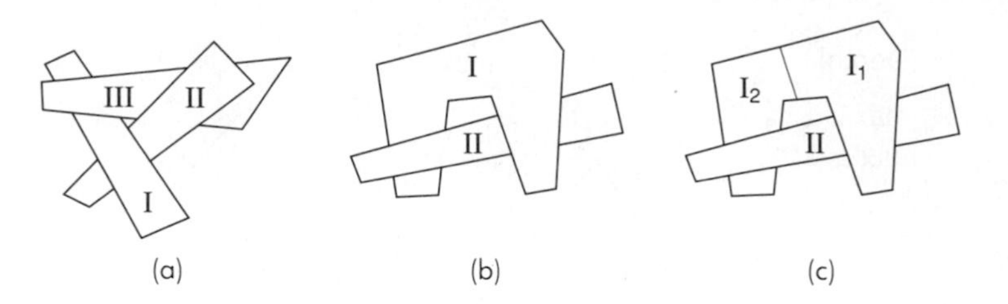

FIGURE 4.23 Ambiguous cases in priority sorting

obscures I. In Figure 4.23(b), I obscures II, and II obscures I. If depth sorting is adopted as the only criterion for determining priority of the polygons, the algorithm for priority sorting will get caught in an infinite loop in the above situations.

In order to avoid the occurrence of such a failure, the polygons once moved (as a result of minimax test, etc.) in the sort list are marked. If an attempt is made later to move a marked polygon, a loop is assumed to have formed. The marked polygon in such cases is divided into two parts. In Figure 4.23(c), polygon I has been divided into I_1 and I_2 by the plane of II. Thus, I_1 is in front of II and I_2 behind it. The ambiguity in priorities is thus obviated.

❑ SCAN-LINE ALGORITHMS

In its simplest form, a scan-line algorithm for hidden line/surface removal works like the z-buffer algorithm described earlier. The following steps are carried out:

1. For every pixel on a scan-line (i.e., y = constant), variables $z_0(x) = 1.0$ and $i_0(x)$ = background value are set.

2. (a) For every pixel on the above scan-line that lies on the projection of a polygon—say, j—depth $z_j(x)$ of the polygon is found.

(b) If $z_i(x)$ is found to be less than the previously set value of $z_0(x)$ then the switch $z_0(x) = z_j(x)$ is made. Also, intensity of display for the pixel (x, y) is set equal to that for the polygon j, that is $i_0(x) = i_j(x)$.

3. Steps 2(a) through 2(b) are repeated for all j. As a result, final value for the display intensity at each pixel, in the composite picture, is now contained in the array i_0. These are copied to the frame buffer and the algorithm proceeds on to the next scan-line, that is step 1.

❑ SPAN-COHERENCE ALGORITHMS

These algorithms utilize the coherence property in scan-lines for sorting out invisible portions of an image [8]. Consider the polygons shown in Figure 4.24(a). Traces of these polygons on the scan-plane are shown in Figure 4.24(b). Five different spans on the scan-line may be identified:

1. Spans 1 and 5, which do not belong to any polygon. The intensity of all the pixels on these spans is set equal to the background value.

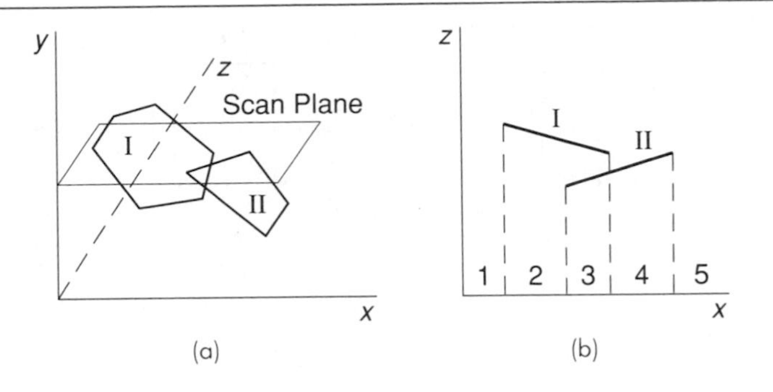

FIGURE 4.24
Selection of spans

2. Spans 2 and 4, which belong to only one polygon. The intensity of all the pixels on each span is set to the value chosen for the particular polygon.
3. Span 3, which belongs to both polygon I and polygon II. The intensity of the pixels on this span is set equal to that of polygon II since it obscures polygon I.

❑ AREA-COHERENCE ALGORITHM

In the hidden surface algorithm suggested by Warnock [9], coherence of pixels in certain areas of the image is utilized to reduce the number of calculations. The algorithm makes use of the fact that the intensity of pixels within a polygon or in certain areas outside the polygon may be determined in a single calculation.

Warnock's algorithm proceeds recursively. At first, the algorithm looks at the entire display surface as a single window and tries to set the intensities of all the pixels. If the image in the window is found to be too complex, the window is subdivided into four smaller windows of equal size. One or more of the windows may become simple enough at this stage for the algorithm to sort out the pixel intensities. If a simple window has been found by the algorithm, it is left untouched, whereas the other windows are further subdivided. The process continues until the intensities in all windows have been determined or until the size of the window has been reduced to that of a single pixel.

The term *simple window*, as used above, implies that the details in the window are such that the algorithm can determine the intensities of all pixels within the window. The algorithm uses some very simple logic to determine these intensities. Polygons are classified as disjoint, intersecting, contained, or surrounding. Depending on the type of polygon, the following steps are adopted.

Disjoint Polygons ■ These are the polygons that are completely outside the given window. If all polygons are disjoint with respect to a given

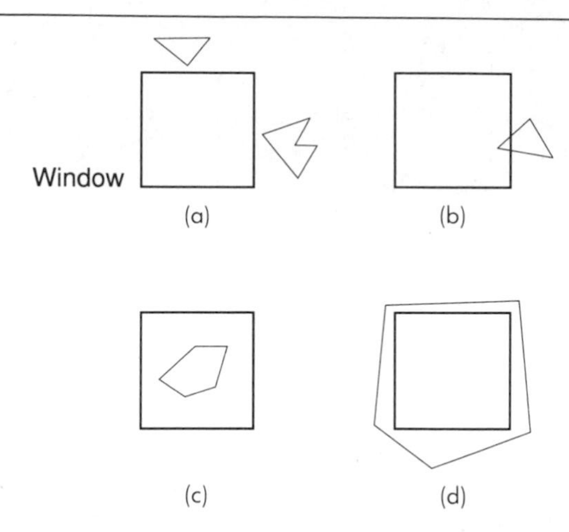

FIGURE 4.25 Area-coherence algorithm

window [see Figure 4.25(a)], then the intensity of the entire window is set to the background value, and the window is dropped from the active list (i.e., no further calculations are performed on this window).

Intersecting Polygons ▪ These are the polygons that intersect the window. If there is a single intersecting polygon in a window [see Figure 4.25(b), then the pixels belonging to the polygon are set to the intensity equal to that of the polygon. The intensity of the other pixels is set to the background value, and the window is dropped from the active list.

Contained Polygons ▪ These are the polygons that are completely contained within the window. If there is a single contained polygon within a window [see Figure 4.25(c)], then the intensity of the pixels within the polygon is set as required for the polygon. The intensity of all other pixels is set to the background value.

Surrounding Polygons ▪ These are the polygons that surround the window. If there is a single surrounding polygon [see Figure 4.25(d)], then the intensity of all pixels is set to the value of the intensity for the polygon. At times, there are contained and/or intersecting polygons along with one or more surrounding polygons in a window. In these cases, if the surrounding polygon obscures all other polygons, then the intensity in the window is set to the level desired for the surrounding polygon.

If the situation in a window does not conform to any of the situations described above, it is further subdivided. No further calculations are required for all other windows. In testing the subdivided windows, use is made of the fact that polygons that are disjoint or surrounding with respect to the larger window remain so for the smaller windows as well.

In Figure 4.26, a few steps from this algorithm are shown. The image consists of two overlapping polygons. Initially, an attempt is made to solve

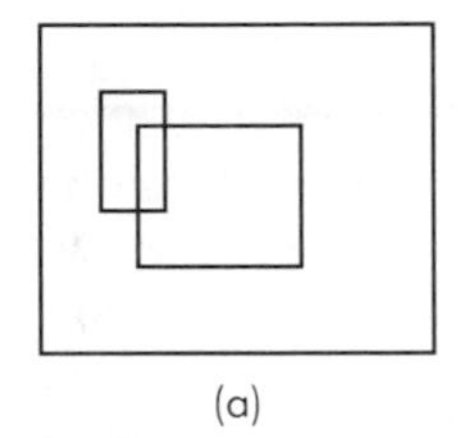
(a)

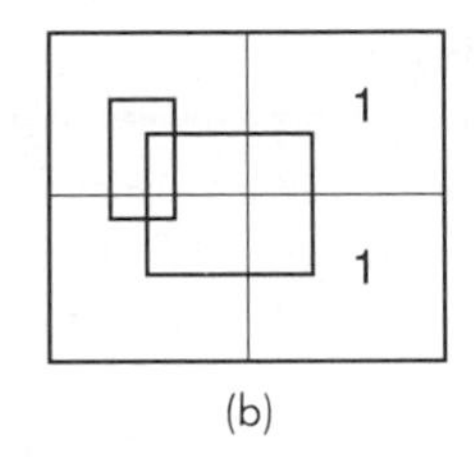

(b)

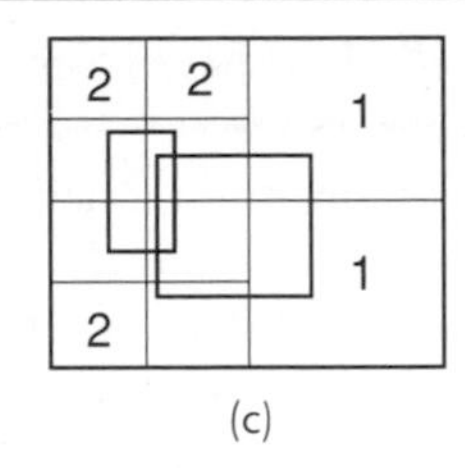

(c)

FIGURE 4.26 Steps in the area-coherence algorithm

the problem for the entire display surface [see Figure 4.26(a)]. Since the situation does not fall into any of the types described above, no solution is possible at this stage. The algorithm then subdivides the window into four parts [see Figure 4.26(b)]. As can be seen, there are intersecting polygons in the two windows (marked 1) on the right. The intensity of these polygons (part of the larger square) may now be set to the desired value. The intensity of the remaining parts of the window is set to the background value. No further examination of these windows is necessary. The algorithm then subdivides the windows on the left. It can be seen that all polygons are disjoint with respect to three windows (marked 2) [see Figure 4.26(c)]. The intensity of these windows is set to the background value. No further calculation need be performed on these windows. The algorithm would now subdivide the remaining five windows as many times as necessary until the entire problem is solved.

Another area-coherence algorithm requiring far fewer calculations has been suggested by Weiler and Atherton [10]. In this method, the windows are not necessarily rectangular or of the same size—rather, they vary depending on the polygons in question. Suppose that the polygons have been depth sorted and that they are situated as shown in Figure 4.27. It is determined through the depth sorting that polygon I is in front of polygon II. Polygon I is therefore made the first window. The intensity of this window is set according to that required for polygon I. The next window in the algorithm is polygon II_a. As can be seen, there is only one surrounding polygon in this window, and, therefore, the intensity of this window is set for polygon II. The entire scan-conversion process is completed in just two steps. This is a significant improvement when compared with Warnock's algorithm. Of course, defining the windows in this method is not as simple as in Warnock's algorithm.

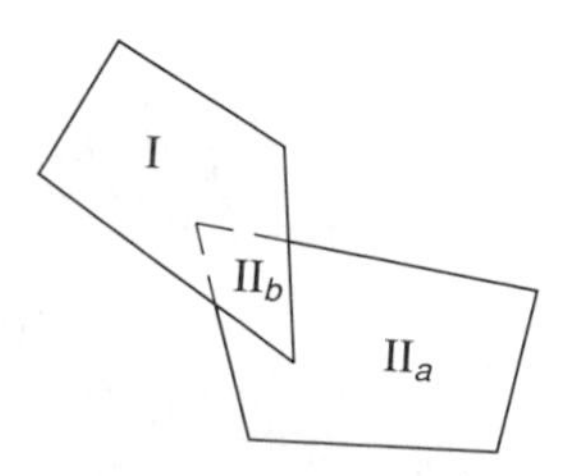

FIGURE 4.27 Weiler-Atherton algorithm

4.9 ▪ Surface Patches

Geometric modeling can be simplified considerably by employing mathematically defined free-form surfaces (and solids). In this section, the reader is introduced to some of these surfaces and the mathematical formulations associated with them.

Complex surfaces and solids may be generated by combining two or more curves. Since the reader is already familiar with the Bézier and B-spline curves, the discussion below is limited to the use of these curves.

❑ BÉZIER SURFACE PATCH

A Bézier surface may be generated from the tensor product of two curves as follows:

$$P(s, t) = \sum_{i=0}^{n} \sum_{j=0}^{m} p_{i,j} B(s)_{i,n} B(t)_{j,m} \tag{4.14}$$

where there are a total of $(n + 1)$ $(m + 1)$ control points, and s and t are single-valued parameters varying from 0 to 1. The blending functions (Bernstein polynomials) are given by the following equations:

$$B(s)_{i,n} = \frac{n!}{i!(n-i)!} s^i (1 - s)^{n-i} \tag{4.15}$$

$$B(t)_{j,m} = \frac{m!}{j!(m-j)!} t^j (1 - t)^{m-j} \tag{4.16}$$

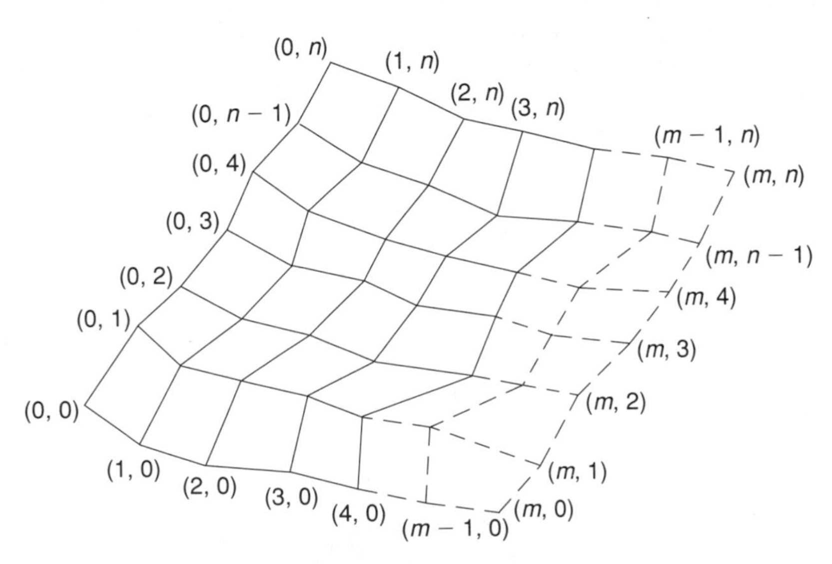

FIGURE 4.28 Control points and control net

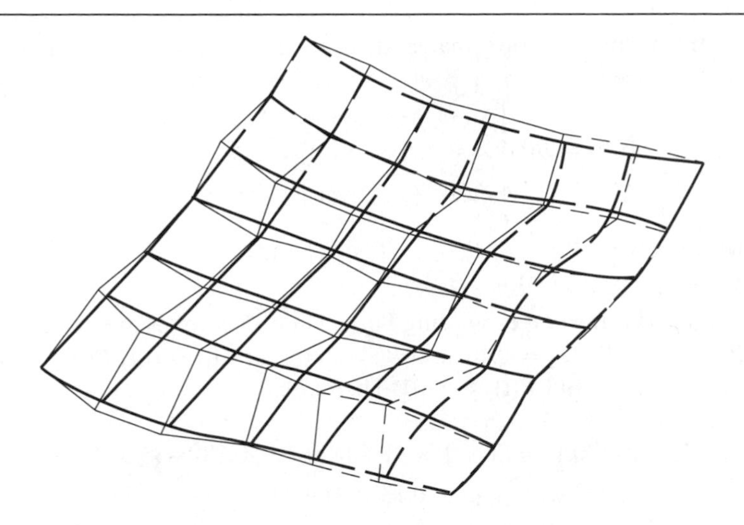

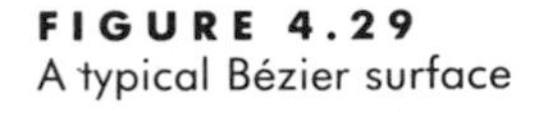

FIGURE 4.29
A typical Bézier surface

The vector $p_{i,j} = \{x_{i,j} \quad y_{i,j} \quad z_{i,j}\}$ contains the coordinates of the control point (i, j) (see Figure 4.28). The vector $P(s, t)$ contains the coordinates of points on the Bézier surface, i.e., $P(s, t) = \{x(s, t) \quad y(s, t) \quad z(s, t)\}$. A typical Bezier surface is shown in Figure 4.29. The grid pattern can be obtained by drawing two sets of Bézier curves. In one set, t is held constant at different levels between 0 and 1, while s is varied from 0 to 1. In the second set, s is held constant, while t is varied from 0 to 1.

EXAMPLE ■ 4.5

A Bézier surface is to be obtained for the following nine control points:

$p_{0,0}(3, 2, 2) \quad p_{0,1}(4, 2, 1.5) \quad p_{0,2}(5, 2, 1.5)$

$p_{1,0}(3, 1, 1.5) \quad p_{1,1}(4, 1, 2) \quad p_{1,2}(5, 1, 1.5)$

$p_{2,0}(3, 0, 2) \quad p_{2,1}(4, 0, 2) \quad p_{2,2}(5, 0, 1.5)$

Find $x(0.1, 0.1)$ at the surface.

In the above example, $n = 2$ and $m = 2$. The formula for $x(0.1, 0.1)$ is given below.

$$
\begin{aligned}
x(0.1, 0.1) &= \sum_{i=0}^{2} \sum_{j=0}^{2} x_{i,j} B_{i,2} B_{j,2} \\
&= \{x_{0,0} B_{0,2}(s) B_{0,2}(t) + x_{0,1} B_{0,2} B_{1,2}(t) \\
&\quad + x_{0,2} B_{0,2}(s) B_{2,2}(t)\} + \{x_{1,0} B_{1,2}(s) B_{0,2}(t) \\
&\quad + x_{1,1} B_{1,2}(s) B_{1,2}(t) + x_{1,2} B_{1,2}(s) B_{2,2}(t)\} \\
&\quad + \{x_{2,0} B_{2,2}(s) B_{0,2}(t) + x_{2,1} B_{2,2}(s) B_{1,2} \\
&\quad + x_{2,2} B_{2,2}(s) B_{2,2}(t)\}
\end{aligned}
$$

In matrix form, the coordinates at any value of s and t are expressed

$$x(s, t) = B(s)\begin{bmatrix} x_{0,0} & x_{0,1} & x_{0,2} \\ x_{1,1} & x_{1,1} & x_{1,2} \\ x_{2,0} & x_{2,1} & x_{2,2} \end{bmatrix} B^T(t)$$

where $B(s) = [B(s)_{0,2} \quad B(s)_{1,2} \quad B(s)_{2,2}]$ and $B(t) = [B(t)_{0,2} \quad B(t)_{1,2} \quad B(t)_{2,2}]$.

For $s = 0.1$, the corresponding Bernstein polynomials are $B(s)_{0,2} = 1$, $B(s)_{1,2} = 0.18$, and $B(s)_{2,2} = .01$. For $t = 0.1$, $B(t)_{0,2} = 1$, $B(t)_{1,2} = 0.18$, and $B(t)_{2,2} = .01$. Therefore,

$$\begin{aligned} x(0.1, 0.1) &= (3 \times 1 \times 1) + (4 \times 1 \times 0.18) + (5 \times 1 \times 0.01) \\ &\quad + (3 \times 0.18 \times 1) + (4 \times 0.18 \times 0.18) \\ &\quad + (5 \times 0.18 \times 0.01) + (3 \times 0.01 \times 1) \\ &\quad + (4 \times 0.01 \times 0.18) + (5 \times 0.01 \times .01) \\ &= 4.4863 \end{aligned}$$

❑

In most applications, several Bézier patches may have to be joined together to model a complex shape. As can be expected, centain continuity conditions should be satisfied for a smooth representation. Zero-order continuity at the junction of two Bézier patches may be maintained by selecting a common set of control points for the coincident edges. First-order continuity across the common edges is maintained by superimposing on the above condition an additional constraint that ensures that the ratio of distances between the control points, near the common edge, on the two surfaces is the same in each row of control points. For example, in Figure 4.30, the two patches will have slope continuity across the common edge if and only if $ab/ab' = cd/cd'$, etc., and the points a, b, and b' and the points c, d, and d', etc., are collinear.

❑ B-SPLINE SURFACE PATCH

A B-spline surface patch may be formed by tensor multiplication of two B-spline curves as follows:

$$P(s, t) = \sum_{i=1}^{k} \sum_{j=1}^{l} p_{i,j} B(s)_{i,k} B(t)_{j,l} \tag{4.17}$$

where k and l represent the order of the constituent curves in s and t, respectively.

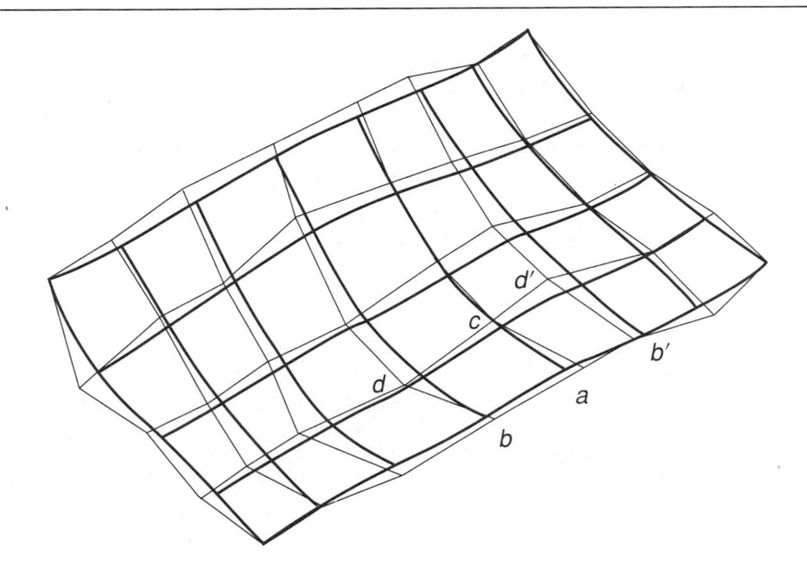

FIGURE 4.30
Two Bézier surfaces joined together

The knot sequence vectors—say, t_s and t_t—are formed in the manner explained in Section 3.2. The B-spline functions $B(s)_{i,k}$ and $B(t)_{j,l}$ can be obtained recursively from equation (3.13). The coordinates for any point on the patch may be determined in a manner similar to that illustrated in Example 4.2.

4.10 ▪ Geometric Properties of Graphics Models

Geometric models are often utilized for finding geometric properties of objects. Given below are some mathematical formulations that are used for determining these properties.

❑ LENGTHS OF CURVES

In Chapter 3, equations for Bézier curves and B-splines are given. The lengths of these curves may be found from an integral equation such as

$$L = \int dl \tag{4.18}$$

Assume that the length of a Bézier curve is to be determined. The parametric equation for this type of curve is given in equation (3.9). In its expanded form, the equation of a cubic Bézier curve is given as

$$\begin{aligned} P(s) &= (1-s)^3 p_0 + 3s(1-s)^2 p_1 + 3s^2(1-s)p_2 + s^3 p_3 \\ &= b_0 p_0 + b_1 p_1 + b_2 p_2 + b_3 p_3 \end{aligned} \tag{4.19}$$

where

$$b_0 = (1 - s)^3, b_1 = 3s(1 - s)^2, \text{etc.}$$

The vector $P(s)$ is differentiated with respect to the parameter s as follows:

$$P'(s) = b'_0 P_0 + b'_1 p_1 + b'_2 p_2 \tag{4.20}$$

where

$$b'_0 = db_0/ds, b'_1 = db_1/ds, \text{etc.}$$

The elemental length dl of the curve is given as

$$dl = |P'(s)|\, ds \tag{4.21}$$

where

$$|P'(s)| = [P'(s) \cdot P'(s)]^{0.5} \tag{4.22}$$

Using the coefficients from equation (4.20), it is found that

$$\begin{aligned} P'(s) \cdot P'(s) = {} & (b'_0)^2 p_0 \cdot p_0 + (b'_1)^2 p_1 \cdot p_1 + (b'_2)^2 p_2 \cdot p_2 \\ & + (b'_3)^2 p_3 \cdot p_3 + 2(b'_0 b'_1) p_0 \cdot p_1 + 2(b'_0 b'_2) p_0 \cdot p_2 \\ & + 2(b'_1 b'_2) p_1 \cdot p_2 \end{aligned} \tag{4.23}$$

The length of the curve is then computed from the following equation:

$$L = \int_0^1 \sqrt{P'(s) \cdot P'(s)}\, ds \tag{4.24}$$

EXAMPLE ■ 4.6

A Bézier curve is drawn with the control points (0, 0), (1, 1), and (2, 0) lying on a semicircle. Estimate the length of the curve.

A quadratic Bézier curve may be drawn from the data given. The equation for the curve is given as

$$P(s) = (1 - s)^2 p_0 + 2s(1 - s) p_1 + s^2 p_2$$

Therefore,

$$P'(s) = -2(1 - s) p_0 + 2(1 - 2s) p_1 + 2s p_2$$

and the dot product

$$\begin{aligned} P'(s) \cdot P'(s) = {} & [-2(1 - s)]^2 p_0 \cdot p_0 + [2(1 - 2s)]^2 p_1 \cdot p_1 \\ & + [2s]^2 p_2 \cdot p_2 - 8(1 - s)(1 - 2s) p_0 \cdot p_1 \\ & - 8s(1 - s) p_0 \cdot p_2 + 8s(1 - 2s) p_1 \cdot p_2 \end{aligned}$$

From the coordinates of the control points, it is seen that $p_0 = \{0 \quad 0\}$ $p_1 = \{1 \quad 1\}$, and $p_2 = \{2 \quad 0\}$. The products are $p_0 \cdot p_0 = \{0 \quad 0\}\{0 \quad 0\}^T = 0$, $p_0 \cdot p_1 = 0$, $p_0 \cdot p_2 = 0$, $p_1 \cdot p_2 = 2$, $p_1 \cdot p_1 = 2$ and $p_2 \cdot p_2 = 4$. Therefore,

$$\begin{aligned} P'(s) \cdot P'(s) &= 8(1 - 2s)^2 + 16s^2 + 16s(1 - 2s) \\ &= 16s^2 - 16s + 8 \end{aligned}$$

$$L = 4 \int_0^1 \sqrt{(s^2 - s + 0.5)}\, ds$$

Substitute $s = (x + 1)/2$ and $ds = dx/2$ in the equation above to get

$$L = \int_{-1}^1 \sqrt{(x^2 + 1)}\, dx$$

Using the Gaussian quadrature formula and $n = 4$,

$$\begin{aligned} L &= 0.34785\sqrt{(0.86113^2 + 1)} + 0.65214\sqrt{(0.33998^2 + 1)} \\ &\quad + 0.34785\sqrt{(0.86113^2 + 1)} + 0.65214\sqrt{(0.33998^2 + 1)} \\ &= 2.2957 \end{aligned}$$

The arc length of a semicircle passing through the control points is 3.1416. ❑

❑ SURFACE AREA

As explained earlier in this chapter, a surface patch comprised of two parametric variables (s, t) can be created using Bézier curves, B-splines, etc. The area of such a surface patch may be found in the following manner:

Step 1 Find the partial derivatives.

$$P_s(s, t) = \frac{\partial P(s, t)}{\partial s}$$

$$P_t(s, t) = \frac{\partial P(s, t)}{\partial t}$$

Step 2 Find the equation for an element of area from the following cross-product:

$$dA = |P_s(s, t) \times P_t(s, t)|\, ds\, dt \tag{4.25}$$

Step 3 Find the surface area by a numerical quadrature method.

$$A = \int_0^1 \int_0^1 dA(s, t) \tag{4.26}$$

❑ VOLUME

A solid *hyperpatch* can be formed by a combination of three parametric curves. A hyperpatch is given by an equation of the following type:

$$P(s, t, u) = \sum_{i=0}^{n} \sum_{j=0}^{m} \sum_{k=0}^{l} p_{i,j,k} B(s)_{i,n} B(t)_{j,m} B(u)_{k,l} \tag{4.27}$$

Figure 4.31 shows a hyperpatch generated by the tensor product of three Bézier curves. A total of 125 control points have been used.

The volume of a parametric hyperpatch may be determined by triple integration of the equation of a volume element given below:

$$dV = [(P_s(s, t, u) \times P_t(s, t, u)) \cdot P_u(s, t, u)]\, ds\, dt\, du \tag{4.28}$$

If the density of the material is known—say, $\sigma(s, t, u)$—the mass of the differential element above is expressed as

$$dM = \sigma(s, t, u)\, dV \tag{4.29}$$

For the case of a homogeneous material, σ is a constant.

The first moments may be found from the following equations:

$$M_{yz} = \int_0^1 \int_0^1 \int_0^1 \sigma(s, t, u) X(s, t, u)\, dV \tag{4.30}$$

$$M_{zx} = \int_0^1 \int_0^1 \int_0^1 \sigma(s, t, u) Y(s, t, u)\, dV \tag{4.31}$$

$$M_{xy} = \int_0^1 \int_0^1 \int_0^1 \sigma(s, t, u) Z(s, t, u)\, dV \tag{4.32}$$

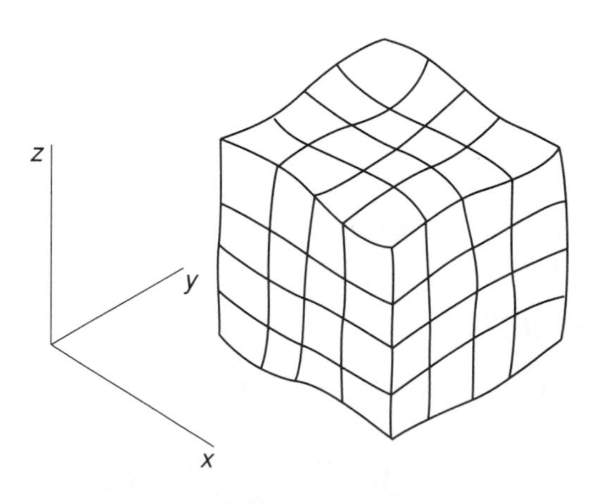

FIGURE 4.31
A three-dimensional hyperpatch

The centroid of the hyperpatch is then found as follows:

$$(\bar{X}, \bar{Y}, \bar{Z}) = \left(\frac{M_{yz}}{M}, \frac{M_{zx}}{M}, \frac{M_{xy}}{M}\right)$$

In these equations, $X(s, t, u)$, $Y(s, t, u)$, and $Z(s, t, u)$ are obtained from $P(s, t, u)$.

Second moments may be found using equations of the following type:

$$I_x = \int_0^1 \int_0^1 \int_0^1 \sigma(s, t, u)[Y^2(s, t, u) + Z^2(s, t, u)]\, dV \quad (4.33)$$

Second moments are used for determining stress, deflection, rotational response, etc. of structural members.

4.11 ▪ Computer Simulation and Animation

Computers are now being widely used to generate graphical models of physical objects and to simulate their time-dependent responses to various types of physical stimuli. Some of the changes that take place in time-dependent phenomena include change in size and/or shape, change in position or orientation, change in color or shade, etc. If it were possible to depict these changes in real time (i.e., at the rate at which the physical phenomena really take place) and to display them graphically, the designer could study problems of congestion on highways, optimal rates of flow of workers and materials on a factory floor, etc. Computer hardware technology has not yet reached the point where it can simulate most time-varying physical phenomena in real time. But, by adopting techniques similar to those found in cinematography, it is possible for a computer to animate rapidly occurring physical phenomena.

Figure 4.32 shows a portion of a cam mechanism in four different angular positions. The images are produced by a program developed by one of the authors. The program determines the appropriate cam profile from the data provided and passes the graphics display function on to a postprocessor. In the postprocessor, the geometry of the cam is recalculated for a predetermined rotation, and then the point of contact between

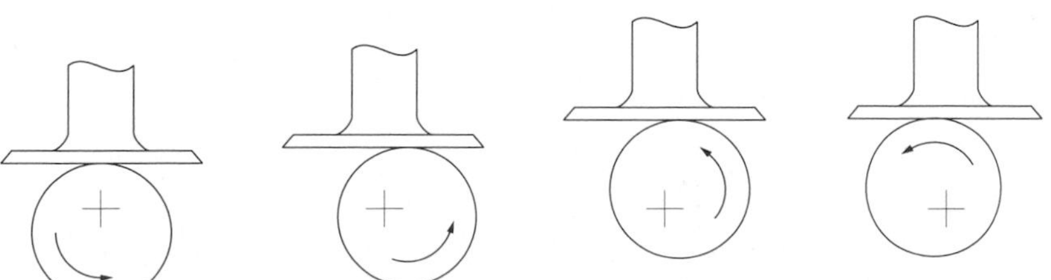

FIGURE 4.32
Animation of a cam and follower

the cam and the follower is calculated. As soon as these calculations are completed, the graphics display is updated. Depending on the speed of the processor, the rate of the graphics update varies from 1 to 2 seconds. Thus, by looking at the computer screen for a while, the observer gets a feel for the interaction between the cam and the follower (i.e., the shift in the point of contact, the extent of movement of the cam, etc.). As any movie fan knows, the above rate of frame display is not sufficient for producing apparently continuous motion. The rate required for that is 20 to 30 frames per second. To depict inherently faster phenomena (such as the collision of two cars, for instance), an even faster rate of display is, in fact, used.

One very common technique for producing smooth animated motion is to generate all the frames in advance, store them on a mass storage device, and display them on the computer screen in a rapid sequence. Here, too, one may be restricted by the persistence of the screen phosphor or the rate of data transfer between secondary storage (the disk drive) and display memory (the graphics display card) in the host computer. In such cases, it is advisable that the frames be recorded on film or on videocassette.

In another technique used to produce realistic animation, a large number of frames are superimposed one over the other. One frame at a time is turned "on," while the other frames are "off." The technique is similar to that of flashing neon signs in which rotating automobile wheels or water cascading down a falls is shown by turning the neon tubes off and on. If you imagine the pictures drawn in darker ink as the lighted frame, the operation of this technique is apparent from Figure 4.33.

The method of animation used in the cam design program described above is known as *modeled animation.* In modeled animation, the computer generates the images based on some mathematical formulation. In a majority of engineering applications, modeled animation may be the only choice available. In artwork such as animated cartoons, however, another method, known as *keyframe* [11] animation, is quite popular. In the keyframe method, the animator creates a number of basic pictures (cels) interactively. By combining several basic pictures, a composite cel known as the *keyframe* is created. The keyframes are terminal pictures (i.e., they depict considerable changes in the picture details). In between the keyframes are several other frames that are obtained by interpolation. The interpolation algorithm often allows for varying such details in the frames as angles, sizes, position, etc.

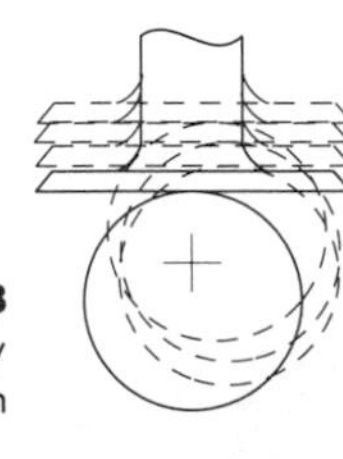
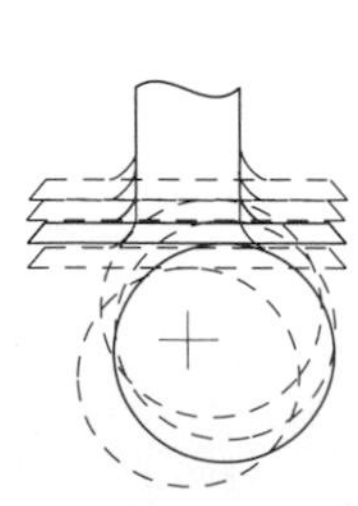
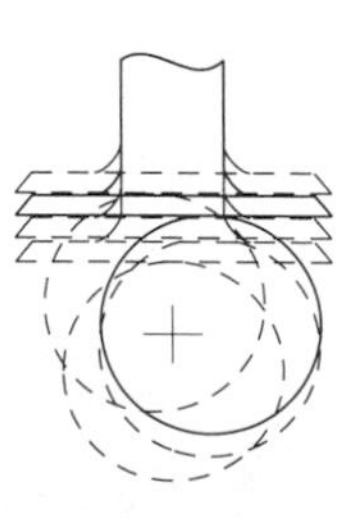
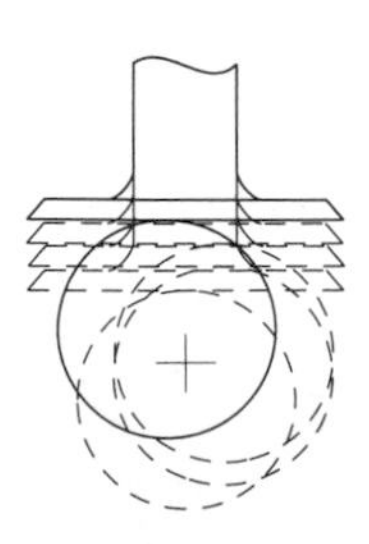

FIGURE 4.33 Animation by superimposition

In both methods of animation (i.e., modeled and keyframe), use of segments is often a good idea. As explained in Appendix AII.4, a segment contains only a part of a complex picture. Different frames are created by altering the segments that have undergone change. Other segments are copied as is.

4.12 ▪ Windows, Viewports, and Viewing Transformations

A draftsperson, before starting to draw, needs to do some planning *vis-à-vis* the scale of the drawing and the exact location and composition of the picture. Thus, if the object is large in comparison to the size of drawing sheet available, a reduction scale is chosen. If the object is small, the draftsperson may decide to draw it on an enlarged scale. Sometimes drawings of several small parts are reduced in size to fit on one sheet.

Obviously, a computer-oriented draftsperson needs to do exactly the same type of planning as described above. All graphics packages therefore provide facilities for such planning. Described below are windows, viewports, and other features that are used for such purposes.

Suppose that a hydraulics engineer wishes to plot a graph of daily consumption of water in his district for the month of June. The variables in his graph are the days of the month and the corresponding millions of gallons. The engineer may plot the days along a horizontal axis and the water consumption quantities along a vertical axis. A point on the graph will be defined by a pair of coordinates (day, millions of gallons). Coordinates such as those described above are known as *world coordinates*. World coordinates are not used directly by the computer. Instead, the computer identifies a point in terms of the coordinates defined on the display device. On a raster display device, the coordinates of a point are specified in terms of a pixel number and a row number. Therefore, it is necessary, that the computer be given some means of converting world coordinates into device coordinates so that a point from the world coordinate system is mapped onto the display device. We will discuss the method of mapping points momentarily. It is necessary, however, that we first understand the concepts of window and viewport.

❑ WINDOWS AND VIEWPORTS

In computer graphics, the user is given the option of delineating the boundaries of the space in the real world from which data are extracted for a graphics display. The space so delineated is known as the *window*. Thus, in the example above, if the engineer wishes to plot data for the entire month, he may specify a window whose lower left corner is at (1, 0) and whose upper

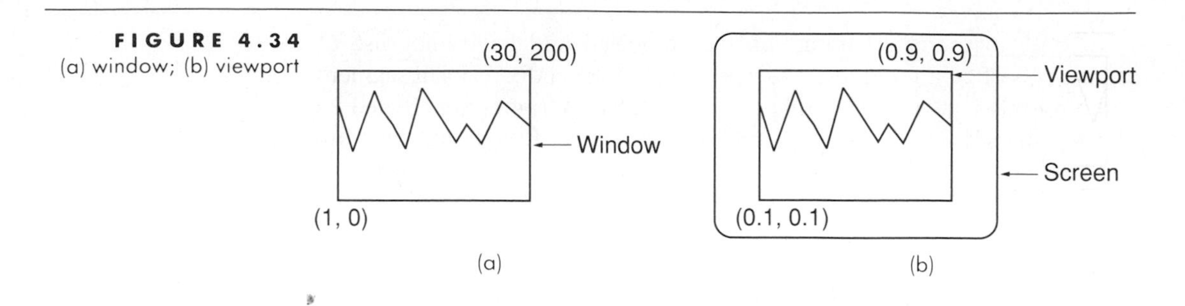

FIGURE 4.34
(a) window; (b) viewport

right corner is at (30, 200) in terms of the world coordinates. The window in this case is defined as a rectangular box, as shown in Figure 4.34(a).

The engineer should now decide where exactly on the display surface the graph should be placed. For this purpose, he selects a *viewport*. Assume that the lower left corner of the display surface has coordinates (0, 0) and the upper right corner has coordinates (1, 1) (i.e., the width and the height of the display device are both equal to one, which is true if the dimensions of the screen are *normalized*). When the coordinates of a point are expressed as fractions of the width and height of the display device, these are then known as the *normalized device coordinates* (NDC). Now, if it is desired that the graph be placed centrally on the display surface, leaving a uniform margin all around, then the coordinates of the lower left corner of the viewport that the engineer selects may be, say, (0.1, 0.1) and those of the upper right corner of the viewport may be (0.9, 0.9). The graph is then plotted on the display surface, as shown in Figure 4.34(b).

❑ VIEWING TRANSFORMATIONS

The process of mapping points from within the window (object space) onto the viewport area is known as the *viewing transformation*. Mathematically, this is achieved by evolving a transformation matrix that is operated on the coordinates of every point in the object space. The transformation matrix may be obtained by assuming that the viewing transformation is carried out in three distinct stages [8]. First, the window is translated such that its lower left corner coincides with the origin of the device coordinates. Next, a scaling is applied so that the size of the window is the same as that of the viewport. In the third and final stage, a translation is again applied so that the window now sits with its lower left corner coincident with the lower left corner of the viewport.

To explain the mechanism of the viewing transformation with a numerical example, return to the engineer's graph once again. Assume that the window defined by the engineer is now given by points (1, 0) and (30, 200) and the viewport by points (0.5, 0.5) and (1, 1). As a first step, the window is moved such that its lower left corner (1, 0) coincides with the origin of

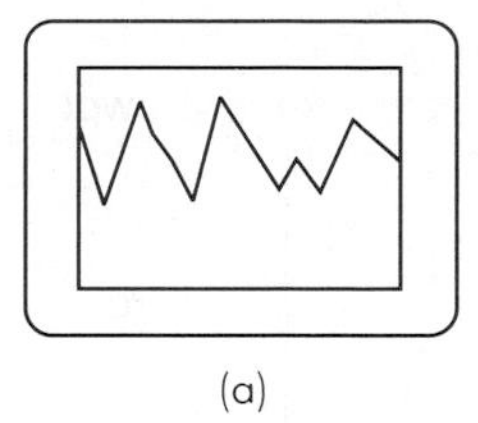

(a)

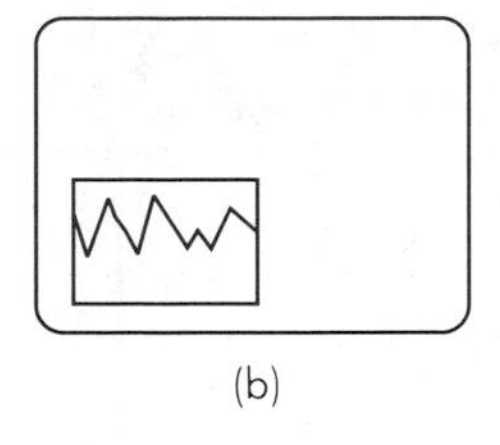

(b)

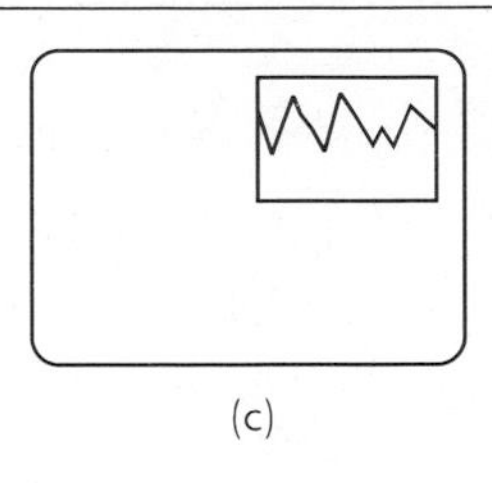

(c)

FIGURE 4.35 (a) a window translated to the origin of the NDC system; (b) a window scaled to the size of the viewport; (c) a window fitted into the viewport

the device coordinates (0, 0). See Figure 4.35(a). Obviously, the translation required in this case is -1 in the horizontal direction and 0 in the vertical direction. The translation matrix in this case therefore reads:

$$T(-1, 0) = \begin{bmatrix} 1 & 0 & 0 \\ 0 & 1 & 0 \\ -1 & 0 & 1 \end{bmatrix}$$

Since it is desired to plot the graph over a length of 0.5 horizontally and 0.5 vertically, it is obvious that a reduction of 0.5/29 must be applied horizontally and 0.5/200 vertically. The scaling matrix is thus

$$S(0.5/29, 0.5/200) = \begin{bmatrix} 0.5/29 & 0 & 0 \\ 0 & 0.5/200 & 0 \\ 0 & 0 & 1 \end{bmatrix}$$

Once the scaling transformation is completed, the window might appear as shown in Figure 4.35(b). Finally, to translate the window to the viewport area, the following matrix is used:

$$T(0.1, 0.1) = \begin{bmatrix} 1 & 0 & 0 \\ 0 & 1 & 0 \\ 0.5 & 0.5 & 1 \end{bmatrix}$$

The window is then moved to the position desired by the user, as shown in Figure 4.35(c).

The complete viewing transformation, as explained above, is carried out by concatenating the following matrix transformations:

$$\{x' \quad y' \quad 1\} = \{x \quad y \quad 1\}T(-1, 0)S(0.5/29, 0.5/200)T(0.5, 0.5)$$

It can be easily verified that points (1, 0) and (30, 200) from the object space map into points (0.5, 0.5) and (1, 1) on the display surface due to the above concatenated transformations.

In order that points expressed in normalized device coordinates be mapped into picture elements, the normalized device coordinates are converted into device coordinates. Let's assume that a computer screen has

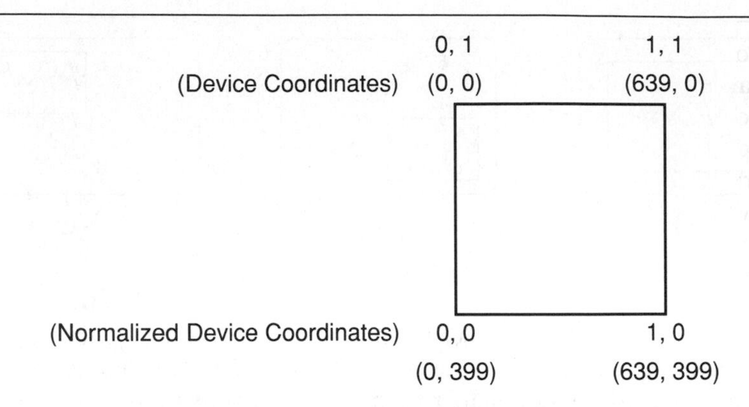

FIGURE 4.36 NDC and device coordinates

400 (0–399) horizontal lines of picture elements with 640 (0–639) pixels in each line. The normalized device coordinates are then scaled by the following matrix:

$$S(639, 399) = \begin{bmatrix} 639 & 0 & 0 \\ 0 & 399 & 0 \\ 0 & 0 & 1 \end{bmatrix}$$

In many display devices, the origin is placed at the top left corner, which is different from the location of the origin in the NDC; a translation may therefore be necessary. The transformation matrix, if needed in the above case, should move the origin of the NDC by an amount of 399 as follows:

$$T(0, 399) = \begin{bmatrix} 1 & 0 & 0 \\ 0 & 0 & 0 \\ 0 & 399 & 1 \end{bmatrix}$$

In a device coordinate system such as the one above, the y axis is generally positive downward, as shown in Figure 4.36. A reflection about the x axis is therefore applied.

$$M(x) = \begin{bmatrix} 1 & 0 & 1 \\ 0 & -1 & 0 \\ 0 & 0 & 1 \end{bmatrix}$$

The device coordinates are then obtained in the following manner:

$$\{x'' \quad y' \quad 1\} = \{x'' \quad y' \quad 1\}S(639, 399)T(0, 399)M(x)$$

4.13 ▪ Clipping

In discussing the viewing transformations, the example cited is such that every detail (points, lines, etc.) that needs to be displayed lies within the window. There are instances, however, when the user does not wish to display the entire contents of the object space. The user then selects a window that contains only those details that are to be displayed; details outside of the window are discarded. The method by which a graphics program determines which objects or parts of objects should be retained or thrown away is known as *clipping*.

A very convenient clipping method for clipping lines was developed by Cohen and Sutherland [1, 8]. Since most graphics images consist of line segments anyway, the algorithm may therefore also be applied to any plane object. In the Cohen and Sutherland algorithm, the object space is divided into nine separate regions, as shown in Figure 4.37. Each of these regions is identified with a pair of letters. The window is identified with the letters (*M*, *M*). Consider line segment 1. Obviously, it should be retained and displayed since it lies entirely within the specified window. The clipping algorithm checks the positions of the ends of the line and determines that they both lie inside the window. It therefore trivially accepts line segment 1. In the case of line segment 2, the algorithm determines that both ends of the line lie in the regions above the window. No part of the line segment may therefore pass through the window. As a result, it is rejected trivially. In a similar manner, line segments 3, 4, and 5 are also rejected. As for line segment 6, one end of the line lies in the region (*T*, *R*) above the window, and the other lies in the region (*B*, *L*) below the window. Although not guaranteed yet, it is possible that in cases where a line crosses over from one side of the window to the other (as line 6 does), it may pass through the window. The algorithm then proceeds further to investigate such cases.

In order to determine the visible portions of line segments, such as line 6 above, the points of intersection of the line with the vertical boundaries of the window are first determined. In Figure 4.37, the line intersects the left side of the boundary at point *b* and the right side of the boundary at point *d*. The algorithm now treats *b* and *d* as the end points of line 6. As can be seen, point *d* lies above the top horizontal boundary, which signifies that the right

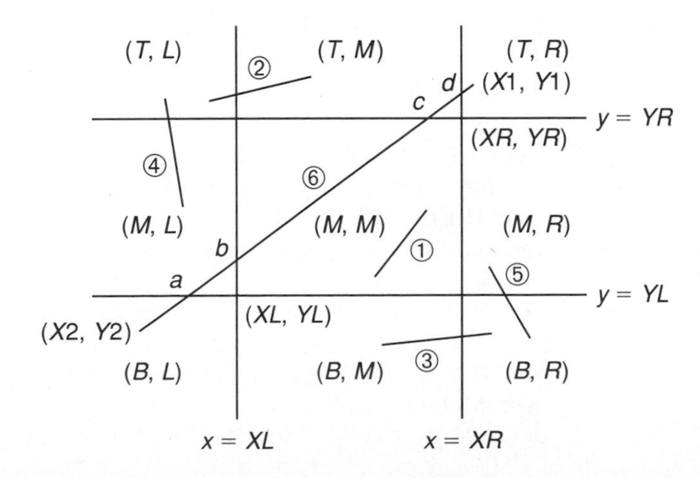

FIGURE 4.37
Nine distinct regions

end of the line is still outside the window. The intersection of line 6 with the top horizontal boundary (point *c*) is found next. The visible portion of line segment 6 is *bc*. Of the six lines shown in Figure 4.37, only line 1 and segment *bc* would be drawn.

Character clipping may also be carried out using the above Cohen-Sutherland algorithm. A common approach is to perform the visibility test on the diagonal of the cell containing the character. If the diagonal is completely outside the window, the character is trivially rejected. For polygons filled with one solid color or shaded, a simpler algorithm is available whereby the visibility test is applied to the entire polygon at once, rather than repeating it on each line of the polygon.

4.14 ▪ Summary

This chapter discusses those aspects of computer graphics and geometric modeling that are directly or indirectly employed in geometric design. The discussions on animation and simulation, three-dimensional modeling, and the method of determining geometric deal with direct applications. Other topics discussed in the chapter, such as surface and hyperpatches, projections, viewing transformations, etc., are included to provide a better understanding of computer graphics for engineers.

References

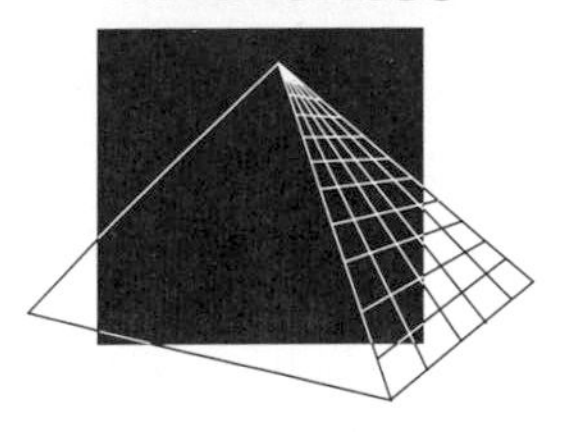

1. Foley, J.D. and A. Van Dam. *Fundamentals of Interactive Computer Graphics.* Reading, Mass.: Addison-Wesley, 1983.
2. Mortenson. M.E. *Geometric Modeling.* New York: John Wiley, 1985.
3. Giesecke, F.E., A. Mitchell, H.C. Spencer, I.L. Hill, and J.T. Dygdon. *Technical Drawing.* New York: Macmillan, 1986.
4. Sutherland, I.E., R.F. Sproull, and R.A. Schumacker. *A Characterization of Ten Hidden-Surface Removal Algorithms.* vol. 6, 1974.
5. Catmull, E. "Computer Display of Curved Surfaces," in *Tutorial and Selected Readings in Interactive Computer Graphics.* Edited by H. Freeman. New York: IEEE, 1980: 309–15.
6. Park, Chan S. *Interactive Microcomputer Graphics.* Reading, Mass.: Addison-Wesley, 1985.
7. Newell, M.E., R.G. Newell, and T.L. Sancha. "A New Approach to the Shaded Picture Problem." Proc. ACM Natl. Conf. 1972:443.
8. Newman, W.M., and R.F. Sproull. *Principles of Interactive Computer Graphics.* New York: McGraw-Hill, 1979.

9. Warnock, J.E. "A Hidden-Surface Algorithm for Computer Generated Half-tone Pictures," University of Utah Computer Science Department Report, TR 4–15, June 1969.
10. Weiler, K., and P. Atherton. " Hidden-Surface Removal Using Polygon Area Sorting", Comp. Graphics, vol. II no. 2, 1977:214.
11. Magnenat-Thalmann, N., and D. Thalmann. *Computer Animation: Theory and Practice*. New York: Springer-Verlag, 1985.

Problems

4.1 Is it possible to use polyhedral surface models for the objects shown below? If yes, draw the model, and prove Euler's relationship in each case.

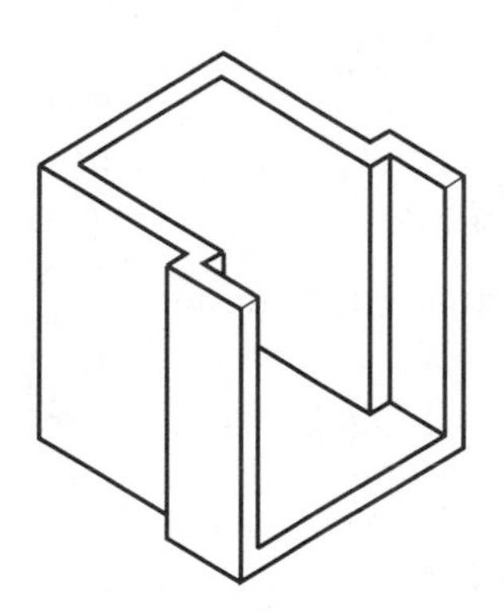

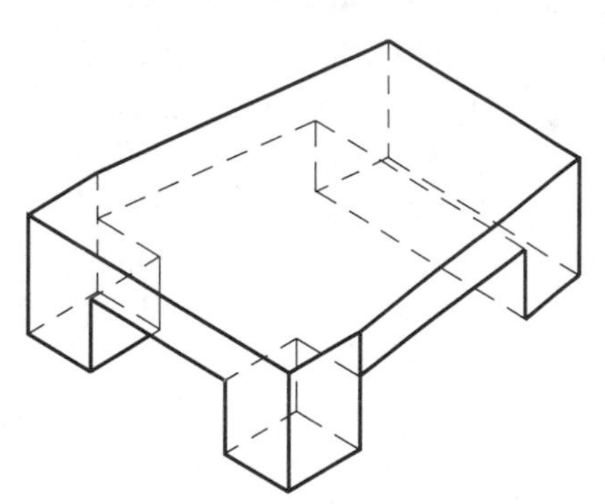

4.2 For the objects shown below, obtain binary tree representations for CSG modeling using the primitives given in Figure 4.5.

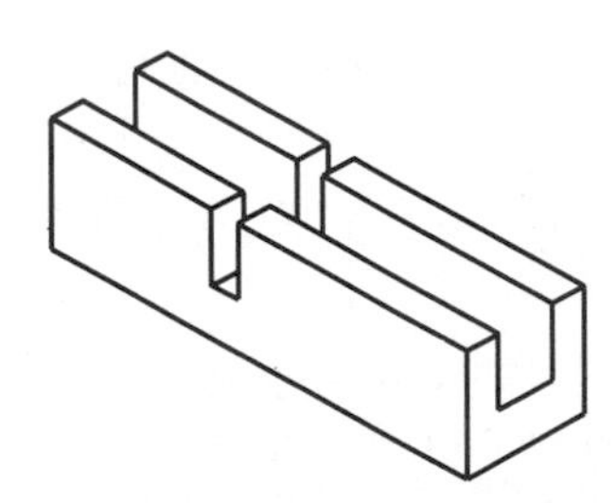

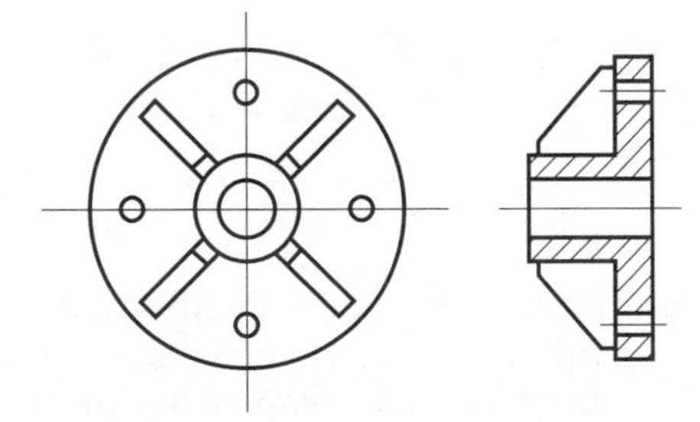

4.3 In isometric projection, the three principal axes of a cube are spaced 120° apart, as shown in Figure 4.11. Also, its principal dimensions (i.e., height, width, and depth) are foreshortened to 82 percent of their true length. From the data given in Example 4.2, check the accuracy of the above statements.

4.4 One type of dimetric projection is obtained by rotating the object -20.7° about the y axis and then $+19.4^\circ$ about the x axis. Obtain a transformation matrix for this dimetric transformation.

4.5 In Example 4.3 a two-point perspective of a cube is drawn. Obtain a three-point perspective, assuming a rotation of 10° about the horizontal axis. The locations of the screen and the viewer may be assumed to be the same as in the example.

4.6 Repeat the problem in Example 4.4 with a 15° rotation about the x axis in addition to the 30° rotation about the y axis.

4.7 The vertices of one of the faces of a solid block are at (1, 1, 1), (3, 1.5, 2), (3.5, 2, 5), and (1.5, 1, 1.4) as one traverses the edges of the face in a counterclockwise manner. Will this face be visible to a person standing at (2, 2, 10)?

4.8 Use the minimax and intersection tests to determine if the following two polygons overlap in the xy plane:
Polygon I: vertices at (3, 3.2), (4.5, 8), (7, 10), and (3.5, 7)
Polygon II: vertices at (4, 5), (5, 6), (6, 12), and (3, 6)

4.9 A point has the coordinates (3.2, 3.5). Determine if it is inside or outside polygons I and II given in problem 4.8.

4.10 In Figure 4.26 two steps in the Warnock algorithm are shown. How many steps (i.e., subdivisions) are required to completely solve the hidden surface problem. Justify your answer by means of diagrams. Use photocopies of Figure 4.26, if necessary.

4.11 In Example 4.2, it is desired that the vertex of the cube originally at the point (5, 6, 3) remain at the same point after the isometric transformation. Determine a suitable matrix for such a transformation.

4.12 From the data given in Example 4.5, find y (0.1, 0.1) and z (0.1, 0.1) for the Bézier surface.

4.13 From the data given in Example 4.5, find x (0.1, 0.1), y (0.1, 0.1), and z (0.1, 0.1) for a biquadratic B-spline surface patch.

4.14 A Bézier curve is utilized to approximate a semi-ellipse using the following control points: (0, 0), (1.5, 1), and (0, 3). Find the length of the curve, and compare it with the actual length of the semi-ellipse.

4.15 Two rectangular circuit boards measuring 18 in. × 30 in. and 40 in. × 60 in. are to be drawn on a screen with a resolution of 1024 × 960. Find the appropriate viewing transformation matrix such that the larger board is placed across the top half of the screen and the smaller board in the bottom half.

4.16 In a particular graphing problem the two-dimensional window is defined by points (51, 13) and (105, 75)—the lower left and upper right corners, respectively. Find mathematically the visible portion of a line whose extremeties lie at points (120, 69) and (42, 19).

4.17 In the figure below, the circular cam rotates about point *O*. In order to animate the working of the cam mechanism, it is necessary that the height *h* be related to the angle (α). Find a suitable equation.

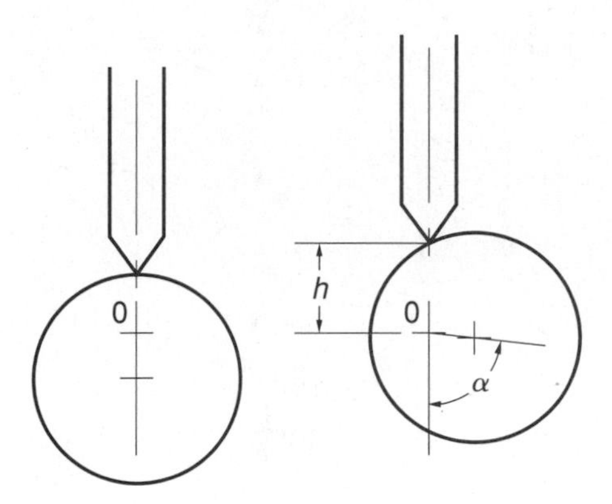

CHAPTER 5

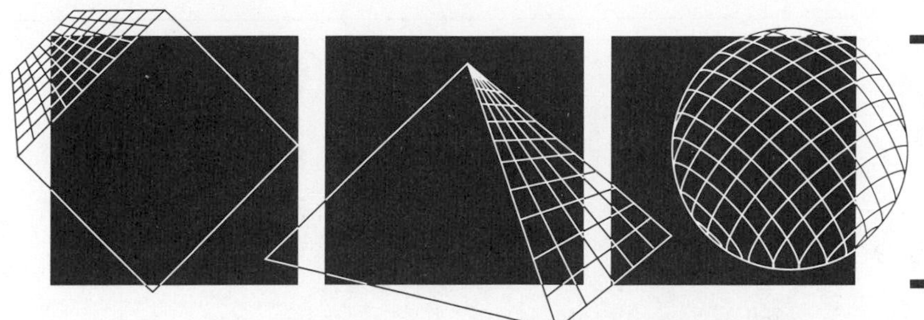

Introduction to Design Databases

5.1 Introduction

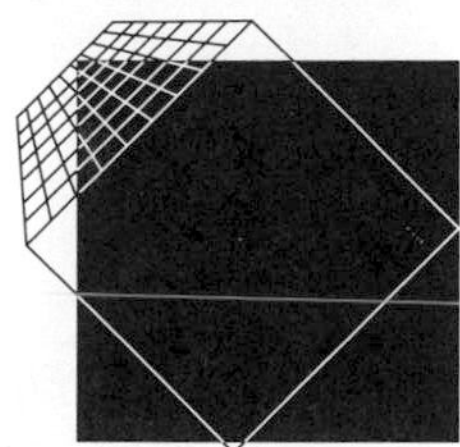

Database is one of the most frequently mentioned terms in the modern technical and business communities. Databases constitute an essential link among various components in CAD, CAM, or CIMS, as described in Chapter 1.

A *database* can be defined as a logical collection of related information. The collection of related information is a common act in a civilized world and has been practiced by humans for many years. Some examples of these collections include a telephone directory, a schoolteacher's student record book, a chef's recipe book, and an accountant's monthly ledger, as well as the subject index at the end of this book. Many more such examples can be found in daily life. The term *database* became prominent, however, only after the advent of electronic computers in the 1950s.

The amount of information in a database can vary from a handful to billions of entries. Typically, each set of information in a telephone book consists of the name, address, and telephone number of a subscriber. For a medium-sized city of half a million people, a telephone book may have an average of 700 pages with 500 sets of information on each page. A total of 350,000 sets of information may thus be included in a single book. It does not take much imagination to figure out the astronomical amount of information that has to be collected, organized logically, and stored in data files for the U.S. census.

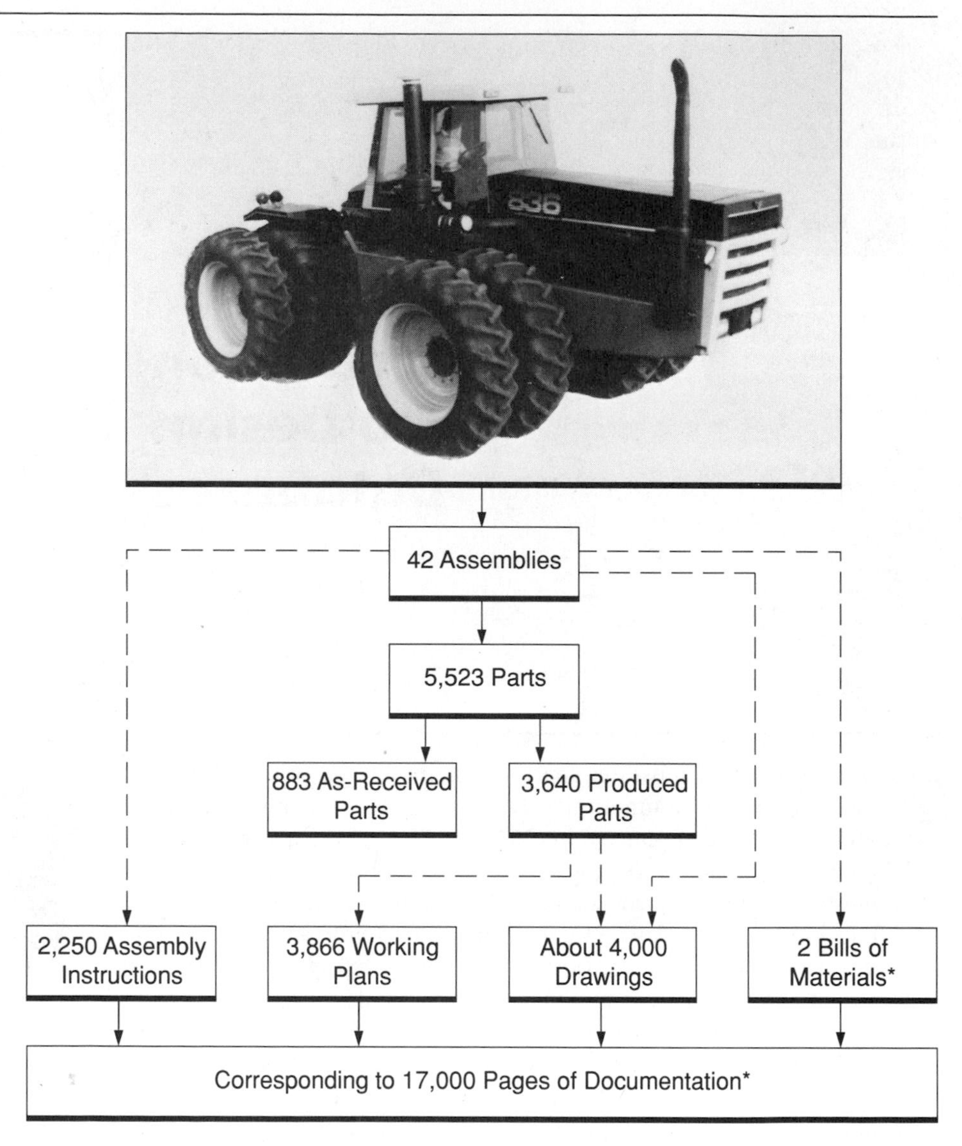

* This includes computer records (i.e., both bills of materials) that do not properly represent actual pages of documentation. Hard copies of everything would amount to this number.

FIGURE 5.1 Amount of engineering data required for a four-wheel drive tractor. Courtesy of Versatile Farm Equipment Company

The volume of information related to the production of industrial products can be just as horrendous as that encountered in business and administration. Figure 5.1 illustrates the amount of relevant technical information associated with a farm tractor. It can readily be seen from this figure that a total of 17,000 pages of documentation, or an equivalent of 25 telephone books, is required to describe this product. A similar example for a trench digger is also available in reference [1].

The colossal amount of information, or data files, that is required to be collected, stored, and manipulated in modern businesss and technical operations has made the use of electronic digital computers essential. Database management systems (DBMSs), which provide for high-speed and efficient data organization, storage, and retrieval, have become an important subject in the curricula of modern business schools. The subject of DBMSs in relation to business and administration is thus relatively well established, and ample books and publications on DBMSs are available [2–4].

The development of DBMSs for design and manufacturing engineering has been evolving rapidly with the advancement in CAD and CAM. However, the scope and methodology of constructing this type of DBMS are still far from being uniform. In this chapter, we will first review the essential elements of databases and DBMSs. The three common data models for business applications will be briefly described. We will then review the fundamental differences between CAD databases and those for business and administration. Such a comparison will provide the reader with some insight into the basic requirements of CAD databases.

5.2 ▪ Database Management Systems

A database may consist of millions of different sets of information. It is highly desirable to develop a system that can accept, organize, store, and retrieve information quickly. Efficient operations can be achieved by using powerful hardware (i.e., computers with large storage memory and high computational speeds) in conjunction with intelligent software systems. The latter is normally referred to as the *database management system.*

The construction of a database begins with the *field* or *attribute*. A field is the smallest entry or block of information in a database. Each field can be a different data type. Take, for instance, the set of information in a typical telephone book, which involves (1) the subscriber's name, (2) the subscriber's address, and (3) the subscriber's telephone number; see Table 5.1.

Each of the above entries (i.e., the name, address, and telephone number) represents a field. A specific collection of these fields or attributes constitutes what is called a *record*. One can thus conclude that information

TABLE 5.1 Information in a typical telephone book

Subscriber's Name	Address	Telephone No.
Hsu, T.R.	27 Sunbury Place	555-1218
Sinha, D.K.	23 Josepha Avenue	555-9724
Smith, D.J.	123 46th Street E.	555-7890
Watt, J.T.	756 Main St.	555-8994

on each subscriber in a telephone book represents a record. A telephone book that consists of many records is referred to as a *data file* or *database*.

Proper organization of data files is the key to an efficient DBMS. Various data models have been developed for such a purpose. The suitability of a particular data model, of course, depends on the intended applications. For example, one type of data model may be suitable for applications that require numerous queries, but little updating. The same data model, however, may not be suitable for applications that require frequent updating. Some data models are more suitable for specially trained personnel, and some are intended for the general public. However, three distinct types of data models often form the basis for commercial DBMS packages: (1) the relational model, (2) the hierarchical model, and (3) the network model. A comprehensive description of these models can be found in reference [5].

5.3 ▪ Data Models

The three types of data models mentioned in the last section are widely used in business and administration. They are designed to handle information that can be described by words, symbols, or numbers and are often referred to as *semantic models*. We will soon realize that semantic models are useful in engineering applications as well. It is therefore necessary for us to learn the basic principles of these models, and in particular those of relational models which are the most popular of these three models.

❑ RELATIONAL MODEL

Of the three types of data models, the relational database model appears to have the simplest structure. Related data are stored in the form of tables, with each row representing a record consisting of fields under each column. Thus, a telephone book is a typical relational database. Table 5.2 represents a relational database for the bill of materials required to produce the table vise illustrated in Figure 5.2.

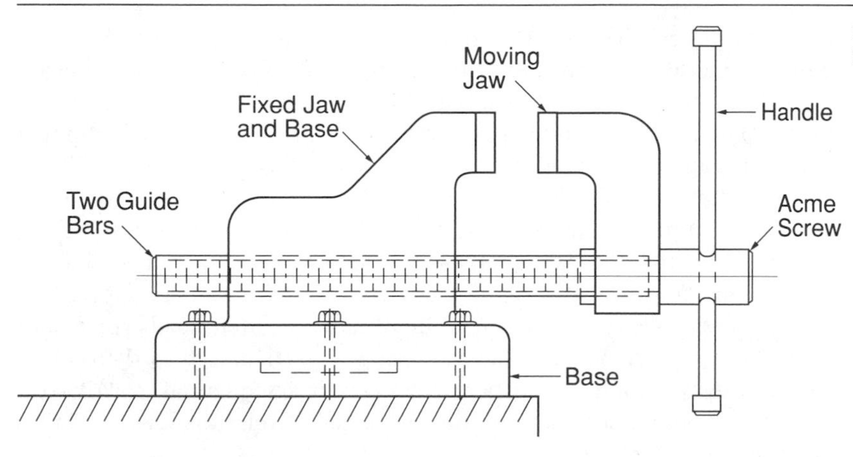

FIGURE 5.2
Schematic of a table vise

The data in Table 5.2 obviously can be represented by a matrix, a_{ij}, with the row number $i = 1, 2, 3, 4, 5, 6$, and the column number $j = 1, 2, 3, 4$. The total number of fields in this database is thus equal to $6 \times 4 = 24$.

Special characteristics of the relational model can be summarized, according to Cook [5], as follows:

1. Each cell in the matrix holds only one field.
2. No two rows can be identical.
3. In at least one of the columns there must be at least one entry (or field) that is unique for each row (e.g., entries in either column 1 or column 2 in Table 5.2 can be used to distinguish a particular record in the particular row).
4. Records or rows need not be in any particular order. The user may add any number of new records at the bottom of the table or insert records in between any two selected rows.
5. Since no two rows are identical and fields are related by columns, but not by rows, the user may delete or modify any row (record) without having to adjust any other part of the database.

TABLE 5.2 Bill of materials for a table vise

Component No. (Index L)	Component Name (Index J)	Material (Index N)	Fabrication Method (Index P)
010	Moving jaw	Steel	Forged
020	Fixed jaw and base	Steel	Forged
030	Acme screw	Steel	Machined
040	Handle	Aluminum	Machined
050	Base	Steel	Forged
060	Guide bar	Steel	Machined

Mathematical manipulation of this type of data model was first proposed by Codd [6], and significant efforts have since been made toward the further development of analytical formulations and computer algorithms. Detailed descriptions of theoretical formulation and computer programming of relational data models are available in references [2–4].

The simplicity in the structure of the relational data model has made it extremely popular among users in the business world. A number of commercial database packages based on this model are available for business and administration. Unfortunately, this unique advantage (i.e., the structural simplicity of the relational model) often leads to major difficulties in searching for and retrieving a particular record from the database. Any search has to be carried out in a sequential, row-by-row manner until the desired information is located. A relational model is thus the least efficient.

EXAMPLE ■ 5.1

Write a simple program in FORTRAN 77 for the following database used in a manufacturing process:

Index	Part No.	Part	Machine to Be Used
1	0011	Cam	Milling machine
2	0012	Cam	Drill press
3	3011	Shaft	Lathe

The program in Figure 5.3 is written to illustrate how a simple relational database can be constructed. The program obviously is not efficient enough to handle a large number of records, as commercial database systems can. The latter systems use high-level machine languages other than FORTRAN.

The program is designed to handle the following specific functions of a relational database:

1. Create the database.
2. Delete from the database.
3. Add to the database.
4. Open an existing database.
5. List the current database.

The attributes in the above data file are defined in the program, which is specifically written for use on a personal computer, and consist of a mixture of integers and characters. A similar program for a database used for payroll tabulation and computations is available in reference [7]. ❑

FIGURE 5.3 Computer program FORTRAN 77 database program

```
c .........
c     Example of a simple database program in FORTRAN
c                 - all records are held in an array in memory
c                   during processing !!
c                 - the array is unsorted and searching is done
c                   inefficiently in a linear fashion !!
c
c     Definition: partno = Part No.               (an integer field)
c                 parts = Parts                   (a 15 char. field)
c                 machin = Machine to be used.    (a 20 char. field)
c .........
      program simpdb
      character*39 record(100), tmprec
      integer*2 partno
      character*15 parts
      character*20 machin
      logical*1 opened
c
      opened = .false.
      nrec = 0
    5 write (*,10) 'Do you wish to: '
   10 format (///1x,a)
      write (*,15) '  1) Create data base'
   15 format (1x,a)
      write (*,15) '  2) Delete from data base'
      write (*,15) '  3) Add to data base'
      write (*,15) '  4) Open existing data base'
      write (*,15) '  5) List current data base'
      write (*,15) '  6) Exit'
      read (*,*,err=5) icode
      if (icode .lt. 1 .or. icode .gt. 6) go to 5
      if (icode .eq. 6) then
c        ... save database and quit
         if (opened) then
            call save (nrec, record)
         end if
         go to 20
      else if (icode .eq. 1) then

c        ... create database
         call build (nrec, record, partno, parts, machin)
         opened = .true.
      else if (icode .eq. 2) then
c        ... delete from database
         call delete (nrec, record, partno, parts, machin)
      else if (icode .eq. 3) then
c        ... add to database
         call insert (nrec, record, partno, parts, machin)
      else if (icode .eq. 4) then
```

FIGURE 5.3
(*Continued*)

```
c         ... open existing database
            call start (nrec, record)
            opened = .true.
         else
c         ... list contents of open data base
            call list (nrec, record, partno, parts, machin)
         end if
         go to 5
   20    stop '... end of program'
         end
c >>>
         subroutine save (nrec, record)

         character*39 record(100)
c
         open (3,file='datbas01.dat',status='unknown')
         write (3,10) nrec
   10    format (i4)
         do 30 i = 1, nrec
            write (3,20) record(i)
   20       format (a39)
   30    continue
         close (3)
         return
         end
c >>>
         subroutine build (nrec, record, partno, parts, machin)
         character*39 record(100), tmprec
         integer*2 partno
         character*15 parts
         character*20 machin
c
         partno = -1
         parts = ' '
         machin = ' '
         call pack (tmprec, partno, parts, machin)
         do 100 i = 1, 100
            record(i) = tmprec
  100    continue
         nrec = 0
         return
         end
```

FIGURE 5.3
(*Continued*)

```
c >>>
      subroutine delete (nrec, record, partno, parts, machin)
      character*39 record(100), tmprec
      integer*2 partno
      character*15 parts
      character*20 machin
      logical*1 found
c
      write (*,5) 'Part Deletion'
    5 format (1x,a)
    7 write (*,10) 'Enter partno: '
   10 format (1x,a\)
      read (*,*,err=30) partno
      if (partno .lt. 0) go to 30
      write (*,10) 'Enter parts: '
      read (*,15) parts
   15 format (a15)
      write (*,10) 'Enter machin: '
      read (*,20) machin
   20 format (a15)
      call pack (tmprec, partno, parts, machin)
      found = .false.
      jrec = -1
      do 25 i = 1, nrec
         if (tmprec .eq. record(i)) then
c           ... found record, ok to delete
            found = .true.
            jrec = i
            partno = -1
            parts = ' '
            machin = ' '

            call pack (tmprec, partno, parts, machin)
            record(i) = tmprec
            call reorg (nrec, record, tmprec)
         end if
   25 continue
      if (.not. found) then
         write (*,5) 'Could not find record!!! - cannot delete'
      end if
      return
   30 write (*,5) 'Invalid partno!!! - re-enter'
      go to 7
      end
```

FIGURE 5.3
(Continued)

```
c >>>
      subroutine insert (nrec, record, partno, parts, machin)
      character*39 record(100), tmprec
      integer*2 partno
      character*15 parts
      character*20 machin
      logical*1 found
c
      write (*,5) 'Part Deletion'
    5 format (1x,a)
    7 write (*,10) 'Enter partno: '
   10 format (1x,a\)
      read (*,*,err=30) partno
      if (partno .lt. 0) go to 30
      write (*,10) 'Enter parts: '
      read (*,15) parts
   15 format (a15)
      write (*,10) 'Enter machin: '
      read (*,20) machin
   20 format (a15)
      call pack (tmprec, partno, parts, machin)
      found = .false.
      jrec = -1
      do 25 i = 1, nrec
         if (tmprec .eq. record(i)) then
c           ... found a similar record, cannot add
            found = .true.
            jrec = i
         end if
   25 continue
      if (.not. found) then
         if (nrec .lt. 100) then
            nrec = nrec+1
            record(nrec) = tmprec
         else
            write (*,5) 'Database overflow!!! - cannot add'
         end if
      else
         write (*,5) 'Duplicate record!!! - cannot add'
      end if
      return
   30 write (*,5) 'Invalid partno!!! - re-enter'
      go to 7
      end
```

```
c >>>
      subroutine start (nrec, record)
      character*39 record(100)
c
      open (3,file='datbas01.dat',status='old')
      read (3,10) nrec
   10 format (i4)
         do 30 i = 1, nrec
         read (3,20) record(i)
   20    format (a39)
   30 continue
      close (3)
      return
      end
c >>>

      subroutine list (nrec, record, partno, parts, machin)
      character*39 record(100), tmprec
      integer*2 partno
      character*15 parts
      character*20 machin
c
      do 20 i = 1, nrec
         tmprec = record(i)
         call unpack (tmprec, partno, parts, machin)
         if (partno .ge. 0) then
            write (*,10) 'partno:', partno, 'parts: ', parts,
     &         ' machin: ', machin
   10       format (1x,a,i3,a,a15,a,a20)
         end if
   20 continue
      close (3)
      return
      end
c >>>
      subroutine pack (tmprec, partno, parts, machin)
      character*39 tmprec
      integer*2 partno
      character*15 parts
      character*20 machin
c
      write (tmprec,99) partno, parts, machin
   99 format (i4,a15,a20)
c ... note: in the above statement, tmprec acts like a character buffer,
c ...       not a disk file
      return
      end
```

FIGURE 5.3
(Continued)

FIGURE 5.3
(*Continued*)

```
c >>>
      subroutine unpack (tmprec, partno, parts, machin)
      character*39 tmprec
      integer*2 partno
      character*15 parts
      character*20 machin
c
      read (tmprec,99) partno, parts, machin
   99 format (i4,a15,a20)
c ... note: in the above statement, tmprec acts like a character buffer,
c ...        not a disk file
      return
      end
c >>>
      subroutine reorg (nrec, record, tmprec)
      character*39 record(100), tmprec
c
      i =.1
      5 if (i .gt. nrec) go to 15
        if (record(i) .eq. tmprec) then
           do 10 j = i, nrec-1
              record(j) = record(j+1)
   10      continue
           nrec = nrec-1
        end if
        i = i+1
        go to 5
   15 return
        end
```

HIERARCHICAL MODEL

A close look at the relational database presented in Table 5.2 will reveal that a significant number of repetitive attributes are contained in the file. For example, the attribute "Steel" is repeated in matrix locations a_{13}, a_{23}, a_{33}, a_{53}, and a_{63}; the attribute "Forged" is repeated in a_{14}, a_{24}, and a_{54}; and the attribute "Machined" is repeated in a_{34}, a_{44}, and a_{64}. This repetition may result in a very significant amount of waste in terms of memory storage, as well as the time required to search for and retrieve a specific record from a database that contains a large number of records. It has thus become necessary to develop other types of data models with less repetition and thus higher efficiencies in the storage and retrieval of information. The reader will see in a moment that the hierarchical model can provide partial improvement over this deficiency, which is inherent in the relational model.

The layout of a hierarchical model is similar to a tree structure, as illustrated in Figure 5.4. Various fields and records are linked through

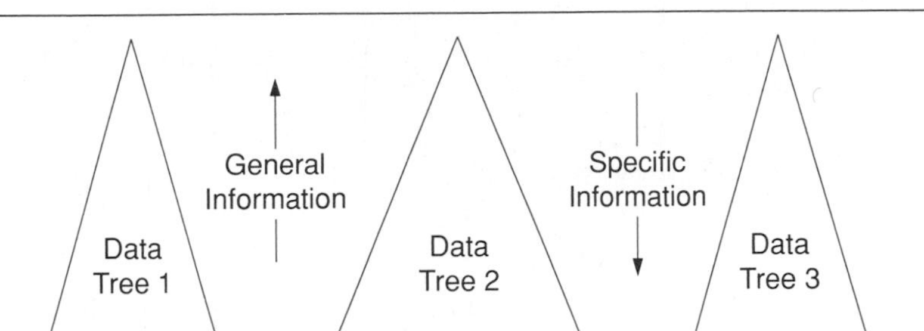

FIGURE 5.4
Schematic of a hierarchical data model

"branches," such as those illustrated in Figure 5.5 for the bill of materials for a table vise. At the top of the information tree is placed the broadest possible description of the contents of the database. As one moves down to the next level, the descriptions become more specific, though the information contained at this level still falls into more than one category. However, as one may readily observe, information stored in subsequent lower levels becomes more and more specific in nature. In the process of searching for a particular set of data, the user may pick up the particular category that most accurately describes the nature of the data being sought. The user then moves a step farther down the information tree through the selected category. By repeating this process at every level of the information tree, one can locate the pertinent data rather quickly.

The database in Figure 5.5 consists of two "trees." A query to the database starts at the top of the trees, with "Material," at which stage the indices steel and aluminum are the keys to further search of the data. The search for a complete set of information can be conducted by moving from the top to the bottom of the tree.

The advantage of using the hierarchical model obviously lies in the expediency in searching for and retrieving particular sets of information. A major shortcoming in this model, however, is the rigidity of the database structure. It is difficult to incorporate modifications into the database, especially in cases in which the new record involves substantial changes in the field data.

❑ NETWORK MODEL

We have noticed from Figure 5.5 that "Fabrication Method" is used as a category for grouping the data on the steel ACME screw and guide bar and the aluminum handle. The key element "Machined" appears three times in the database, twice in the "Steel" family and another time in the "Aluminum" family. This duplication can be avoided by cross-referencing the two materials families, as illustrated in Figure 5.6. This compact data model is usually referred to as the network model.

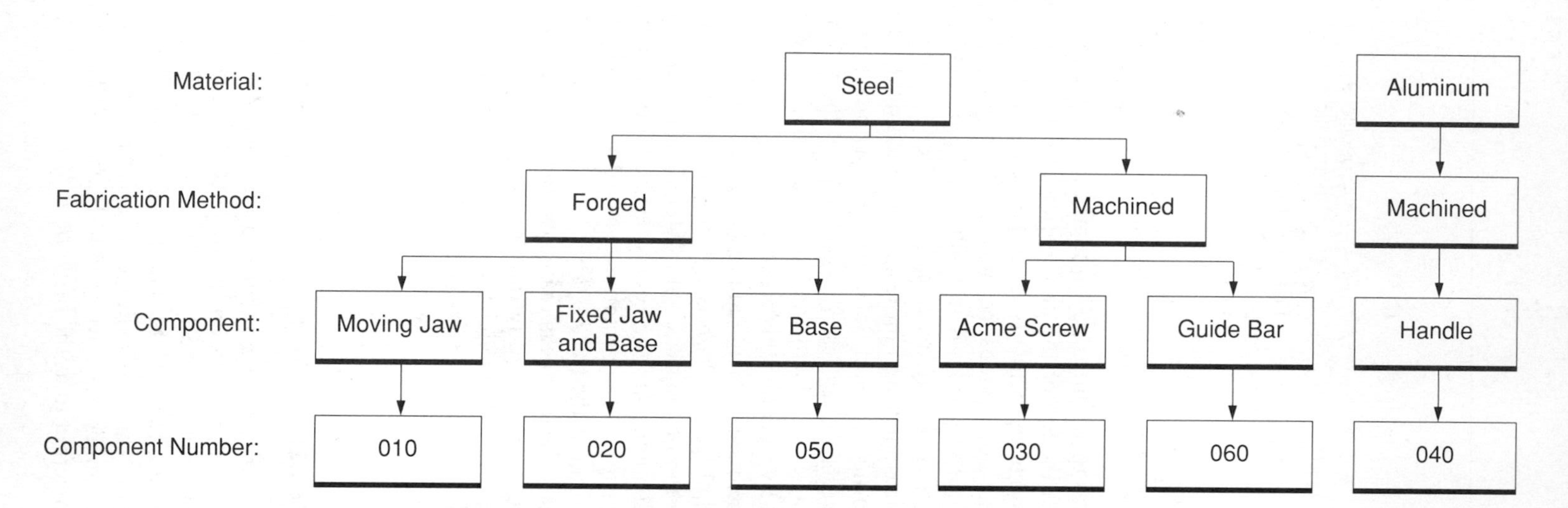

FIGURE 5.5 Structural diagram of a hierarchical data model

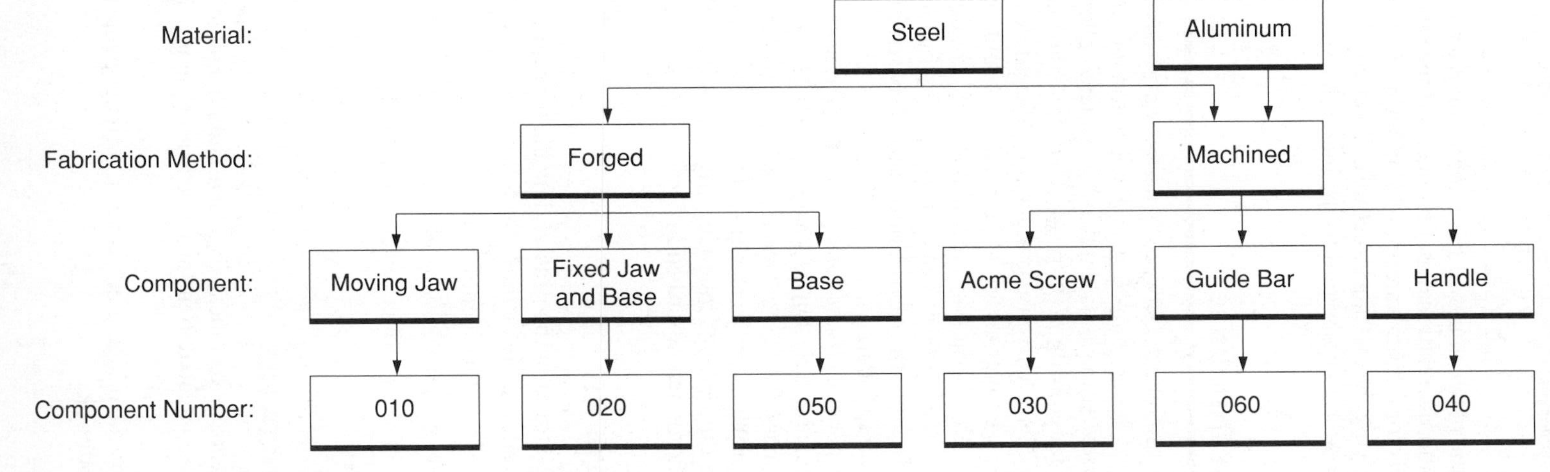

FIGURE 5.6 Structural diagram of a network model

The fact that the network model has a more compact structure than do the hierarchical and the relational models makes it the most efficient of all these three models for a given amount of computing power. However, the difficulty in modifying the existing structure of a network model can be just as great, if not greater, as compared to the hierarchical model.

5.4 ▪ Design Database

The fundamental difference between a design database and a business database is that a design database is a collection of product-oriented information, whereas a business database is a mere collection of related information used for consulting purposes. Therefore, the main thrust in the construction of a design database is to include efficient descriptions of the configurations (or geometries) of all components and to provide production and assembly instructions and detailed information necessary for manufacturing the final product. Such descriptions include not only semantic information, as in the case of business databases, but also geometric details of the components.

Following the general discussions given in references [1] and [8], it appears that, in general, a design database consists of five groups: (1) general data, (2) technical data, (3) geometric data, (4) product structure data, and (5) data on previous projects. Separate files may be constructed for each of the above data groups following any one or a combination of the three types of data models described in Section 5.3.

❑ GENERAL DATA FILE

Information to be stored in the general data file is nontechnical in nature. Typically, a general data file includes the following items:

- Design file and product identification
- Customer information: name, address, and other pertinent data
- Specification of the product, special considerations, and instructions
- Cost estimates

❑ TECHNICAL DATA FILE

As the title suggests, a technical data file should include all pertinent technical information related to the product. Major items to be included are as follows:

- Description of the design method used, including charts, design codes, or handbooks

- Design criteria and constraints used
- Material data for each component
- Other information such as mass, volume, moment of inertia, and overall dimensions

❑ GEOMETRIC DATA FILE

Information collected in this file relates to the description of the product's geometry. Dimensions and configurations of all components should be stored in the file. Proper description of the product configuration and its storage and retrieval constitute a major function of design databases, and much of the ongoing development effort is being made in these areas.

Semantic description of single two-dimensional objects is a relatively simple task because configurations of these objects can be described by graphic primitives such as dots, arcs, circles, straight lines, and curves. These primitives can be stored by recording the coordinates of their respective strategic points. A detailed description of the procedure for constructing a database for two-dimensional objects will be presented in Section 5.5.

❑ PRODUCT STRUCTURE DATA FILE

This data file can be viewed as the necessary "bridge" between the data files for the design and the manufacturing of a product. Two major sets of information are involved in this data file.

Parts Information ■ Generally, parts that make up the product can be acquired from two sources, as illustrated in Figure 5.1 (i.e., standard parts are purchased from outside suppliers, and other parts are produced in-house). In the case of a standard part, information related to the sources of supply must be fully documented. For parts produced in-house, pertinent information on tools, tool paths for NC machines, tolerances, and surface finishing ought to be included. Material information for each part in terms of quantity (or a bill of materials, such as that given in Table 5.2) and inventory information for the parts are also necessary. On rare occasions, parts are purchased as rough castings and then reworked in-house. In such cases, either a new category can be created for these parts, or they can be classified as "in-house produced" parts.

Assembly Procedure ■ Proper procedures must be prescribed for assembling the various parts into the final product. Information such as which part goes into what should be explicitly stored in this data file.

❑ DATA FILE ON PREVIOUS PROJECTS

In cases where a current product has evolved from existing products, duplication of data files should be avoided. The designer can simply build current design data files on the basis of relevant existing ones.

A sample flowchart of a typical design database that encompasses most of the data files described above is available in reference [1].

5.5 ▪ Geometric Databases for Two-Dimensional Objects

As we have learned in Section 5.4, the distinction between a business or administration database and a design database is that the design database contains data files that describe the geometry of solid structures. These files can be stored and retrieved efficiently. While a three-dimensional geometric description of solid objects involves complex mathematical formulations and the organization of the data is too complicated to be described here, we will illustrate in this section how a much simpler two-dimensional geometric database can be constructed. A relational data model will be used for this purpose.

Since, in general, any two-dimensional object of plane geometry can be described by a combination of points, lines, arcs, and curves, this type of geometry can be described by organizing entities involving these four basic configurations. The creation of the configuration of the entire object can be accomplished by generating the four files, each containing one of these entities. However, a fifth file is necessary to describe the relationships among the various entities that are required to create the object geometry.

The proper descriptions of the four entities are as follows:

1. *Points* (entity type 1) are entered to the entity file with a pair of coordinates, (x, y).
2. *Lines* (entity type 2) are entered to the entity file with two sets of coordinates: one set of coordinates, (x_1, y_1), for the starting point and the other set of coordinates, (x_2, y_2), for the ending point.
3. *Arcs* (entity type 3) are entered to the entity file with the coordinates of the center of the arc [i.e., (x_0, y_0)] and the coordinates of the starting point at (x_1, y_1) and of the ending point at (x_2, y_2).
4. *Curves* (entity type 4) are entered to the entity file with sets of coordinates of a minimum of three points or *knots* on the curve. Coordinates of knots are necessary to generate spline functions, as described in Section 3.2.

Each of the above four entity files thus includes a list consisting of the entity type, the entity identifier, and the associated coordinates. The *relational file* is simply a sequential list of the entities used as the designer creates the geometric data file for the object. The programming structure of this particular type of database is illustrated in Figure 5.7. The source listing of this program is included in the listing of the MICROCAD code described in Chapter 10.

The following example illustrates how the entity files and relational file may be constructed for a two-dimensional configuration. Take, for example, the lifting lug of a pressure vessel, as depicted in Figure 5.8. The geometry of this object can be created in the following sequence:

1. *One line from (1, 1) to (4, 1):* entered as entity type 2, entity identifier 1
2. *One curve passing through three knots at (4, 1), (6.02, 3), (4, 5):* entered as entity type 4, entity identifier 1
3. *One line from (4, 5) to (1, 5):* entered as entity type 2, entity identifier 2 (for the second line segment input)
4. *One line from (1, 5) to (1, 1):* entered as entity type 2, entity identifier 3 (for the third line segment)
5. *One arc (a circle in this case) with the center at (4, 3) with both the starting and the ending points at (4, 4):* entered as entity type 3, entity identifier 1

The relational and entity files may be created as follows:

Relational File: [Entity type, number of entities, entity identifier, number of points for curve (i.e., entity type 4)]

Sequence 1, a line: [2, 1, 1]
Sequence 2, a curve: [4, 1, 1, 3]
Sequence 3, a line: [2, 1, 2]
Sequence 4, a line: [2, 1, 3]
Sequence 5, an arc: [3, 1, 1]

Entity File:

Entity type 2, lines: [Entity identifier, coordinates (x_1, y_1), (x_2, y_2)]
1, (1, 1), (4, 1)
2, (4, 5), (1, 5)
3, (1, 5), (1, 1)

Entity type 3, arcs: [Entity identifier, coordinates (x_0, y_0), (x_1, y_1), (x_2, y_2)]
1, (4, 3), (4, 4), (4, 4)

Entity type 4, curves: [Entity identifier, number of knots, coordinates (x, y) of knots]
1, 3, (4, 1), (6.02, 3), (4, 5)

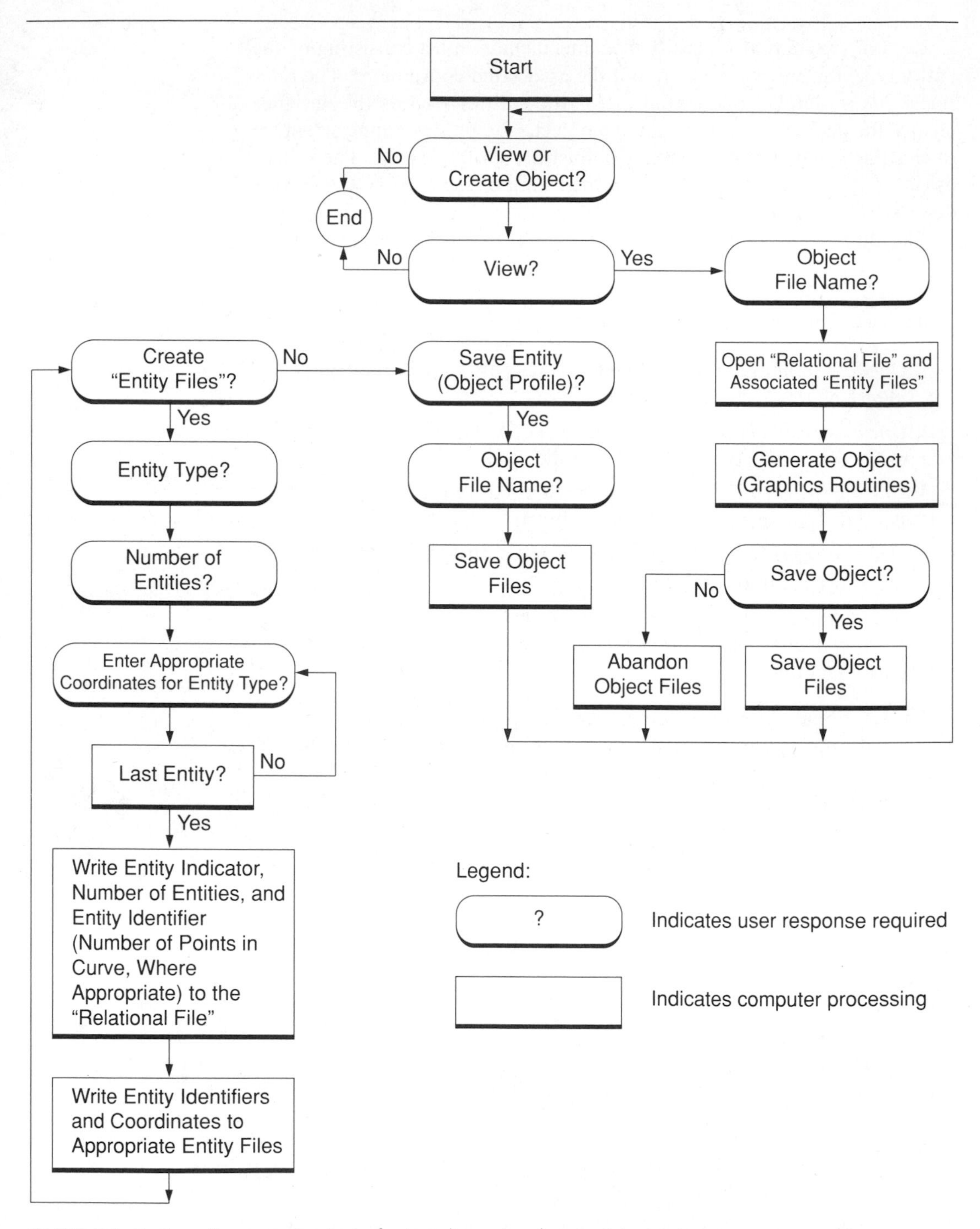

FIGURE 5.7 Program structure of a two-dimensional geometric database

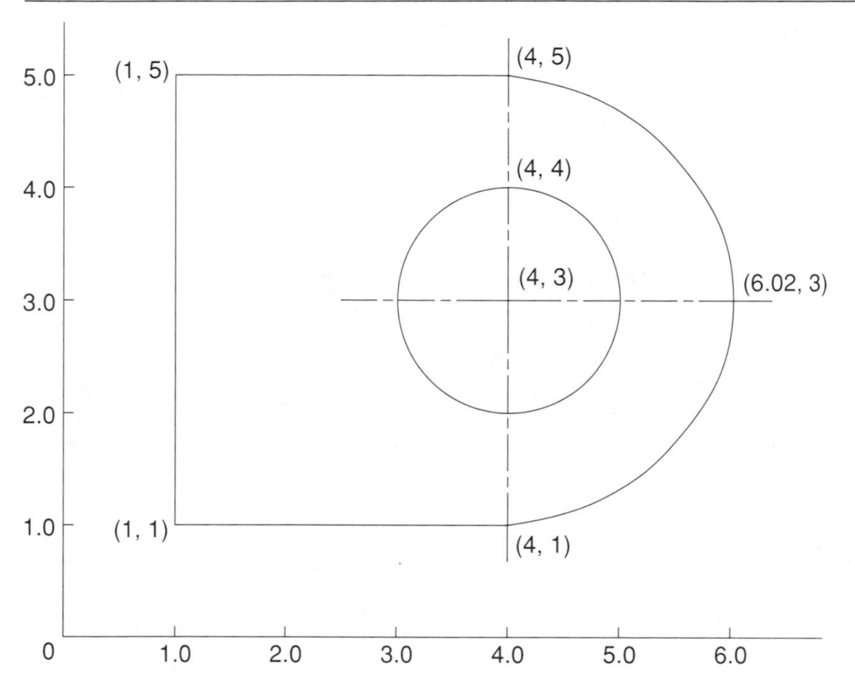

FIGURE 5.8
Profile of a lifting lug

EXAMPLE ■ 5.2

The PROFILE computer program was developed for two-dimensional geometric databases [9] and can construct two-dimensional profiles consisting of lines, arcs, and curves. The program was written in FORTRAN 77, with graphic primitives and operations handled by the GKS, as described in Chapter 3.

In this program, a point is drawn as a line with both ends having the same coordinates. Similarly, a circle is drawn as an arc with both ends having the same coordinates. A wide variety of curves may be drawn. The number of points required for constructing a curve may vary from 3 to 20. One constraint, however, is that the specified curve must be described by single-valued functions (i.e., one function value, y, corresponding to each x value). For the cases in which the specified curve cannot be described by a single-valued function, the user may split the curve into two or more sections so that each section can be described by a single-valued function. A concentration of points near the regions with sharp direction changes can result in better correlation between the function-generated curve and the original curve. Figure 5.9 shows an example of the profile of a cam that consists of two lines, a point, an arc, a circle, and a curve that was split into two sections.

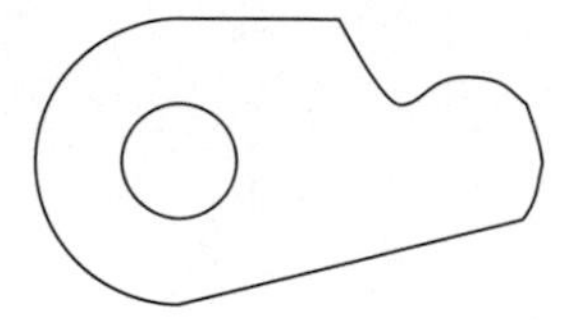

FIGURE 5.9
Profile of a cam

TABLE 5.3
Geometric entities for the profile of a cam

Entity Type	Pertaining Coordinates
2—line	End points of lines
3—arc	Center of curvature, end points of arc
4—curve	Consecutive points along the curve

Table 5.3 summarizes the information necessary to draw the configuration of the cam. In addition to that information, the number of points describing the curve must be entered. The user should also enter the file name if it is to be saved, and the desired minimum and maximum coordinates for drawing the picture must be entered as well.

The flowcharts used to develop the PROFILE program and the listing of the code are presented in Appendix III.

Input data for the construction of the profile of the cam shown in Figure 5.9 include the following items:

Entity Type	No. Pts	Pertaining Coordinates (x, y)			
3		(10, 10)	(10, 12)	(10, 12)	
2		(10, 10)	(10, 10)		
2		(21, 8)	(10, 5)		
3		(10, 10)	(10, 15)	(10, 5)	
2		(10, 15)	(15, 15)		
4	7	(15, 15)	(17, 12)	(18, 12.7)	(19, 13)
		(21, 12)	(21.3, 11)	(21.5, 10)	
4	3	(21, 8)	(21.3, 9)	(21.5, 10)	

(x min, y min), (x max, y max) = (3, 0) (23, 20) for the location of the origin of the screen coordinates.

The above information was stored in a number of files in the form of one- and two-dimensional arrays. The relational file is two-dimensional and stores such information as the total number of entity types, the types of entities, and the entity identifiers. The entity identifiers simply keep successive track of the number of each type used. These identifiers are also kept in their own one-dimensional files (i.e., one file each for lines, arcs, and curves). Each type of entity has its own two-dimensional file that contains coordinate values. The program also keeps track of the total number of entities.

Once the data are entered, the profile of the object can be drawn. Algorithms included in the program will perform all necessary computations. For arcs and curves, these algorithms calculate intermediate points. Lines are then drawn between the points, providing a good approximation of the original curve. Obviously, as more points are used, the better the approximation becomes. However, there is an optimum number of points at which the time consumed for the extra calculations is not worth the marginal improvement in the quality of the results.

The basic logic used for the PROFILE program is illustrated in the summary flowchart presented in Appendix III. The main program allows the user to do three things: (1) create a database, (2) draw the picture, and (3) quit the program. If the option of creating a database is chosen, the subroutine CREATE will be called up. This is the subroutine that prompts for the input of the required information. It creates the database for the input information. This database may then be stored on a memory disk. If the user wishes to draw the profile of the object, the subroutine DRAW is called up. According to the input data, the program will call up appropriate subroutines that draw lines, arcs, and curves, as well as the profile. After the profile is drawn, the user can either re-enter the main program for more action by repeating the above procedures or quit.

This program illustrates the basic concept of databases used for two-dimensional geometric modeling purposes. It also shows the nature of the information required to generate the profile of an object and how this information is used to provide a graphical representation of an object's geometry. ❑

5.6 ▪ Geometric Databases for Three-Dimensional Objects

The description of three-dimensional objects is much more complicated than that of two-dimensional objects. In order to produce databases for three-dimensional objects, it is often necessary to include the following three sets of information in a geometric data file [8]:

1. *Geometric information*: This information includes the coordinates of the vertices and the mathematical equations for all the edges or faces.
2. *Topological information*: For solids involving convex/concave surfaces, the topological description must be given in addition to the geometric information.

3. *Ancillary information*: This information includes other descriptions of the object, such as the color of the surface and the degree of transparency of the model (see Plate 2).

Similar to two-dimensional objects, three-dimensional solids can be described by assembling certain geometric primitive types such as boxes, cylinders, and cones. A preferable method is to have the solid be made up of a number of edges, faces, and vertices, with proper descriptions of their positions in space and their topological connections. The task of describing correct topological connections for these primitives appears to be the most difficult task in the construction of a three-dimensional geometric database. Much of this difficulty is attributable to the fact that edges, vertices, and faces must be stored as entities with unique identities within the particular object.

Three-dimensional objects also introduce many new constraints because, although vertices and faces are entities in the database, they may not exist other than as part of an object. Pointers such as the ones that describe the relationship among various edges must be valid. The topological arrangement of faces, edges, and vertices must be consistent with a closed solid. The geometry must also be consistent so that the object does not intersect itself. Finally, there are design constraints that the solid model must satisfy.

When the geometry of the object is retrieved for viewing or transferring to a hard disk for storage, it is necessary to ensure that the output of the edges and vertices, plus any other related information such as the surface finish, is correct. A few useful tips for efficiently describing such relationships and the algorithms for constructing three-dimensional solid geometries are available in references [8], [10], and [11].

5.7 ▪ The IGES Standard

The principles of geometric databases for two-and three-dimensional objects have been outlined in the two preceding sections. Geometric information for objects can be stored, modified, and retrieved by databases constructed on these principles by software engineers. The PROFILE program described in Section 5.5 illustrates such an application. In reality, however, the task involved in such construction is by no means trivial. There is clearly a need for a standard geometric database that can be readily adapted to various application (or CAD) programs with proper pre- and postprocessors. The concept is very much like that of GKS for graphic primitives, as presented in Chapter 3.

A standard package evolved over the years for geometric databases is the *Initial Geometric Exchange Specification* (*IGES*) *standard*. It allows the transferring of geometric data between different graphic systems.

The basic structure of the IGES is built on the information on entities as defined in Section 5.5. Three principal types of entities are contained in the data file: geometry entities, annotation entities, and structural and definition entities.

1. *Geometry entities*: These entities include the categories of points, lines, arcs, and curves for two-dimensional objects and the geometric primitives required for three-dimensional objects such as those described in the foregoing section. Algorithms for creating these primitives, including those for two-dimensional objects, are provided either by the user, or by a graphic package such as the GKS standard.

2. *Annotation entities*: All information pertinent to engineering drawings, such as dimensions, labels, and notes, should be included in these entities.

3. *Structural and definition entities*: These entities include the definitions of all individual components and their relationships to the overall structure that is being modeled.

Because IGES is intended to be used as a standard geometric database for CAD applications, efficient storage and retrieval of data comprise a prime consideration. It is thus conceivable that the file structure of the IGES is built toward a network data model, as described in Section 5.3.

The IGES is intended to be used as an essential element for geometric data links in many commercial CAD/CAM systems. It also serves as a vital geometry interface between application programs for automatic mesh generation in finite element analyses and CAD involving finite element analysis [12]. The latter two topics will be described in detail in Chapter 6.

5.8 ▪ Basic Requirements for Design Database Management Systems

In view of the differences between design and business databases, straightforward adoption of the three common data models, described in Section 5.3, as the basis for the design database management system (DDBMS) is inadequate. More sophisticated and complex DBMSs have to be constructed for these databases.

Three unique characteristics of the DDBMS should be considered when constructing such a system. A DDBMS should be (1) dynamic in information handling, (2) multiuser oriented, and (3) multilayer structured.

❑ DYNAMIC INFORMATION HANDLING

Information to be handled in a DDBMS is continuously developed, supplied, and refined during various stages of the design process, as described in Chapter 1. Information such as customers' specifications, materials suppliers and specifications, and design and fabrication methods may change or be modified from time to time. A good DDBMS should accommodate such changes with considerable readiness.

❑ MULTIUSER ORIENTED

Often, a large portion of the information stored in a design database may be common to several products. One obvious example is the materials data which are likely to be used by more than one user at any given time. A design database at Chrysler Corporation was shared by more than 3,000 users from 550 work stations at various sites [13]. Therefore, the DDBMS should be constructed in such a way that access to the system by multiple users does not jeopardize the integrity of the data structure, and yet the system can still provide for fast retrieval of necessary information.

❑ MULTILAYER STRUCTURE

A design database is most likely to involve information related to both graphical and textual properties of products. It must be able to provide several levels of information in data files of distinct natures. For example, the geometric description of an object may be made up of a number of elements, and each of these elements needs to be described. The relationships between these distinct information data files are not one-to-one, but many-to-many (e.g., face to edges, shaft to wheels, etc.).

In order to satisfy the basic requirements associated with the aforementioned unique characteristics, a DDBMS must include the following basic modules:

- Access control modules for efficient access to the design database by multiple users
- Concurrent control modules for temporary storage of data being used by each individual user
- Temporary working modules for individual users

Various safety checks, of course, are to be built into the DDBMS to secure the integrity of both semantic and geometric information shared by multiple users. Easy maintenance and diagnostic checks are also among many other desirable elements of the system.

5.9 ▪ Databases for Integrated Engineering Systems

A close examination of Figure 5.1 relating to the production of four-wheel-drive tractors will reveal that almost all the data required to produce this machine are associated with the production process and its control or with the manufacturing aspects of the product. While design data files are required for those parts produced in-house, these constitute a relatively minor portion of the overall database. A similar proportion between the amounts of design data and manufacturing data exists in the production of most other complex machines. It is thus desirable to develop intelligent DBMSs for an integrated CAD and CAM system in order to substantially enhance productivity. Integrated CAD and CAM database systems will allow direct and fast transmission of data from design to manufacturing, without having to handle enormous numbers of drawings and pages of production documentation as in the traditional way.

The scope of an integrated CAD/CAM database is by no means clearly defined at this moment in time, just as with the DDBMS. However, a summary of the major elements in such systems is given below [6, 14].

Before we deal with the subject of an integrated CAD and CAM database, a close look at the major elements of a CAM system alone is quite useful. We will then try to integrate this system with the CAD database presented in earlier sections.

The major data modules in a common CAM database may be categorized into four models: (1) material model, (2) production model, (3) parts model, and (4) assembly model.

❑ MATERIAL MODEL

Data to be stored in this model include identifications and inventories of the materials used to produce various parts. Also included is the status of materials as related to orders and shipments by various suppliers.

❑ PRODUCTION MODEL

This data model should include all the information necessary for manufacturing the parts for the specified product. Information on the geometry and dimensions, as well as cutting tools and cutting paths, should be contained in the production model. Generally, these are determined at the design stage and stored in the design database. Retrieval of relevant information should be possible when the need arises. Additional information such as the production process and scheduling should also be included in the production model.

❑ PARTS MODEL

This model consists of information on all as-received parts. Similar to the material model, inventories of these parts and the status of ordered parts must be clearly specified in the file.

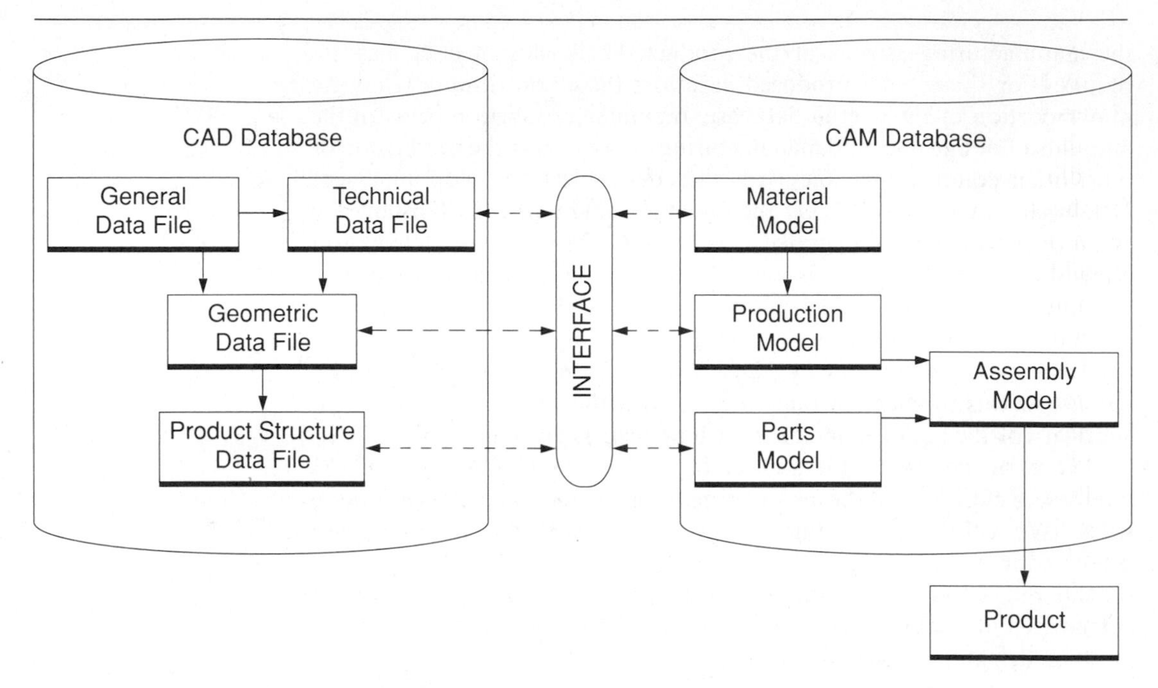

FIGURE 5.10 Structure of an integrated CAD/CAM database

❑ ASSEMBLY MODEL

As illustrated in Figure 5.1, a final product is made up of many parts. All parts, whether produced in-house or as-received, must be assembled to complete the product. Detailed assembly instructions must be available to the assembler, whether the assembly is carried out by human or by machine. The latter, of course, applies to an automated manufacturing system. In an automated assembly plant, a more sophisticated data model is necessary.

The relationship between CAD and CAM databases as an integrated system can be better explained by the diagram in Figure 5.10. The transmission of data between the two databases follows the dotted lines in the figure.

5.10 ▪ Summary

The prime objective of a computer-aided engineering system is obviously productivity enhancement. In the case of CAD, sophisticated computer software and hardware have been developed to handle the tasks of conceptual design, design synthesis and optimization, geometric modeling, analysis, and drafting, which are traditionally performed by manual labor. Many large-scale complex design tasks can now be carried out by computers with highly accurate results and at great speed. However, one factor that limits further improvement of CAD efficiency is the difficulty in handling the massive amounts of data that are supplied into and generated by the design process. Database management systems (DBMSs) that are both intelligent and effective can alleviate the problem of data flow and should therefore be treated as an essential component of CAD. As will be demonstrated in Chapter 10, intelligent design databases constitute an essential element of the next generation of CAD involving expert systems.

Theories for database management systems in business and administration are well established. Most of these systems are constructed on three common models: relational, hierarchical, and network models. Commercial DBMS packages for these applications are available.

Design database management systems (DDBMSs) are much more complex than DBMSs for business and administration. Design DBMSs are required to contain both semantic information and graphical description of products. Data to be handled by these systems are dynamic in nature as they are continuously generated and modified with the evolution of the design. These systems are also expected to be shared simultaneously by many users (i.e., designers). Each user has his or her own distinct purpose and application. The relational model is often used in conjunction with the hierarchical model in these systems.

A growing trend in the DBMS is toward the integration of CAD and CAM databases. The role that a design database can play in a simple integrated CAD/CAM system will be demonstrated in Chapter 10. The efficient transmission of data between these two types of databases is an essential element in an automated manufacturing system.

References

1. Eberlein, W., and H. Wedekind. "A Methodology for Embedding Design Databases into Integrated Engineering Systems." In *File Structures and Data Bases for CAD*, edited by J. Encarnacao and F.L. Krause, International Federation for Information Processing. Amsterdam; North-Holland, 1982.
2. Date, C.J. *An Introduction to Database Systems*, vol. 1. 3d ed. Reading, Mass.: Addison-Wesley, 1982.
3. R.A. Frost, ed. *Database Management Systems.* New York: McGraw-Hill, 1984.
4. Borkin, S.A. *Data Models: A Semantic Approach for Database Systems.* Cambridge, Mass.: MIT Press, 1980.
5. Cook, R. "Conquering Computer Clutter." *High Technology* (December 1984): pp. 60–70.

6. Codd, E.F. "A Relational Model of Data for Large Shared Data Banks." *Communications of the Association for Computing Machinery* 13, no. 6 (June 1970).
7. Zwass, V. *Programming in FORTRAN. New York:* Barnes & Noble, 1981. pp. 172–184.
8. Gardan, Y., and M. Lucas. *Interactive Graphics in CAD.* New York: Unipub, 1984.
9. Dutchak, P. "Database Management with Emphasis on Engineering Design." B.Sc. thesis, Department of Mechanical Engineering, University of Manitoba, 1989.
10. Dassler, R., J.H. Germer, F.L. Krause, and G. Pohlmann. "Databases for Geometric Modelling and Their Application." In *File Structures and Databases for CAD*, edited by J. Encarnacao and F.L. Krause. Amsterdam: North-Holland, 1982.
11. Dewey, B.R. *Computer Graphics for Engineers.* New York: Harper & Row, 1988.
12. Jayaram, S., and A. Myklebust. "Automatic Generation of Geometry Interfaces Between Applications Programs and CADCAM Systems." *Computer-Aided Design* 22, no. 1 (Jan–Feb 1990): pp. 50–56.
13. Krouse, J.K. "Engineering Without Paper." *High Technology* (March 1986): pp. 38–46.
14. Udagawa, Y., and T. Mizoguchi. "An Advanced Database System ADAM—Toward Integrated Management of Engineering Data." In *Proceedings of International Conference on Data Engineering* (April 1984): pp. 3–11.

Problems

5.1 Modify the computer program presented in Figure 5.3 in order to store and retrieve the information given in Table 5.2 for the production of a table vise.

5.2 Make any necessary revisions to the computer program given in Figure 5.3 in order to store and retrieve the following records:

Part No.	Part	Supplier	Cost
0102	3/4 nut	ABC Inc.	$ 0.50 ea.
0105	3/4 bolts	Atlas Co.	$ 0.80 ea.
0211	Lift arm	Sam Foundries	$125.00 ea.
0328	Sockets	Long Beach Shop	$ 1.20 ea.

5.3 Write a simple program to store the set of test results shown below.

Time (min.)	Pressure (Pa)	Temperature (°C)
1.0	300	275
2.0	350	325
3.0	375	400
4.0	400	600

A fourth column for the volume of the gas (i.e., volume = temperature/pressure) should be computed and added to the data file by the program.

5.4 Write a simple program to store the geometry of a thin plate with curved edges with the dimensions given below.

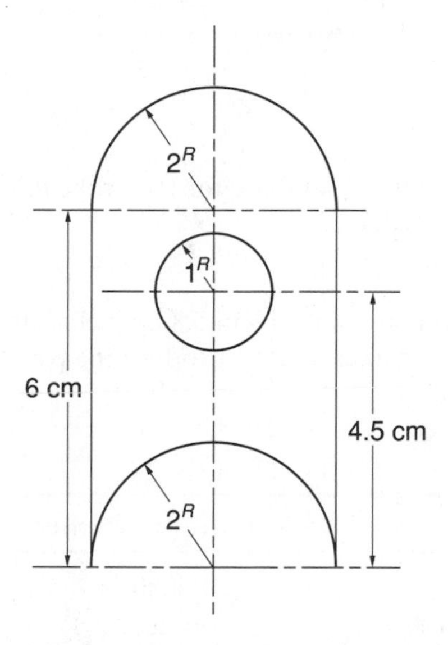

5.5 Design and construct a database for storing and retrieving graphic primitives such as line segments, arcs, and curves.

5.6 Establish a hierarchical file in place of the relational file used in the case described in Section 5.5.

5.7 Prepare a bill of materials for the following double shoe drum brake illustration. All parts are made of steel except where marked.

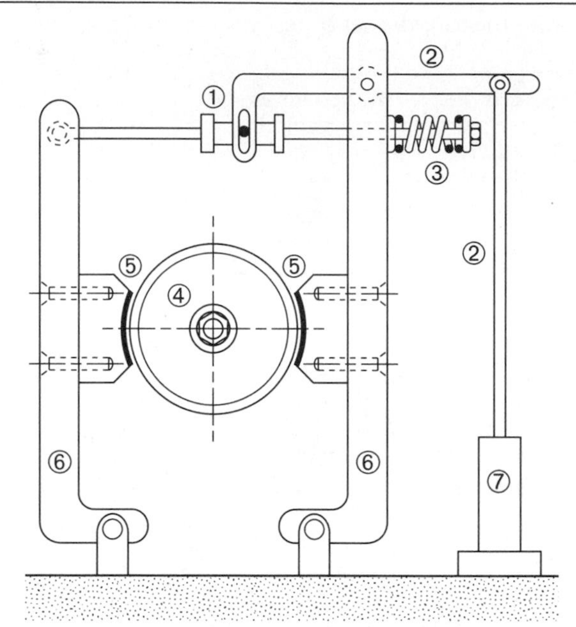

5.8 Construct a materials database for the brake in problem 5.7 using

(a) A relational model
(b) A hierarchical model

5.9 Modify the database in problem 5.8 by including the fabrication methods and machines used to produce the parts as given in the following table:

Part	Fabrication Method	Machine Used
Collar	Machined	Lathe
Connecting rod	Forged	None
Compression spring	Purchased	None
Cast-iron drum	Cast	None
Cast-iron brake lining	Cast	None
Rocking arm	Forged	Grinder
Solenoid cylinder	Cast	Lathe

5.10 Prepare input data to the PROFILE program in Example 5.2 for the construction of the geometry of the gear tooth illustrated in Figure 3.7.

CHAPTER 6

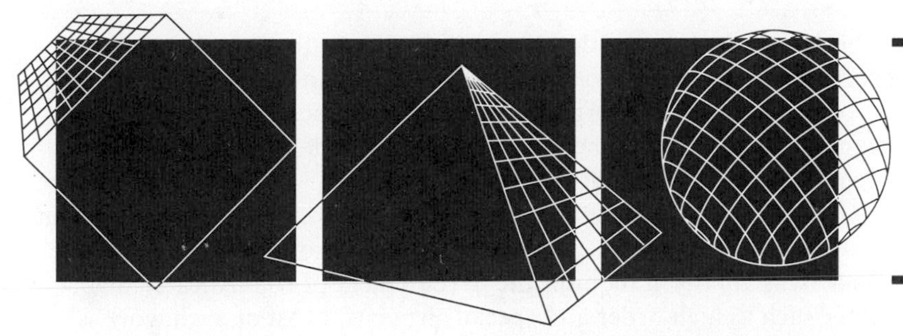

Overview of the Finite Element Method*

6.1 Introduction

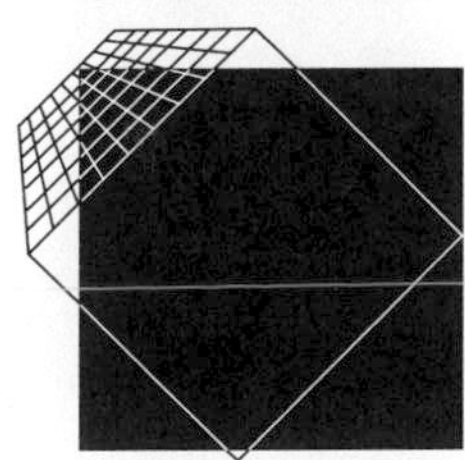

The rapid advances in the nuclear and aerospace technologies in the past two decades have made reliable stress analysis a critical factor in the design process. In recent years, there are increasing needs for the use of nontraditional materials (e.g., polymers and composites) for primary load-bearing components of aerospace crafts and vehicles in order to reduce structural weight and cost of production. Sophisticated design analysis has become more a necessity than ever.

The analysis required for the above applications clearly is multidisciplinary in nature, as will be described later, and sometimes involves the additional complexity of multiphase (e.g., solid and fluid interactions and phase change phenomena, such as welding and solidification of metals. These extremely complicated factors preclude the use of classical methods, and numerical techniques such as the finite element method (FEM) appear to be the most viable alternative solutions.

The versatility of the finite element method for the solution of practical problems involving complex geometries and loading/boundary conditions has been recognized by researchers since 1956 [1]. Several excellent books describing the basic principles of this subject have been

* A significant portion of the first half of this chapter has been published in Chapter 1 of reference [15].

published in the last two decades [2–11]. Books that are devoted to special purposes have also been made available in recent years [12–16]. Also, of course, many worthwhile conference proceedings and monographs dealing with the theory and application of this technique have been published on an ongoing basis.

Generally speaking, there are two ways of developing a finite element analysis. The first approach is to place emphasis on the accuracy of the result. This approach requires the development of sophisticated algorithms, such as high-order and special elements. Most of such work is tailored to some special purpose. The second approach is to develop general-purpose programs and commercial codes such as ANSYS, NASTRAN, and MARC analysis. Special features of these codes can be found in a survey of finite element codes [17]. These codes are versatile enough to handle just about any engineering problem, but require the users (engineers) to have substantial training in order to become proficient in their use. The proprietary nature of these commercial packages allows the engineers to use them only as "black boxes," and often the important physical sense of this powerful tool is lost.

The seemingly unlimited capability of the FEM for handling design analyses of machine components having extremely complex geometries and loading and boundary conditions makes it an important component in the CAD technology. Some commercial CAD packages have already integrated the FEM and solid modeling with a common interface and common database for both disciplines.

The objective of this book is to provide the reader with an overview, but not an in-depth description, of this major CAD component—the FEM. It is thus necessary to skip some of the fundamental subjects and principles of the finite element method. Such topics as the variational principle, integration schemes, and convergence criteria will be omitted in the text. These topics are adequately covered in the texts cited in references [2–14].

This chapter will thus present only a general description of the finite element method. In addition to bringing the reader up to date on the application of the FEM in engineering analyses, the concept of discretization described in the following section will provide the reader with an appreciation of the engineering sense of this method. The steps of the general finite element analysis will be presented in such a way as to include the formulation of some key equations.

6.2 ▪ The Concept of Discretization

Many engineering analyses involve physical quantities, such as stress (or strain) everywhere in a solid caused by applied forces (or pressure), temperature at every point in a solid caused by heat sources (or sinks),

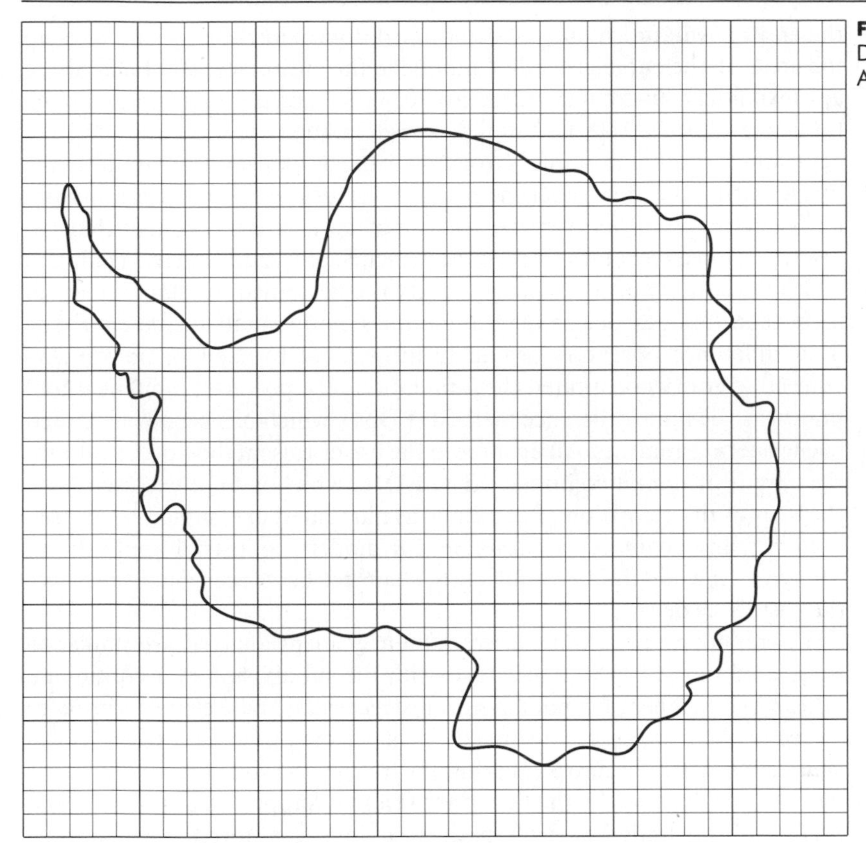

FIGURE 6.1
Discretized map of Antarctica

velocities at every location in a fluid due to difference in heads or pressure, and amplitudes of vibration at different parts of a solid caused by sources of excitation. One will recognize the fact that the substances (either solids or fluids) being considered are assumed to be continuous or continua. Therefore, we are really asked to solve problems involving an infinite number of solution points, which implies infinite degrees of freedom (DOF).* Common sense suggests to us that the degree of difficulty of the solution to a problem is proportional to the number of DOF involved. The FEM that is based on the concept of discretization of continua for converting a problem involving infinite DOF to one with a finite number of DOF is obviously a vital alternative way of solving such a problem.

The use of the discretization concept in science and technology is not new. One such common practice was to measure the areas enclosed by close curves before the invention of the planimeter. As illustrated in Figure 6.1,

* A degree of freedom can be regarded as one desired unknown quantity.

the area of Antarctica may be determined approximately by summing up the areas of the cubic grids that subdivide the entire domain. Each of the approximately 404.5 cubes represents 40,000 sq. km, giving an approximate total area of 16,180,000,000 sq. km, which is about 13.6 percent over the actual size. Another example is to estimate the work produced by an engine by measuring the area under the pressure versus the volume diagram from an indicator [16]. Many readers may not be aware of the fact that the discretization concept was indeed the principle of integration in calculus, in which the entirety of a physical quantity is determined by summing up all incremental values determined at infinitesimally small increments. The finite difference operators originally formulated by Newton in the seventeenth century constituted the foundation of a powerful numerical tool known as the finite difference method (FDM), which has been widely used in engineering analyses. An upsurge in the use of this method occurred with the advent of digital computers. The FDM, which is built exclusively on the basis of discretization of continua, as illustrated in Chapter 2, has been used in many engineering disciplines for modern industrial applications. Many worthwhile monographs on this subject have been published over the past three decades.

Discretization procedures used in finite element analysis are similar to those used in the FDM. Each starts with the subdivision of a continuum into a finite number of subdomains called *elements*. These elements are connected at the corners, or in some cases at selected points on edges, called *nodes*, as shown in Figure 6.2. In this illustration, the solid has a finite shape that can be described by the boundary Ω and is supported at points A and B. The solid is under a set of specified actions (e.g., forces, heat sources) which can be described by

$$\{P\} = P_1, P_2, P_3, \ldots$$

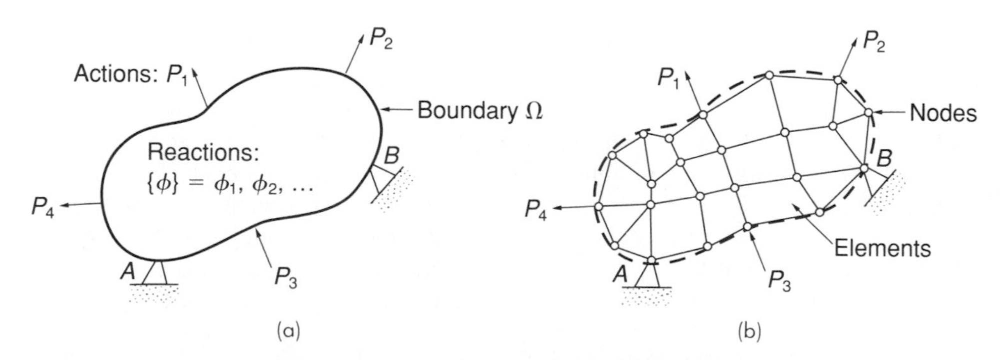

FIGURE 6.2 Discretization of a solid: (a) original body; (b) discretized body

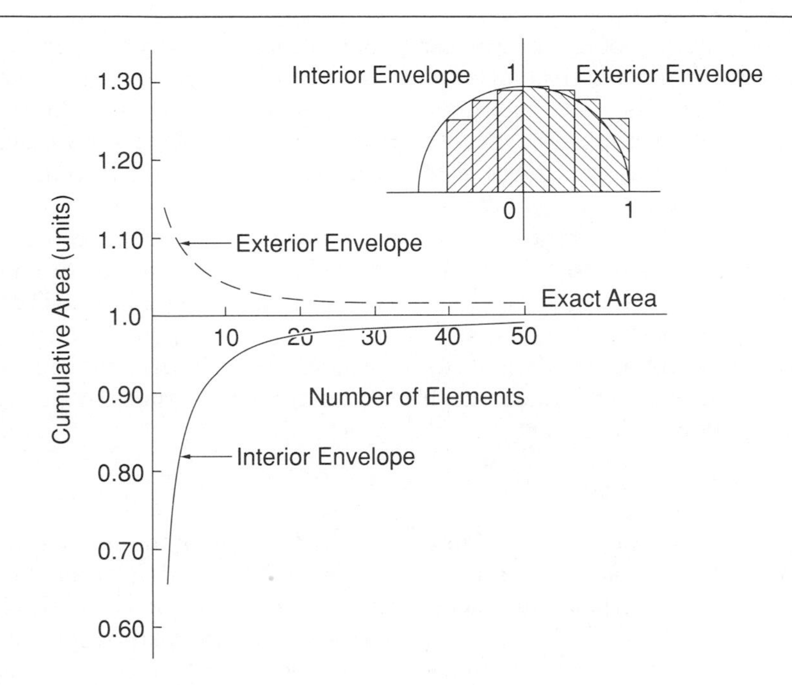

FIGURE 6.3 Convergence of an area computation by discretization

By virtue of these actions, there exist the required reactions (e.g., deflections, strains, stresses, temperatures) to be described by

$$\{\phi\} = \phi_1, \phi_2, \phi_3, \ldots$$

Since there are an infinite number of points in the solid bounded by Ω and each point is associated with a set of values $\{\phi\}$, there are obviously an infinite number of ϕ values to be determined.

The original solid in Figure 6.2(a) is subdivided into a finite number of subdomains (elements) with certain shapes (triangular and quadrilateral plates, in this case), and these elements are interconnected at the nodes. After such discretization, the original continuum is no longer a continuous body, but contains a number of discrete pieces connected at the corners and/or edges in some cases. Another noticeable change is that the originally continuous curved boundary, Ω, has now become segments of straight lines.* It is obvious that the discretized body is geometrically similar, but not identical, to the original solid. The degree of similarity of the two bodies, of course, depends on the number of elements used in the discretized body (model). The more elements used in the model, the closer it gets to the original geometry, and thus the result is closer to the exact solution, as illustrated by a simple example in Figure 6.3. The reader will

* Curved "edges" are allowed for certain special elements. In that case, the curved surface of the original solid may be preserved after the discretization.

find that the area of a semicircle can be approximated by summing up the areas of the dividing rectangles; one set is entirely enclosed within the circular boundary, and another set is approached from the exterior boundary. Both sets can be shown to asymptotically converge to the exact area as an increasing number of the elements are used in the discretized model.

The unique advantage of discretization is that after such a process, analysis needs only to be applied to the individual elements of certain simple geometries (e.g., triangular plates), rather than to the entire solid of complex geometry.

The physical quantities, $\{\phi\}$, to be determined in the element can be obtained from these values at the nodes through certain shape or interpolation functions, or

$$\{\phi\}^e = [N]\{\phi\}^n$$

where $\{\phi\}^e$ and $\{\phi\}^n$ are the respective physical quantities in the element and the corresponding nodes, and $[N]$ is the shape or interpolation function. The interpolation functions are usually derived on the basis of local coordinates of the element. However, a transformation of this function from the local coordinates to the global coordinate system* is performed afterward.

Since the above equation requires solutions only for the individual elements or at the nodes and there are now only a finite number of these elements and nodes in the discretized model, the original problem involving an infinite number of DOF has thus been reduced to a finite number of DOF by means of the discretization process. Solutions of $\{\phi\}$ in the discretized body are hence substantially more attainable.

6.3 ▪ Application of the Finite Element Method in Engineering Analysis

The concept of discretization on which the FEM is based is by no means new, as described in the foregoing section. The term *finite element method*, however, was not used until the 1950s [1]. It is not surprising that this method was first used for the stress analysis of aircraft structures because sophisticated stress analysis is traditionally emphasized in the aerospace industry. The potential applications of this method were soon recognized by researchers and engineers involved in other disciplines and industries,

* Global coordinates means that the origin of the coordinates is set outside the element.

and the rate at which industry has since adopted this method has been staggering.

While a complete list of the applications of this method is not possible by any standard, a few examples including reference sources are presented here. Those who are interested in exploring these subjects in depth may refer to these references.

1. Industrial Applications.

- *Aerospace:* stress analysis of aircraft structural and engine components; aerodynamic and performance analyses; mechanism and linkage analyses; simulation of aircraft response to various applied loads
- *Automobile manufacturing:* stress analysis of vehicle structures; dynamic and impact analyses and simulations
- *Shipbuilding:* stress analysis of structures for components and assembled products; hydrodynamic performance analysis
- *Nuclear power:* thermomechanical stress analysis of reactor components; thermohydraulic performance analysis of systems; simulation of normal operating and accident conditions
- *Steel and metal processing:* stress and thermal analyses of process equipment; prediction of residual stresses and deformation on processed products; macroscopic evaluation of heat treatment procedure
- *Construction:* stress analysis of building structures, concrete foundations, underground tunnels, bridges, etc.
- *Resource development and mining:* stress analysis of excavation equipment, geotechnical materials; analysis of response of geological materials to static (e.g., hydraulic) or dynamic (e.g., explosive) loads; thermal mechanical-hydraulic analyses in cases such as geothermal energy development, structural integrity analysis of mine shafts.

2. Scientific and Engineering Disciplines. The use of the FEM in its earlier years in disciplines other than the stress analysis of solid structures has been demonstrated in several excellent books, including those by Desai [9] and Zienkiewicz [3]. Following is a list of some more recent publications in various disciplines:

- Elastic-plastic stress analysis [4, 13]
- Heat and mass transfer [7, 18, 19]
- Thermal stress analysis [15]
- Dynamics and vibrations [8]
- Fluid mechanics [12, 20–22]
- Diffusion and mass transport [23]
- Soil mechanics [24, 25]
- Concrete engineering [26]
- Geomechanics [14, 27–31]

- Biomechanics and bioengineering [32–36]
- Material science [37]
- Mechanisms and linkages [38, 39]
- Physical science [40, 41]

The above list, of course, is by no means complete, and it is only natural for one to assume that it will grow longer and diversify into many more branches of science and engineering applications as time goes by.

6.4 ▪ Steps in the Finite Element Method

The wide range of applications of the FEM in engineering analysis has been illustrated in the foregoing section. It is unrealistic for anyone to attempt to establish a set of standard procedures for all the computations for the problems described above. However, as a general guideline, most finite element analyses follow eight steps.

❑ STEP 1: DISCRETIZATION OF THE REAL STRUCTURE

Depending on the nature of the structure, a variety of types of elements are available. Figure 6.4 shows typical and commonly used elements. The *bar elements* are frequently used in modeling trusses and frames, and the *plane elements* are suitable when structures have planar geometries or when the variation of the physical quantities is limited to a plane (i.e., the plane stress or plane strain case). For structures with an axis of geometric symmetry (e.g., cylinders and pressure vessels), the *torus elements* can be used. Finally, for structures with extremely complicated geometry, when a three-dimensional model becomes a necessity, the *tetrahedron* or *hexahedron elements* are commonly used. Many other types of elements (e.g., *beam* and *shell elements*) are developed for special applications. Some elements have additional nodes on the sides, which can either be curved or straight. Many isoparametric elements* follow this description. Modern finite element computer programs allow various combinations of different element types, as will be presented in Chapter 7, and hence make finite element modeling highly versatile.

* The interpolation functions associated with isoparametric elements are formulated on the basis of the parametric coordinates in the element.

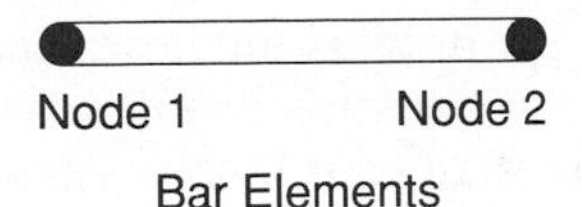

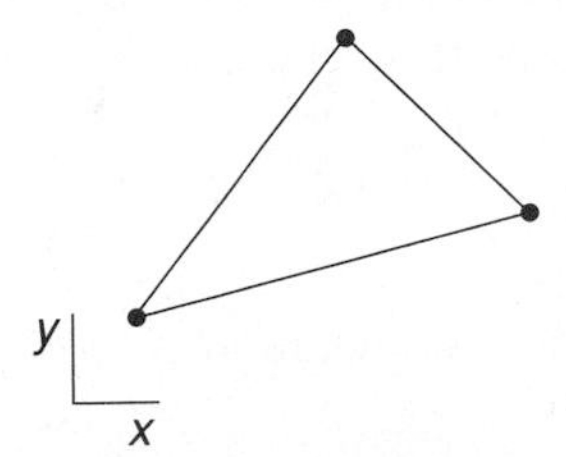

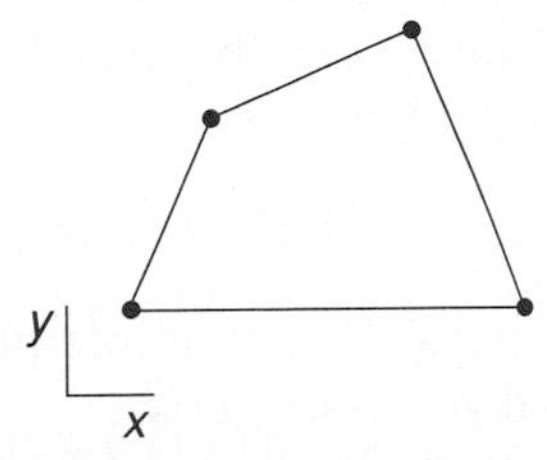

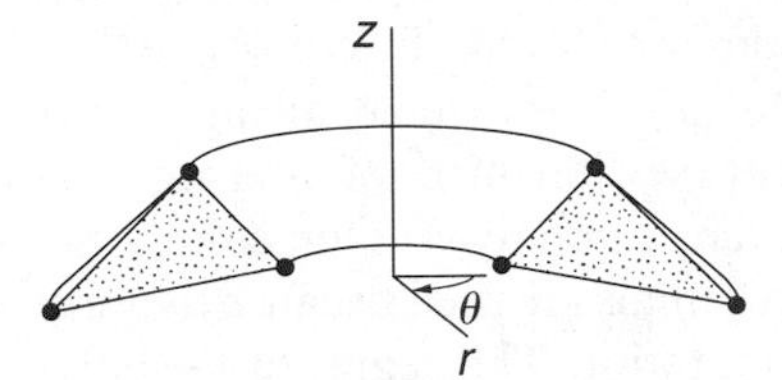

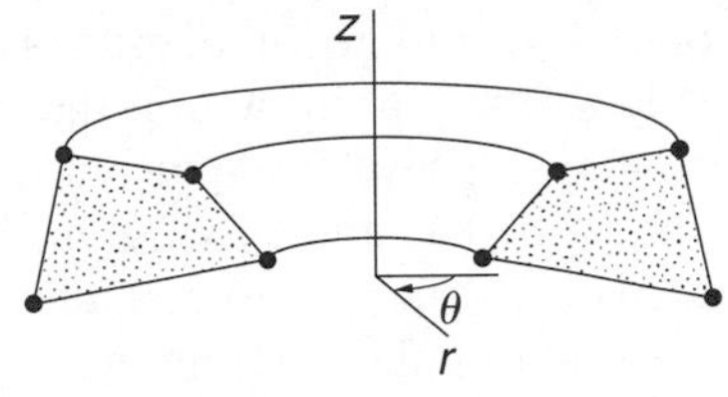

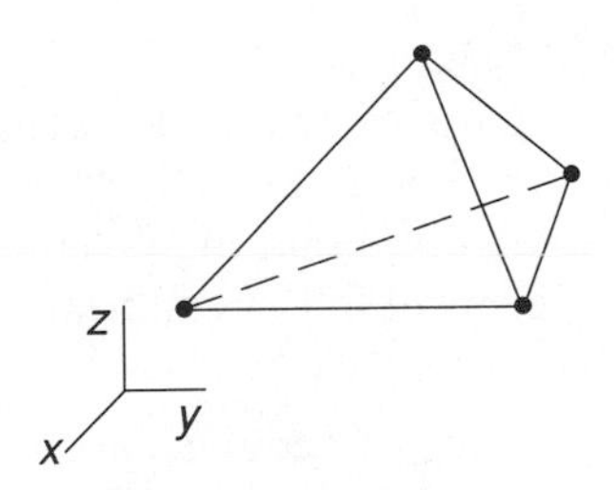

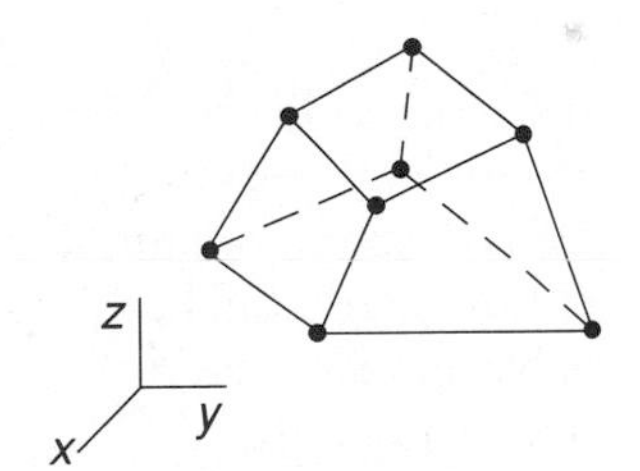

FIGURE 6.4
Typical simplex elements

❑ STEP 2: IDENTIFY THE PRIMARY UNKNOWN QUANTITIES AND AN APPROPRIATE INTERPOLATION FUNCTION

Again, depending on the nature of the problem, the primary unknown quantity involved in the solution varies from case to case. One may choose the displacement component as the primary unknown quantity in a stress analysis problem, temperature in a heat conduction analysis, and velocity in a fluid flow problem.

The primary unknown quantity in an element usually is represented by a vector quantity, $\{\phi(\vec{r})\}$, in which $\vec{r}$ represents coordinates [e.g., (x, y, z) in a Cartesian coordinate system or (r, θ, z) in a cylindrical polar coordinate system] in the elements. This quantity must be related to the corresponding values at the associated nodes via an interpolation (or shape or trial) function chosen by the analyst. Mathematically, this relationship can be expressed as

$$\{\Phi(\vec{r})\} = [N(\vec{r})]\{\phi\} \tag{6.1}$$

in which $[N(\vec{r})]$ is the interpolation function and $\{\phi\}$ are the corresponding nodal values of $\{\Phi(\vec{r})\}$.

The reader should take note from the above equation that there is a relationship between the primary unknown quantities in the element and the equivalent values of these quantities at the associate nodes. Conversely, once those unknown quantities at the nodes are solved, the corresponding values at any point within the element can be computed by means of the relationship given in equation (6.1). There are several forms of interpolation functions that one may use in the finite element analysis. One commonly used type is the polynomial function. Polynomials are used because they are easily manipulated in the subsequent computation. The degree of the polynomial chosen is related to the number of nodes in the element and, to some extent, to the nature of the problems, as will be discussed in Chapter 7.

EXAMPLE ■ 6.1

Referring to the bar element illustrated in Figure 6.5, two nodes were assigned at $x = x_1$ and $x = x_2$ for node 1 and node 2, respectively. Derive the interpolation function on the assumption that the variation of the longitudinal displacement in the bar element follows a linear polynomial function.

If we let the displacement in the bar element be $U(x)$ [equivalent to $\{\Phi(\vec{r})\}$ in equation (6.1)], then the simplest form of polynomial func-

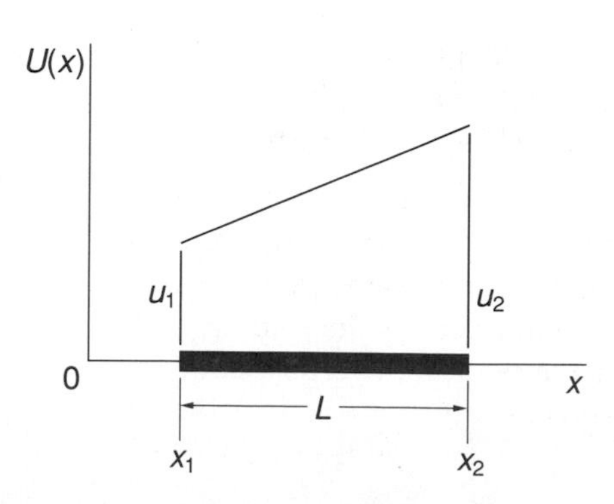

FIGURE 6.5 Linear interpolation function for a bar element

tion describing this physical quantity becomes

$$U(x) = \alpha_1 + \alpha_2 x \tag{6.2}$$

where α_1 and α_2 are two arbitrary constants that can be evaluated from the nodal conditions.

We further let u_1 and u_2 be the respective corresponding displacements at node 1 and node 2. These are unknown quantities at the nodal locations. The reader should be reminded that they are discrete values. The following two simultaneous equations can be derived by substituting u_1 and u_2 for $U(x)$ in equation (6.2) with corresponding nodal coordinates x_1 and x_2:

$$u_1 = \alpha_1 + \alpha_2 x_1$$

$$u_2 = \alpha_1 + \alpha_2 x_2$$

from which one may solve for α_1 and α_2:

$$\alpha_1 = -\frac{x_2}{x_1 - x_2} u_1 + \frac{x_1}{x_1 - x_2} u_2$$

and

$$\alpha_2 = \frac{1}{x_1 - x_2} u_1 - \frac{1}{x_1 - x_2} u_2$$

The following relationship between the element quantity, $U(x)$, and the nodal quantities, u_1 and u_2, can be derived by substituting α_1 and α_2 in the above expressions into equation (6.2):

$$U(x) = \left\{ \frac{x - x_2}{x_1 - x_2} \quad -\frac{x - x_1}{x_1 - x_2} \right\} \begin{Bmatrix} u_1 \\ u_2 \end{Bmatrix} \tag{6.3}$$

By comparing equation (6.3) with equation (6.1), we conclude that the interpolation function has the form

$$[N(x)] = \left\{ \frac{x - x_2}{x_1 - x_2} \quad -\frac{x - x_1}{x_1 - x_2} \right\}$$ ❑

EXAMPLE ■ 6.2

Derive the interpolation function for the triangular plate element illustrated in Figure 6.6. The primary unknown quantity, $\Phi(x, y)$, in the element can mean a temperature field, pressure variation, etc. Assume the coordinates of the three associate nodes are specified as (x_1, y_1), (x_2, y_2), and (x_3, y_3) with respective nodal values of ϕ_1, ϕ_2, and ϕ_3. Once again, we assume the variation of $\Phi(x, y)$ in the element follows a simple linear polynomial function.

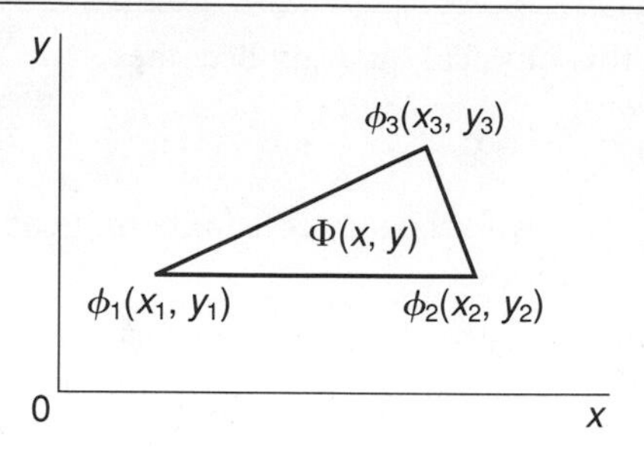

FIGURE 6.6
Interpolation function in a two-dimensional domain

Following the element configuration illustrated in Figure 6.6, the function $\Phi(x, y)$ can be assumed to vary within the specified domain according to a linear function:

$$\Phi(x, y) = \alpha_1 + \alpha_2 x + \alpha_3 y = \{1 \quad x \quad y\} \begin{Bmatrix} \alpha_1 \\ \alpha_2 \\ \alpha_3 \end{Bmatrix}$$

$$= \{R(x, y)\}^T \{\alpha\} \tag{6.4}$$

The coefficients α_1, α_2, and α_3 are constants that can be determined by substituting the coordinates of the nodes with specified nodal values ϕ_1, ϕ_2 and ϕ_3 to give

$$\phi_1 = \alpha_1 + \alpha_2 x_1 + \alpha_3 y_1$$
$$\phi_2 = \alpha_1 + \alpha_2 x_2 + \alpha_3 y_2$$
$$\phi_3 = \alpha_1 + \alpha_2 x_3 + \alpha_3 y_3$$

or in a matrix form:

$$\{\phi\} = [A]\{\alpha\} \tag{6.5}$$

and
$$\{\alpha\} = [A]^{-1}\{\phi\} = [h]\{\phi\} \tag{6.6}$$

The matrix $[A]$ in the above expressions contains the specified coordinates of the given nodes as

$$[A] = \begin{bmatrix} 1 & x_1 & y_1 \\ 1 & x_2 & y_2 \\ 1 & x_3 & y_3 \end{bmatrix}$$

The inversion of the above matrix, $[A]^{-1} = [h]$ can be readily performed to give

$$[h] = \frac{1}{|A|}\begin{bmatrix} x_2y_3 - x_3y_2 & x_3y_1 - x_1y_3 & x_1y_2 - x_2y_1 \\ y_2 - y_3 & y_3 - y_1 & y_1 - y_2 \\ x_3 - x_2 & x_1 - x_3 & x_2 - x_1 \end{bmatrix} \tag{6.7}$$

where

$$\begin{aligned} |A| &= \text{the determinant of the elements in } [A] \\ &= (x_1y_2 - x_2y_1) + (x_2y_3 - x_3y_2) + (x_3y_1 - x_1y_3) \\ &= \text{twice the area of the triangle bounded by } \phi_1, \phi_2, \text{ and } \phi_3 \end{aligned}$$

By substituting equation (6.7) into (6.6) and then (6.4), the function $\Phi(x, y)$ can be evaluated by the three nodal quantities ϕ_1, ϕ_2, and ϕ_3, or $\{\phi\}$, to be

$$\Phi(x, y) = \{R(x, y)\}^T[h]\{\phi\} \tag{6.8}$$

The interpolation function for this case can be expressed by comparing the above expression with that shown in equation (6.1) and results in the following form:

$$[N(x, y)] = \{R(x, y)\}^T[h] \tag{6.9}$$

where the matrix $\{R(x, y)\}^T = \{1 \quad x \quad y\}$ and $[h]$ is given in equation (6.7).

The interpolation function $[N(x, y)]$ given in equation (6.9) is a linear function of x and y, which is the simplest among all such functions known to exist. ❑

EXAMPLE ■ 6.3

A triangular plane element is situated in the xy plane with nodes I, J and K, as shown in Figure 6.7.

(a) Derive the interpolation function $[N(x, y)]$ based on an assumed linear variation of the primary unknown quantity, $T(x, y)$, in the element.

(b) If the temperature T has the values $T_i = 500°\text{C}$, $T_j = 400°\text{C}$, and $T_k = 300°\text{C}$ at the three nodes I, J, and K, what is the temperature at point P in the element with a coordinate of (4, 1)?

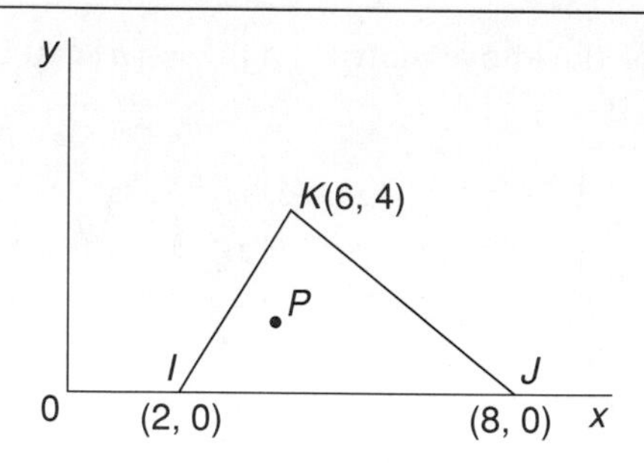

FIGURE 6.7
A triangular element

(a) We will first evaluate the area of the element to be $A = 12$ units. The interpolation function $[N(x, y)]$ can be expressed by using equation (6.9) with the $[h]$ matrix to be evaluated following equation (6.7) as

$$[h] = \frac{1}{2A}\begin{bmatrix} x_j y_k - x_k y_j & x_k y_i - x_i y_k & x_i y_j - x_j y_i \\ y_j - y_k & y_k - y_i & y_i - y_j \\ x_k - x_j & x_i - x_k & x_j - x_i \end{bmatrix}$$

in which (x_i, y_i), (x_j, y_j), and (x_k, y_k) are the respective coordinates of nodes I, J, and K.

Thus, by substituting appropriate nodal coordinates into the above expression, one may obtain

$$[h] = \frac{1}{12}\begin{bmatrix} 16 & -4 & 0 \\ -2 & 2 & 0 \\ -1 & -2 & 3 \end{bmatrix}$$

The interpolation function for the element can be evaluated to be

$$\begin{aligned} [N(x, y)] &= \{R(x, y)\}^T[h] \\ &= \{1 \quad x \quad y\}\frac{1}{12}\begin{bmatrix} 16 & -4 & 0 \\ -2 & 2 & 0 \\ -1 & -2 & 3 \end{bmatrix} \\ &= \left\{\left(\frac{4}{3} - \frac{x}{6} - \frac{y}{12}\right) \quad \left(-\frac{1}{3} + \frac{x}{6} - \frac{y}{6}\right) \quad \frac{y}{4}\right\} \end{aligned}$$

(b) The temperature distribution in the element, $T(x, y)$, can be expressed as

$$T(x, y) = [N(x, y)]\{T\}$$

in which $\{T\}$ represents the temperatures at the nodes. We thus have the temperature in the element as

$$T(x, y) = \left\{ \left(\frac{4}{3} - \frac{x}{6} - \frac{y}{12}\right) \quad \left(-\frac{1}{3} + \frac{x}{6} - \frac{y}{6}\right) \quad \frac{y}{4} \right\} \begin{Bmatrix} T_i = 500 \\ T_j = 400 \\ T_k = 300 \end{Bmatrix}$$

$$= 500\left(\frac{4}{3} - \frac{x}{6} - \frac{y}{12}\right) + 400\left(-\frac{1}{3} + \frac{x}{6} - \frac{y}{6}\right) + 300\frac{y}{4}$$

The temperature at point P with coordinates of $x = 4$ and $y = 1$ thus becomes 433.3°C. ❑

❑ STEP 3: DEFINE THE RELATIONSHIP BETWEEN ACTIONS AND REACTIONS

The laws of physics usually provide certain relationships between the *actions* on a substance and the induced *reactions* from the same substance. Table 6.1 indicates some of these physical quantities present in common engineering analysis. The physical law that relates $\{F\}$ and $\{u\}$ in the static stress analysis case is the minimization of potential energy, which is a function of these two quantities. The Fourier law can be used to relate $\{Q\}$ and $\{T\}$ in the heat conduction analysis.

❑ STEP 4: DERIVE THE ELEMENT EQUATIONS

Element equations relate the reaction in terms of the primary unknown and the action that causes the reaction. There are generally two distinct methods used in the FEM for the derivation of these equations—namely, the *Rayleigh-Ritz method* based on the variational principle and the *Galerkin weighted residual method.* The latter method proves to be more practical for problems that can be completely described by a set of

TABLE 6.1
Actions and reactions

Applications	Actions: $\{P\}$	Reactions: $\{\Phi\}$
Stress analysis	Forces $\{F\}$	Displacements $\{u\}$, Strains $\{\varepsilon\}$, Stresses $\{\sigma\}$
Heat conduction	Thermal forces $\{Q\}$	Temperatures $\{T\}$
Fluid flow	Pressure or heads $\{p\}$	Velocities $\{V\}$

differential equations. Many field problems, such as heat conduction-convection and fluid flow problems, can be handled in this way. Mathematical formulations of these two distinct methods are beyond the scope of this book. The reader is referred to reference [16] for detailed descriptions.

A general form of the *element equations* can be shown as follows:

$$[K_e]\{q_e\} = \{Q\} \tag{6.10}$$

where

$[K_e]$ = element stiffness matrix
$\{q_e\}$ = vector of primary unknown quantities at the nodes
$\{Q\}$ = vector of nodal forcing parameters

❑ STEP 5: DERIVE OVERALL STRUCTURE STIFFNESS EQUATIONS

This step of the analysis assembles all individual element equations to provide *stiffness equations* for the entire structure, or mathematically

$$[K]\{q\} = \{R\} \tag{6.11}$$

in which

$[K]$ = overall stiffness matrix = $\sum_{e=1}^{m} [K_e]$

m = total number of elements in the discretized model

$\{R\}$ = assemblage of resultant vector of *applied* nodal forcing parameters

$\{q\}$ = vector of nodal quantities of the entire structure

It should be noted that the entries of the $[K_e]$ matrices common to other elements through nodal connection should be summed up algebraically during the assembly process. This procedure can be demonstrated by the following case illustration.

Referring to the plane structure shown in Figure 6.8, the solid plate is discretized into four triangular elements interconnected at five nodes. If we allow the plate to deform in the xy plane only, then each node will be associated with two unknown displacement components along the respective x and y coordinates. The total number of unknown displacement components in the whole system without any nodal conditions imposed on it is thus equal to $5 \times 2 = 10$, which is also the total number of degrees of

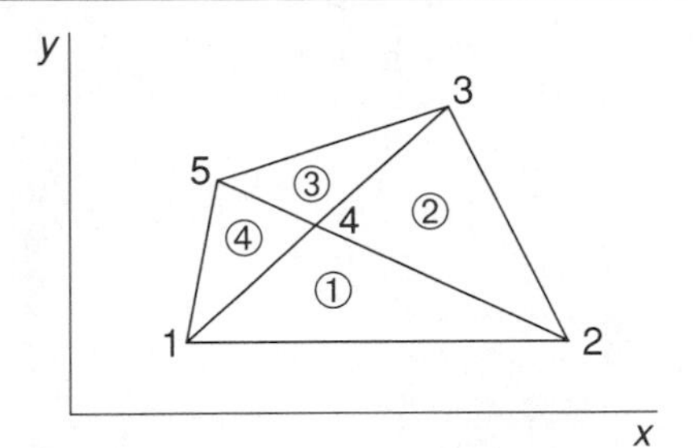

FIGURE 6.8
An assembly of elements

freedom in this case. Since the sizes of the overall stiffness and force matrices are identical to the total number of degrees of freedom involved in the problem, we have thus established the sizes of these matrices for the present case as 10 × 10 for the matrix $[K]$ and 10 × 1 for $\{R\}$.

We shall further recognize the fact that since each node has two degrees of freedom, it is conceivable that the entries under rows 1 and 2 and columns 1 and 2 of the $[K]$ matrix should be filled with the corresponding entries of the $[K_e]$ matrix related to node 1. Likewise, the entries under rows 3 and 4 and columns 3 and 4 of the $[K]$ matrix correspond to the entries of the $[K_e]$ matrix associated with node no. 2. The same rule applies to the remaining parts of the $[K]$ matrix, as well as the $\{R\}$ matrix. The layouts of the $[K]$ and $\{R\}$ matrices for this particular case are illustrated in Figure 6.9. Closed circles (dots) in this diagram represent the proper positions of the entries of the element stiffness matrices. The overall stiffness and the resultant force matrices are constructed by simple summation of these entries at the same locations.

The specified boundary conditions on $\{q\}$ or $\{R\}$ are imposed on the assembled equations in equation (6.11) before solving the equations.

❑ STEP 6: SOLVE FOR THE PRIMARY UNKNOWNS

It is apparent that equation (6.11) represents a set of simultaneous algebraic equations. The total number of equations is identical to the total number of primary unknowns, $q_i (i = 1, 2, \ldots, n)$, at the nodes. Depending on the size and degree of symmetry of the $[K]$ matrix, there are generally two methods that can be used to solve for $\{q\}$ [i.e., Gaussian elimination (Section 2.3) and matrix inversion (as described in Appendix I)].

In practical applications, it is desirable to partition the $[K]$ matrix and rearrange equation (6.11) into the following form.

$$\begin{bmatrix} K_{aa} & K_{ab} \\ K_{ba} & K_{bb} \end{bmatrix} \begin{Bmatrix} q_a \\ q_b \end{Bmatrix} = \begin{Bmatrix} R_a \\ R_b \end{Bmatrix}$$

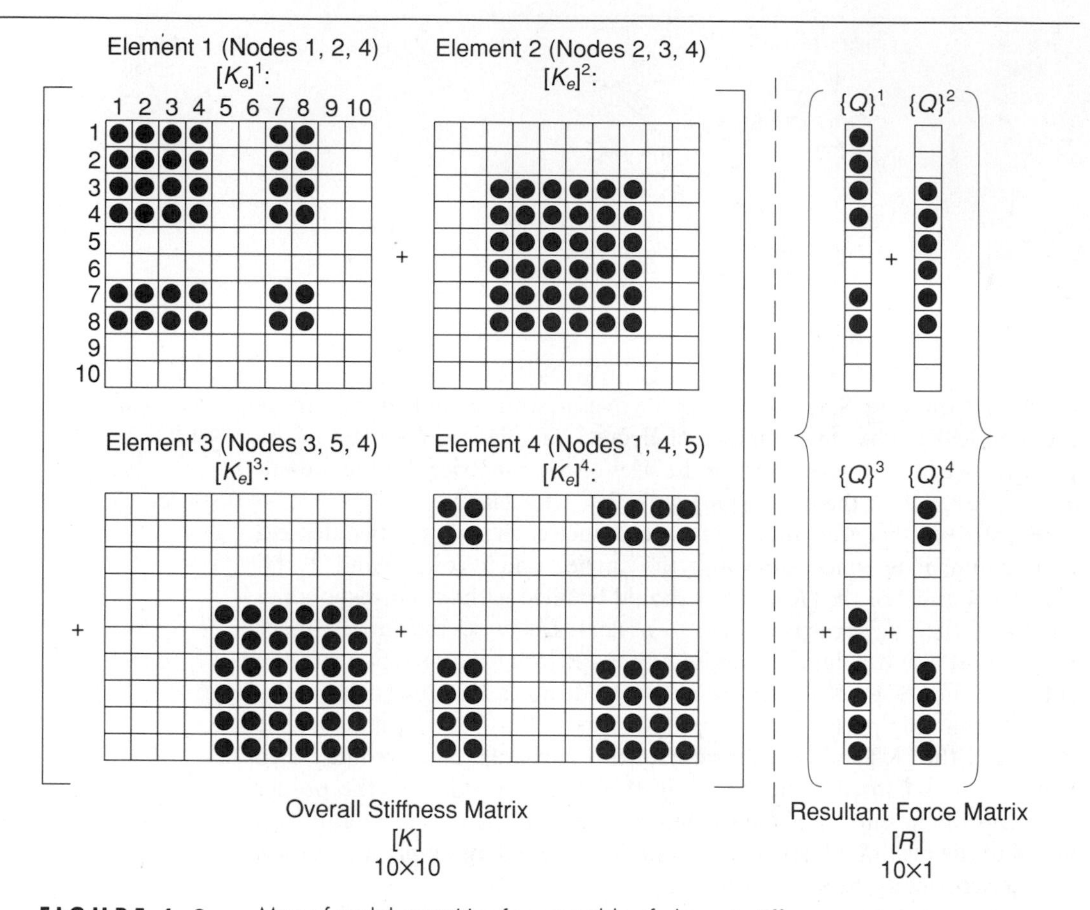

FIGURE 6.9 Map of nodal quantities for assembly of element stiffness matrix

where

$$\{q_a\} = \text{specified (known) nodal quantities}$$

$$\{R_b\} = \text{specified (known) } \textit{applied} \text{ nodal resultant forces}$$

The reader is reminded that any interchange of rows of the $[K]$ matrix must be accompanied by the interchange of respective columns.

The unknown nodal quantities, $\{q_b\}$, can thus be computed from the specified values by solving the above equation, resulting in the following expression:

$$\{q_b\} = [K_{bb}]^{-1}(\{R_b\} - [K_{ba}]\{q_a\}) \tag{6.12}$$

EXAMPLE ■ 6.4

Use the matrix partition method to solve the following set of equations:

$$\begin{bmatrix} 5 & 4 & -3 & 5 & -6 \\ 4 & -5 & 4 & -4 & 3 \\ 2 & 3 & -4 & 5 & -6 \\ -1 & -2 & 3 & -4 & 5 \\ -7 & 6 & -5 & 4 & -3 \end{bmatrix} \begin{Bmatrix} u_1 \\ u_2 \\ u_3 \\ u_4 \\ u_5 \end{Bmatrix} = \begin{Bmatrix} -15 \\ 39 \\ -37 \\ 33 \\ -55 \end{Bmatrix}$$

with specified values of $u_3 = 7$, $u_4 = 2$, and $u_5 = 6$.

By following the procedure presented above, the original matrix equation can be partitioned into the following form:

$$\left[\begin{array}{ccc|cc} -4 & 5 & 3 & -1 & -2 \\ 4 & -3 & -5 & -7 & 6 \\ 5 & -6 & -4 & 2 & 3 \\ \hline 5 & -6 & -3 & 5 & 4 \\ -4 & 3 & 4 & 4 & -5 \end{array}\right] \left\{\begin{array}{c} 2 \\ 6 \\ 7 \\ \hline u_1 \\ u_2 \end{array}\right\} = \left\{\begin{array}{c} 33 \\ -55 \\ -37 \\ \hline -15 \\ 39 \end{array}\right\}$$

The submatrices that appeared in equation (6.12) become

$$[k_{aa}] = \begin{bmatrix} -4 & 5 & 3 \\ 4 & -3 & -5 \\ 5 & -6 & -4 \end{bmatrix} \qquad [k_{ab}] = \begin{bmatrix} -1 & -2 \\ -7 & 6 \\ 2 & 3 \end{bmatrix}$$

$$[k_{ba}] = \begin{bmatrix} 5 & -6 & -3 \\ -4 & 3 & 4 \end{bmatrix} \qquad [k_{bb}] = \begin{bmatrix} 5 & 4 \\ 4 & -5 \end{bmatrix}$$

$$\{q_a\} = \begin{Bmatrix} 2 \\ 6 \\ 7 \end{Bmatrix} \quad \{q_b\}\begin{Bmatrix} u_1 \\ u_2 \end{Bmatrix} \qquad \{R_a\} = \begin{Bmatrix} 33 \\ -55 \\ -37 \end{Bmatrix} \qquad \{R_b\}\begin{Bmatrix} -15 \\ 39 \end{Bmatrix}$$

The unknown quantities in $\{q_b\}$ can be solved by using the relationship shown in equation (6.12) as

$$\{q_b\} = \begin{Bmatrix} u_1 \\ u_2 \end{Bmatrix} = \begin{bmatrix} 5 & 4 \\ 4 & -5 \end{bmatrix}^{-1} \left(\begin{Bmatrix} -15 \\ 39 \end{Bmatrix} - \begin{bmatrix} 5 & -6 & -3 \\ -4 & 3 & 4 \end{bmatrix} \begin{Bmatrix} 2 \\ 6 \\ 7 \end{Bmatrix} \right)$$

$$= \begin{bmatrix} 0.12195 & 0.09756 \\ 0.09756 & -0.12195 \end{bmatrix} \begin{Bmatrix} 32 \\ 1 \end{Bmatrix} = \begin{Bmatrix} 3.99996 \\ 2.99997 \end{Bmatrix} = \begin{Bmatrix} 4 \\ 3 \end{Bmatrix}$$ ❑

❑ STEP 7: SOLVE FOR THE SECONDARY UNKNOWNS

Once the primary unknown, $\{q\}$, is solved, other unknowns may be calculated from the available physical relationships. For example, in the case of elastic stress-deformation analysis, the primary unknowns are the displacement components at the nodes. One may readily envisage that the corresponding element displacement components may be determined by equation (6.1) with the computed nodal values. The strain components in the element can then be calculated from the displacement-strain relationships following the theory of elasticity. The stress components, of course, can be evaluated by means of the well-known generalized Hooke's law from the computed strain components.

☐ STEP 8: INTERPRETATION OF RESULTS

Results obtained from finite element computer codes used to be in tabulated form, indicating the calculated primary and all secondary physical quantities at specified nodes and elements under given applied loads. The analyst could then select critical sections of the body and evaluate these results with respect to the established design criteria as in most engineering design processes. In some instances, the computed results can be expressed graphically. Figure 6.10 shows the contours of isoclinic lines in a gear tooth made of steel. The contours produced by the finite element analysis [42] are shown to give close correlation to those

FIGURE 6.10 Stress contours in a gear tooth by the finite element analysis and photoelastic investigation: (a) isoclinics from finite element analysis; (b) isoclinics from photoelastic experiment

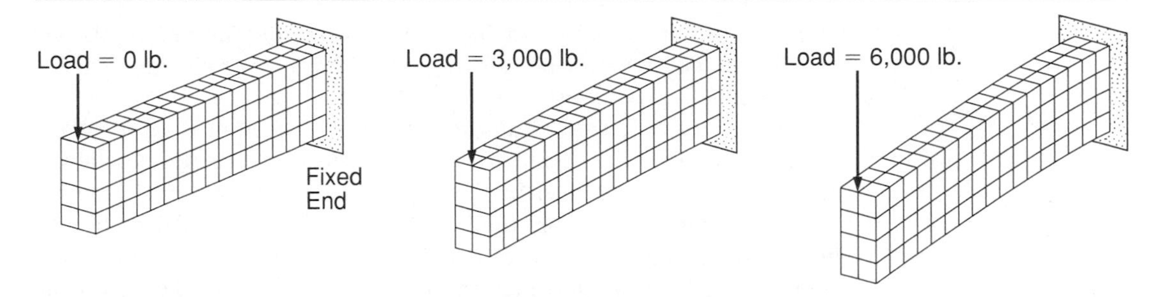

FIGURE 6.11 Graphic representation of deflection in a cantilever beam

produced by the photoelastic investigation [43]. These contours may be expressed in different colors to distinguish stresses in tension or compression, as well as critcal areas as shown in Plate 16. The critical locations and magnitudes of stress concentrations in the gear tooth are readily visible.

Expressions of finite element analysis results now take the form of visual displays of the models under different loading conditions. Figure 6.11 illustrates the shape of a cantilever beam under two different levels of loading. Many commercial finite element codes can produce much more sophisticated graphic output, such as that shown in Plates 5 and 6, with animated output as well as comprehensive expression of critical thermal and stress zones in the structure with color codings (Plates 10 and 13).

6.5 ▪ Automatic Mesh Generation

The power and versatility of the finite element method as an effective tool for engineering analyses have been described in Section 6.3. A major drawback of using this method, however, is the laborious and time-consuming task involved in setting up the meshes and entering the information associated with these meshes into the computer. This task becomes even more insurmountable in the cases that involve structures of complex geometries. Automatic mesh generation has become a desirable feature for most commercial finite element programs.

Mesh generation in a finite element analysis involves setting the locations of nodes in the model for the structure, numbering the nodes and elements, and specifying the nodal coordinates and nodes that form individual elements. A number of different mesh generators have been developed since the early 1970s. A comprehensive review of many of these methods is available in reference [44]. Mathematical formulation of these algorithms is complicated and beyond the scope of this book. We will provide the reader with only an overview of the principles involved in automatic mesh generation.

The primary objective of automatic mesh generation is to minimize manual input to the finite element analysis. It constitutes a major portion of the *preprocessor* of a finite element program. Computer-generated meshes usually result in fewer errors than what can be accomplished by human effort. Most automatic mesh generators require the user to input mesh topology first. Relevant information such as the topology of the region and the mesh to be generated in the region has to be specified. Desirable mesh shapes in triangular or quadrilateral plane geometries for two-dimensional models, or tetrahedron or hexahedron elements for three-dimensional models, as shown in Figure 6.4, need to be entered by the user. Meshes can then be generated based on the user-specified density of nodes, density of element areas, or density of element volumes. The latter scheme is used for three-dimensional analyses. After the mesh generation, the user is expected to make minor local adjustment to rectify any irregularity in the overall mesh. This step may not be necessary if an effective mesh smoothing algorithm is built in the mesh generator.

❑ MESH GENERATION BY NODAL DENSITY DISTRIBUTION

Many algorithms for mesh generation require the user to specify the density or number of nodes in a specific region of the model. Once the density of nodes is specified, the algorithm can assign proper locations for all individual nodes with sets of coordinates. Elements in the region are then constructed by Boolean connectivity matrices with associated nodes. The numbering of nodes and elements is established by following chronological sequence. The following simple mesh generation scheme is used to illustrate the above procedure.

The region is a plane bounded by vertices *A*, *B*, *C*, and *D*, as shown in Figure 6.12. A finite element mesh can be generated by an algorithm, with the user specifying the density of nodes along the four edges. Let us assume that five nodes are needed along edges *AB* and *CD*, and four nodes along edges *AC* and *BD*. The algorithm should automatically locate five nodes along each of the edges *AB* and *CD*, with equal spacing. Likewise, four nodes are equally spaced along edges *AC* and *BD*. Since the coordinates of the terminal nodes at vertices *A*, *B*, *C*, and *D* are fixed once the coordinate system for the region is set, the intermediate nodes (e.g. nodes 2, 3, and 4 on *AB*, nodes 17, 18, and 19 on *CD*, nodes 6, and 11 on *AC*, and nodes 10 and 15 on *BD*) can be readily identified. All interior nodes in the region can be automatically generated by interpolating the coordinates from those of the terminal nodes, as shown in Table 6.2. A typical user's input to the program on nodal coordinates is also illustrated in the table.

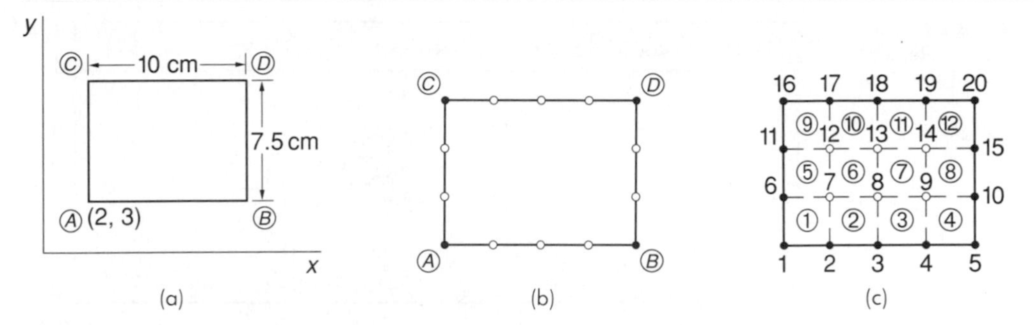

FIGURE 6.12 Mesh generation by nodes: (a) region to be modeled; (b) generated boundary nodes; (c) generated mesh

TABLE 6.2
Input of terminal nodal coordinates

Node No.	x coordinate	y coordinate
1	2	3.0
5	12	3.0
6	2	5.5
10	12	5.5
11	2	8.0
15	12	8.0
16	2	10.5
20	12	10.5

Once the coordinates of all numbered nodes are interpolated and stored in the computer, the description of elements can be accomplished by the identification of associated nodes (e.g., element 1 in Figure 6.12(c) is identified with nodes 1, 2, 7, and 6; likewise, nodes 8, 9, 14, and 13 are associated with element 7). The user's input of element information in this case can be kept to a minimum, as indicated in Table 6.3.

The reader may have noticed that nodal information for only 6 of the total 12 elements in the region needs to be entered. The node numbers of the remaining elements can be established by interpolation from the numbers of each pair of terminal element numbers, as shown in the first column in the table.

This simple technique of mesh generation can be used effectively in two-dimensional plane geometries. One can expand the use of the above scheme to a more general case of plane geometry, as illustrated in Figure 6.13. In

TABLE 6.3 Input of element information

Element No.	Node *I*	Node *J*	Node *K*	Node *L*
1	1	2	7	6
4	4	5	10	9
5	6	7	12	11
8	9	10	15	14
9	11	12	17	16
12	14	15	20	19

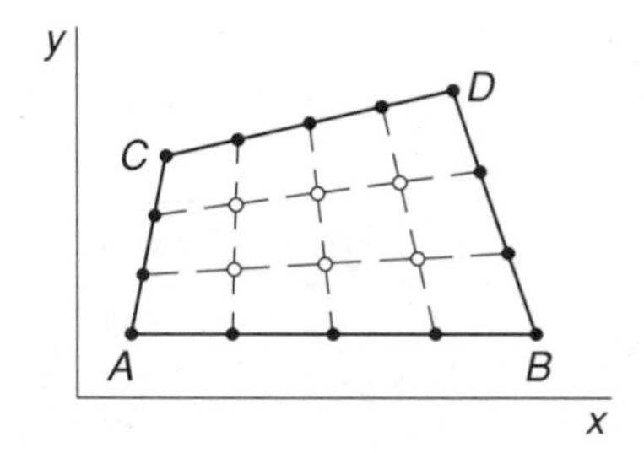

FIGURE 6.13 Mesh generation in a quadrilateral region

such cases, the interpolation of the coordinates of intermediate and interior nodes requires a slightly more complicated formula that takes into account the slopes of the lines on which all these nodes lie.

Many mesh generators produce meshes of triangular shapes, rather than quadrilateral elements as illustrated in the above example. Various methods called *triangulation* are available for that purpose, as presented in reference [44]. A good triangulation algorithm maximizes the sum of the smallest angles of the triangles for optimal analytical results.

❑ MESH GENERATION BY AREA DENSITY DISTRIBUTION

Mesh generation by nodal density distribution works well for two-dimensional analyses involving relatively simple geometries, such as those illustrated in Figures 6.12 and 6.13. A more efficient method is the specification of the density of *element areas* in the region. The principle of this mesh generation scheme can be demonstrated by a simple example of generating two elements of equal areas in a plane region, as shown in Figure 6.14(a).

The reader will find that one way the region *ABC* can be discretized by two elements with equal areas is depicted in Figure 6.14(b). The locations of

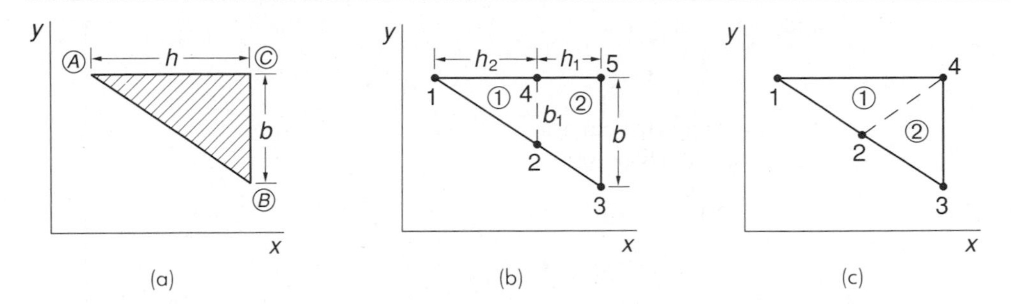

FIGURE 6.14 Mesh generation by area density distribution: (a) region to be modeled; (b) discretization by area—triangular and quadrilateral elements; (c) discretization by area—triangular elements

nodes 2 and 4 can be established by solving for h_1 or h_2 from the following equations:

$$(b + b_1)h_1/2 = b_1 h_2/2 = (bh/2)/2 \tag{6.13}$$

$$h_2/b_1 = h/b \tag{6.14}$$

$$h_1 + h_2 = h \tag{6.15}$$

The solution of $h_1 = 0.293h$ can be used to establish the coordinates of nodes 2 and 4.

There are other ways in which this region can be discretized into two elements with equal areas, as illustrated in Figure 6.14(c). The area of triangular plane elements with specified nodal coordinates of its vertices can be determined by the determinant of the matrix [A] in equation (6.5).

EXAMPLE ■ 6.5

Determine the coordinates of node 2 in a region similar to what is shown in Figure 6.14(c) with the coordinates of the three vertices given as follows:

Vertex A (node 1): $x_1 = 2$ $y_1 = 8$ cm
Vertex B (node 3): $x_3 = 10$ $y_3 = 2$ cm
Vertex C (node 4): $x_4 = 12$ $y_4 = 8$ cm

The area of the region is computed to be $A = 30\ \text{cm}^2$. If we let A_1 and A_2 be the respective areas for elements 1 and 2 and let the coordinates of node 2 be (x, y), the areas of both elements can be determined by the determinant of the matrix [A] in equation (6.5), or

$$2A_1 = (x_1 y - x y_1) + (x y_4 - x_4 y) + (x_4 y_1 - x_1 y_4)$$

and

$$2A_2 = (xy_3 - x_3y) + (x_3y_4 - x_4y_3) + (x_4y - xy_4)$$

The following two equations are obtained after substituting the numerical values of $x_1, y_1, \ldots, x_4, y_4$ into the above expressions:

$$2A_1 = -10y + 80 \tag{6.16}$$

$$2A_2 = -6x + 2y + 56 \tag{6.17}$$

Since the area of element 1 is the same as the area of element 2, (i.e., $A_1 = A_2$), one can obtain the following equation by equating equations (6.16) and (6.17).

$$6x - 12y = -24 \tag{6.18}$$

The condition of $A = 2A_1 = 2A_2 = 30\ \text{cm}^2$ can lead to the solution of $y = 5$ cm from equation (6.16), and the solution of $x = 6$ can then be found from equation (6.18) with the value of y. The coordinates of node 2 are thus (6, 5). ❑

❑ MESH GENERATION BY VOLUME DENSITY DISTRIBUTION

The concept of specifying the density distribution of element areas for two-dimensional finite element meshes can be extended to three-dimensional models. In such cases, the density distribution of *element volume* is used. The volume of the basic three-dimensional elements (i.e., tetrahedron elements, depicted in Figure 6.4) can be determined by the following determinant:

$$V = \frac{1}{6}\begin{vmatrix} 1 & 1 & 1 & 1 \\ x_1 & x_2 & x_3 & x_4 \\ y_1 & y_2 & y_3 & y_4 \\ z_1 & z_2 & z_3 & z_4 \end{vmatrix} \tag{6.19}$$

where (x_1, y_1, z_1), (x_2, y_2, z_2), (x_3, y_3, z_3), and (x_4, y_4, z_4) are the coordinates of the four nodes of the element, as shown in Figure 6.15.

This mesh generation scheme is illustrated by the following example involving the discretization of a wheel with a variable cross-section, as depicted in Figure 6.16(a). The region for the finite element mesh can be approximated by a section of a pie shape with variable thickness, as shown in Figure 6.16(b). One may first divide the region into two subvolumes by the plane designated by $JKHG$. The finite element mesh in each subvolume can then be generated by the user's specification of the number of tetrahedron elements in each subvolume. We may visualize that the

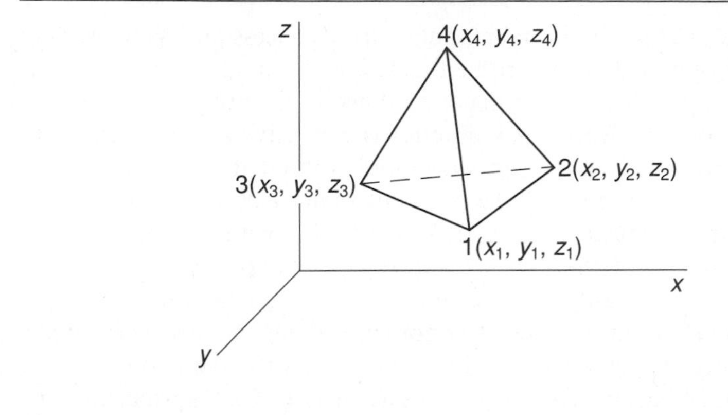

FIGURE 6.15
Nodal coordinates of a tetrahedron element

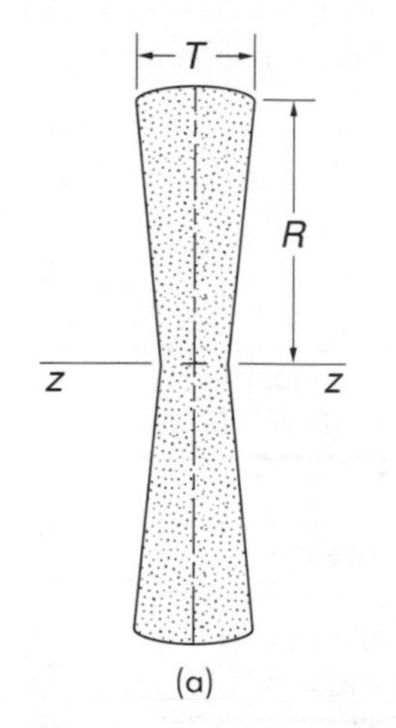

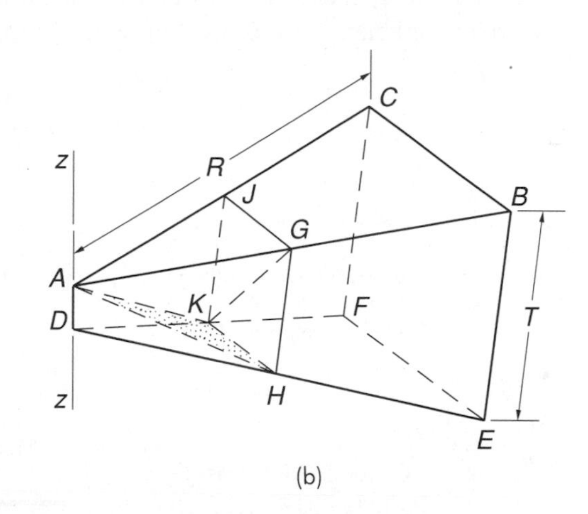

FIGURE 6.16
A three-dimensional mesh generation: (a) cross-section of a wheel; (b) region for finite element model

subvolume near the axle *AD* of the wheel can be subdivided into three tetrahedrons, *ADKH*, *AGJK*, and *AGHK*, as illustrated in Figure 6.16(b). It is conceivable that a minimum of four tetrahedrons can be generated in the other subvolume. The locations, and thus the coordinates, of all intermediate nodes in the mesh can be determined by using the formula for the computation of the volume of a tetrahedron, given in Equation (6.19), and by following a procedure similar to that used in Example 6.5 for two-dimensional mesh generation.

6.6 ▪ Integration of CAD and Finite Element Analysis

Of all the principal benefits of CAD as presented in Section 1.3, a major cost-saving benefit is the possibility of eliminating the need for building prototypes of the product. Prototype building is a common practice in industry because the traditional manufacturing process for a new product requires prototypes after the product is designed. These prototypes allow engineers to conduct tests and make modifications until all design specifications are met. However, this practice is expensive and time consuming, and there usually is not enough time to build these prototypes as design cycles are shortened by the competitive marketplace. Skipping this step in the manufacturing process is now possible with the sophisticated *geometric modeling* of the product and design analyses, which use such reliable techniques as the finite element method and design optimization. Close interactions among these technologies in engineering design have been illustrated in Figure 1.6. In this section, we will take a look at how geometric modeling, or *solid modeling*, as needed in a CAD system, and the powerful finite element analysis can be integrated to offer designers an effective and efficient tool for dealing with complex design problems.

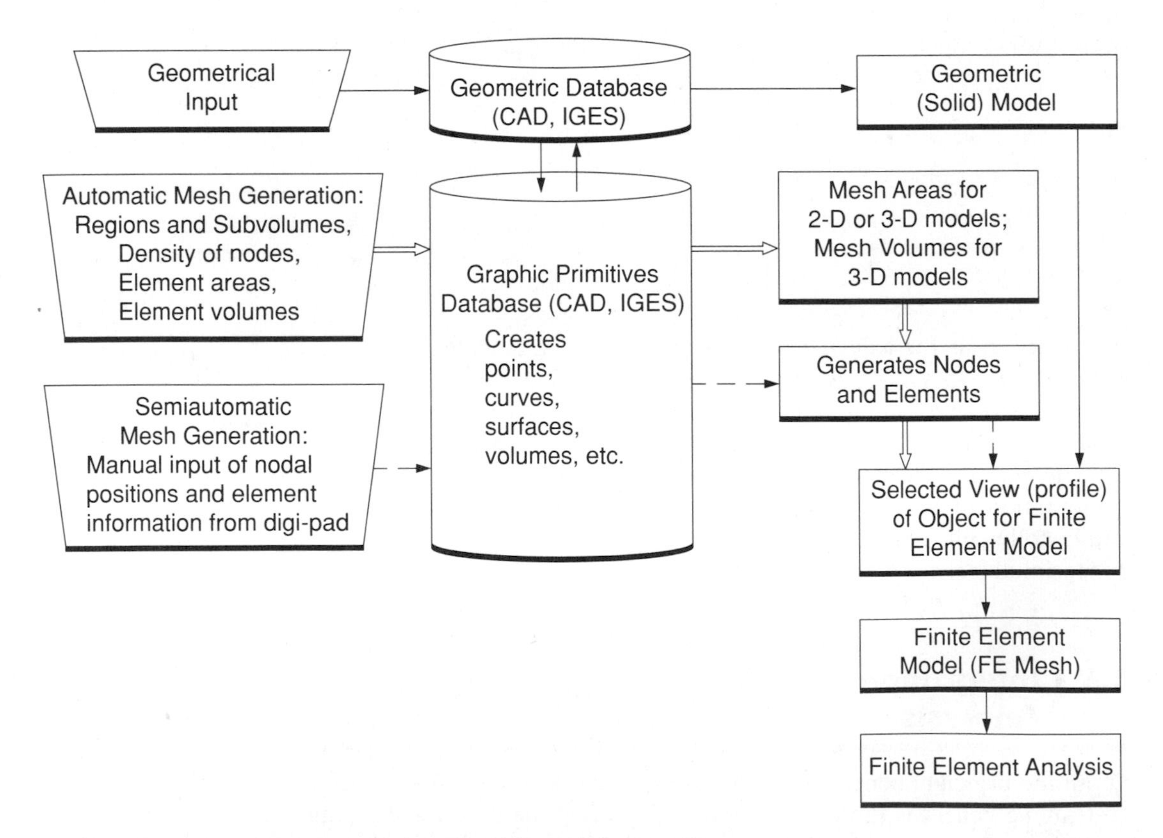

FIGURE 6.17 Integration of CAD and finite element analysis

CAD can play an important role in a finite element analysis (FEA) in the preprocessing stage of the analysis. As mentioned in the previous section, generating a finite element model is a laborious, time-consuming, and error-prone process. Automatic mesh generation through the use of interactive CAD modeling has become a desirable feature of FEA programs. The principle of interactive CAD and FEA is illustated in Figure 6.17.

The process begins with the user's input of geometric data of the product. The tentative geometry of the product (i.e., solid models) is created and stored in the geometric database. The principle of constructing a geometric database is presented in Chapter 5, and the IGES package described in Section 5.7 can be used for this purpose. This geometry can be retrieved with a selected view for the finite element model. The user may choose a specific mesh generation scheme for the creation of the finite element model by following the block diagram in Figure 6.17.

A case study is presented here to demonstrate how CAD can be used as an effective preprocess of an FEA. The case is related to the finite element stress analysis of a cam-shaft assembly. The software program used in the study is SDRC I-DEAS Version 4.0. Two different mesh generation algorithms are available in this particular software: (1) free node generator and (2) mapped mesh generator.

❑ FREE NODE GENERATOR

The profile of the solid cam was created first, as shown in Figure 6.18(a), in which the circular portion (to the left of the vertical axis) was created by eight points along the arc and the remaining curved portion was created by cubic spline functions using 12 knots. A wireframe model of the cam was then constructed with a specified thickness, as shown in Figure 6.18(b). The location of the center of the axle for the shaft is marked by "x." The geometry of the shaft, shown in Figure 6.18(c), was then created, using the primitive of cylinder available in the database. Again, the location of the axis of the shaft is marked by "x." The cam and the shaft were then assembled by matching the axes marked by "x," using the Boolean union algorithm in the program. This assembly is shown in Figure 6.18(d). The symmetry of the geometry of the assembled about the horizontal plane required the finite element mesh to be generated only on half of the assembled, as illustrated in Figure 6.18(e). At this stage, the user was required to enter the desired number of nodes in specified subvolumes of the solid. As one may visualize from Figure 6.18(f), eight subvolumes were assigned—six on the shaft and two on the cam. The density distribution of nodes in each subvolume was entered, and the finite element mesh for the half-assembly was then completed by the automatic mesh generator, as shown in Figure 6.18(g). The mesh for the entire assembly, shown in Figure 6.18(h), can be produced by using the mirror reflection algorithm, which is available in most commercial CAD and finite element programs.

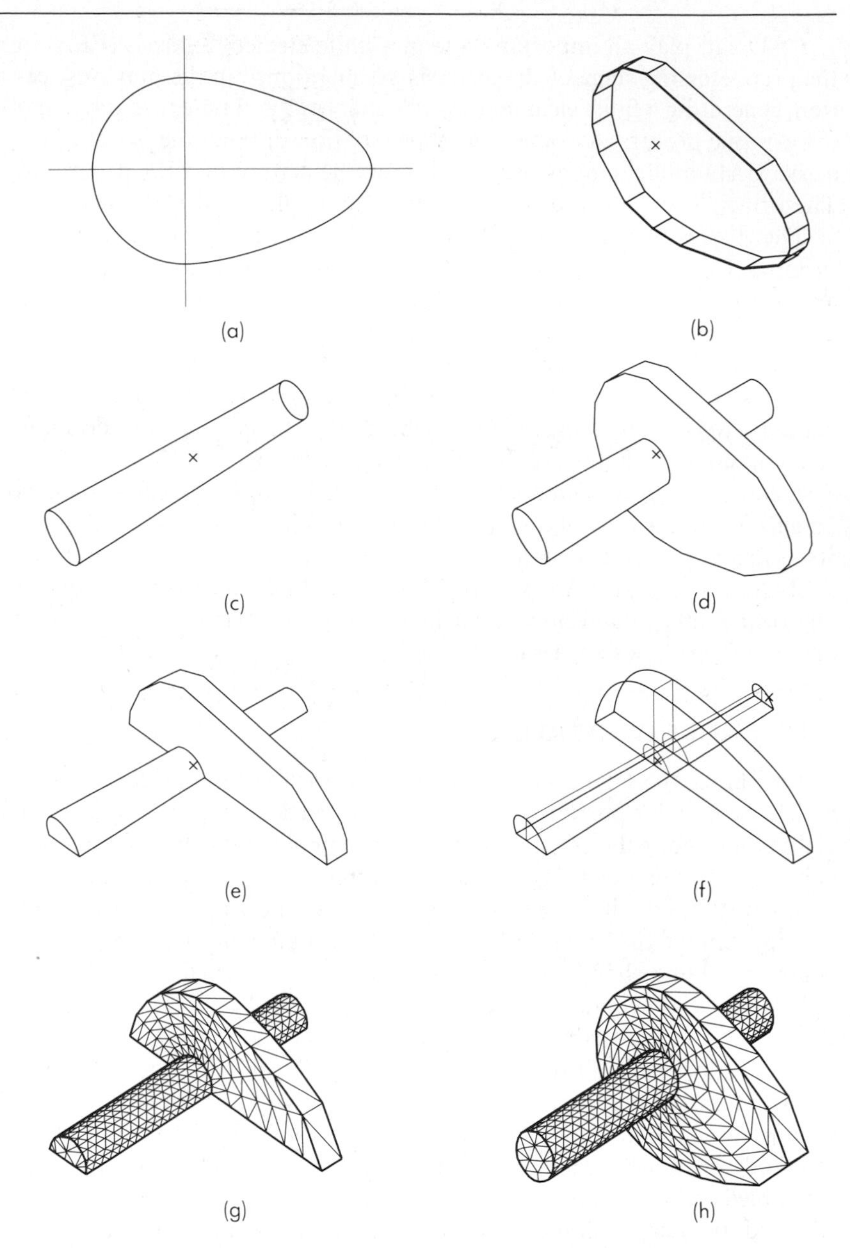

FIGURE 6.18
Finite element mesh by free node generator

❑ MAPPED MESH GENERATOR

The discretization process began with the retrieval of the profile of half of the cam, as shown in Figure 6.19(a). Two regions were identified: the cam region and the shaft region. The mesh in the cam region was to follow density distributions of 7 nodes along the radii and 12 nodes on the

Color Plates

PLATES 1—4
Geometric model of a rear-axle and frame assembly of a truck (Courtesy of WABCO Construction and Mining Equipment, Peoria, Illinois)

PLATES 5—6
Animated motion of a rear-axle suspension system of a truck (Courtesy of WABCO Construction and Mining Equipment, Peoria, Illinois)

PLATES 7—10
Molding process of a plastic keyboard (Courtesy of the SDRC Corporation)

PLATE 11
Solid model of a rocker arm (Courtesy of Swanson Analysis Systems Inc. and Onan Corporation)

PLATE 12
Finite element meshes for initial rocker arm geometry and the optimized geometry (Courtesy of Swanson Analysis Systems Inc. and Onan Corporation)

PLATE 13
Distribution of stress in rocker arms before and after design optimization analysis (Courtesy of Swanson Analysis Systems Inc. and Onan Corporation)

PLATE 14
Orthographic projections of a gear tooth with oblique view

PLATE 15
Graphical output: Superimposed deformed and undeformed meshes of a gear tooth subjected to a concentrated force applied at the upper left face

PLATE 16
Graphical output: Contours of stress component in R-direction in a gear tooth

PLATE 17
Engineering drawing of a gear tooth with enhanced image filling produced by MICROCAD code on a microcomputer

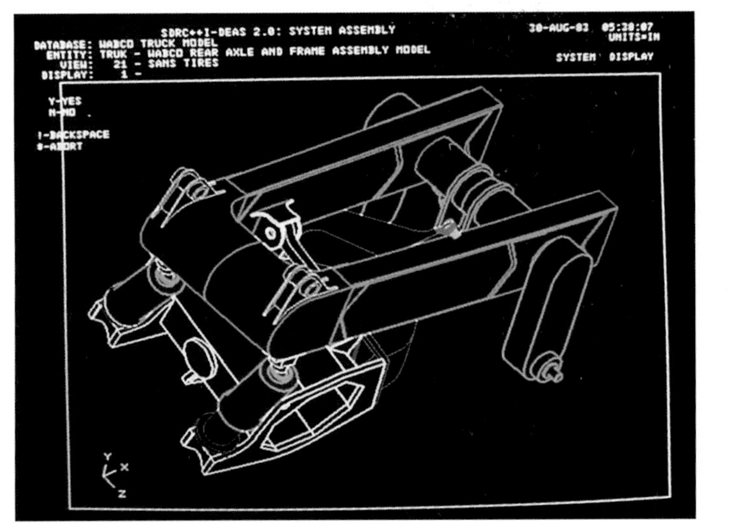

PLATE 1 ■ A line sketch

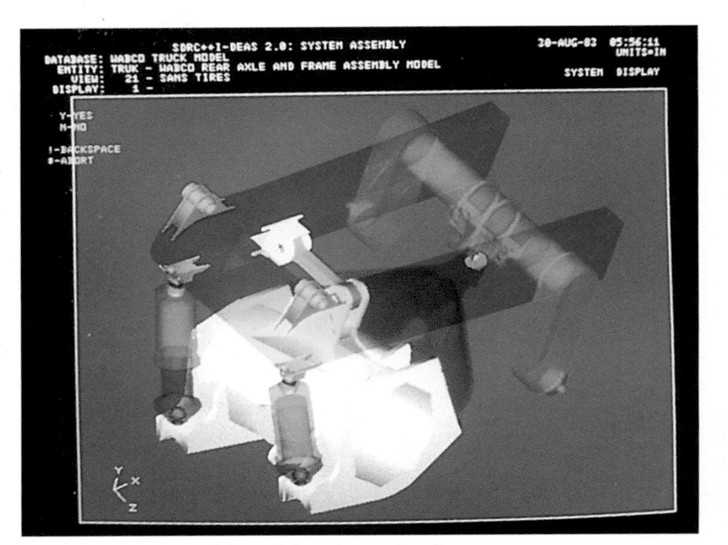

PLATE 2 ■ Solid model without wheel

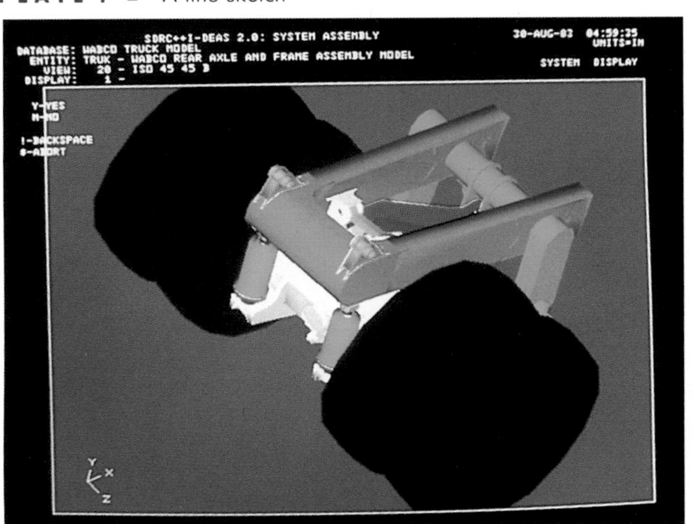

PLATE 3 ■ Solid model with wheels

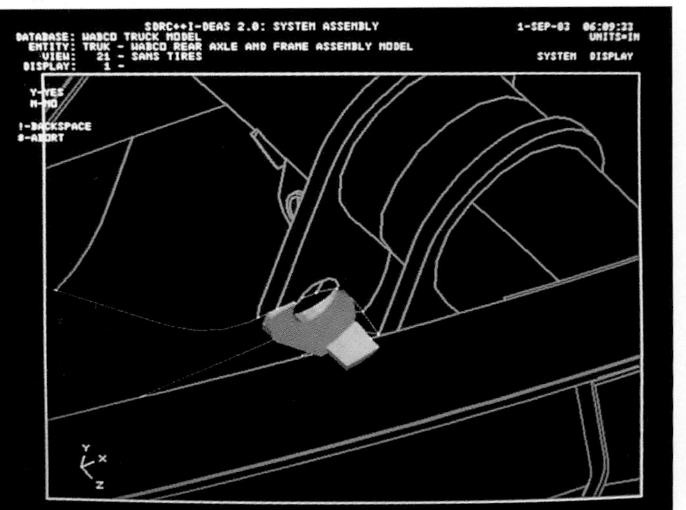

PLATE 4 ■ Close-up view on the latch device

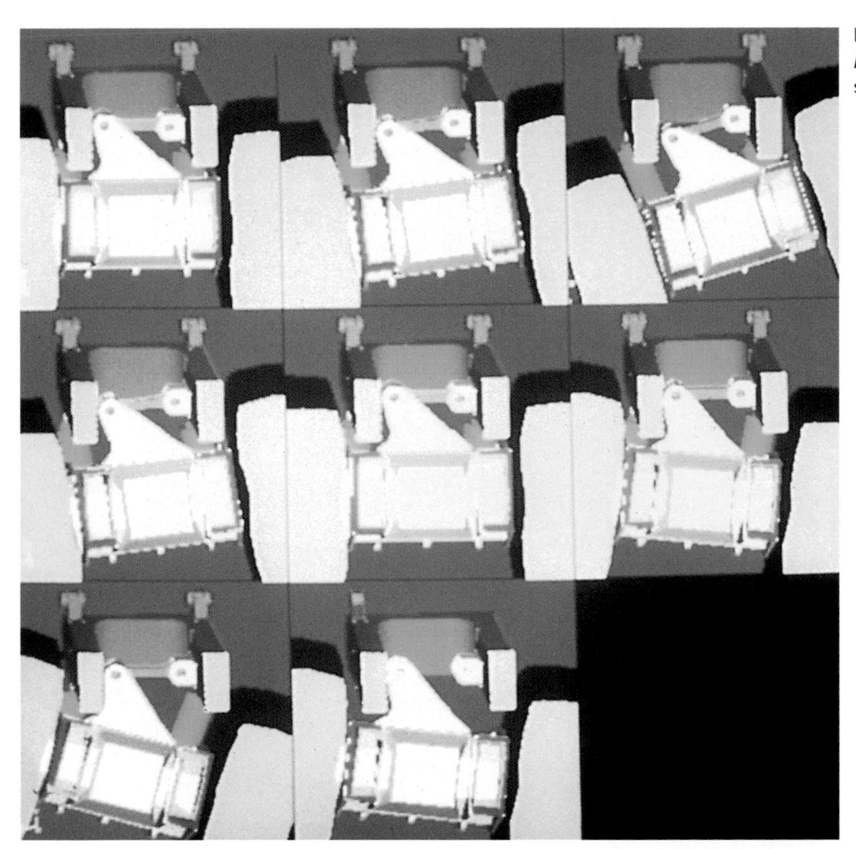

PLATE 5
Movement at eight selected instances

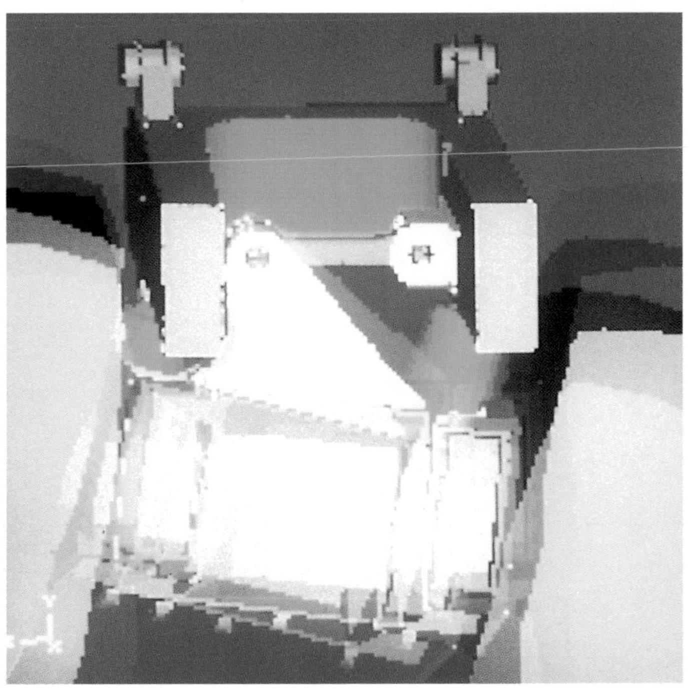

PLATE 6
Superimposed positions at two instances

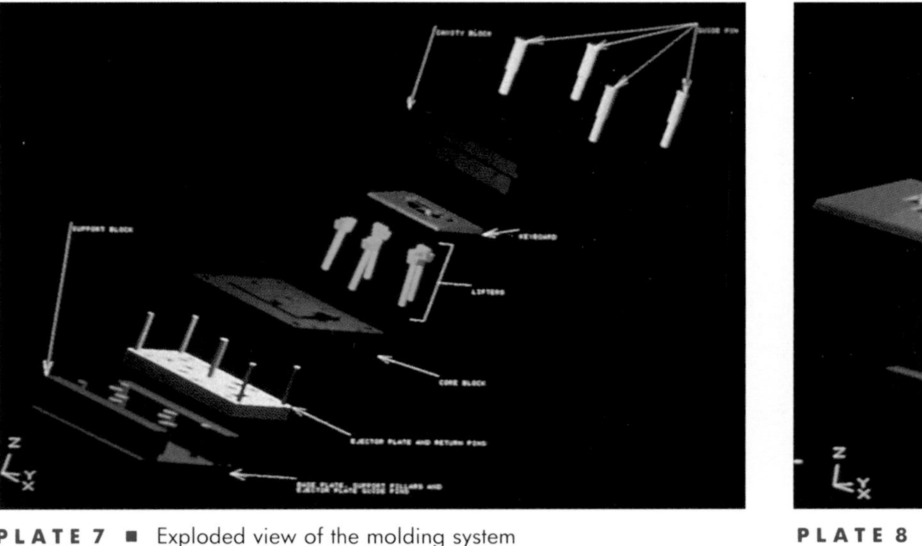

PLATE 7 ■ Exploded view of the molding system

PLATE 8 ■ Finished product with mold open

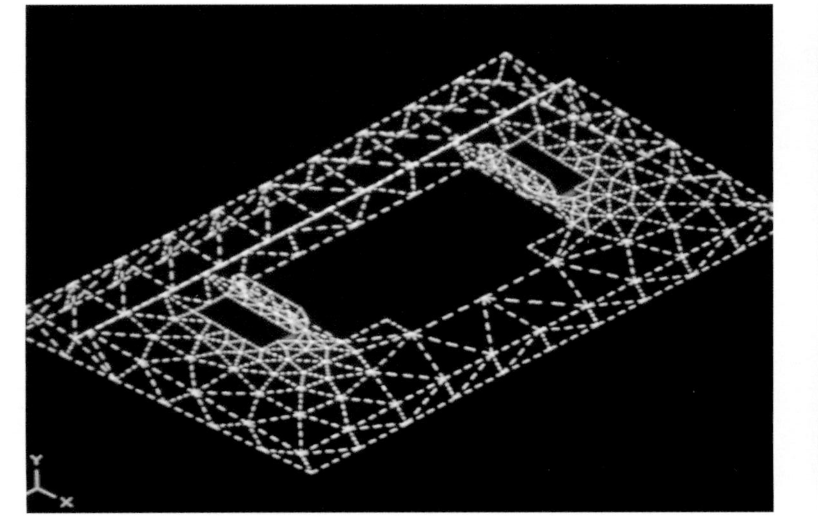

PLATE 9 ■ Finite element meshes and stress variation in the keyboard

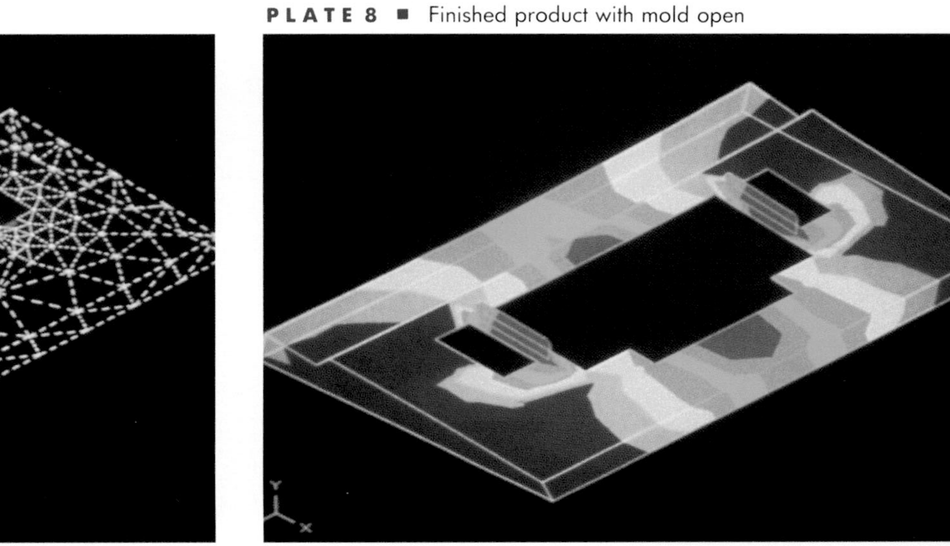

PLATE 10 ■ Temperature distribution in the keyboard

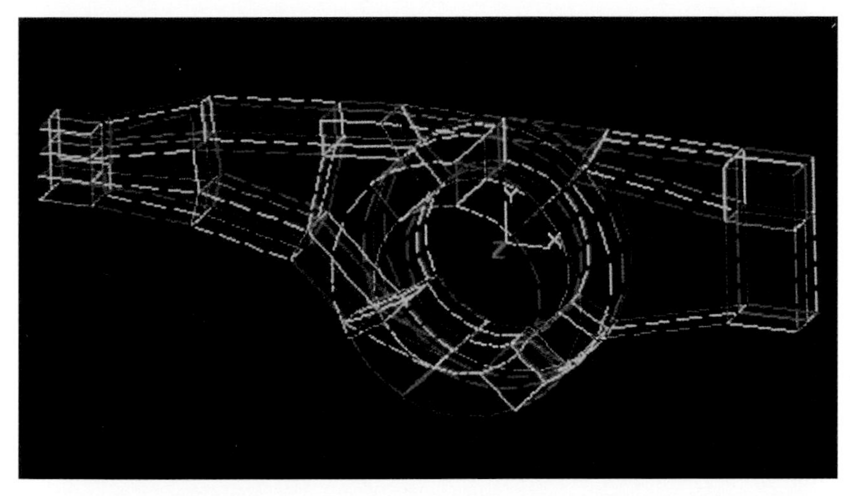

PLATE 11a
Line segments

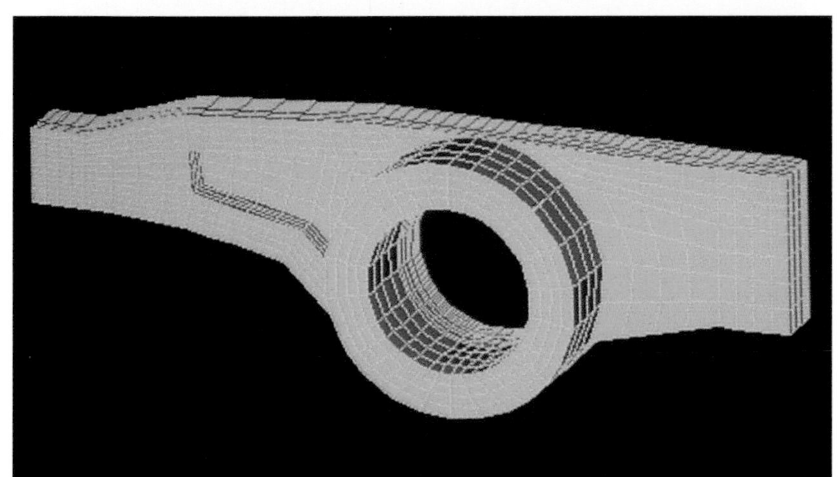

PLATE 11b
Areas made up by line segments

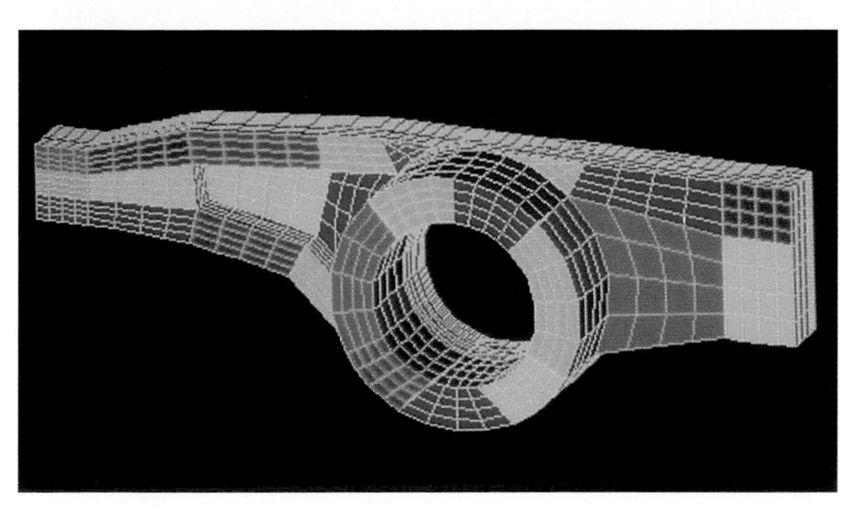

PLATE 11c
Volume blocks made up by areas

PLATE 12
Finite element meshes for initial rocker arm geometry (top) and the optimized geometry (bottom)

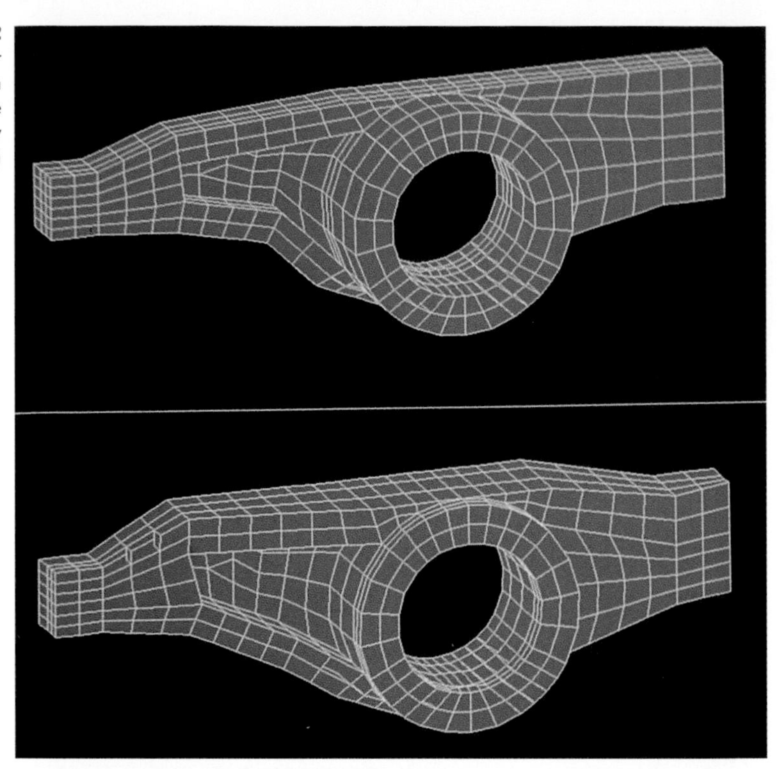

PLATE 13
Distributions of stress in rocker arms before and after design optimization analysis

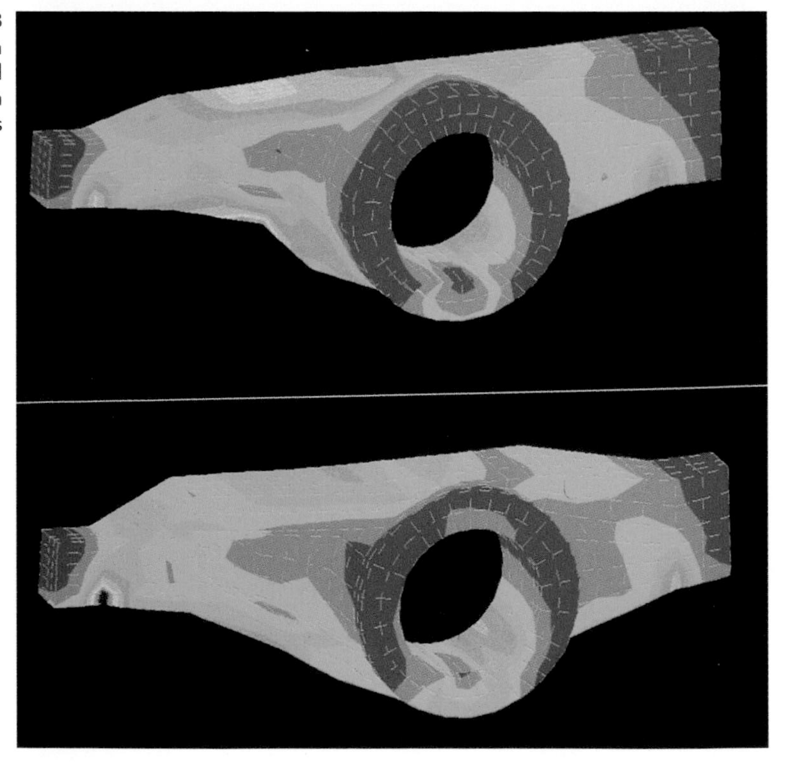

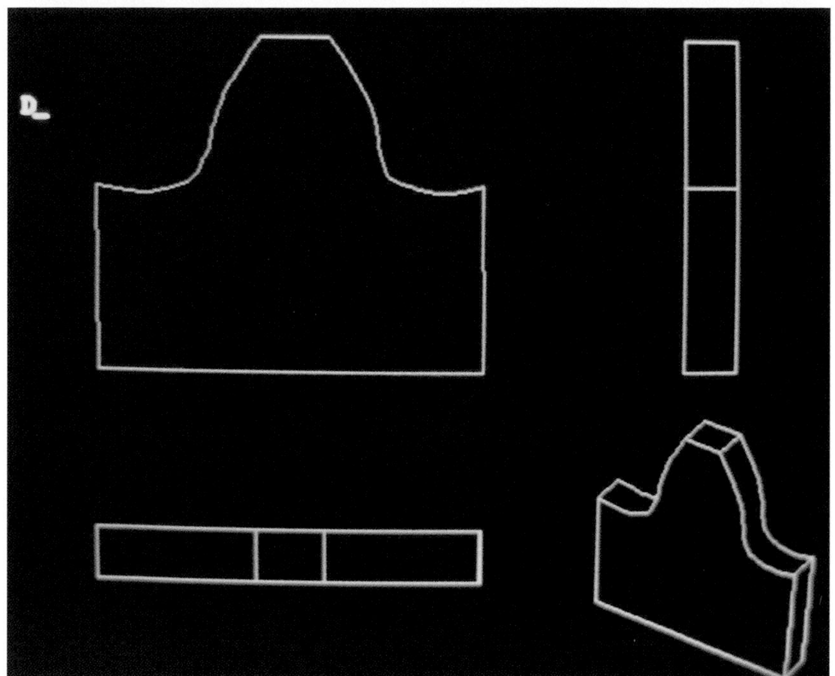

PLATE 14
Orthographic projections of a gear tooth with oblique view

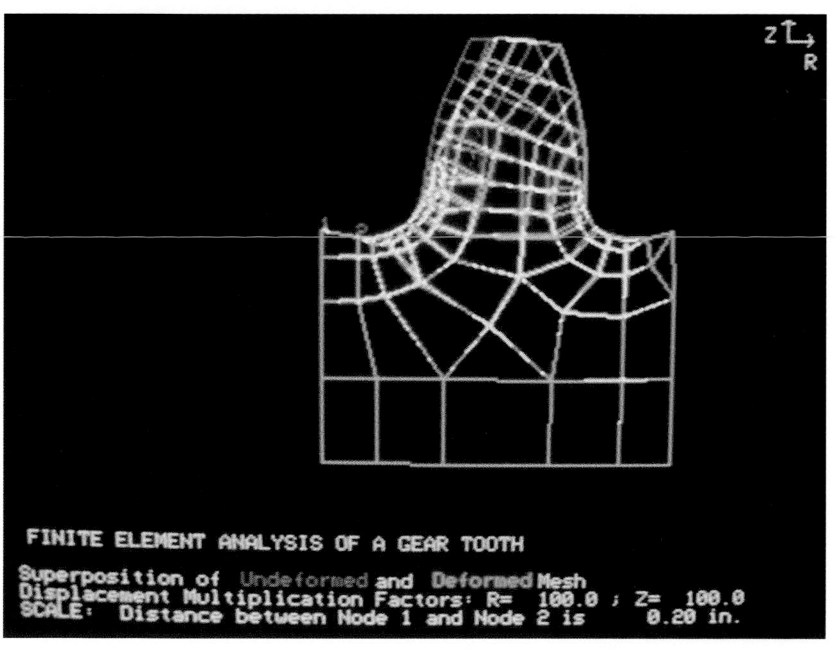

PLATE 15
Graphical output: Superimposed deformed and undeformed meshes of a gear tooth subjected to a concentrated force applied at the upper left face

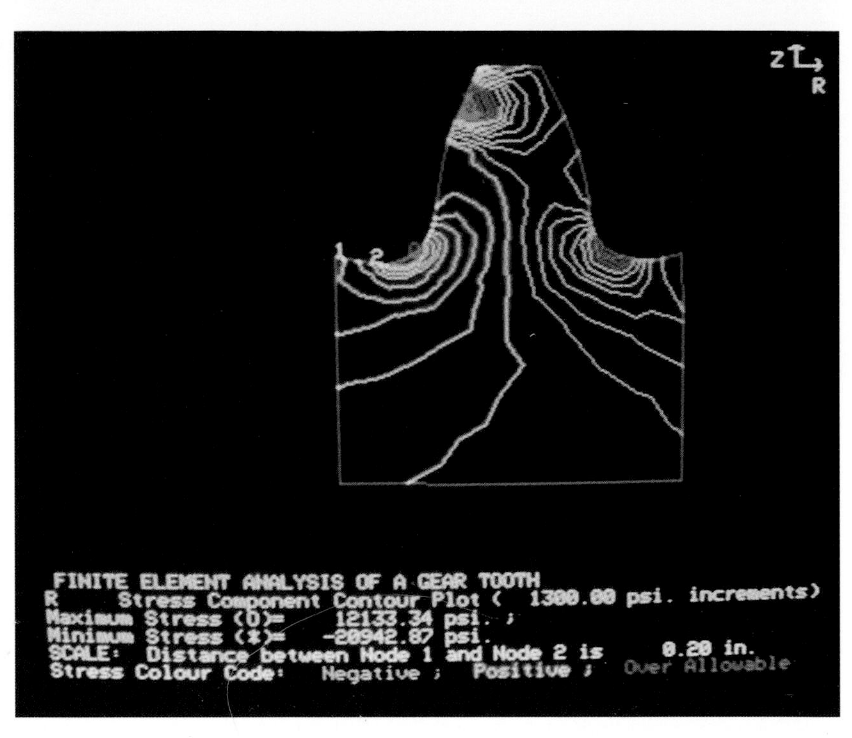

PLATE 16
Graphical output: Contours of stress component in R-direction in a gear tooth

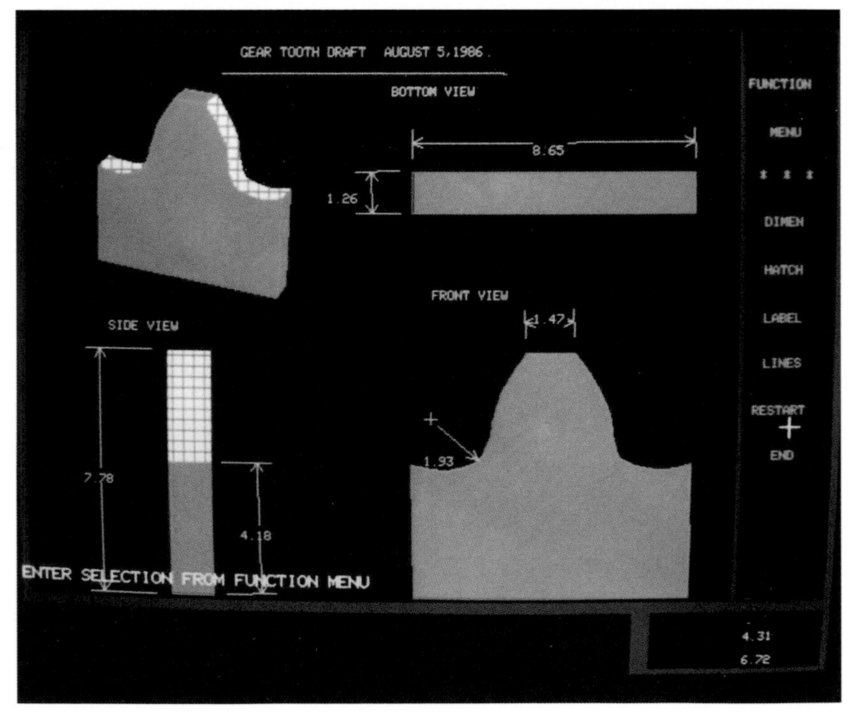

PLATE 17
Engineering drawing of a gear tooth with enhanced image filling produced by MICROCAD code on a microcomputer

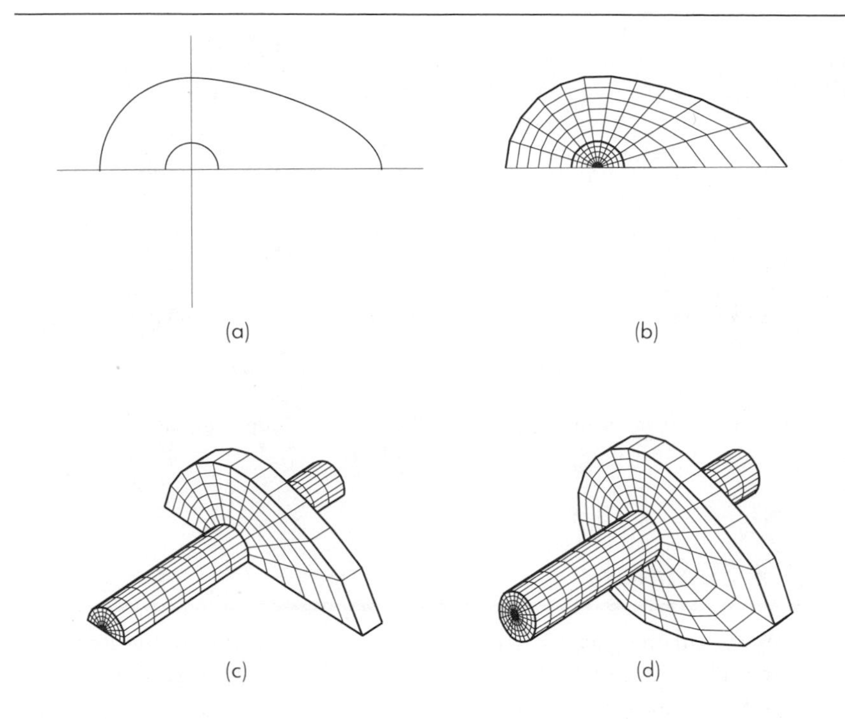

FIGURE 6.19
Finite element mesh by mapped mesh generator

perimeters, whereas the nodal density distributions in the shaft region involved 5 nodes along the radii and 9 on the perimeters. Minor manual adjustment of the mesh near the interface of the two regions was necessary for the compatibility of nodal locations at the interface of the two regions. Generated meshes in both regions are shown in Figure 6.19(b). The mesh in the cam-shaft assembly can be constructed by the extrusion of the plane mesh in Figure 6.19(b) to the desirable thickness of the cam and the length of the shaft, as depicted in Figure 6.19(c). The mesh for the full geometry of the assembly can be readily constructed using the mirror reflection algorithm in the CAD program, as shown in Figure 6.19(d).

Both automatic mesh generators provided by the SDRC I-DEAS program gave satisfactory results. The free nodes generator required more user input, but allowed the user better control of density distributions of elements in various regions. The mapped mesh generator, on the other hand, required less intervention from the user, but allowed the user little control of the density of elements in certain parts of the region. One obvious shortcoming was the rather coarse mesh distribution toward the tip of the cam, as can be seen in Figure 6.19(b). Figure 6.20 shows the solid models of the cam-shaft assembly. One may readily observe that the solid geometry of the assembly produced by the free node generator indeed appears more smooth and refined than does the one produced by the mapped mesh generator.

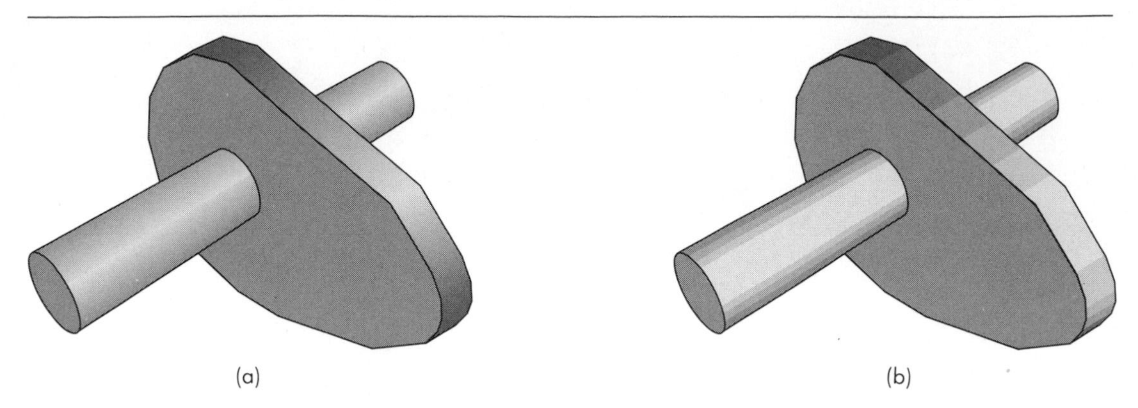

FIGURE 6.20 Solid model of a cam-shaft assembly produced by (a) free node generator; (b) mapped mesh generator

6.7 ▪ Summary

Finite element analysis (FEA) has become an essential part of modern CAD systems. This chapter is intended to provide the reader with an overview of this powerful analytical tool.

The versatility of FEA for effective analyses and applications in science and engineering was outlined. This was followed by the illustration of the principle of discretization of continuous solids on which the FEA is built. Most engineering analyses using FEA follow procedures involving eight steps. These steps were presented with numerical illustrations. Mathematical formulations at the elementary level were presented to illustrate the principle of the popular automatic mesh generation function, which is offered by many commercial FEA programs. Three methods, nodal density distribution, area density distribution, and volume density distribution, were presented. Automatic mesh generation has made integration of FEA and CAD a possibility. The chapter concluded with the description of the working principle of such an integration, with examples using two distinct approaches: the free node generator and the mapped mesh generator approaches available from a commercial CAD system.

— References

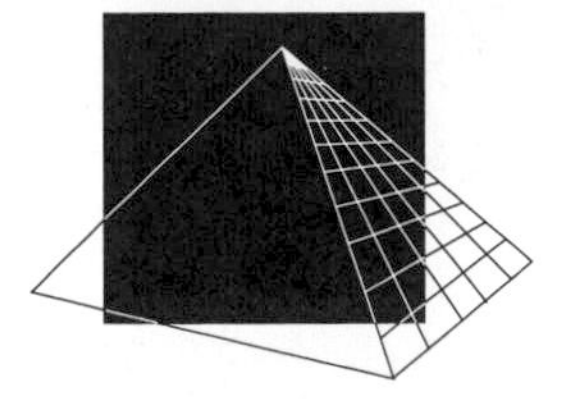

1. Turner, M.J., R.W. Clough, H.C. Martin, and L.J. Topp. "Stiffness and Deflection Analysis of Complex Structure." *Journal of Auronautical Science*, 23 (1956): 805–23.
2. Zienkiewicz, O.C., and Y.K. Cheung. *The Finite Element Method in Structural and Continuum Mechanics*. New York: McGraw-Hill, 1967.

3. Zienkiewicz, O.C. *The Finite Element Method in Engineering Science.* New York: McGraw-Hill, 1971.
4. Desai, C.S., and J.F. Abel. *Introduction to the Finite Element Method.* Van Nostrand Reinhold, 1972.
5. Cook, R.D. *Concepts and Applications of Finite Element Analysis.* 3d ed. New York: John Wiley, 1974.
6. Gallagher, R.H. *Finite Element Analysis—Fundamentals.* Englewood Cliffs, N.J.: Prentice-Hall, 1975.
7. Segerlind, L.J. *Applied Finite Element Analysis.* 2d ed. New York: John Wiley, 1976.
8. Bathe, K.J., and E.L. Wilson. *Numerical Methods in Finite Element Analysis.* Englewood Cliffs, N.J.: Prentice-Hall, 1976.
9. Desai, C.S. *Elementary Finite Element Method.* Englewood Cliffs, N.J.: Prentice-Hall, 1979.
10. Hinton, E., and D.R.J. Owen. *An Introduction to Finite Element Computations.* Pineridge, U.K.: Pineridge Press, 1979.
11. Akin, J.E. *Application and Implementation of Finite Element Methods.* New York: Academic Press, 1982.
12. Gallagher, R.H., J.T. Oden, C. Taylor, and O.C. Zienkiewicz, eds. *Finite Elements in Fluids.* Vols. 1–3. New York: John Wiley, 1975.
13. Owen, D.R.J., and E. Hinton. *Finite Elements in Plasticity.* Pineridge, U.K.: Pineridge Press Ltd., 1980.
14. Naylor, D.J., and G.N. Pande. *Finite Elements in Geotechnical Engineering.* Pineridge, U.K.: Pineridge Press, 1981.
15. Hsu, T.R. *Finite Element Method in Thermomechanics.* Boston, Mass.: George Allen & Unwin, 1986.
16. Hsu, T.R., and D.K. Sinha. *Finite Element Analysis by Microcomputer.* Duxbury, Mass.: Kern International, 1988.
17. Noor, A.K. "Survey of Computer Programs for Solution of Nonlinear Structural and Solid Mechanics Problems." *Computer and Structures,* 13 (1981): 425–465.
18. Lewis, R.W., K., Morgan, and O.C. Zienkiewicz, eds. *Numerical Methods in Heat Transfer.* New York: John Wiley, 1981.
19. Foulser, R.W.S. "Mass Diffusion Through a Moving Interface Using a Moving Finite Element Mesh." *Journal of Computational Physics,* 55, no. 3 (1984): 408–425.
20. Huffenus, J.P., and D. Khaletzky. "A Finite Element Method to Solve the Navier-Stokes Equations Using the Method of Characteristics." *International Journal for Numerical Method in Fluids,* 4, nos. 3 & 4 (1984): 247–269.
21. Mizukami, A., and M. Tsuchiya. "A Finite Element Method for the Three-Dimensional Non-Steady Navier-Stokes Equations." *International Journal for Numerical Method in Fluids,* 4, nos. 3 & 4 (1984): 349–357.

22. Srinivas, K., and C.A.J. Fletcher. "Finite Element Solutions for Laminar and Turbulent Compressible Flow." *International Journal for Numerical Method in Fluids*, 4, nos. 5 & 6 (1984): 421–439.
23. Abdel-Hadi, E.A.A., T.R. Hsu, and K.S. Bhatia. "Upwind Finite Element Analysis of Advection-Diffusion Equation." *Proceedings of 4th International Conference on Numerical Methods in Thermal Problems.* U.K.: University College of Swansea, July 1985.
24. Runesson, K., and J.R. Booker. "Finite Element Analysis of Elastic-Plastic Layered Soil Using Discrete Fourier Series Expansion." *International Journal for Numerical Method in Engineering*, 19, nos. 3 & 4 (1983): 473–478.
25. Aalto, J. "Finite Element Seepage Flow Nets." *International Journal for Numerical and Analytical Methods in Geomechanics*, 8, nos. 3 & 4 (1984): 297–303.
26. Owen, D.J.R., J.A. Figueiras, and F. Damjanic. "Finite Element Analysis of Reinforced and Prestressed Concrete Structures Including Thermal Loading." *Computer Methods in Applied Mechanics and Engineering*, 41, nos. 1, 2 & 3 (1983): 323–366.
27. Taylor, L.M., and D.V. Swenson. "A Finite Element Model for the Analysis of Tailored Pulse Stimulation of Boreholes." *International Journal for Numerical and Analytical Methods in Geomechanics*, 7, nos. 3 & 4 (1983): 469–484.
28. Olson, A.H., J.A. Orcutt, and G.A. Frazier. "The Discrete Wavenumber/Finite Element Method for Synthetic Seismograms." *Geophysical Journal of the Royal Astronomical Society*, 77 (1984): 421–460.
29. *Ratigan, J.L "A Finite Element Formulation for Brine Transport in Rock Salt." International Journal for Numerical and Analytical Methods in Geomechanics*, 8, nos. 3 & 4 (1984): 225–241.
30. Baca, R.G., R.C. Arnett, and D.W. Langford. "Modelling Fluid Flow in Fractured-Porous Rock Massess by Finite Element Techniques." *International Journal for Numerical Method in Fluids* 4, nos. 3 & 4 (1984): 337–348.
31. Gomez-Masso, A., and I. Attalla. "Finite Element Versus Simplified Methods in the Seismic Analysis of Underground Structures." *Earthquake-Engineering and Structural Dynamics*, 12, nos. 3 & 4 (1984): 347–367.
32. Huiskes, R., and E.Y.S. Chao. "A Survey of Finite Element Analysis in Orthopedic Biomechanics: The First Decade." *Journal of Biomechanics*, 16, nos. 5 & 6 (1983): 385–409.
33. Oonishi, H., H. Isha, and T. Hasegawa. "Mechanical Analysis of the Human Pelvis and Its Application to the Artificial Hip Joint—By Means of the Three-Dimensional Finite Element Method." *Journal of Biomechanics*, 16, nos. 5 & 6 (1983): 427–444.

34. Furlong, D.R., and A.N. Palazotto. "A Finite Element Analysis of the Influence of Surgical Herniation on the Viscoelastic Properties of the Intervertebral Disc." *Journal of Biomechanics*, 16, no. 10 (1983): 785–795.
35. Diller, K.R., and L.J. Hayes "A Finite Element Model of Burn Injury in Blood-Perfused Skin." *Journal of Biomechanical Engineering*, 105, no. 3 (1983): 300–307.
36. Lee, G.C., N.T. Tseng, and Y.M. Yuan. "Finite Element Modeling of Lungs Including Interlobar Fissures and the Heart Cavity." *Journal of Biomechanics*, 16, nos. 8 & 9 (1983): 679–690.
37. Basombrio, F.G. "Finite Element Formulation of Radioisotope Diffusion in Metal Grain Textures." *Journal of Computational Physics*, 54, nos. 2 & 3 (1984): 237–244.
38. Cleghorn, W.L., R.G. Fenton, and B. Tabarrok. "Finite Analysis of High-Speed Flexible Mechanisms." *Mechanisms and Machine Theory*, 16, no. 4 (1981): 407–424.
39. Cleghorn. W.L., R.G. Fenton, and B. Tabarrok. "Steady-State Vibrational Response of High-Speed Flexible Mechanisms." *Mechanism and Machine Theory*, 19, no. 4/5 (1984): 417–423.
40. McDaniel, T.W., R.B. Fernadez, R.R. Root, and R.B. Anderson. "An Accurate Scalar Potential Finite Element Method for Linear, Two-Dimensional Magnetostatics Problems." *International Journal for Numerical Method in Engineering*, 19, nos. 5 & 6 (1983): 725–737.
41. Singh, B., and J. Lal. "Finite Element Method for Unsteady MHD Flow Through Pipes with Arbitrary Wall Conductivity." *International Journal for Numerical Method in Fluids*, 4, nos. 3 & 4 (1984): 291–302.
42. Hsu, T.R. "Application of Finite Element Technique to the Technology Transfer Design." ASME Paper 76-DET-74, 1976.
43. Dolan, T.J., and E.L. Broghamer. *A Photoelastic Study of Stresses in Gear Tooth Profiles*. Bulletin no. 335. Urabana, Ill.: Engineering Experimental Station, University of Illinois, 1942.
44. Ho-Le, K. "Finite Element Mesh Generation Methods: A Review and Classification." *Computer-Aided Design*, 20 no. 1 (Jan–Feb, 1988): 27–38.

Problems

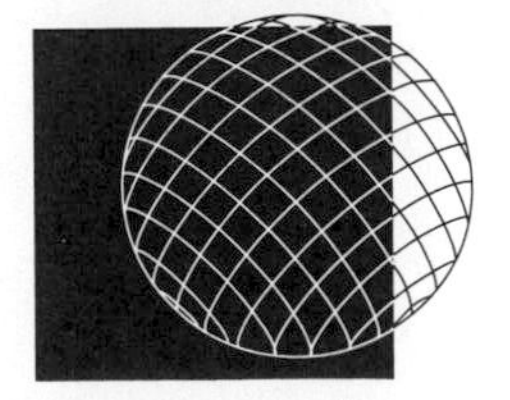

6.1 Give at least one example of the application of the discretization concept in a common engineering problem.

6.2 Identify the primary and secondary unknown quantities for the finite element analysis for at least one engineering discipline other than those given in Table 6.1.

6.3 Find the interpolation functions for the bar element illustrated in Figure 6.5, with the primary quantity varying along the x direction, by assuming

(a) $U(x) = \alpha_1 x + \alpha_2 x^2$ (b) $U(x) = 1 + \alpha_1 x + \alpha_2 x^2$
(c) $U(x) = 1 + \alpha_1 x^2 + \alpha_2 x^3$

6.4 Find the numerical value of a function $\Phi(x, y)$ at the centroid of a triangular plane enclosed by three nodes located at (0, 0), (4, 1), and (3, 3.5), using a linear polynomial function. Assume ϕ_1, ϕ_2, and ϕ_3 are the corresponding values of Φ at these three nodes.

6.5 What will be the value of the function $\Phi(x, y)$ in problem 6.4 if a higher order polynomial (e.g., $\Phi(x, y) = \alpha_1 + \alpha_2 x^2 + \alpha_3 y^2$) is used.

6.6 Prove that the area of a triangular element enclosed by three apices at (x_1, y_1), (x_2, y_2), and (x_3, y_3) is equal to $|A|/2$ in equation (6.7).

6.7 Derive the interpolation function for a triangular element with an extra node on one of its three edges, as illustrated below.

$$x_4 = \frac{1}{4}(x_2 - x_1) \qquad y_4 = \frac{1}{4}(y_2 - y_1)$$

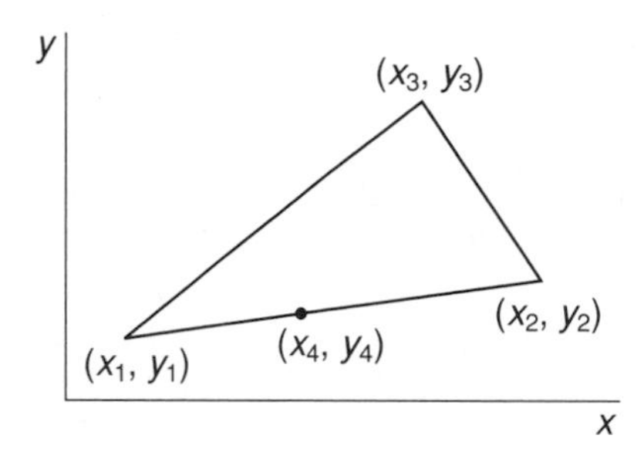

6.8 Solve the following system of equations following the method described in step 6 of Section 6.4 and equation (6.12):

$$\begin{bmatrix} 5 & 2 & -3 & -4 \\ 2 & 7 & -1 & -2 \\ 3 & 5 & -2 & 3 \\ 4 & -3 & 4 & -1 \end{bmatrix} \begin{Bmatrix} x_1 \\ x_2 \\ x_3 \\ x_4 \end{Bmatrix} = \begin{Bmatrix} b_1 \\ b_2 \\ b_3 \\ b_4 \end{Bmatrix}$$

$$x_2 = 2 \qquad x_3 = 3 \qquad b_1 = -16 \qquad b_2 = 5$$

6.9 Find the overall stiffness matrix in the discretized solid shown in the figure below with the following element matrices:

$$[K_e^1] = \begin{bmatrix} 1 & 2 & 3 \\ 2 & 4 & 5 \\ 3 & 5 & 6 \end{bmatrix} \quad \text{for nodes 1, 2, and 5}$$

$$[K_e^2] = \begin{bmatrix} 2 & 1 & 4 \\ 1 & 3 & 5 \\ 4 & 5 & 5 \end{bmatrix} \quad \text{for nodes 2, 3, and 5}$$

$$[K_e^3] = \begin{bmatrix} 4 & 7 & 2 \\ 7 & 3 & 2 \\ 2 & 2 & 1 \end{bmatrix} \quad \text{for nodes 3, 4, and 5}$$

$$[K_e^4] = \begin{bmatrix} 1 & 4 & 7 \\ 4 & 2 & 5 \\ 7 & 5 & 8 \end{bmatrix} \quad \text{for nodes 1, 5, and 4}$$

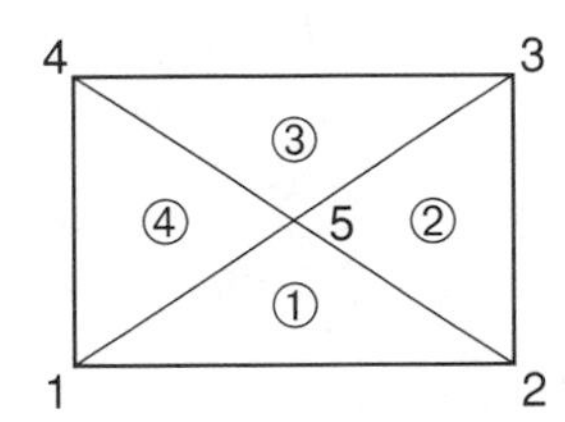

6.10 The simplex element for a three-dimensional solid is of tetrahedron geometry, as shown in Figure 6.4. Derive the interpolation function $[(N(x, y, z)]$ for this element with unknown temperature field $T(x, y, z)$ and corresponding nodal temperature $\{T\}^T = \{T_1 \quad T_2 \quad T_3 \quad T_4\}$ at the respective nodal coordinates (x_i, y_i), $i = 1, 2, 3, 4$. (*Hint:* A simplex element is the element in which the primary unknown quantity varies in a linear function of the local coordinates.)

6.11 Discretize a region of triangular shape with apices A, B, and C, as shown in Figure 6.14(b), into three elements of equal areas, and determine the intermediate nodes on edges AB and AC.

6.12 Determine the coordinates of node 4 in (a) and (b) of the figure shown at the top of the next page, with equal areas for all divided elements.

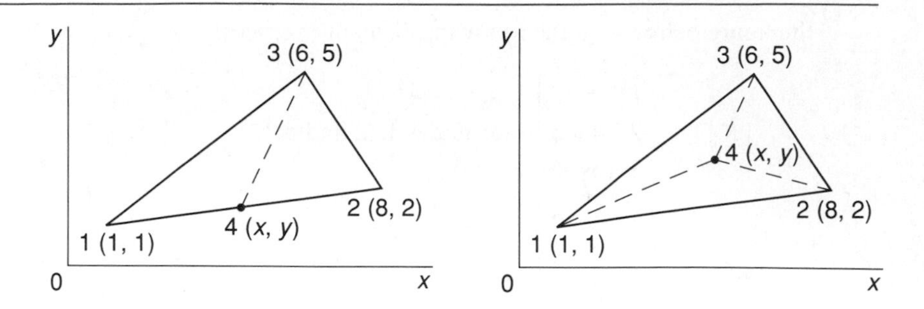

6.13 Determine the coordinates of the interior node in a region containing four triangular elements with equal areas. The shape of the region is defined below.

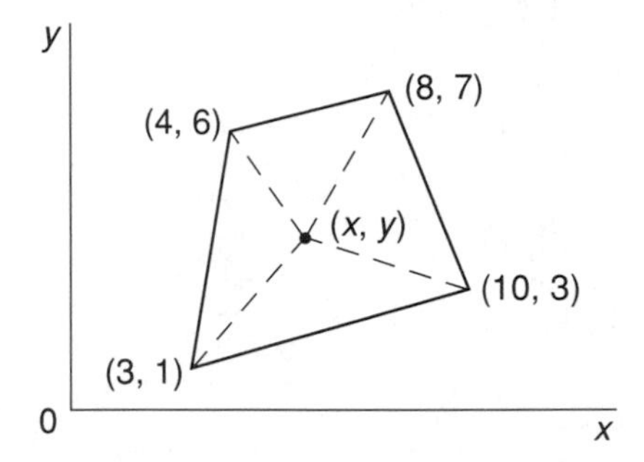

6.14 Determine the x and y coordinates for a corner node N, shown below, such that the three tetrahedron elements will have the same volume.

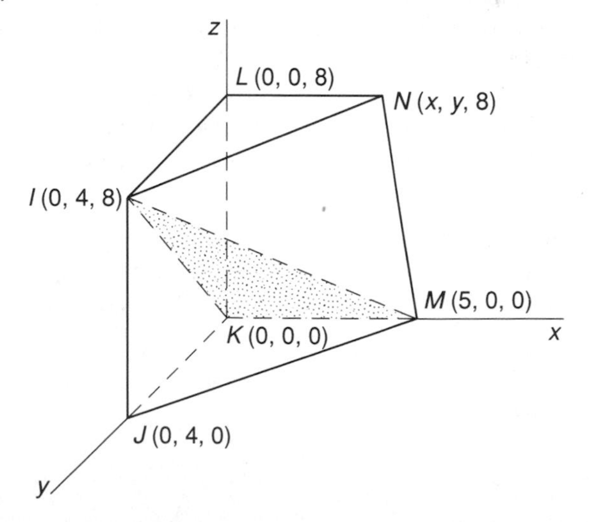

CHAPTER 7

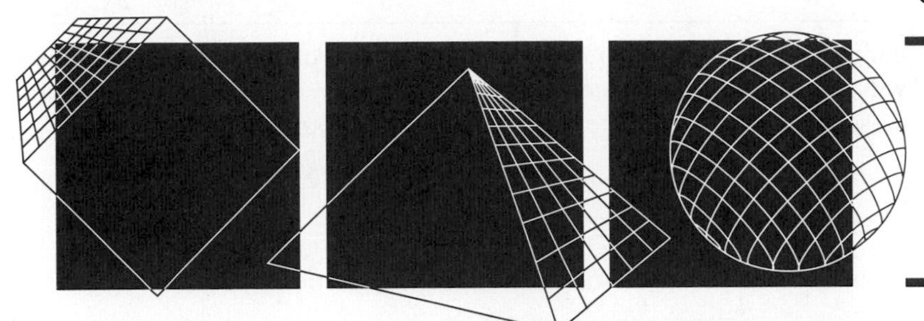

Elastic Stress Analysis by the Finite Element Method

7.1 Introduction

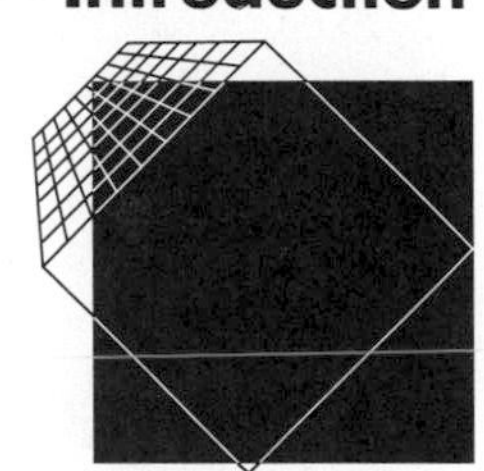

The importance of stress analysis in engineering design has been emphasized in Chapter 1. It is obvious that the integrity of structure has to be the primary consideration in any design problem. Reliable techniques for the assessment of stresses are necessary to achieve such assurance of integrity. The principle of the finite element method has been presented in the foregoing chapter, and here we present the formulation for elastic stress analysis of one- and two-dimensional structures to illustrate the methodology of this powerful technique.

An overview of general-purpose finite element programs will also be presented. This will be followed by a description of analytical capabilities of a popular commercial code, ANSYS®. It is similar to many other commercial codes. Discussion of ANSYS should provide the reader with an opportunity to appreciate these codes that are being used in many industrial establishments.

7.2 ▪ Review of Basic Formulations in Linear Elasticity Theory

It is obvious that the subject of linear elasticity theory cannot be adequately covered in a single section such as this. What we intend to do here is to review some of the fundamental assumptions regarding the behavior of common engineering materials, as well as some of the terminologies that will be used in the subsequent derivation. Only those subjects that are closely related to the finite element formulation are included.

❑ FUNDAMENTAL ASSUMPTIONS

As in all mathematical derivations, a few assumptions will have to be made here:

1. The material is treated as a homogeneous continuous medium, or a continuum. The effects induced by the inherent defects and voids in the material are neglected.
2. The material is isotropic (i.e., its properties are uniform in all directions).
3. Only a small deformation is allowed (e.g., by limiting the strain to less than 0.1 percent for steel).
4. Stress and strain, or force and the induced deformation, follow a linear relationship.
5. Strains (or deformations) are fully recovered after the applied loads are removed.

These assumptions justify the use of a simple linear elastic relationship between the deformations of a solid and the applied forces that produce such deformations. It is a well-known fact that the deflection of a spring, ∂, can be determined by a simple formula of $\partial = F/k$ in which F and k are the applied force and the stiffness constant of the spring, respectively. The latter is a specified material property. This simple formula is valid as long as the deflection is small. In reality, one can imagine that a deformable solid is made up by an infinite number of interconnected "springs." Deformations of these pseudo-springs can be evaluated by a set of linear functions related to the corresponding applied forces. In finite element analysis, one can treat each discrete element as a "spring" whose deformations at interconnected nodes, $\{u\}$, can be correlated to their respective forces, $\{F\}$, by a linear function such that

$$[K_e]\{u\} = \{q\}$$

in which $[K_e]$ is called the element stiffness matrix, which is analogous to the spring constant in the simple case of a stretched spring.

❑ TERMINOLOGIES

Stresses ■ When a solid body is subjected to a system of externally applied loads, there exist two notable responses to these loads: *deformation* of the solid and the induced *internal resistance*. It is the latter that enables

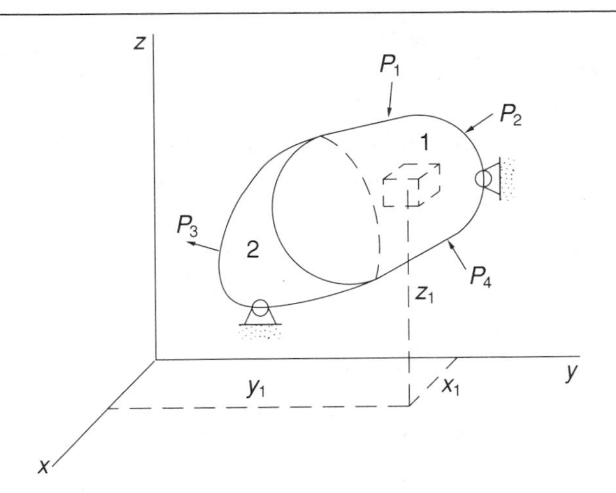

FIGURE 7.1
A deformed solid under equilibrium condition

the deformed solid to resist further deformation and reach a new state of equilibrium.

Figure 7.1 shows a deformed solid situated in an x, y, z coordinate system. In order to achieve a new equilibrium state after the application of the forces $P_1, P_2, \ldots, P_n$, internal resistance is induced everywhere in the solid. It is the intensity of this resistance that is called *stress*. One can visualize that both the magnitude and the direction of stresses can vary from point to point in the deformed solid. Stress is thus classified as a *tensor* quantity since a complete definition of stress requires not only the magnitude and direction, but also the position in the solid.

After realizing that stresses are tensor quantities, it is imperative that we find ways to designate these quantities. If we "zoom" into an infinitesimally small cubic element situated in the solid in an equilibrium condition and let (x, y, z) be the reference coordinate system (Figure 7.1), we can then "magnify" the cube, as shown in Figure 7.2. One can imagine that some

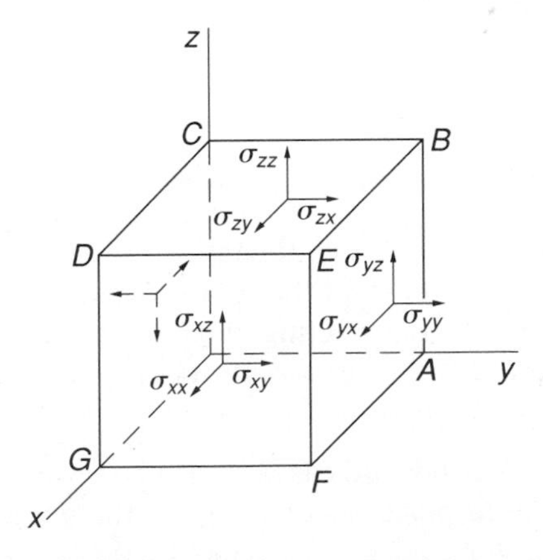

FIGURE 7.2
Stress components

stresses may exist along the faces of this tiny cubic element and some are acting normal to these faces. These components are commonly designated using the following rules:

1. The position of the stress components is completely defined by the coordinates of the centroid of the element [e.g., (x_1, y_1, z_1) in Figure 7.1].
2. Each component of stress consists of two subscripts.
3. The first subscript denotes the axis that is normal to the plane (face) containing the stress component.
4. The second subscript denotes the direction of the stress component.

As shown in Figure 7.2, there can be a total of nine stress components in a small element in a deformed solid:

$$\begin{bmatrix} \sigma_{xx} & \sigma_{xy} & \sigma_{xz} \\ \sigma_{yx} & \sigma_{yy} & \sigma_{yz} \\ \sigma_{zx} & \sigma_{zy} & \sigma_{zz} \end{bmatrix} = [\sigma]$$

It is not difficult to prove that for a solid in equilibrium condition, the following relationships exist:

$$\sigma_{xy} = \sigma_{yx} \qquad \sigma_{yz} = \sigma_{zy} \qquad \sigma_{xz} = \sigma_{zx}$$

Hence, the total number of *independent* stress components is reduced to six:

$$\begin{bmatrix} \sigma_{xx} & \sigma_{xy} & \sigma_{xz} \\ & \sigma_{yy} & \sigma_{yz} \\ & \text{SYM} & \sigma_{zz} \end{bmatrix} = [\sigma]$$

In finite element analysis, however, the stress components in an element are usually expressed as a *column* matrix:

$$\{\sigma\}^T = \{\sigma_1 \quad \sigma_2 \quad \sigma_3 \quad \sigma_4 \quad \sigma_5 \quad \sigma_6\} \tag{7.1}$$

with

$$\sigma_1 = \sigma_{xx} \qquad \sigma_2 = \sigma_{yy} \qquad \sigma_3 = \sigma_{zz}$$

$$\sigma_4 = \sigma_{xy} \qquad \sigma_5 = \sigma_{yz} \qquad \sigma_6 = \sigma_{xz}$$

One may notice that the stress components with identical subscripts are the ones acting normal to the faces of the cube, whereas the ones with different subscripts are the components acting along these faces. The former types of stresses are referred to as the *normal stress components*, and the latter types are called *shearing* or *shear stress components*.

Displacement ▪ A solid body is said to be deformed due to external loads when the *relative* positions of points in the body are changed. The

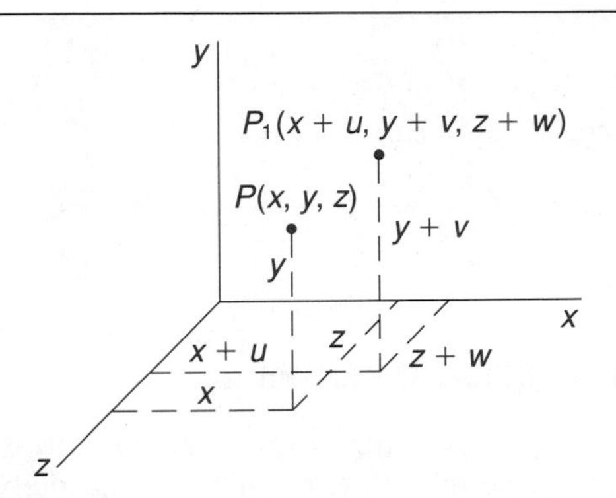

FIGURE 7.3 Displacement components

displacement of a point is defined as the vector distance from the initial to the final location of the point. If we let

u = displacement component of a point along the x axis

v = displacement component of a point along the y axis

w = displacement component of a point along the z axis

then the coordinates of the point $p(x, y, z)$ in its final position P_1 will be $(x + u, y + v, z + w)$, as shown in Figure 7.3.

The three distinct components u, v, and w can be lumped into one vectorial quantity [i.e., $\{U(x, y, z)\}$ or $\{U\}$ in the finite element formulation]. One such typical expression is

$$\{U\}^T = \{U_1 \quad U_2 \quad U_3\} \tag{7.2}$$

with

$$U_1 = u \qquad U_2 = v \qquad U_3 = w$$

Strains ■ As we define stress to be the intensity of internal resistance in the solid induced by applied forces, it may be appropriate to define strain to be the intensity of deformation of the solid. Like stresses, strain components can be classified into both normal strain components and shear strain components. The reader, however, should bear in mind that normal strain components cause changes in the dimensions (and thus volume) of the solid, whereas the physical effect of shear strain components on the solid is merely a change in the shape, but not in the size. Normal strain components carry no unit; however, shear strain components carry a unit of radians.

Strain components are usually expressed in column matrices in finite element analysis as

$$\{\varepsilon\}^T = \{\varepsilon_1 \quad \varepsilon_2 \quad \varepsilon_3 \quad \varepsilon_4 \quad \varepsilon_5 \quad \varepsilon_6\} \tag{7.3}$$

with

$$\varepsilon_1 = \varepsilon_{xx} \qquad \varepsilon_2 = \varepsilon_{yy} \qquad \varepsilon_3 = \varepsilon_{zz}$$

$$\varepsilon_4 = \varepsilon_{xy} \qquad \varepsilon_5 = \varepsilon_{yz} \qquad \varepsilon_6 = \varepsilon_{xz}$$

❑ FUNDAMENTAL RELATIONSHIPS

The three essential physical quantities—namely, stresses, strains, and displacements—involved in the theory of linear elasticity can be interrelated through certain mathematical expressions and material properties such as Young's modulus and the Poisson ratio. Here, we will present only the formulations relevant to the subsequent derivation for finite element analysis.

Strain-Displacement Relations ▪ As mentioned at the beginning of this chapter, the responses of a solid to applied loads (forces, pressures, temperature fields, etc.) include the generation of internal resistances responsible for the establishment of the new state of equilibrium and also an associated deformation field. For a solid situated in an (x, y, z) space, its deformation may vary from point to point. In other words, a more complete description of displacements should be expressed as

$$u = u(x, y, z) \qquad v = v(x, y, z) \qquad w = w(x, y, z)$$

for the three respective components along x, y, and z directions.

If one defines the strain components with a complete description of their magnitudes, directions, and positions in the solid as follows:

$$\varepsilon_{xx} = \varepsilon_{xx}(x, y, z),\ \varepsilon_{yy} = \varepsilon_{yy}(x, y, z), \ldots, \text{etc.}$$

then, by following the derivations presented in Chapter 1 of reference [1], we can correlate these two sets of physical quantities by

$$\varepsilon_{xx}(x, y, z) = \frac{\partial u(x, y, z)}{\partial x} \tag{7.4}$$

$$\varepsilon_{yy}(x, y, z) = \frac{\partial v(x, y, z)}{\partial y} \tag{7.5}$$

$$\varepsilon_{zz}(x, y, z) = \frac{\partial w(x, y, z)}{\partial z} \tag{7.6}$$

$$\varepsilon_{xy}(x, y, z) = \frac{\partial v(x, y, z)}{\partial x} + \frac{\partial u(x, y, z)}{\partial y} \tag{7.7}$$

$$\varepsilon_{yz}(x, y, z) = \frac{\partial w(x, y, z)}{\partial y} + \frac{\partial v(x, y, z)}{\partial z} \tag{7.8}$$

$$\varepsilon_{xz}(x, y, z) = \frac{\partial w(x, y, z)}{\partial x} + \frac{\partial u(x, y, z)}{\partial z} \tag{7.9}$$

The above relationship can be expressed in matrix form as follows:

$$\begin{Bmatrix} \varepsilon_{xx}(x, y, z) \\ \varepsilon_{yy}(x, y, z) \\ \varepsilon_{zz}(x, y, z) \\ \varepsilon_{xy}(x, y, z) \\ \varepsilon_{yz}(x, y, z) \\ \varepsilon_{xz}(x, y, z) \end{Bmatrix} = \begin{bmatrix} \frac{\partial}{\partial x} & 0 & 0 \\ 0 & \frac{\partial}{\partial y} & 0 \\ 0 & 0 & \frac{\partial}{\partial z} \\ \frac{\partial}{\partial y} & \frac{\partial}{\partial x} & 0 \\ 0 & \frac{\partial}{\partial z} & \frac{\partial}{\partial y} \\ \frac{\partial}{\partial z} & 0 & \frac{\partial}{\partial x} \end{bmatrix} \begin{Bmatrix} u(x, y, z) \\ v(x, y, z) \\ w(x, y, z) \end{Bmatrix} \tag{7.10}$$

If we let $\{\varepsilon\}$ and $\{U\}$ represent respective strain and displacement components, as shown in equations (7.4) through (7.9) and (7.2), then the following simple relation is obtained:

$$\{\varepsilon\} = [D]\{U\} \tag{7.11}$$

in which

$$[D] = \begin{bmatrix} \frac{\partial}{\partial x} & 0 & 0 \\ 0 & \frac{\partial}{\partial y} & 0 \\ 0 & 0 & \frac{\partial}{\partial z} \\ \frac{\partial}{\partial y} & \frac{\partial}{\partial x} & 0 \\ 0 & \frac{\partial}{\partial z} & \frac{\partial}{\partial y} \\ \frac{\partial}{\partial z} & 0 & \frac{\partial}{\partial x} \end{bmatrix} \tag{7.12}$$

Stress-Strain Relations ■ For the solid situated in an (x, y, z) space, as shown in Figure 7.1, the relationships among the six independent stress components in equation (7.1) and the corresponding strain components in equation (7.3) can be summarized by the generalized Hooke's law [1] as

$$\begin{Bmatrix} \sigma_{xx}(x, y, z) \\ \sigma_{yy}(x, y, z) \\ \sigma_{zz}(x, y, z) \\ \sigma_{xy}(x, y, z) \\ \sigma_{yz}(x, y, z) \\ \sigma_{xz}(x, y, z) \end{Bmatrix} = \frac{E}{(1+v)(1-2v)}$$

$$\times \begin{bmatrix} 1-v & v & v & 0 & 0 & 0 \\ & 1-v & v & 0 & 0 & 0 \\ & & 1-v & 0 & 0 & 0 \\ & & & \frac{1-2v}{2} & 0 & 0 \\ & \text{SYM} & & & \frac{1-2v}{2} & 0 \\ & & & & & \frac{1-2v}{2} \end{bmatrix} \begin{Bmatrix} \varepsilon_{xx}(x, y, z) \\ \varepsilon_{yy}(x, y, z) \\ \varepsilon_{zz}(x, y, z) \\ \varepsilon_{xy}(x, y, z) \\ \varepsilon_{yz}(x, y, z) \\ \varepsilon_{xz}(x, y, z) \end{Bmatrix}$$

or, in a matrix form,

$$\{\sigma(x, y, z)\} = [C]\{\varepsilon(x, y, z)\} \tag{7.13}$$

The matrix

$$[C] = \frac{E}{(1+v)(1-2v)} \begin{bmatrix} 1-v & v & v & 0 & 0 & 0 \\ & 1-v & v & 0 & 0 & 0 \\ & & 1-v & 0 & 0 & 0 \\ & & & \frac{1-2v}{2} & 0 & 0 \\ & \text{SYM} & & & \frac{1-2v}{2} & 0 \\ & & & & & \frac{1-2v}{2} \end{bmatrix} \tag{7.14}$$

is called the elasticity matrix, with E and v denoting the respective Young's modulus and Poisson's ratio of the material.

Strain Energy in Elastic Solids ▪ The reader may recall the assumption we made earlier that strains (or deformations) in a solid are fully recoverable after the removal of all applied loads. Indeed, if one stretches an elastic band with a force, the band will return to its original length once the applied force is removed. This commonly observed phenomenon can be explained by the fact that certain energy has been generated and stored in the solid as a result of the deformation. This energy is released upon the release of applied forces to the solid and prompts the solid to return to its original shape. As this energy is induced in the solid by the deformation (or strain) of the solid, it is commonly referred to as the *strain energy*.

The strain energy stored in a deformed solid, such as that depicted in Figure 7.1, can be expressed in the following form:

$$U = \frac{1}{2}\int_v (\sigma_{xx}\varepsilon_{xx} + \sigma_{yy}\varepsilon_{yy} + \sigma_{zz}\varepsilon_{zz} + \sigma_{xy}\varepsilon_{xy} + \sigma_{xz}\varepsilon_{xz} + \sigma_{yz}\varepsilon_{yz})\,dv \tag{7.15}$$

in which v is the volume of the solid.

The reader is reminded that strain energy is a scalar quantity (i.e., it has a single value). One should not confuse the notation of U for strain energy with the vectorial quantity $\{U(x, y, z)\}$ for the displacement field used in the previous expressions. The expression of U in the matrix form can be readily accomplished by replacing the stress and strain components in equation (7.15) with the respective matrix expressions. Thus, the strain energy in equation (7.15) can be expressed as

$$U(\{\varepsilon\}, \{\sigma\}) = \frac{1}{2}\int_v \{\varepsilon\}^T\{\sigma\}\,dv \tag{7.16}$$

with $\{\sigma\}$ and $\{\varepsilon\}$ shown in equation (7.1) and equation (7.3), respectively.

7.3 ▪ Finite Element Formulation

As mentioned in Step 4 of Section 6.4, the element equation for elastic stress analysis of a solid using the finite element method can be derived by the Rayleigh-Ritz method based on the variational principle. A detailed description of this method can be found in several references [1–3]. The essence of this method is to keep the potential energy in all elements, π, in the discretized structure at the minimum state. This state ensures the equilibrium condition of the deformed structure.

The derivation of element equations thus involves the formulation of the potential energy function in the elements. Since element displacements $\{U(x, y, z)\}$ are normally used as the primary unknown quantities and the

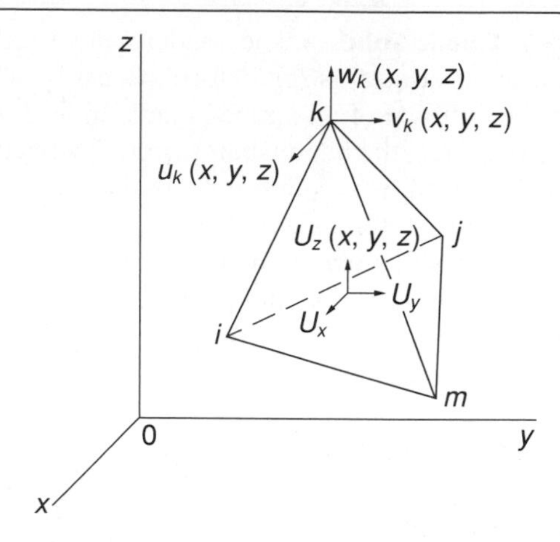

FIGURE 7.4
A tetrahedron element

discrete values of these quantities are their corresponding values at the nodes, the potential energy function should be expressed in terms of nodal displacement components, $\{u\}$, through appropriate interpolation functions. Minimization of these potential energy functions by means of a variational process with respect to $\{u\}$ will be carried out to establish a set of simultaneous algebraic equations for all displacement components.

Take, for an example, the tetrahedron element in Figure 7.4. Following equation (6.1), the displacement components in the element $\{U(x, y, z)\}$ can be expressed in terms of the nodal components as

$$\{U(x, y, z)\} = \begin{Bmatrix} U_x(x, y, z) \\ U_y(x, y, z) \\ U_z(x, y, z) \end{Bmatrix} = [N(x, y, z)]\{u\} \tag{7.17}$$

where $\{u\}$ are the displacement components at nodes i, j, k, and m.

Another way to express the relationship in equation (7.17) is to partition the elements of the interpolation function, $[N(x, y, z)]$, into groups associated with displacement components of the same nodes. For the case illustrated in Figure 7.4, the discretization equation for the tetrahedron element can be expressed as

$$\{U(x, y, z)\} = \{N_i(x, y, z) \quad N_j(x, y, z) \quad N_k(x, y, z) \quad N_m(x, y, z)\} \begin{Bmatrix} u_i \\ u_j \\ u_k \\ u_m \end{Bmatrix} \tag{7.18}$$

Thus, the element displacement components can be expressed in terms of their respective components at the nodes as

$$\{U(x, y, z)\} = N_i\{u_i\} + N_j\{u_j\} + N_k\{u_k\} + N_m\{u_m\} \quad (7.19)$$

One may readily visualize that the N_i and N_j in a uniaxially loaded bar element, as described in Example 6.1, have the form of

$$N_i = N_1(x) = \frac{x - x_2}{x_1 - x_2} \quad \text{and} \quad N_j = N_2(x) = -\frac{x - x_1}{x_1 - x_2}$$

A check for the correct derivation of the interpolation function can be made by examining the following relationship:

$$N_i(x, y, z) + N_j(x, y, z) = 1 \quad (7.20)$$

Substitution of equation (7.17) into equation (7.11) will result in the relationship between the elements strains–nodal displacements:

$$\{\varepsilon(x, y, z)\} = [B(x, y, z)]\{u\} \quad (7.21)$$

where the $[B]$ matrix has the form

$$[B(x, y, z)] = [D][N(x, y, z)] \quad (7.22)$$

with $[D]$ given in equation (7.12).

The element stress-nodal displacements relationship can be expressed by substituting equation (7.21) into equation (7.13), resulting in

$$\{\sigma(x, y, z)\} = [C][B(x, y, z)]\{u\} \quad (7.23)$$

Upon substituting equation (7.21) and equation (7.23) into equation (7.16), the strain energy in the element can be expressed in terms of nodal displacements as

$$U(\{u\}) = \frac{1}{2}\int_v \{u\}^T[B(x, y, z)]^T[C][B(x, y, z)]\{u\}\, dv \quad (7.24)$$

The potential energy in a deformed solid (or element, in the present case) can be evaluated by summing up the strain energy in equation (7.24) and the work done to the element by the applied forces. Mathematically, the potential energy, π, can be expressed as

$$\pi = U + W_p$$

in which W_p is the potential of the applied loads. As the work done to the element by the loads, W, is the negative of their potential, or $W_p = -W$, the potential energy in the element thus equals $\pi = U - W$. The work done to the solid by the applied loads has the general form of

$$\begin{aligned} W &= \int_v \{U(x, y, z)\}^T\{f\}\,dv + \int_s \{U(x, y, z)\}^T\{t\}\,ds \\ &= \int_v \{u\}^T[N(x, y, z)]^T\{f\}\,dv + \int_s \{u\}^T[N(x, y, z)]^T\{t\}\,ds \end{aligned} \tag{7.25}$$

where $\{f\}$ is the body force vector (e.g., gravitational or inertia forces) and $\{t\}$ is the surface traction vector (e.g., pressure or other forms of distributed loads on the boundary s).

Summing up the strain energy in equation (7.24) and the work in equation (7.25), the potential energy in the deformed element can be expressed in terms of nodal displacements as

$$\begin{aligned} \pi(\{u\}) = \frac{1}{2}\int_v &\{u\}^T[B(x, y, z)]^T[C][B(x, y, z)]\{u\}\,dv \\ &- \int_v \{u\}^T[N(x, y, z)]^T\{f\}\,dv \\ &- \int_s \{u\}^T[N(x, y, z)]^T\{t\}\,ds \end{aligned} \tag{7.26}$$

By following the argument presented at the beginning of this section regarding the equilibrium condition for the deformed element, one can ensure such a condition by keeping the potential energy in the minimum state. Mathematically, this can be accomplished by letting

$$\frac{\partial\pi(\{u\})}{\partial\{u\}} = 0$$

In a strict sense, a deformed structure is in an equilibrium condition only when the potential energy in the entire structure is at a minimum. In a discretized solid, however, one may assume that the potential energy of the entire system is the sum of potential energies in all discrete elements. Minimizing the total potential energy function of the structure may be achieved by minimizing this function in each individual element, as implied in the last expression.

Realizing the fact that $\{u\}$ in equation (7.26) represents displacement components at the nodes of an element, with all these nodes fixed in the (x, y, z) space, it is justifiable to remove $\{u\}$ outside the integrals. Thus, one can readily arrive at the following set of equations by performing the variation of π in equation (7.26) with respect to $\{u\}$:

$$\left(\int_v [B(x, y, z)]^T[C][B(x, y, z)]\,dv\right)\{u\} - \int_v [N(x, y, z)]^T\{f\}\,dv - \int_s [N(x, y, z)]^T\{t\}\,ds = 0$$

The above equations can be rearranged to give a simple form:

$$[K_e]\{u\} = \{q\} \tag{7.27}$$

where

$$[K_e] = \text{stiffness matrix} = \int_v [B(x, y, z)]^T[C][B(x, y, z)]\,dv \tag{7.28}$$

$$\{q\} = \text{nodal force matrix} = \int_v [N(x, y, z)]^T\{f\}\,dv + \int_s [N(x, y, z)]^T\{t\}\,ds \tag{7.29}$$

Equation (7.27) represents a set of simultaneous equations involving all unknown nodal displacement components in the element. The remaining procedure for the finite element analysis follows what was presented in Section 6.4.

7.4 ▪ One-Dimensional Stress Analysis of Solids

The finite element analysis of one-dimensional solids involves the uniaxially loaded bars illustrated in Figure 7.5. This problem will be presented as a special case of the general formulation presented in the foregoing section.

Let us consider the bar element shown in Figure 7.5. The bar has a length of L with a uniform cross-sectional area A. Since the forces applied to the bar at the two ends (nodes 1 and 2) are along the x axis, one may expect the bar to deform only in the longitudinal direction. If we let the displacement along the bar be $U(x)$, the equivalent nodal displacements are $\{u\}$, to represent u_1 at node 1 and u_2 at node 2. The discretization in equation (7.17) becomes

$$U(x) = [N(x)]\{u\}$$

FIGURE 7.5
A bar element

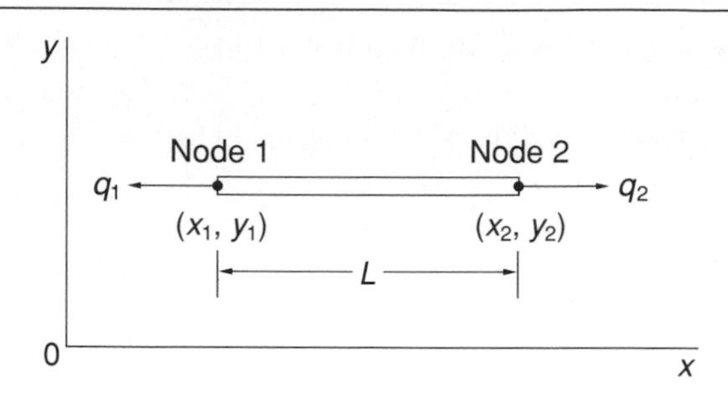

The interpolation function $[N(x)]$ based on a simple linear polynomial variation of $U(x)$ in the above expression has been derived as given in equation (6.3), or

$$[N(x)] = \left\{ \frac{x - x_2}{x_1 - x_2} \quad - \frac{x - x_1}{x_1 - x_2} \right\}$$

From this, we can derive the $[B]$ matrix in equation (7.22) for the present case as

$$[B(x)] = \frac{\partial}{\partial x}[N(x)] = \left\{ \frac{1}{x_1 - x_2} \quad - \frac{1}{x_1 - x_2} \right\}$$

Since the only stress and stain components that need to be considered are $\sigma_{xx}(x)$ and $\varepsilon_{xx}(x)$ in a one-dimensional analysis, the elasticity matrix $[C]$ in equation (7.13) is reduced to a single value of E, the Young's modulus of the material.

By substituting the above expressions of $[N(x)]$ into equation (7.28) with $dv = Adx$ and $x_1 - x_2 = -L$ in which A is the cross-sectional area of the bar, the stiffness matrix can be readily derived to take the form

$$[K_e] = \int_0^L \left\{ -\frac{1}{L} \quad \frac{1}{L} \right\}^T E \left\{ -\frac{1}{L} \quad \frac{1}{L} \right\} Adx = \frac{EA}{L} \begin{bmatrix} 1 & -1 \\ -1 & 1 \end{bmatrix}$$

Because the forces acting on the bar element are point loads, the force matrix in equation (7.29) reduces to a simple matrix of

$$\{q\}^T = [F_1 \quad -F_2\}$$

The element equation for the bar element thus takes the form of

$$\frac{EA}{L} \begin{bmatrix} 1 & -1 \\ -1 & 1 \end{bmatrix} \begin{Bmatrix} u_1 \\ u_2 \end{Bmatrix} = \begin{Bmatrix} F_1 \\ -F_2 \end{Bmatrix}$$

The reader should realize that not all one-dimensional stress analysis problems can be handled in as straightforward a manner as the above case.

The following example of the uniaxial stretching of a tapered rod may significantly increase the complexity of the analysis. It will also provide the reader with a sense of selecting a higher degree of polynomials for the interpolation function, as well as various ways in which to improve the accuracy of the results.

EXAMPLE ■ 7.1

A tapered rod is hung vertically from a solid base, as depicted in Figure 7.6. A force F is applied at the free end. Dimensions and material properties of the rod are given as follows:

Base diameter, D	127 mm
Length, b	483 mm
Projected length, L	508 mm
Young's modulus, E	68,940 MPa
Poisson's ratio, ν	0.33
Mass density, ρ	2,519 kg/m^3
Applied force, F	8,896 N

The analytical solutions for this case are summarized below. The stress:

$$\sigma_{yy}(y) = \frac{F + \dfrac{\pi}{12}\rho g D^2 L(1 - y/L)^3}{\dfrac{\pi}{4} D^2(1 - y/L)^2} \tag{7.30}$$

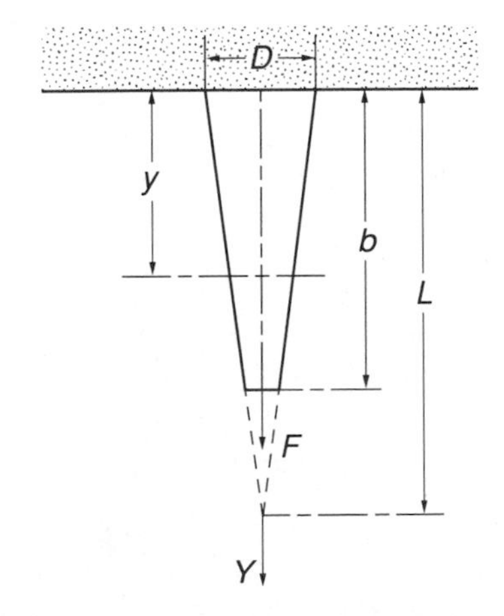

FIGURE 7.6
A tapered rod

The strain:

$$\varepsilon_{yy}(y) = \sigma_{yy}(y)/E \tag{7.31}$$

The displacement:

$$V(y) = \frac{4F}{\pi D^2 E} \frac{y}{(1 - y/L)} + \frac{\rho g}{3E}(Ly - y^2/2) \tag{7.32}$$

Case 1. Single element with linear interpolation function

Consider the entire tapered rod as one element with two nodes, as shown in Figure 7.7. The variation of the displacement in the rod is assumed to follow a linear polynomial function (i.e., $V(y) = \alpha_1 + \alpha_2 y$). The displacement at node 1 is $V_1 = 0$, and V_2 is designated as the unknown displacement at node 2. A force, $F_2 = 8{,}896$ N, is applied at that node. No force is applied at node 1 (i.e., $F_1 = 0$, despite the fact that there is a reaction at this node).

The assumption of linear variation of $V(y)$ in the element enables us to use the interpolation function given in equation (6.4), or

$$\{N(y)\} = \left\{1 - \frac{y}{b} \quad \frac{y}{b}\right\}$$

The $[B]$ matrix can be obtained by differentiating $\{N(y)\}$ with respect to y:

$$[B(y)] = \frac{\partial}{\partial y}\{N(y)\} = \left\{-\frac{1}{b} \quad \frac{1}{b}\right\}$$

The stiffness matrix can thus be determined by using equation (7.28) with $dv = A_0(1 - y/L)^2 dy$ in which A_0 is the cross-sectional area of the rod at the base.

$$\begin{aligned}[K_e] = [K] &= \int_v \left\{-\frac{1}{b} \quad \frac{1}{b}\right\}^T E \left\{-\frac{1}{b} \quad \frac{1}{b}\right\} dv \\ &= \frac{EA_0L}{3b^2}[1 - (1 - \beta)^3]\begin{bmatrix} 1 & -1 \\ -1 & 1 \end{bmatrix}\end{aligned} \tag{7.33}$$

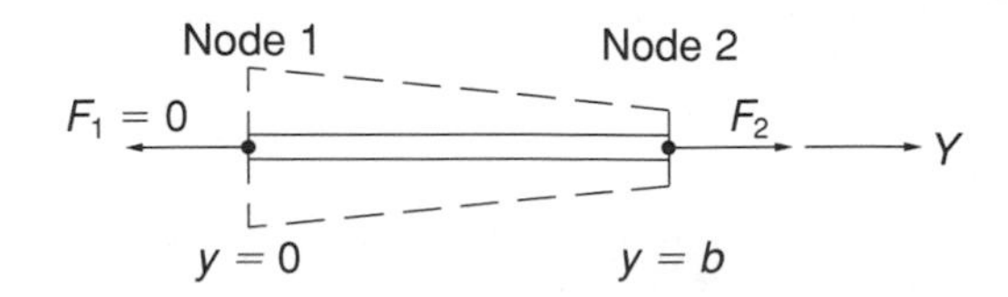

FIGURE 7.7 Single-element model for the tapered rod (*Note*: F_1 and F_2 are applied nodal forces, not nodal reactions)

The force matrix in equation (7.29) can be expressed for the present case as

$$\{q\} = \int_v [N(y)]^T\{W\}\,dv + \{F\}$$

$$= WA_0 \begin{Bmatrix} L[1-(1-\beta)^3]/3 + b[-1/2 + 2\beta/(3L) - \beta^2/(4L^2)] \\ b[1/2 - 2\beta/(3L) + \beta^2/(4L^2)] \end{Bmatrix} + \begin{Bmatrix} 0 \\ F \end{Bmatrix}$$

where $\beta = b/L$ and W is the weight density of the rod.

Substituting the $[K]$ and $\{q\}$ matrices into equation (7.27), one can readily solve for $V_2 (V_1 = 0)$:

$$V_2 = \frac{F + WA_0 b[1/2 - 2\beta/(3L) + \beta^2/(4L^2)]}{\dfrac{EA_0 L}{3b^2}[1-(1-\beta)^3]}$$

The numerical value of V_2 from the above expression is 1.412×10^{-5} m, which is more than one order of magnitude less than the corresponding analytical solution of 1.0382×10^{-4} m derived from equation (7.32). The variation of the displacement in the element follows a straight line, as depicted by curve (1) in Figure 7.8. An attempt will be made to improve the accuracy of the result by using

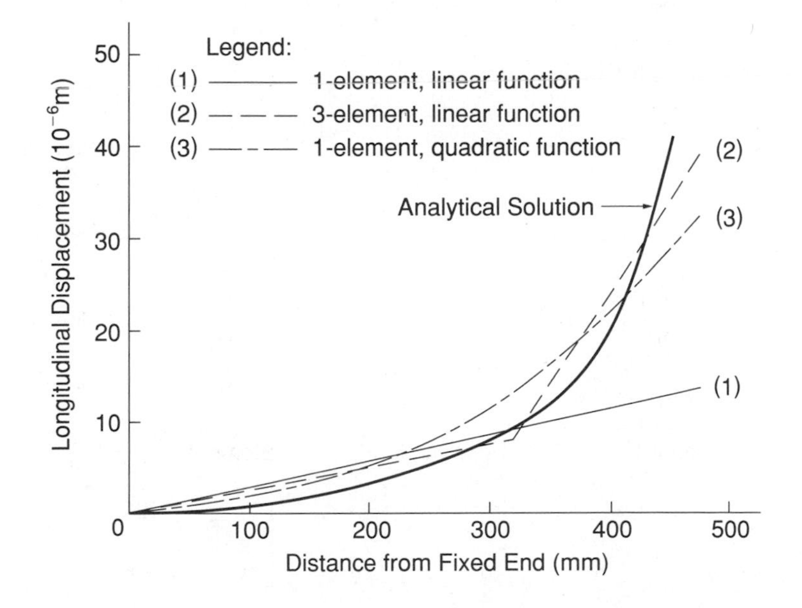

FIGURE 7.8
Correlation of results

more elements in the discretized model, as will be presented in the next case.

Case 2. Three elements with linear interpolation function

Instead of using just one element for the entire rod, three elements of equal length (i.e., $b/3$) are used in this case, as illustrated in Figure 7.9. The use of equation (6.4) for the derivation of the interpolation functions for each of the elements in Figure 7.9 requires the substitution of proper local nodal coordinates. The table below will be useful in keeping track of the proper coordinates for the interpolation function:

$$[N(y)] = \left\{ \frac{y - y_2}{y_1 - y_2} \quad -\frac{y - y_1}{y_1 - y_2} \right\}$$

Element No.	Node i	Node j	y_1 for Node i	y_2 for Node j
1	1	2	0	$b/3$
2	2	3	$b/3$	$2b/3$
3	3	4	$2b/3$	b

The stiffness matrices for each element can be derived by the following general expression:

$$[K_e] = \int_{y_1}^{y_2} \begin{Bmatrix} -3/b \\ 3/b \end{Bmatrix} E\{-3/b \quad 3/b\} A_0(1 - y/L)^2\, dy$$

Expressions for these matrices are

$$[K_1] = \frac{3A_0E}{b\beta}[1 - (1 - \beta/3)^3]\begin{bmatrix} 1 & -1 \\ -1 & 1 \end{bmatrix} \qquad \text{for element 1}$$

$$[K_2] = \frac{3A_0E}{b\beta}[(1 - \beta/3)^3 - (1 - 2\beta/3)^3]\begin{bmatrix} 1 & -1 \\ -1 & 1 \end{bmatrix} \qquad \text{for element 2}$$

$$[K_3] = \frac{3A_0E}{b\beta}[(1 - 2\beta/3)^3 - (1 - \beta)^3]\begin{bmatrix} 1 & -1 \\ -1 & 1 \end{bmatrix} \qquad \text{for element 3}$$

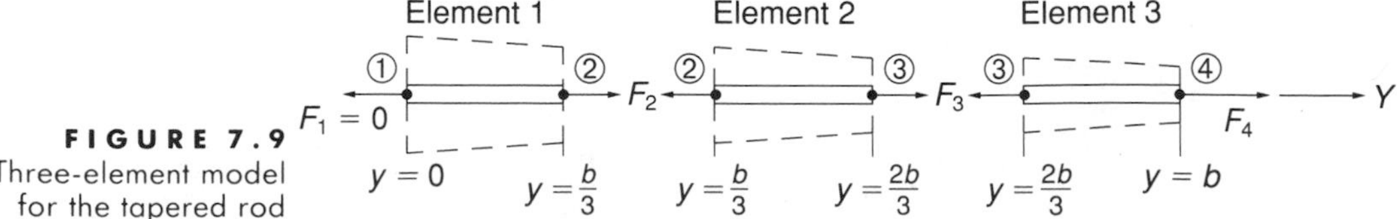

FIGURE 7.9 Three-element model for the tapered rod

The nodal force matrices for each element can be evaluated by the expression

$$\{q\} = \int_v [N(y)]^T\{W\}\,dv + \{F\}$$

$$= WA_0 \int_{y_1}^{y_2} \begin{Bmatrix} (1 - 3y/b)(1 - y/L)^2 \\ 3y(1 - y/L)^2/b \end{Bmatrix} dy + \{F\}$$

from which force matrices for each element can be established:

$$\{q_1\} = WA_0 \begin{Bmatrix} L[1 - (1 - \beta/3)^3]/3 - b[1/2 - 2\beta/(9L) + \beta^2/(36L^2)]/3 \\ b[1/2 - 2\beta/(9L) + \beta^2/(36L^2)]/3 \end{Bmatrix} + \begin{Bmatrix} -F_1 \\ F_2 \end{Bmatrix}$$

$$\{q_2\} = WA_0 \begin{Bmatrix} L[(1 - \beta/3)^3 - (1 - 2\beta/3)^3]/3 - b[3/2 - 14\beta/(9L) + 15\beta^2/(36L^2)]/3 \\ b[3/2 - 14\beta/(9L) + 15\beta^2/(36L^2)]/3 \end{Bmatrix} + \begin{Bmatrix} -F_2 \\ F_3 \end{Bmatrix}$$

$$\{q_3\} = WA_0 \begin{Bmatrix} L[(1 - 2\beta/3)^3 - (1 - \beta)^3]/3 - b[5/2 - 38\beta/(9L) + 9\beta^2/(12L^2)]/3 \\ b[5/2 - 38\beta/(9L) + 9\beta^2/(12L^2)]/3 \end{Bmatrix} + \begin{Bmatrix} -F_3 \\ F_4 \end{Bmatrix}$$

The assembled stiffness and force matrices with the appropriate numerical values of the parameters take the forms

$$[K] = \sum_1^3 [K_i] = 0.572 \times 10^{10} \times \begin{bmatrix} 0.6809 & -0.6809 & 0 & 0 \\ & 0.9507 & -0.2698 & 0 \\ & & 0.3190 & -0.0492 \\ & \text{SYM} & & 0.0492 \end{bmatrix}$$

$$\{R\} = \sum_1^3 \{q_i\} = \begin{Bmatrix} 11.45 \\ -26.61 \\ -47.89 \\ 116.2 \end{Bmatrix} + \begin{Bmatrix} 0 \\ 0 \\ 0 \\ 8896 \end{Bmatrix}$$

The displacements at nodes 1 to 4 (i.e., $\{V\}^T = \{V_1 \quad V_2 \quad V_3 \quad V_4\}$) can be solved by the following equation:

$$[K\}\{V\} = \{R\}$$

with $V_1 = 0$ as the specified boundary condition. Numerical solutions from the above equation give

$$V_2 = 0.2308 \times 10^{-5} \text{ m}$$

$$V_3 = 0.8151 \times 10^{-5} \text{ m}$$

$$V_4 = 0.4021 \times 10^{-4} \text{ m}$$

The corresponding strains and stresses in each element are

$\varepsilon_1 = 0.1435 \times 10^{-4}$ $\quad\varepsilon_2 = 0.3632 \times 10^{-4}$ $\quad\varepsilon_3 = 0.1968 \times 10^{-3}$

$\sigma_1 = 989.3$ KPa $\quad\sigma_2 = 2503.9$ KPa $\quad\sigma_3 = 13567.4$ KPa

The reader should notice that the variation of the displacement within each of the three elements remains linear due to the assumed linear interpolation functions. Results obtained by using the three-element model are graphically represented by curve (2) in Figure 7.8. A significantly improved correlation with the analytical results can be seen from these illustrations.

Case 3. Single element with higher-order interpolation function
The question of selecting the proper interpolation function for the finite element formulation was raised earlier (see Section 6.4). Here, we will demonstrate that more accurate results can be achieved by using fewer elements but higher-order interpolation functions in the formulation. The use of Gaussian integration, as presented in Section 2.6, in finite element analysis will also be demonstrated in this case study. A close examination of the analytical solution of the displacement in equation (7.32) will reveal that the solution follows a polynomial function of the form

$$V(y) = C_0 y(1 + y/L + y^2/L^2 + \cdots + y^n/L^n + \cdots)$$

in which C_0 is a constant.
The selection of an interpolation function such that

$$V(y) = \alpha_1 + \alpha_2 y + \alpha_3 y^2 \tag{7.34}$$

with α_1, α_2, and α_3 being arbitrary constants appears to be logical as this function conforms with part of the analytical solution shown above.

The determination of the above quadratic displacement function in equation (7.34) involves the solution of α_1, α_2, and α_3. A minimum of three sets of nodal coordinates need to be specified for this purpose. A third node is thus necessary. Although the choice of the location of the third node in the rod element is arbitrary, a halfway location was chosen in this case for the sake of convenience. The single element with three nodes is illustrated in Figure 7.10.

By following procedures similar to those outlined in Examples 6.1 and 6.2, one can readily solve for α_1, α_2, and α_3 in terms of the three nodal displacements:

$\alpha_1 = V_1 \quad \alpha_2 = (-3V_1 + 4V_1 - V_3)/b \quad \alpha_3 = 2(V_1 - 2V_2 + V_3)/b^2$

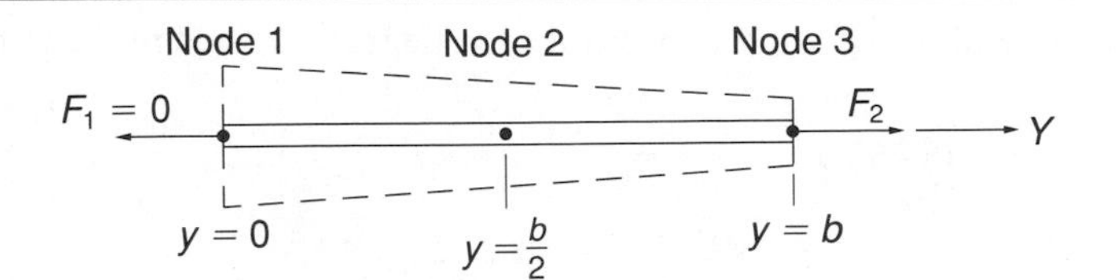

FIGURE 7.10
A single element with three nodes

Upon substituting the above expressions into equation (7.34), the interpolation function can be shown to take the form

$$[N(y)] = \{(1 - 2y/b)(1 - y/b) \quad 4y(1 - y/b)/b \quad -y(1 + 2y/b)/b\} \tag{7.35}$$

The $[B]$ matrix in equation (7.21) can be derived by differentiating the $[N]$ in equation (7.35) with respect to the coordinate y to give

$$\begin{aligned}[B(y)] &= \frac{\partial}{\partial y}[N(y)] \\ &= \{(-3/b + 4y^2/b^2) \quad (4/b - 8y/b^2) \quad (-1/b + 4y/b^2)\}\end{aligned}$$

The element stiffness matrix, $[K_e]$, can be determined by following equation (7.28):

$$\begin{aligned}[K_e] &= \int_0^b [B(y)]^T E[B(y)] A_0(1 - y/L)^2\,dy \\ &= EA_0 \int_0^{0.483} [B(y)]^T[B(y)](-y/0.508)^2\,dy \end{aligned} \tag{7.36}$$

The integral in equation (7.36) can be evaluated either by direct integration, as we did in the previous cases, or by the Gaussian integration method described in Section 2.6. The latter method will be used in this case because it is frequently used in finite element analyses.

The performance of Gaussian integration in equation (7.36) requires first a transformation of coordinates so that the integral will have the same form as that shown in equation (2.79). Such a transformation of coordinates can be achieved by letting

$$y = \frac{1}{2}(0.483 - 0)\zeta + \frac{1}{2}(0.483 + 0) = 0.2415\zeta + 0.2415$$

by which the integral in equation (7.36) in the y coordinate with the interval between 0 and 0.483 is transformed into the ζ coordinate with an interval between -1 and $+1$. One can also observe from the above relationship that $dy = 0.2415\zeta$.

The integral in equation (7.36) can then be transformed into the form

$$[K_e] = (8.757 \times 10^8)\frac{0.483 - 0}{2}\int_{-1}^{1}[B(\zeta)]^T[B(\zeta)] \times \left[1 - \frac{(0.2415\zeta + 0.2415)}{0.508}\right]^2 d\zeta$$

$$= 2.115 \times 10^8 \sum_{n=1}^{3}[B(\zeta_n)]^T[B(\zeta_n)]\left(\frac{0.2665 - 0.2415\zeta}{0.508}\right)^2 H_n$$

Numerical values of ζ_n and H_n are available in Table 2.2. For the case involving three sample points (i.e., $n = 3$), we have

$$\zeta_1 = -0.77459 \qquad H_1 = 0.5555$$

$$\zeta_2 = \quad 0.0 \qquad H_2 = 0.8888$$

$$\zeta_3 = \quad 0.77459 \qquad H_3 = 0.5555$$

By substituting the above values into the last expression for $[K_e]$, the numerical values of the elements of the $[K_e]$ matrix can be determined to give

$$[K_e] = 10^{10}\begin{bmatrix} 0.2827 & -0.3183 & 0.03566 \\ & 0.3962 & -0.07735 \\ & \text{SYM} & 0.04227 \end{bmatrix}$$

The body force matrix can be evaluated by following a procedure similar to that used in the other two cases. It is, however, neglected in this case study because the magnitude is trivial. The element equation that is used to solve the displacements at the three nodes takes the form

$$10^{10}\begin{bmatrix} 0.2827 & -0.3183 & 0.03566 \\ & 0.3962 & -0.07735 \\ & \text{SYM} & 0.04227 \end{bmatrix}\begin{Bmatrix} V_1 \\ V_2 \\ V_3 \end{Bmatrix} = \begin{Bmatrix} F_1 \\ F_2 \\ 8896 \end{Bmatrix}$$

The appropriate boundary conditions require that $V_1 = 0$ and $F_1 = F_2 = 0$, which leads to the solution of V_2 and V_3: $V_2 = 0.6501 \times 10^{-5}$ m and $V_3 = 0.3309 \times 10^{-4}$ m.

The strain distribution in the element can be determined by the following relationship:

$$\{\varepsilon(y)\} = [B(y)]\{V\} = (-1.467 + 34.443y) \times 10^{-5}$$

The corresponding stress distribution has the form

$$\{\sigma(y)\} = [C]\{\varepsilon(y)\} = -1011.41 + 23746\text{y} \quad (\text{KPa})$$

The variation of the displacement in the entire tapered rod using the interpolation function in equation (7.35) is graphically represented by curve (3) in Figure 7.8. It is readily seen that the results obtained from Case 3, using one element with a higher-order interpolation function, are much superior to those of the similar case using one element, but with a simple linear interpolation function (Case 1). The best correlation of results with the analytical solution obviously occurred in the case that involved more elements (i.e., Case 2). There, significantly more effort was needed in the computation. ❑

7.5 ▪ Two-Dimensional Stress Analysis of Solids (Plane Stress Case)

Many engineering problems involve structures of *thin* geometry, with the loads applying to the plane of the thickness, as illustrated in Figure 7.11. Thin plates, gears, the web of many large I-beams, wheels, etc., all fit this description. Since the variation of stresses with respect to the out-of-plane coordinate z is considered to be constant and negligible, it is reasonable to assume that out of the six components of stress in equation (7.1), three of them—σ_{zz}, σ_{xz}, σ_{yz}—can be neglected in comparison to the remaining three components—σ_{xx}, σ_{yy}, and σ_{xy}. This idealization is called *plane stress*, and the stress components can be expressed as

$$\{\sigma\} = \begin{Bmatrix} \sigma_{xx} \\ \sigma_{yy} \\ \sigma_{xy} \end{Bmatrix}$$

The corresponding components of strain are

$$\{\varepsilon\} = \begin{Bmatrix} \varepsilon_{xx} \\ \varepsilon_{yy} \\ \varepsilon_{xy} \end{Bmatrix}$$

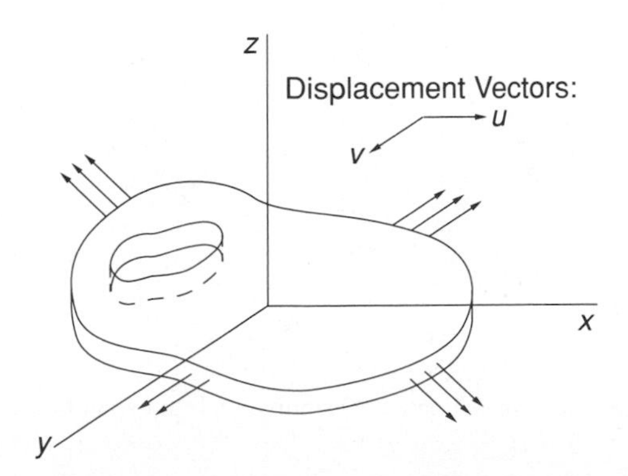

FIGURE 7.11 A typical plane stress geometry

Equation (7.10) becomes

$$\{\varepsilon\} = \begin{Bmatrix} \dfrac{\partial U}{\partial x} \\ \dfrac{\partial V}{\partial y} \\ \dfrac{\partial U}{\partial y} + \dfrac{\partial V}{\partial x} \end{Bmatrix} = \begin{bmatrix} \dfrac{\partial}{\partial x} & 0 \\ 0 & \dfrac{\partial}{\partial y} \\ \dfrac{\partial}{\partial y} & \dfrac{\partial}{\partial x} \end{bmatrix} \begin{Bmatrix} U \\ V \end{Bmatrix} \tag{7.37}$$

where U and V are the respective displacement components in the x and y directions.

The $[C]$ matrix in the stress-strain relationship in equation (7.13) has a simplified form of

$$[C] = \frac{E}{1 - \nu^2} \begin{bmatrix} 1 & \nu & 0 \\ \nu & 1 & 0 \\ 0 & 0 & \dfrac{1-\nu}{2} \end{bmatrix} \tag{7.38}$$

Since the triangular plane element is regarded as the fundamental element in finite element analysis,* the following formulation is performed on this basic element geometry.

Let us consider a triangular plane element with three nodes, i, j, and m, as shown in Figure 7.12. The corresponding nodal positions are specified by the coordinates (x_i, y_i), (x_j, y_j), and (x_m, y_m).

As the solid can deform only in the xy plane, the *element displacement field* thus involves two components, $u(x, y)$ and $v(x, y)$, which can be expressed in a vectorial form as follows:

$$\{U(x, y)\} = \begin{Bmatrix} u(x, y) \\ v(x, y) \end{Bmatrix} = [N(x, y)]\{u\} \tag{7.39}$$

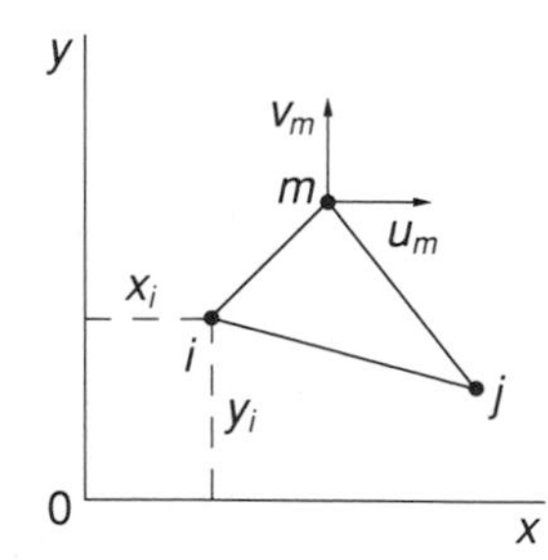

FIGURE 7.12 A typical triangular plate element

* Quadrilateral elements can be treated by subdivision into two or four triangular elements.

where

$\{U(x, y)\}$ = displacement field in the element

$\{u\}$ = displacements at nodes i, j, and m

$[N(x, y)]$ = interpolation function

As described earlier, the derivation of the interpolation function in equation (7.39) usually begins with an assumption of a function representing the variation of the displacement components in the element. Selection of this function often depends on the expected behavior of the material as a result of the applied actions, as demonstrated in Case 3 of Example 7.1. Here, we assume these functions take the simple form of linear polynomials, as shown below.

$$u(x, y) = \alpha_1 + \alpha_2 x + \alpha_3 y \qquad (7.40)$$

for the element displacement along the x direction and

$$v(x, y) = \alpha_4 + \alpha_5 x + \alpha_6 y \qquad (7.41)$$

for the element displacement along the y direction.

These two functions can be expressed in the matrix form as

$$\{U(x, y)\} = \begin{Bmatrix} u(x, y) \\ v(x, y) \end{Bmatrix} = \begin{bmatrix} 1 & x & y & 0 & 0 & 0 \\ 0 & 0 & 0 & 1 & x & y \end{bmatrix} \{\alpha_1 \quad \alpha_2 \quad \alpha_3 \quad \alpha_4 \quad \alpha_5 \quad \alpha_6\}^T \qquad (7.42)$$

This form may be abbreviated as

$$\{U(x, y)\} = [R(x, y)]\{\alpha\}$$

where $\alpha_1, \alpha_2, \ldots, \alpha_6$ are arbitrary constants that can be expressed in terms of the nodal displacements (i.e., u_i, u_j, u_m, v_i, v_j, and v_m) upon substituting the respective nodal coordinates into equations (7.40) and (7.41):

$$u_i = \alpha_1 + \alpha_2 x_i + \alpha_3 y_i \qquad v_i = \alpha_4 + \alpha_5 x_i + \alpha_6 y_i$$

$$u_j = \alpha_1 + \alpha_2 x_j + \alpha_3 y_j \qquad v_j = \alpha_4 + \alpha_5 x_j + \alpha_6 y_j$$

$$u_m = \alpha_1 + \alpha_2 x_m + \alpha_3 y_m \qquad v_m = \alpha_4 + \alpha_5 x_m + \alpha_6 y_m$$

The nodal displacements, $\{u\}$, can be expressed in the following matrix from the above relations:

$$\{u\} = \begin{Bmatrix} u_i \\ u_j \\ u_m \\ v_i \\ v_j \\ v_m \end{Bmatrix} = \begin{bmatrix} 1 & x_i & y_i & 0 & 0 & 0 \\ 1 & x_j & y_j & 0 & 0 & 0 \\ 1 & x_m & y_m & 0 & 0 & 0 \\ 0 & 0 & 0 & 1 & x_i & y_i \\ 0 & 0 & 0 & 1 & x_j & y_j \\ 0 & 0 & 0 & 1 & x_m & y_m \end{bmatrix} \begin{Bmatrix} \alpha_1 \\ \alpha_2 \\ \alpha_3 \\ \alpha_4 \\ \alpha_5 \\ \alpha_6 \end{Bmatrix}$$

or

$$\{u\} = [A]\{\alpha\} \tag{7.43}$$

The matrix $[A]$ involves given nodal coordinates only.

From equation (7.43), one can determine the coefficients, $\alpha_1 . \alpha_2, \ldots, \alpha_6$, from the relationship

$$\{\alpha\} = [A]^{-1}\{u\} = [h]\{u\} \tag{7.44}$$

where $[h] = [A]^{-1}$ involves the specified nodal coordinates. By substituting equation (7.44) into equation (7.42), the element displacements may be expressed as

$$\{U(x, y)\} = [R(x, y)][h]\{u\} \tag{7.45}$$

After comparing the expressions in equations (7.45), (6.1), and (7.17), one can conclude that the interpolation function takes the form

$$[N(x, y)] = [R(x, y)][h] \tag{7.46}$$

with

$$[R(x, y)] = \begin{bmatrix} 1 & x & y & 0 & 0 & 0 \\ 0 & 0 & 0 & 1 & x & y \end{bmatrix} \tag{7.47}$$

The relationship between the element strains and the nodal displacements can be derived by substituting equation (7.42) into equation (7.11):

$$\{\varepsilon\} = \begin{Bmatrix} \varepsilon_{xx} \\ \varepsilon_{yy} \\ \varepsilon_{xy} \end{Bmatrix} = \begin{Bmatrix} \dfrac{\partial u}{\partial x} \\ \dfrac{\partial v}{\partial y} \\ \dfrac{\partial u}{\partial y} + \dfrac{\partial v}{\partial x} \end{Bmatrix} = \begin{Bmatrix} \alpha_2 \\ \alpha_6 \\ \alpha_3 + \alpha_5 \end{Bmatrix}$$

The reader should notice at this point that the strain components expressed in the above expression are constants within the element (i.e., there is no variation in each of these strain components in the element). This is due to the fact that linear polynomial functions are used for the displacements in equations (7.40) and (7.41). Algorithms based on the constant strain element (and thus constant stress) formulation are simple in derivation. However, smaller, and therefore many more, elements are needed in the areas of the structure in which high concentrations of stresses or strains are expected. Such arrangement is necessary to ensure the accuracy of the result.

By substituting equation (7.43) into equation (7.44) and then into the above expression, we can show the element strains–nodal displacements relationship to be identical to that given in equation (7.21), or

$$\{\varepsilon(x, y)\} = [B(x, y)]\{u\} \tag{7.48}$$

in which

$$[B]=\frac{1}{2A}\begin{bmatrix} (y_j-y_m) & (y_m-y_i) & (y_i-y_j) & 0 & 0 & 0 \\ 0 & 0 & 0 & (x_m-x_j) & (x_i-x_m) & (x_j-x_i) \\ (x_m-x_j) & (x_i-x_m) & (x_j-x_i) & (y_j-y_m) & (y_m-y_i) & (y_i-y_j) \end{bmatrix} \tag{7.49}$$

with A equal to the area of the triangle bounded by the three nodes, or

$$A = (x_m - x_i)(y_m - y_j) - \frac{1}{2}(y_m - y_i)(x_m - x_i)$$
$$- \frac{1}{2}(y_i - y_j)(x_j - x_i) - \frac{1}{2}(y_m - y_j)(x_m - x_j)$$

The element equations for a triangular plate element can be derived by following the procedure described in step 4 in Section 6.4 (i.e., by minimizing the potential energy function).

The potential energy function for this case is similar to what is shown in equation (7.26) as

$$\pi = \frac{1}{2}\{u\}^T\left(\int_v [B]^T[C][B]\,dv\right)\{u\} - \{u\}^T\int_v [N]^T(f\}\,dv$$
$$- \{u\}^T\int_s [N]^T\{t\}\,ds$$

Minimizing π with respect to the nodal displacements, $\{u\}$—i.e., letting

$$\frac{\partial \pi}{\partial \{u\}} = 0$$

will lead to the element equation shown in equation (7.27).

$$[K_e]\{u\} = \{q\}$$

with expressions for the stiffness and nodal force matrices given in equations (7.28) and (7.29), respectively.

Once the stiffness matrix for each element is evaluated, the overall stiffness matrix for the entire structure can be assembled following the description in step 5, Section 6.4:

$$[K] = \sum_{m=1}^{M} [K_e^m]$$

The overall equilibrium equation for the entire structure becomes

$$[K]\{u\} = \{R\} \tag{7.50}$$

where $\{R\}$ is the resultant nodal force vector.

The nodal displacement of all the nodes in the whole structure can be calculated from the simultaneous equations in (7.50) by such numerical techniques as the Gaussian elimination technique presented in Chapter 2.

It is easily seen that evaluation of the element stiffness matrix in equation (7.28) requires the performance of integration over the element area with the assumption that the element has a constant thickness w:

$$[K_e] = \int_y \int_x [B(x, y)]^T [C][B(x, y)] w \, dx \, dy$$

The above integration sometimes becomes very complicated. However, if the size of the element is not too large, the above integral can be approximated as follows:

$$[K_e] \doteq [B]^T [C][B] w A \tag{7.51}$$

where A is the area of the element; matrix $[B]$, which relates the element strain and nodal displacements, is expressed in equation (7.49); and matrix $[C]$ is given in equation (7.38).

Most commercial codes evaluate integrals for the stiffness and force matrices using a numerical integration method, such as the Gaussian integration technique illustrated in Example 7.1. For integrations over two-dimensional domains or in a three-dimensional space, relevant formulations for Gaussian quadrature are available in such publications as references [2 and 4].

EXAMPLE ■ 7.2

Use the finite element method to find (a) the displacements at the corners of a triangular plate induced by a force acting at one of the corners, as illustrated in Figure 7.13(a); (b) the displacement in the plate; (c) the stresses and strains in the plate; and (d) the reactions at the fixed corner. The plate has a thickness of 1 inch and is made of an aluminum alloy with the following material properties: Young's modulus, $E = 10^7$ psi, and Poisson's ratio, $\nu = 0.3$.

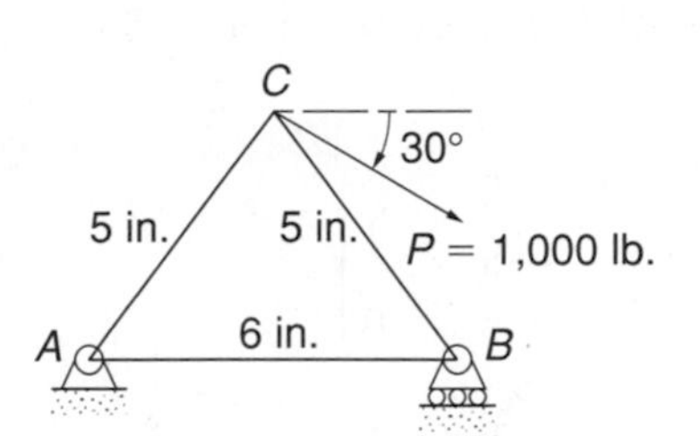

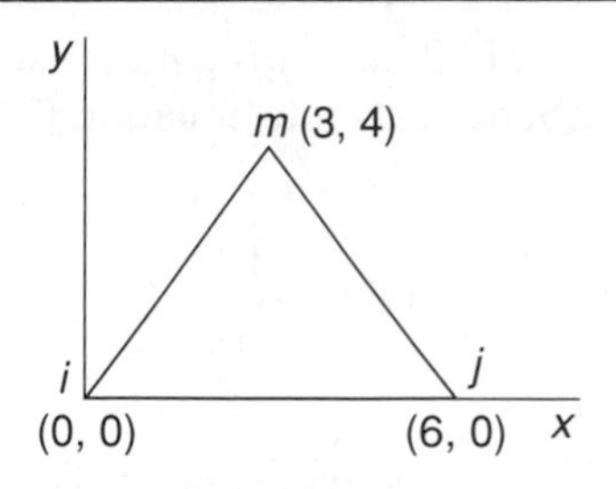

FIGURE 7.13 Finite element analysis of a plane element: (a) dimensions; (b) nodal coordinates

We assume that one triangular element with nodes i, j, and m is adequate for the plate. The origin of the coordinate system is set to coincide with the fixed corner, as shown in Figure 7.13(b). Coordinates of the nodes can then be established as ($x_i = 0$, $y_i = 0$), ($x_j = 6$, $y_j = 0$), and ($x_m = 3$, $y_m = 4$). The area of the element is readily determined to be 12 in^2.

Let us use linear polynomial functions such as those shown in equations (7.40) and (7.41) for the element displacement functions. The $[A]$ matrix in equation (7.43) has the form

$$[A] = \begin{bmatrix} 1 & 0 & 0 & 0 & 0 & 0 \\ 1 & 6 & 0 & 0 & 0 & 0 \\ 1 & 3 & 4 & 0 & 0 & 0 \\ 0 & 0 & 0 & 1 & 0 & 0 \\ 0 & 0 & 0 & 1 & 6 & 0 \\ 0 & 0 & 0 & 1 & 3 & 4 \end{bmatrix}$$

The $[h]$ matrix in equation (7.45) is obtained by inverting the above matrix to give

$$[h] = \begin{bmatrix} 1 & 0 & 0 & 0 & 0 & 0 \\ -0.167 & 0.167 & 0 & 0 & 0 & 0 \\ -0.125 & -0.125 & 0.25 & 0 & 0 & 0 \\ 0 & 0 & 0 & 1 & 0 & 0 \\ 0 & 0 & 0 & -0.167 & 0.167 & 0 \\ 0 & 0 & 0 & -0.125 & -0.125 & 0.25 \end{bmatrix} \quad (7.52)$$

The interpolation function matrix can be expressed by using equations (7.46) and (7.47) with the form

$$[N(x, y)] = \begin{bmatrix} (1 - 0.167x - 0.125y) & (0.167x - 0.125y) & 0.25y & 0 & 0 & 0 \\ 0 & 0 & 0 & (1 - 0.167x - 0.125y) & (0.167x - 0.125y) & 0.25y \end{bmatrix}$$

The $[B]$ matrix is then constructed with the substitution of nodal coordinates into equation (7.49) as

$$[B] = \frac{1}{24}\begin{bmatrix} -4 & 4 & 0 & 0 & 0 & 0 \\ 0 & 0 & 0 & -3 & -3 & 6 \\ -3 & -3 & 6 & -4 & 4 & 0 \end{bmatrix}$$

For the present case, the $[C]$ matrix in equation (7.38) has the form of

$$[C] = 10.99 \times 10^6 \begin{bmatrix} 1 & 0.3 & 0 \\ 0.3 & 1 & 0 \\ 0 & 0 & 0.35 \end{bmatrix}$$

We are now ready to evaluate the element stiffness matrix by substituting the above $[B]$ and $[C]$ matrices into equation (7.51), resulting in the following form:

$$[K_e] = 228{,}958.32 \begin{bmatrix} 19.15 & -12.85 & -6.3 & 7.8 & -0.6 & -7.2 \\ & 19.15 & -6.3 & 0.6 & -7.8 & 7.2 \\ & & 12.6 & -8.4 & 8.4 & 0 \\ & & & 14.6 & 3.4 & -18 \\ & \text{SYM} & & & 14.6 & -18 \\ & & & & & 36 \end{bmatrix}$$

Since there is only one element in the model, the element stiffness matrix is used as the overall stiffness matrix. The solution to the problem can thus be shown below:

(a) The nodal displacements $\{u\}$:

The following boundary conditions are used:

$$u_i = v_i = v_j = 0$$

and the applied nodal forces can be expressed as

$$f_{ix} = f_{iy} = f_{jx} = f_{jy} = 0$$

$$f_{mx} = p\cos 30^\circ = 866 \text{ lb.}$$

$$f_{my} = p\sin 30^\circ = -500 \text{ lb.}$$

The second subscript, x or y, attached to the applied nodal force components denotes the direction of these forces in the respective x and y directions.

The overall equilibrium equation in equation (7.50) takes the following form:

$$228{,}958.32 \times \begin{bmatrix} 19.15 & -12.85 & -6.3 & 7.8 & -0.6 & -7.2 \\ & 19.15 & -6.3 & 0.6 & -7.8 & 7.2 \\ & & 12.6 & -8.4 & 8.4 & 0 \\ & & & 14.6 & 3.4 & -18 \\ & \text{SYM} & & & 14.6 & -18 \\ & & & & & 36 \end{bmatrix}$$

$$\times \begin{Bmatrix} u_i = 0 \\ u_j \\ u_m \\ v_i = 0 \\ v_j = 0 \\ v_m \end{Bmatrix} = \begin{Bmatrix} f_{ix} = 0 \\ f_{jx} = 0 \\ f_{mx} = 866 \\ f_{iy} = 0 \\ f_{jy} = 0 \\ f_{my} = -500 \end{Bmatrix}$$

The solution of the above matrix of equations requires rearrangement of the rows and columns according to the rule established in Section 6.4 and equation (6.12). For the present case, it was necessary to interchange rows 2 and 4 first (accompanied by the interchange of the respective columns in the [K] matrix), followed by the interchange of rows and columns 3 and 5. The resultant equations have the form

$$228{,}958.32 \times \left[\begin{array}{ccc|ccc} 19.15 & 7.8 & -0.6 & -12.85 & -6.3 & -7.2 \\ 7.8 & 14.6 & 3.4 & 0.6 & -8.4 & -18 \\ -0.6 & 3.4 & 14.6 & -7.80 & 8.4 & -18 \\ \hline -12.85 & 0.6 & -7.8 & 19.15 & -6.3 & 7.2 \\ -6.3 & -8.4 & 8.4 & -6.3 & 12.6 & 0 \\ -7.2 & -18 & -18 & 7.2 & 0 & 36 \end{array}\right]$$

$$\times \left(\begin{array}{c} 0 \\ 0 \\ 0 \\ \hline u_j \\ u_m \\ v_m \end{array}\right) = \left(\begin{array}{c} 0 \\ 0 \\ 0 \\ \hline 0 \\ 866 \\ -500 \end{array}\right)$$

The three nonzero nodal displacements can be solved from the above partitioned matrix equations by the following simultaneous equations:

$$228{,}958.32 \begin{bmatrix} 19.15 & -6.3 & 7.2 \\ -6.3 & 12.6 & 0 \\ 7.2 & 0 & 36 \end{bmatrix} \begin{Bmatrix} u_j \\ u_m \\ v_m \end{Bmatrix} = \begin{Bmatrix} 0 \\ 866 \\ -500 \end{Bmatrix}$$

The following numerical values were obtained for these displacement components:

$$u_j = 0.15989 \times 10^{-3} \text{ in.}$$

$$u_m = 0.38012 \times 10^{-3} \text{ in.}$$

$$v_m = -0.09264 \times 10^{-3} \text{ in.}$$

(b) The displacement components in the plate may be expressed by using equation (7.39) with the interpolation function, $[N(x, y)]$, and the nodal displacements:

$$\begin{aligned} u(x, y) &= (1 - 0.167x - 0.125y)u_i \\ &\quad + (0.167x - 0.125y)u_j + 0.25yu_m \\ &= (0.167x - 0.125y) \times 0.15989 \times 10^{-3} \\ &\quad + 0.25 \times 0.38012 \times 10^{-3}y \end{aligned}$$

$$\begin{aligned} v(x, y) &= (1 - 0.167x - 0.125y)v_i \\ &\quad + (0.167x - 0.125y)v_j + 0.25yv_m \\ &= -0.09264 \times 0.25 \times 10^{-3}y \end{aligned}$$

(c) The strain components in the element can be computed by using equation (7.48), as follows:

$$\begin{Bmatrix} \varepsilon_{xx} \\ \varepsilon_{yy} \\ \varepsilon_{xy} \end{Bmatrix} = \frac{1}{24} \times \begin{bmatrix} -4 & 4 & 0 & 0 & 0 & 0 \\ 0 & 0 & 0 & -3 & -3 & 6 \\ -3 & -3 & 6 & -4 & 4 & 0 \end{bmatrix}$$

$$\times \begin{Bmatrix} u_i = 0 \\ u_j = 0.15989 \\ u_m = 0.38012 \\ v_i = 0 \\ v_j = 0 \\ v_m = -0.090264 \end{Bmatrix} \times 10^{-3} = \begin{Bmatrix} 26.648 \\ -23.160 \\ 75.044 \end{Bmatrix} \times 10^{-6}$$

The stress components can be computed by the generalized Hooke's law, (i.e., $\{\sigma\} = [c]\{\varepsilon\}$):

$$\begin{Bmatrix} \sigma_{xx} \\ \sigma_{yy} \\ \sigma_{xy} \end{Bmatrix} = 10.99 \times 10^6 \begin{bmatrix} 1 & 0.3 & 0 \\ 0.3 & 1 & 0 \\ 0 & 0 & 0.35 \end{bmatrix} \begin{Bmatrix} 26.648 \\ -23.160 \\ 75.044 \end{Bmatrix} \times 10^{-6}$$

$$= \begin{Bmatrix} 216.503 \\ -166.670 \\ 288.657 \end{Bmatrix} \text{ psi}$$

(d) The reactions at all the nodes in the element can be obtained by using the expression in problem 7.1 with the form of

$$\{R\} = \int_v [B]^T\{\sigma\}\,dv = [B]^T\{\sigma\}\,\Delta V$$

$$\begin{Bmatrix} R_{ix} \\ R_{jx} \\ R_{mx} \\ R_{iy} \\ R_{jy} \\ R_{my} \end{Bmatrix} = \frac{1}{24}\begin{bmatrix} -4 & 0 & -3 \\ 4 & 0 & -3 \\ 0 & 0 & 6 \\ 0 & -3 & -4 \\ 0 & -3 & 4 \\ 0 & 6 & 0 \end{bmatrix}\begin{Bmatrix} 216.503 \\ -166.670 \\ 288.657 \end{Bmatrix} \times 12 \times 1$$

$$= \begin{Bmatrix} -865.99 \\ 0.021 \\ 865.97 \\ -327.31 \\ 827.32 \\ -500.01 \end{Bmatrix} = \begin{Bmatrix} -866.0 \\ 0.0 \\ 866.0 \\ -327.3 \\ 827.3 \\ -500.0 \end{Bmatrix}$$

The reactions at node 1 therefore have the numerical values of $R_{ix} =$ 866.0 lb. toward the left in the horizontal direction and $R_{iy} = 327.3$ lb. in the downward direction along the y coordinate. ❑

7.6 ▪ General-Purpose Finite Element Programs

The incredible versatility of the finite element method for solving industrial problems involving various engineering disciplines has been illustrated in Section 6.3. A finite element program that can be used to solve a large class of engineering problems obviously is desirable for industrial users. Programs that have this capability are called *general-purpose programs.* A number of such programs are available either through commercial outlets or by private arrangement. Surveys on general-purpose programs have been made by individuals or organizations in recent years, and the results have been published [5, 6].

Since most general-purpose finite element programs are developed for commercial purposes, the following configurations are common in these programs:

1. The program should be capable of solving a large class of engineering problems, ranging from elastic stress analysis of structures to some unusual applications such as piezoelectric analysis for transformers and microphones.

2. It must have a large element library from which users can select element configurations that closely match the configurations of the structures and the nature of the loading and deformation conditions (e.g., bending elements, friction elements, etc.).

3. It must be cost-effective in terms of manpower and computation. Highly efficient solution methods are used for problems involving a large number of DOF (degrees of freedom) (i.e., unknown nodal variables).

4. It must be user friendly. Popular commercial codes normally have easy-to-use preprocessors (e.g., automatic mesh generator) and postprocessors (e.g., color-coded graphical output, such as that shown in Plates 9, 10, 13, 15, and 16).

5. It should be adaptable to most computer hardware, from micro- to supercomputers.

6. It can be readily interfaced with other software—in particular, CAD/CAM/drafting software through special buffer programs called *translators*, as illustrated in Figure 7.14. Brief descriptions of some of the commercial CAD codes are available in reference [7].

Commercial codes are usually available to users on a leasing basis. The user pays the vendor or developer a license fee and additional fees on the volume of usage (e.g., number of DOF). The vendor is responsible for the software's maintenance, as well as its upgrading. A major disadvantage of using these codes, however, is that the user has no access to the source code because almost all these programs are proprietary for obvious reasons.

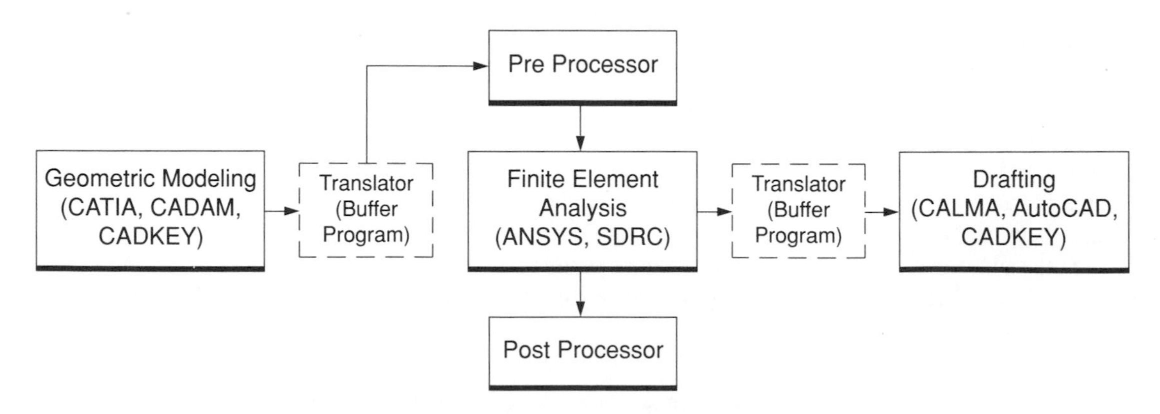

FIGURE 7.14 Integration of finite element analysis with CAD

7.7 ▪ The ANSYS Program

There are obviously a number of well-established general-purpose finite element analysis programs available in the marketplace. Many of these codes have the characteristics and capabilities described in the foregoing section. Here, we will present the reader with an outline of the specific features of one of the most widely used codes, the ANSYS program [8]. Our purpose is merely to familiarize the reader with much of the versatility that a commercial code can provide.

The ANSYS program has been developed and maintained by Swanson Analysis Systems, Inc., since 1970. It is probably the most widely used large-scale general-purpose code in the marketplace today. According to the developer, the customer base includes over 2,800 licenses, including more than 900 university licenses. It is used by engineers worldwide.

❑ PROGRAM CAPABILITIES

1. *Static structural analysis:* Includes elastic, plastic, creep, swelling, and buckling stress analysis of solids with small or large deformations
2. *Dynamic structural analysis:* Includes linear and nonlinear transient vibration, modal analysis, seismic, and random vibration
3. *Heat transfer analysis:* Handles both steady state and transient heat flow in solids, and heat transfer involving phase change and coupled thermal-structural effects
4. *Fluid dynamics and acoustic analysis:* Includes potential flow, free convection, and mass transport
5. *Kinematic analysis:* Handles both rigid-body and flexible linkages and mechanisms
6. *Magnetic analysis:* Includes inductance, flux density, flux lines, forces, power loss, etc.; the main applications are in solenoids, actuators, motors, and transformers
7. *Piezoelectric analysis:* Handles problems involving three-dimensional structures with AC or DC electrical sources, such as transducers, oscillators, resonators, and microphones

Input material properties to all the above analyses can be a function of temperature, as in the case of structural behavior in high-temperature environments.

❑ MODELING CAPABILITIES

1. The code accepts geometries of structures defined by three common coordinate systems: Cartesian, cylindrical, and spherical.

2. It can be run in either interactive or batch mode. Prompting commands are returned by a program in the interactive mode. Free-format data input is allowed in either mode.

❑ ELEMENT LIBRARY

There are altogether 77 different types of elements that can be used to discretize a structure. Some of these elements are illustrated in Figure 7.15.

The choice of elements is largely dependent on the geometry of the structure, as well as on the nature of the analysis (e.g., elastic, heat transfer) to be conducted.

❑ SOLUTION METHODS

The ANSYS program uses the wave-front direct solution method for solving a system of simultaneous equations such as that given in equation (7.50). This technique simultaneously assembles and solves an overall stiffness matrix [$[K]$ in equation (7.50)] from the individual element matrices such as those shown in equation (7.28). This procedure progressively moves through the model, element by element. In the meantime, equations corresponding to the particular element's DOF are solved. Once the DOF are solved, they are immediately removed from the matrix, using the Gaussian elimination method described in Chapter 2. The DOF set present in the assembled matrix is called *wave front*.

Other techniques such as the Guyan reduction method and Jacobi eigenvalue extraction are used for dynamic analyses.

❑ PREPROCESSING CAPABILITIES

1. Automatic mesh generation is available for complex two- or three-dimensional structural models. The user needs only to define the geometry and boundaries of the structure. The input keypoints can create line segments that automatically form two-dimensional meshes on the basis of the specified area discretization or three-dimensional meshes on the basis of the specified volume of meshes.

2. The multiple region mesh generation option allows the user to discretize structures in different portions (e.g., a shell and a nozzle). The program can then "assemble" the meshes of these portions to form the model for the entire structure.

3. The symmetry reflections and system transfers for nodal coordinates provide significant savings of the user's time and effort in the discretization process.

4. Digitizing via cross-hairs or tablet hardware may be used with automatic mesh generating routines.

Spar	Spar	Tension-Only Spar	Elastic Beam	Elastic Beam
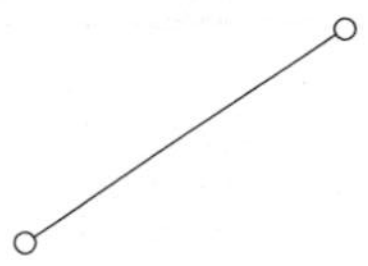		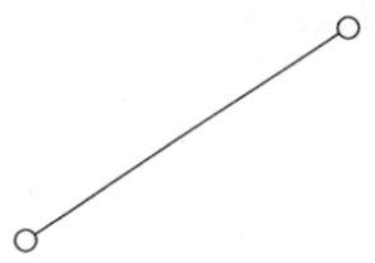	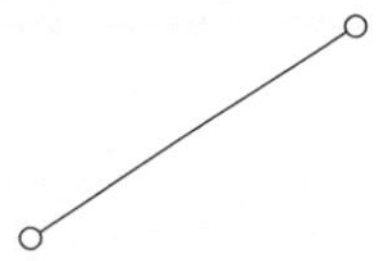	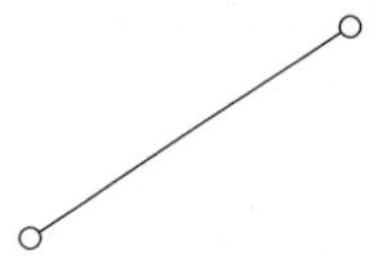
STIF1 2 nodes 2-D space DOF: UX, UY F: (P, R, S) Axial force only (truss).	STIF8 2 nodes 3-D space DOF: UX, UY, UZ F: (P, R, S) Axial force only (truss).	STIF10 2 nodes 3-D space DOF: UX, UY, UZ F: (R, S) Bilinear behavior may be tension-only (hook or cable) or compression-only (gap).	STIF3 2 nodes 2-D space DOF: UX, UY, ROTZ F: (R, S) For bending members with symmetric cross-sections.	STIF4 2 nodes 3-D space DOF: UX, UY, ROTX ROTY, ROTZ F: (R, S) Bending or torsional members with symmetric cross-sections.

Isoparametric Solid	Isoparametric Solid	Hyperelastic Solid	Crack Tip Solid	Tetrahedral Solid
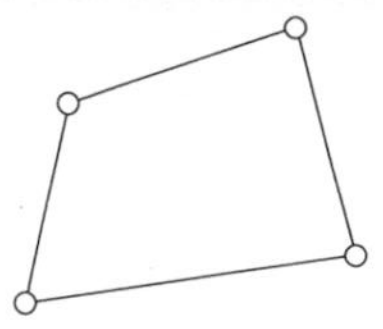	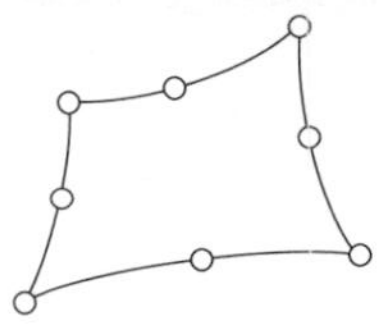	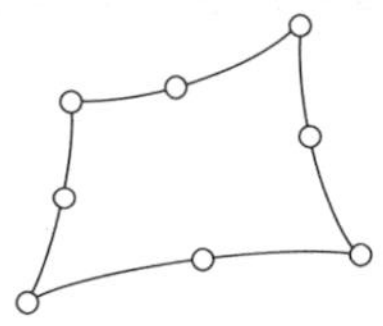	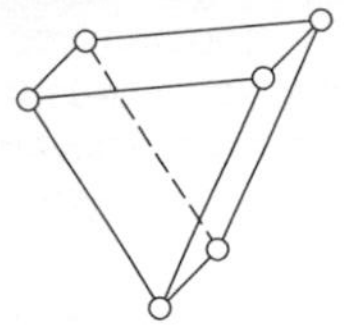	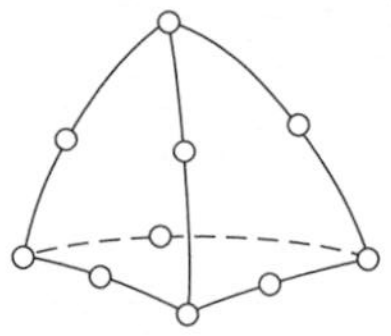
STIF42 4 nodes 2-D space DOF: UX, UY F: (P, R, S) Plane stress, plane strain, or axisym analyses.	STIF82 8 nodes 2-D space DOF: UX, UY F: (P, S) Plane stress, plane strain, or axisym models.	STIF84 8 nodes 2-D space DOF: UX, UY F: For rubber-like Mooney-Rivlin or Blatz-Ko materials. Plane strain or axisym with torsion.	STIF85 6 nodes 3-D space DOF: UX, UY, UZ F: (P) Models stresses at a crack front using an inverse square-root singularity method.	STIF92 10 nodes 3-D space DOF: UX, UY, UZ F: (P, R, S) Particularly suited to automatic meshing of arbitrary volumes.

FIGURE 7.15 Typical element library ANSYS program (partial listing)

(continues)

Source: Reprinted from ANSYS Revision 4.3 Seminar Notes, *Introduction to ANSYS*, Vol. 1, with permission from Swanson Analysis Systems, Inc., 1990)

Isoparametric Solid	Isoparametric Solid	Anisotropic Solid	Reinforced Solid	Hyperelastic Solid
STIF45 8 nodes 3-D space DOF: UX, UY, UZ F: (P, R, S) General application to 3-D solid models. Generalized plane strain option.	STIF95 20 nodes 3-D space DOF: UX, UY, UZ F: (P) General application to 3-D solid models. Generalized plane strain option.	STIF64 8 nodes 3-D space DOF: UX, UY, UZ F: (S) Applicable to anisotropic or crystalline materials. Generalized plane strain option.	STIF65 8 nodes 3-D space DOF: UX, UY, UZ F: (P) Matrix material may crack (in 3 directions), crush, creep, or yield. Reinforcing may be included which creeps and yields. Suitable for concrete, rock, fiber-glass, composite, etc.	STIF86 8 nodes 3-D space DOF: UX, UY, UZ F: For rubber-like Mooney-Rivlin or Blatz-Ko materials.
Quadrilateral Shell	**Isoparametric Shell**	**Plastic Triangular Shell**	**Membrane Shell**	**Plastic Axisymmetric Shell with Torsion**
STIF63 4 nodes 3-D space DOF: UX, UY, UZ, ROTX, ROTY, ROTZ F: (R, S) General thin shell or plate analyses. Membrane-only, bending-only, or both.	STIF93 8 nodes 3-D space DOF: UX, UY, UZ, ROTX, ROTY, ROTZ F: (S) Suited for curved shell structures.	STIF48 3 nodes 3-D space DOF: UX, UY, UZ, ROTX, ROTY, ROTZ F: (P, R, S) Plastic anal-yses of shell structures. Membrane-only, bending-only, or both.	STIF41 4 nodes 3-D space DOF: UX, UY, UZ F: No bending. Allows membrane collapse in one or two orthogonal directions for cloth-type structures.	STIF51 2 nodes 2-D space DOF: UX, UY, UZ, ROTZ F: (P, R, S) Axisym plastic analyses of conical shells. Torsional behavior option.

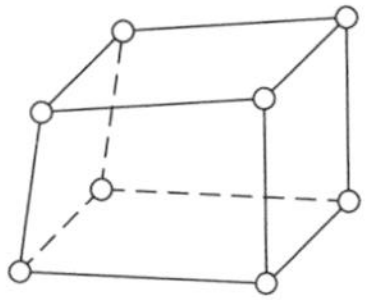

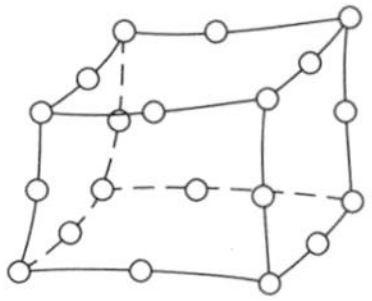

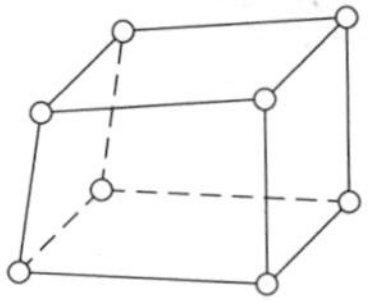

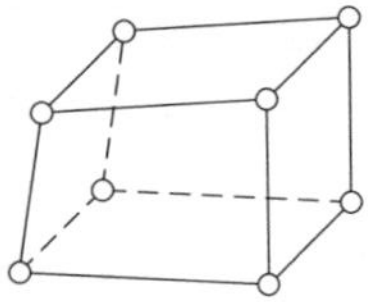

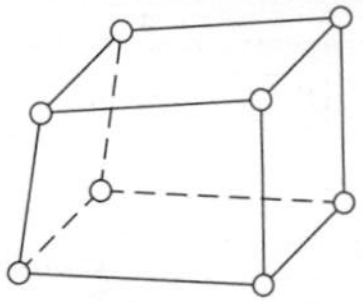

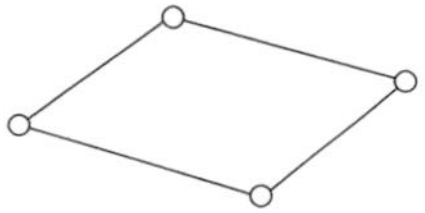

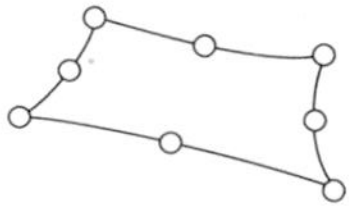

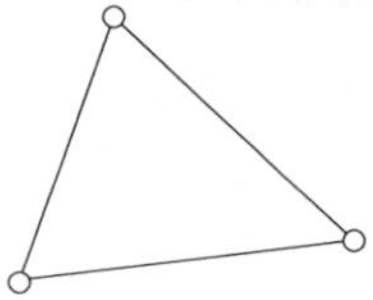

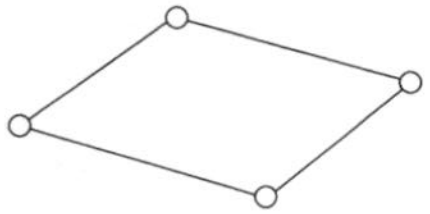

FIGURE 7.15 *(Continued)*

❑ POSTPROCESSING CAPABILITIES

1. Results of ANSYS code analyses can be printed and/or plotted. In the case of plotting, it is possible to plot the following items: nodes and elements; keypoints, line segments, areas, volumes; boundary conditions; displaced shapes; solution data (contour plots of any stored data and plane and topographic).

2. The following plot types are available: wireframe, hidden line removed, section, and perspective views. These plot-out results can be shown in color with map control (e.g., Plates 9, 10, and 13), shaded images (e.g., Plate 2), and animation (e.g., Plates 5 and 6).

❑ OTHER SPECIAL CAPABILITIES

The design optimization feature in the ANSYS program allows any aspect of a design, such as stress, natural frequency, and temperature, to be optimized, in addition to the cost of material and the weight of the structure as handled by usual optimization procedures. The user needs only to prescribe the data for the initial design and to specify the design and state variables with corresponding limiting values and the objective function. The objective function can then be minimized by the program. A practical design case involving the optimization of the geometry of a rocker arm for higher natural frequencies is presented in Chapter 10. The ANSYS program was used to achieve such optimization.

ANSYS results can be transferred to and from some CAD/CAM software systems through translators, as illustrated in Figure 7.14. The reader already has learned from Chapter 5 that data files in CAD programs usually contain geometric information for structures defined by points, lines, and surfaces. These points, lines, etc., may be made to correspond to node locations and element connectivity and be stored in the ANSYS database. Other semantic data such as material properties and boundary conditions can also be transferred back and forth between the two programs. Swanson Analysis Systems, Inc. has developed translators to make it possible for the ANSYS program to interface with eight commercially available CAD/CAM systems.

7.8 ▪ Summary

Finite element formulation for elastic stress analysis of structures was presented in this chapter. Stress analysis, a primary concern in the design process, ensures the integrity of structures. Key formulas were used to derive the expressions for the strain energy in a deformed solid. Element equations were then derived by minimizing the potential energies, which include the strain energy in the elements. Solutions involving displacements, strains, and stresses for a longitudinally loaded tapered rod were obtained by the finite element formulation. This case study was used to provide the reader with guidance in selecting interpolation functions. However, comparison of the numerical results indicated that cases that employ more elements with simple polynomial interpolation functions produced more accurate results than those cases using fewer elements, but higher-order interpolation functions. Finite element formulation for the stress analysis of two-dimensional structures of plane stress geometry was presented with numerical illustrations.

An overview of general purpose finite element programs was also included. Many industrial firms now use these programs for their design analyses. A general description of the capabilities of an established commercial program, the ANSYS code, was also presented. Many features in this code can be found in similar programs. General purpose finite element programs are extremely valuable for CAD applications.

References

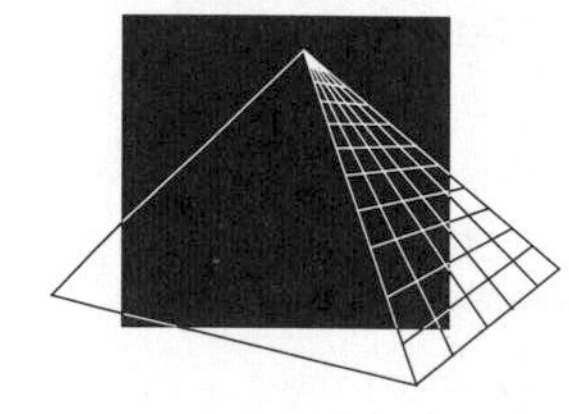

1. Cook, R.D. *Concepts and Applications of Finite Element Analysis.* New York: John Wiley, 1974. Chapter 3.
2. Bathe, K.J., and E.L. Wilson. *Numerical Methods in Finite Element Analysis.* Englewood Cliffs, N.J.: Prentice-Hall, 1976. Chapter 5.
3. Hsu, T.R., and D.K. Sinha. *Finite Element Analysis by Microcomputer.* Duxbury, Mass.: Kern International, 1988.
4. Zienkiewicz, O.C. *The Finite Element Method in Engineering Science.* New York: *McGraw-Hill,* 1971.
5. Noor, A.K. "Survey of Computer Programs for Solution of Nonlinear Structural and Solid Mechanics Problems." *Computers and Structures,* 13 (1981): 425–465.
6. Fredriksson, B., and J. Mackerle. *Structural Mechanics Finite Element Computer Programs: A Decade with Finite Elements 1970–1980.* Linkoping, Sweden: AZC Advanced Engineering Corp., 1980.
7. *A Survey of CAD CAM Systems.* 3d ed. Dallas: Productivity International, 1983.
8. *Introduction to ANSYS.* ANSYS Revision 4.3 Seminar Notes. Houston, Pa.: Swanson Analysis Systems, Inc., August 1987.

Problems

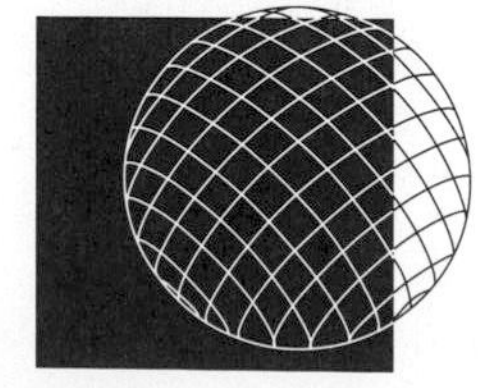

7.1 Prove that the nodal force vector in an element may be expressed by

$$\{F\} = \int_v [B]^T \{\sigma\}\, dv$$

7.2 Use finite element analysis to solve the stresses and strains in the steel, copper, and aluminum sections in the compound rod illustrated below and the displacements at the joints. Assume the sections are welded together, so the rod can deform only in its axial direction. The compound bar has a uniform cross-sectional area of 650 mm^2, with the following moduli of elasticity: $E_{st} = 206$ GPa, $E_{cu} = 103$ GPa, $E_{al} = 69$ GPa. Verify your finite element analysis solution by the classical method from the strength of material theory.

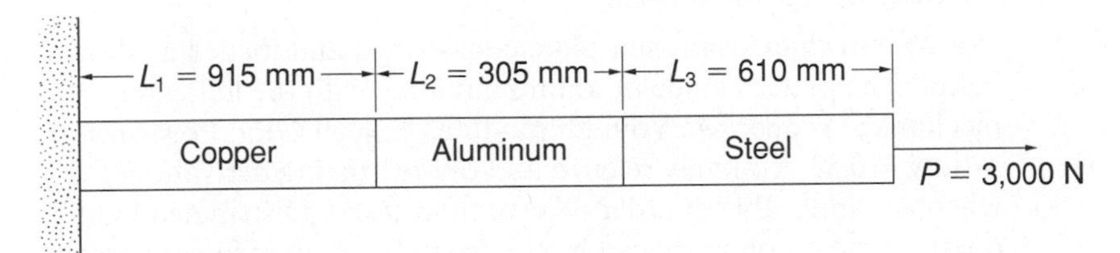

7.3 Use the finite element method to solve for the following quantities in a compound bar similar to the one in problem 7.2, but arranged differently:

(a) Total elongations in each section
(b) Displacement at the midlength point in each section
(c) Stress and stain in each section

The bar has a uniform cross-sectional area of 650 mm^2, with the following moduli of elasticity: 206 GPa for section 1, 103 GPa for section 2, and 69 GPa for section 3.

$L_1 = 915$ mm | $L_2 = 610$ mm | $L_3 = 305$ mm

Steel | Copper | Aluminum | $P = 3{,}000$ N

Section 1 | Section 2 | Section 3

7.4 Use finite element theory to calculate the average stresses and strains in the triangular plate shown below. Also find the displacements at points A and B. The material has the following properties: $E = 69$ GPa, $v = 0.33$. Assume the thickness of the plate to be 25 mm.

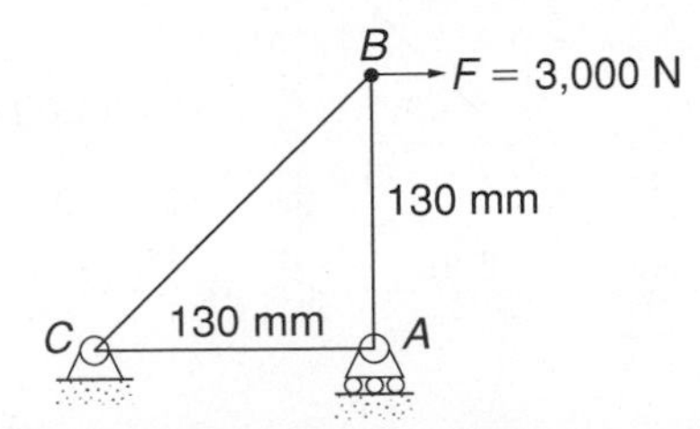

7.5 Solve problem 7.4 with the following changes of conditions:

- Force F is applied in the opposite direction.
- The type and conditions of support at corners A and C are reversed.

7.6 Use the finite element theory to compute

(a) Displacement components at the corners in the x and y directions
(b) Average strains and stresses in the plate
(c) Reactions at the corners

in a 25-mm-thick triangular plate constrained and loaded as shown below. The plate is made of aluminum alloy with the following mechanical properties: Young's modulus $E = 69$ GPa; Poisson's ratio $\nu = 0.33$. You may refer to and quote existing derivations wherever applicable in order to save time. (*Hint:* Distributed load on an element boundary can be treated as equivalent forces applied at two terminal nodes (e.g., the equivalent lumped forces at nodes i and m in the figure below are $f_i = f_m = PA/2$ where A is the area under the pressure P).

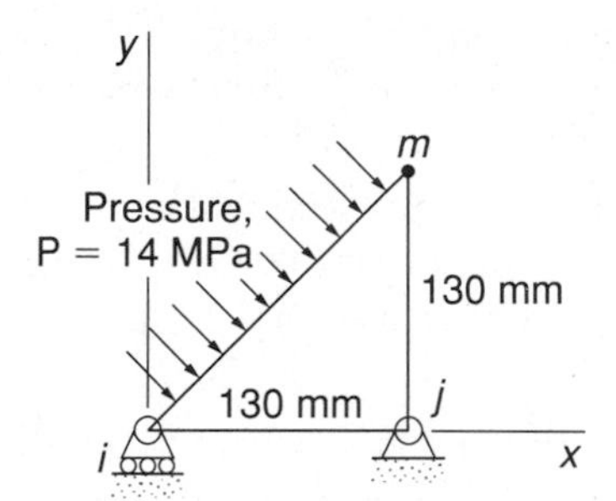

7.7 A right triangular plate 1 inch thick is designed to withstand distributed pressure on its vertical edge, as illustrated in the figure below. The plate is made of aluminum with Young's modulus of

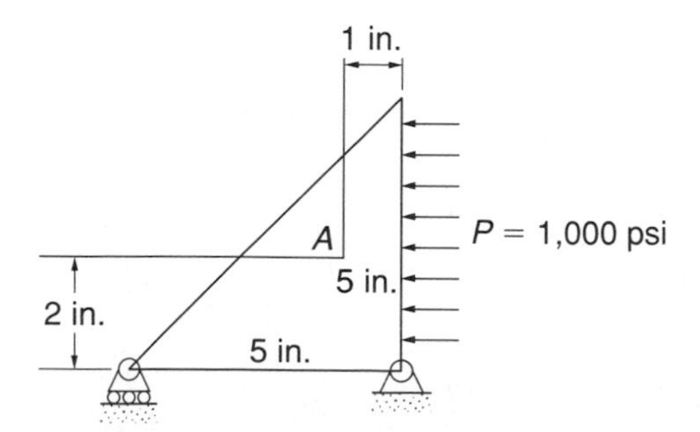

10^7 psi and Poisson's ratio of 0.3. Use the finite element method to find

(a) Displacement at the corners and also at point A
(b) The stresses and strains in the plate
(c) The reactions at the supports

7.8 Derive matrix equations for the unknown nodal displacements in the two-element plane structure illustrated below. Assume the element equations have the following forms:

$$\begin{bmatrix} 1 & 2 & 3 & 4 & 5 & 6 \\ & 5 & 7 & 8 & 9 & 10 \\ & & 10 & 5 & 4 & 3 \\ & & & 9 & 2 & 1 \\ & \text{SYM} & & & 8 & 4 \\ & & & & & 3 \end{bmatrix} \begin{Bmatrix} u_1 \\ v_1 \\ u_2 \\ v_2 \\ u_4 \\ v_4 \end{Bmatrix} = \begin{Bmatrix} f^1_{1x} \\ f^1_{1y} \\ f^1_{2x} \\ f^1_{2y} \\ f^1_{4x} \\ f^1_{4y} \end{Bmatrix}$$

for element 1 and

$$\begin{bmatrix} 10 & 20 & 30 & 40 & 50 & 60 \\ & 50 & 70 & 80 & 90 & 100 \\ & & 100 & 50 & 40 & 30 \\ & & & 90 & 20 & 10 \\ & \text{SYM} & & & 80 & 40 \\ & & & & & 30 \end{bmatrix} \begin{Bmatrix} u_2 \\ v_2 \\ u_3 \\ v_3 \\ u_4 \\ v_4 \end{Bmatrix} = \begin{Bmatrix} f^2_{2x} \\ f^2_{2y} \\ f^2_{3x} \\ f^2_{3y} \\ f^2_{4x} \\ f^2_{4y} \end{Bmatrix}$$

for element 2. The entries in the above element stiffness matrices are fictitious numbers for the sake of simplicity.

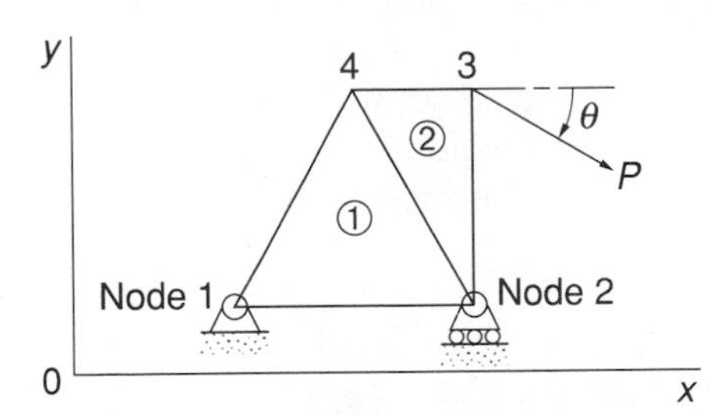

7.9 Solve the problem with the same conditions as presented in Example 7.2 except that the triangular plate is welded together by two separate pieces at C and D, as shown in the following figure.

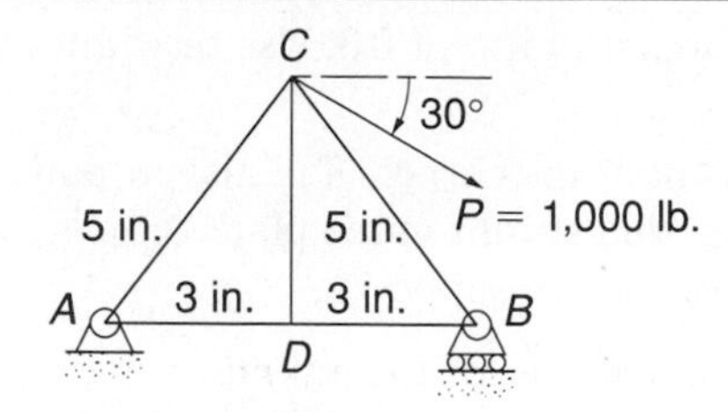

7.10 Use the same technique as in problem 7.4 to find the average stress and strain in two triangular parts that form a square plate when welded at the corners.

Material properties:

Plate (1): $E_1 = 69$ GPa, $\nu_1 = 0.33$
Plate (2): $E_2 = 104$ GPa, $\nu_2 = 0.25$

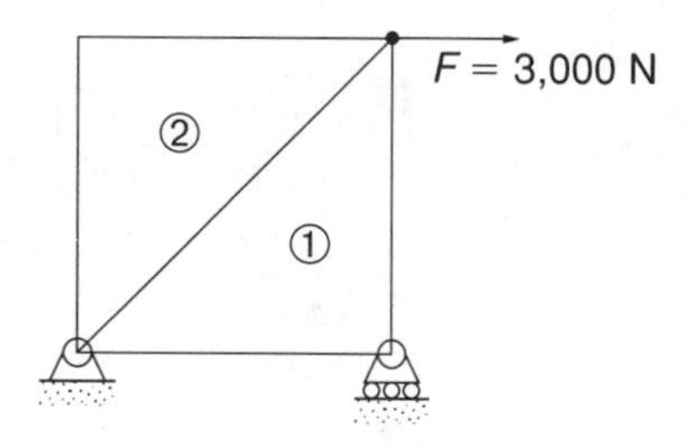

Dimensions of the plates are identical to those in problem 7.4.

7.11 Determine the details of the overall stiffness equation shown in equation (6.11) without boundary conditions by using four finite elements of equal length for the homogeneous circular rod whose cross-sectional area decreases with distance from the fixture shown below. The rod deflects axially due to the static external load F applied at the free end ($x = L$).

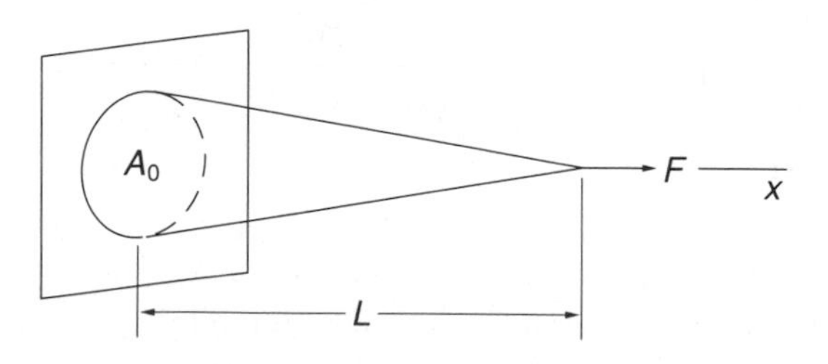

7.12 Apply boundary conditions to the overall stiffness equation generated in the above problem. Write a computer program to determine u, ε_{xx}, and σ_{xx} at all nodal points of the four-element idealization used in problem 7.11 where $A_0 = 6{,}400$ mm^2, $L =$ 300 mm, $E = 69$ GPa, and $F = 600$ N. The library subroutine should be used to solve the final set of linear simultaneous equations. The program should be checked initially for the corresponding uniform beam ($A_0 = 6{,}400$ mm^2 throughout), and these results should also be submitted.

7.13 Derive the expression of the interpolation function for a triangular torus element with nodal coordinates (r_i, z_i), (r_j, z_j), and (r_k, z_k), for the respective nodes I, J, and K in an (r, z) cylindrical coordinate system, with z being the axis of symmetry. Assume the element displacement components in the radial and axial directions follow linear functions: $U_r(r, z) = b_1 + b_2 r + b_3 z$ and $U_z(r, z) = b_4 + b_5 r + b_6 z$ in which $b_1, b_2, \ldots, b_6$ are arbitrary constants.

7.14 Prove that the plane strain case can indeed be treated as a degenerated axisymmetric solid with a large radius in the finite element formulation.

7.15 An overhead traveling crane was designed to lift a load of 3,500 N through a pulley, as illustrated below. Using linear interpolation functions, find the average stresses in the body of the crane (i.e., the plate ijm with a thickness of 25 mm) by the finite element method. Assume the wheels of the pulley are made of perfectly rigid material. The properties of the plate are $E = 210$ GPa and $\nu = 0.3$.

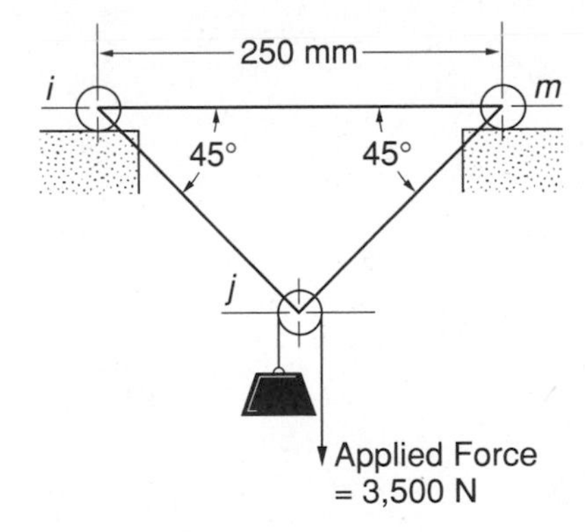

CHAPTER 8

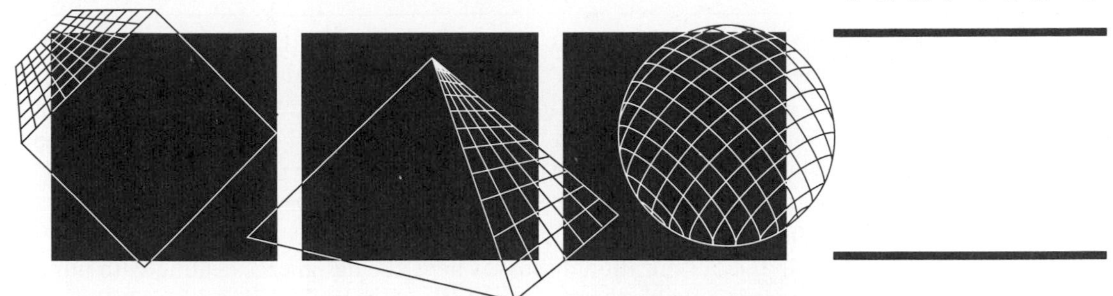

Design Optimization

8.1 Introduction

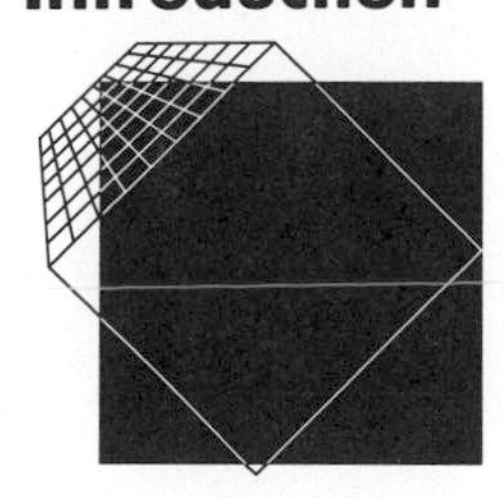

In Chapter 1, an outline of the design procedure is given. It shows that design starts with a set of given specifications. One of the first responsibilities of a design engineer is to synthesize the product based on certain concepts he or she may have regarding its nature. The synthesis usually involves selecting from a number of options, some of which may perform better than the others or may be easier to manufacture and maintain. A designer thus tries to optimize the product even before it is created. This stage of optimization is, more or less, qualitative in nature and is dependent on the ability and experience of the designer. To illustrate the point with a simple example, suppose that a drive is to be designed for a sawmill. The options are an electric motor and gears or an electric motor and belt transmission. Considering the nature of the machinery involved, a belt drive seems to be a more appropriate choice since the belt drive can take greater shock loads and is easier to replace. Thus, qualitatively, the belt drive is an optimal design.

As with all qualitative analyses, mathematical relationships are difficult to derive in situations such as the above. In other words, it is extremely difficult in most cases to obtain mathematical equations whereby the optimal design could be arrived at by plugging certain numbers into a mathematical model. In this chapter, no attempt is made to develop optimization techniques for design synthesis. It is assumed that a design concept has been formulated and the proper product synthesized. The techniques discussed are those that enable the designer to give the product a definite shape (i.e., to find numerical values for, say, diameters, lengths, speed ratios, etc.). The underlying assumption in the whole exercise is that the designer is looking for the best proportioned product—which may be so, from the perspective of weight, cost, performance, convenience, or a myriad of other considerations.

In the optimization discussed here, the designer is dealing with numbers. It is possible therefore to write mathematical algorithms rather easily. Some of these algorithms are given later in the chapter. Before starting, however, it is desirable that a few technical terms relevant to optimization be explained.

8.2 ▪ Design Variables, Parameters, and Constraints

A product is described in terms of certain characteristic quantities. A helical spring, for example, may be described in terms of its coil diameter, number of turns, cross-sectional geometry (round, square, etc.) of the wire, etc. At times, no restrictions are imposed on the values of some or all of these quantities. Those quantities that are not predefined for a product are known as *design variables*, and other characteristic quantities that are predefined are known as *design parameters*.

The role of design variables in design optimization is crucial. These should be correctly identified and their effect on the product properly understood. By altering the values of these variables, one can generate different design options and therefore optimize.

In almost every design problem, there are certain constraints. For example, constraints may arise from the fact that in a hollow tubular column, the outside diameter of the tube may not exceed 5 inches in order to fit in the space available, or that its wall thickness may not be less than 0.1 inch in order to avoid local denting. There may also be limits put on the allowable compressive stress in order to avoid buckling.

Design constraints are either behavior constraints or side constraints. *Side constraints* are those that put limits or bounds on the design variable [i.e., they restrict the range of the design variables on one (maximum or minimum) end or both (maximum and minimum) ends]. The 5-inch-diameter limitation on the column design above represents a side constraint. The limitation that the compressive stress may not exceed a certain value is a *behavior constraint*. Behavior constraints are invariably present in engineering design problems, whereas side constraints may or may not exist. For example, it is inconceivable that a load-carrying member would be designed with complete disregard for its strength, but a column may be of any size in some cases.

8.3 ▪ Objective Function

As mentioned above, any optimization exercise should have a certain objective—be it minimization of weight, maximization of fuel efficiency, or minimization of machining cost. For numerical optimization, it is necessary that the design objective be expressed as a function of the design variables. A mathematical expression of the design objective in the form explained above is known as the *objective function.*

To further elucidate the concept of objective function and its use, let's consider the design of a helical compression spring (with round wire) of minimum weight. Assume that the load-carrying capacity and the maximum deflection in the spring are specified. Also, assume that the designer has an unlimited choice of spring index, but no choice in selecting the spring material. The weight of the spring can be expressed by the following formula:

$$W = \rho(n + e)\pi^2 D d^2/4 \tag{8.1}$$

where

ρ = the density (or specific weight) of the wire material

n = the number of active turns

e = the total number of idle turns at the ends

D = the mean coil diameter

d = the wire diameter

Equation (8.1) may be modified to read

$$W = (\pi^2 D^3 n/4C^2 + \pi^2 d^3 Ce/4)\rho \tag{8.2}$$

where C is the spring index, D/d.

Maximum shear stress (τ_s) and deflection (δ) in the spring are given by the following equations [1]:

$$\tau_s = 8KFD/\pi d^3 \quad \text{or} \quad d = \sqrt[3]{8KFD/\pi\tau_s} \tag{8.3}$$

and

$$\delta = 8FD^3 n/Gd^4 \quad \text{or} \quad D^3 n = G\partial d^4/8F \tag{8.4}$$

where

K = the Wahl's factor

F = the specified load

δ = the maximum specified deflection

τ_s = the allowable shear stress in the wire material

G = the modulus of rigidity of the wire material

Substituting values from equations (8.3) and (8.4) into equation (8.2), an expression for the weight of the spring in terms of C is obtained as follows [2]:

$$W = [(2FG\delta/\tau_s^2)K^2 + (\pi^2/4)(8F/\pi\tau_s)^{1.5}eK^{1.5}C^{2.5}]\rho \quad (8.5)$$

Equation (8.5) is the objective function for the spring weight minimization problem. By varying C, the corresponding changes in W may be studied and the minimum value determined.

The above is a simple illustration of the design optimization process. In most cases, the objective function contains several variables and constraints. The objective function and the constraints are often nonlinear; as a result, the mathematics of the optimization process are far more complex. The above example, nonetheless, represents a typical optimization process.

8.4 ▪ Constrained and Unconstrained Optimization

Unconstrained optimization, as the name implies, relates to the technique of finding optimum solutions in cases where there are no constraints. Intuitively, the reader might realize that there are very few engineering design problems in which there are absolutely no constraints. The question may therefore be asked whether unconstrained minimization has any relevance in engineering design. The answer is a definite yes. This is so because many of the constrained optimization problems can be converted into unconstrained optimization problems and optimal solutions found by the simpler methods. As is shown below, unconstrained optimization is far simpler than constrained optimization.

❑ UNCONSTRAINED MINIMA

Let's assume that there is an objective function $F(X)$ expressed in terms of a vector of design variables $X = \{x_1 \quad x_2 \quad \cdots \quad x_n\}^T$. $F(X)$ will have a stationary value at X_q if

$$\nabla F(X) = \begin{Bmatrix} \dfrac{\partial F(X)}{\partial x_1} \\ \dfrac{\partial F(X)}{\partial x_2} \\ \vdots \\ \dfrac{\partial F(X)}{\partial x_n} \end{Bmatrix}_{X = X_q} = \{0\} \tag{8.6}$$

If $F(X_q)$ is also a minimum, then the Hessain matrix H, as defined in equation (8.7) below, is positive definite at X_q. A positive definite matrix has all eigenvalues greater than 0. The Hessain matrix is given as follows:

$$H = \begin{bmatrix} \dfrac{\partial F(X)}{\partial x_1^2} & \dfrac{\partial F(X)}{\partial x_1 \partial x_2} & \cdots & \dfrac{\partial F(X)}{\partial x_1 \partial x_n} \\ \dfrac{\partial F(X)}{\partial x_2 \partial x_1} & \dfrac{\partial F(X)}{\partial x_2^2} & \cdots & \dfrac{\partial F(X)}{\partial x_2 \partial x_n} \\ \vdots & \vdots & & \vdots \\ \dfrac{\partial F(X)}{\partial x_n \partial x_1} & \dfrac{\partial F(X)}{\partial x_n \partial x_2} & \cdots & \dfrac{\partial F(X)}{\partial x_n^2} \end{bmatrix} \tag{8.7}$$

It should, however, be understood that equations (8.6) and (8.7) only guarantee that $F(X_q)$ is a local minimum. It is quite possible that the objective function has a value lower than the above in a different region of the variable space. In order to find the absolute minimum, all stationary points are checked.

❑ CONSTRAINED MINIMA

Suppose that in the above case it is also specified that $x_1 \geq a$ and $x_3 = b$. Then, based on these two design constraints, constraint functions $g(X)$ and $h(X)$ are formed, specifying that $g(X) \equiv a - x_1 \leq 0$ and $h(X) \equiv x_3 - b = 0$. The minimization problem is then expressed as follows:

$$\begin{aligned} &\text{minimize: } F(X) \\ &\text{subject to: } g(X) \leq 0 \\ &\qquad\qquad\quad h(X) = 0 \end{aligned}$$

Typically, a constrained optimization problem is stated as follows:

$$\begin{aligned} &\text{minimize: } F(X) \\ &\text{subject to: } g_i(X) \leq 0 \\ &\qquad\qquad\quad h_j(X) = 0 \end{aligned}$$

where $g_i(X)$ are the inequality constraint functions and $h_j(X)$ are the equality constraint functions. It is important to classify the constraint functions as above since their treatment in the algorithms that follow is often quite different.

Suppose that there were two variables in $F(X)$ (i.e., $X = \{x_1 \quad x_2\}^T$) and two constraints [i.e., $g(X)$ and $h(X)$] in the problem. It is possible that the plots of the functions $g(X)$ and $h(X)$ appear in the design space ($x_1 x_2$ plane), as shown in Figure 8.1. Function $g(X) = 0$ divides the design space into two regions—feasible and infeasible. In the feasible region, $g(X) \leq 0$, and in the infeasible region, $g(X) > 0$. As far as the equality constraint is concerned, feasible designs are only those that lie on the curve $h(X) = 0$. Both $h(X) > 0$ and $h(X) < 0$ represent violations of constraint. Thus, considering the two constraints together, it can be said that the feasible designs are those that lie on the darker portion of $h(X)$. Objective function isoclines (i.e., contours of fixed values) are also shown in the figure, from which it is obvious that the optimal solution is given by the point X^*.

It can be shown that the optimal design point, such as X^* above, in a multiple (k inequality and l equality) constraint situation satisfies the following Kuhn-Tucker conditions:

(i) X^* is a feasible design point

(ii) $\lambda_i g_i(X) = 0 \quad i = 1, k; \lambda_i \geq 0$

(iii) $\nabla F(X^*) + \sum_{i=1}^{k} \lambda_i \nabla g_i(X^*) + \sum_{m=1}^{l} \lambda_{k+m} \nabla h_j(X^*) = 0$
$\lambda_i \geq 0$; λ_{k+m} unrestricted in sign

In (ii) above, if the search point is at the constraint surface itself [i.e., $g_i(X) = 0$], then λ_i should be greater than zero. If, on the other hand, $g_i(X) \leq 0$, then $\lambda_i = 0$.

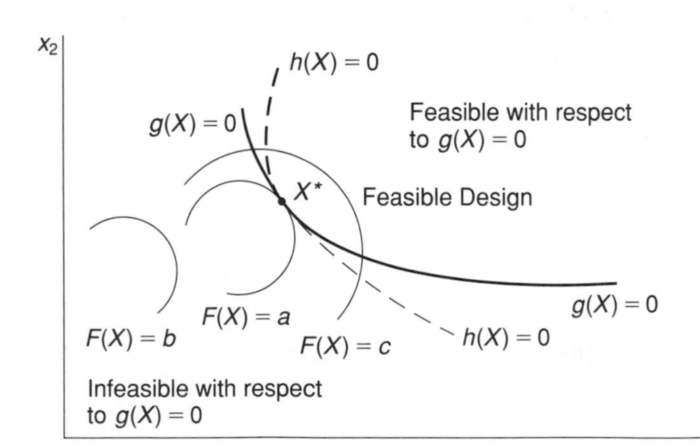

FIGURE 8.1 Objective and constraint functions

While Kuhn-Tucker conditions are necessary conditions for a constrained optimal design, they are not always sufficient conditions. The situation is akin to the unconstrained optimization problem, where a vanishing gradient and a positive definite Hessain matrix do not imply a global minimum. However, in the case of a convex programming problem, Kuhn-Tucker conditions are necessary and sufficient. A *convex programming problem* is one in which the objective function and the constraint function are all convex in the region of search. A *convex objective function* has only one minimum. A *convex constraint function* is one in which a straight line joining two points on the constraint surface lies entirely in the feasible region as far as that particular constraint is concerned. Figure 8.1 depicts a convex problem.

8.5 ▪ Unconstrained Optimization with One Design Variable

❑ COMPUTER GRAPHICS METHOD

In equations (8.6) and (8.7), the necessary conditions for unconstrained optimum are given. Thus, it seems reasonable to expect that by differentiating the objective function, it should be possible to find the optimal solution. The above logic does not work in most practical problems because the objective function is not easily differentiable. The attempt is often made therefore to develop mathematical methods for optimization without differentiation. A few such methods are discussed later. Objective functions of one or two variables are quite easily optimized with computer graphics because such problems may be geometrically represented in two-dimensional space. The advantage of using computer graphics in optimization, as is evident later, is that the method not only yields an optimal solution, but also provides detailed information concerning the response of the objective function to the changes in the design variables.

To illustrate the usefulness of computer graphics, let's start with optimization of a helical spring. Equation (8.5) is used for this purpose. The equation is derived in such a manner that in every design the stress in the spring wires is equal to the allowable limit and the diameter of the spring wire is never negative. Thus, two constraints have been eliminated. There is another constraint: $C > 1$. This constraint can be obviated easily by considering only $C > 1$. It is known that in a properly designed helical spring $C = 6 - 10$. Therefore, there is certainly no loss in the generality of the search by starting at a value of $C = 2$. Thus, by the above deliberations, an unconstrained engineering design optimization problem is created. For the solution, see Example 8.1.

EXAMPLE ■ 8.1

Design a compression spring of minimum weight, given the following data:

Maximum deflection	30 mm
Axial load	500 N
Allowable shear stress	207 MPa
Modulus of rigidity	79 GPa
Density	7800 kg/m^3

The ends of the spring are squared and ground (i.e., $e = 2$).

A graph ($W \sim C$) for $C = 2$ to $C = 14$ is shown in Figure 8.2. From the curve, it is learned that the weight of the spring is fairly constant around $C = 7$. It is thus probably immaterial whether $C = 7$ or $C = 6.8$ is adopted. This bit of information is quite vital in manufacturing because it is imprudent to attempt to design springs with arbitrary spring indices and wire diameters. A wise designer would thus select a design for which the correct diameter wire is available from stock and the required spring index achievable with the existing manufacturing facilities. The reader should appreciate that such valuable information would have been missed in a purely mathematical optimization.

The knowledge gained from Figure 8.2 is now used for further refining the optimal solution. In Figure 8.3, ($W \sim C$) for $C = 6.0$ to $C = 8.0$ is plotted. Looking at Figure 8.3, it appears that the absolute minimum occurs at $C = 7.2$.

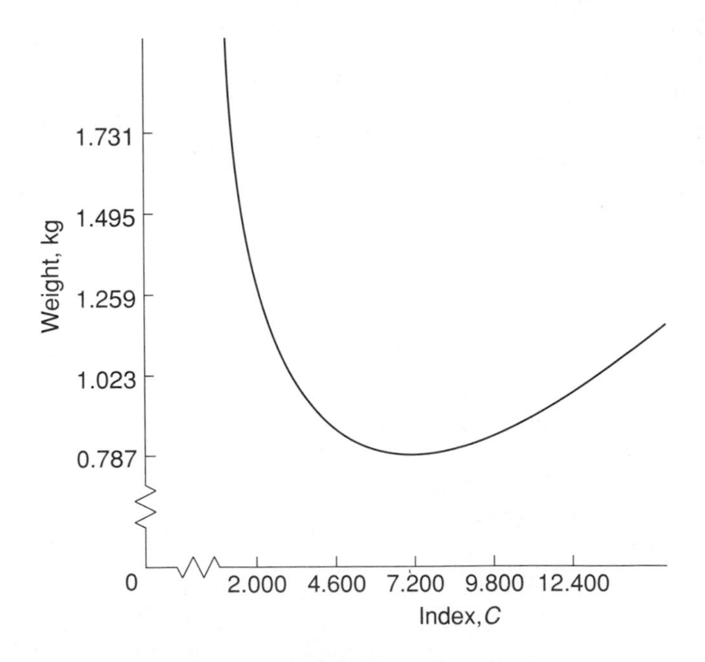

FIGURE 8.2 Weight versus spring index

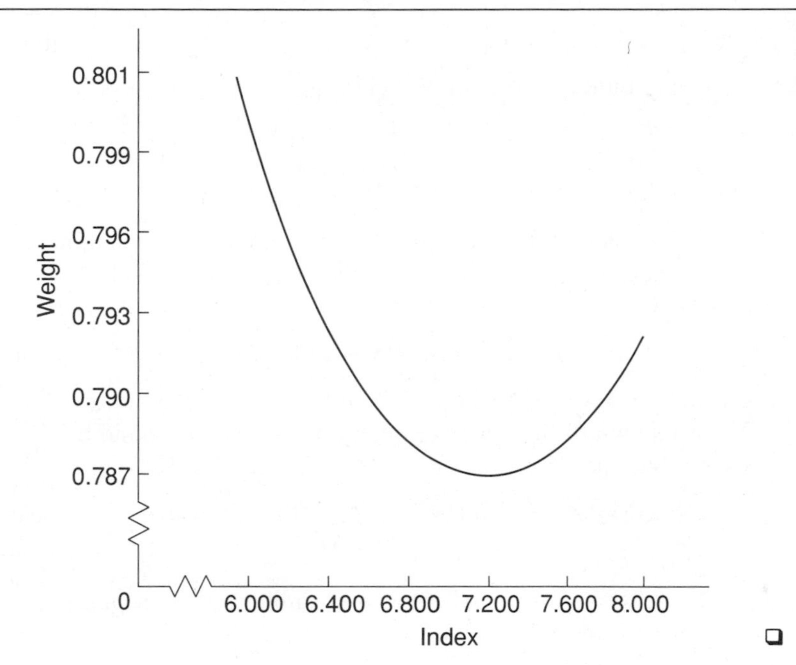

FIGURE 8.3
Weight versus spring index

OPTIMIZATION BY CURVE-FITTING

There are instances where it is not possible to write explicit mathematical objective functions even though the design variables are known. In such cases, an approximate objective function is derived from experimental or statistical data by the techniques of curve-fitting. A case in point may be the selection of optimal speed of an I.C. engine. It is known that the fuel consumption rate is a function of the speed of the engine, but finding a mathematical relationship between the two is extremely difficult.

Suppose that an objective function $F(x)$ depends on a single variable x. To establish an approximate objective function, the designer obtains values of the objective function at different values of x—spreading the observations over the entire range of the variable x. Depending on the number of sampling points available and the nature of the objective function (if known), a straight line or a quadratic or cubic curve is fitted through the data points and the minimum determined by differentiating or by plotting graphs. In Example 8.2, the problem of Example 8.1 is solved once again by the method of curve fitting.

EXAMPLE 8.2

Assuming that the precise equation for the weight of a helical spring is not known, design a spring of minimum weight, using the data given below:

Spring Index	Spring Weight (kg)
6.0	0.8014
7.0	0.7896
8.0	0.7921

Since three sampling points are given, it is possible to find a quadratic approximation to the objective function by the method of curve-fitting,

$$F(x) = 1.1725 - 0.10475x + 0.00715x^2$$

where $F(x)$ = weight of the spring and x = the spring index. The minimum of the objective function may then be found by differentiation:

$$dF(x)/d(x) = -0.10475 + 2 \times 0.00715x = 0$$

giving $x = C = 7.325$.

As can be seen, the optimal solution is similar to the one found in Example 8.1 above. ❑

8.6 ▪ Constrained Optimization with One Design Variable

Constrained optimization problems involving a single variable are also easily handled with computer graphics. To illustrate a possible computer graphics approach, a constrained optimization problem is solved in Example 8.3 below. Any other problems with single or multiple constraints may be solved in a similar manner.

EXAMPLE ▪ 8.3

Design a circular torsion bar of minimum weight, using the following data:

Applied torque	2,000 lb.-in.
Length of bar	40 in.
Modulus of rigidity	11,500,000 lb./in^2
Specific weight	0.284 lb./in^3

It is required that the twist in the bar be at least 0.25 radian.

Assume that the inside diameter of the hollow bar is given as xD, where D is the outside diameter and x is a ratio in the range $1 > x \geq 0$. The weight of the bar may be expressed as $W = \pi L\rho(1 - x^2)\dfrac{D^2}{4}$, where ρ is the specific weight of the bar material. By using the well-known torsion equations, the weight equation is modified as follows:

$$W = (\pi/4)L(1 - x^2)[(16/\pi)(T/\tau)/(1 - x^4)]^{2/3}\rho \qquad (8.8)$$

In the above equation, the constraint on the allowable shearing stress has been removed by requiring that the highest stress (τ) in the bar be equal to the allowable stress. The constraint on x can be overlooked if we decide to confine our search to the area from, say, $x = 0$ to $x = 0.9$. We are then left with a single constraint—namely, that on the twist. Let's say that the twist is represented by the symbol θ; then the constraint function is formed as shown below.

$$g(x) \equiv 0.25 - \theta \leq 0 \qquad (8.9)$$

A plot of the objective (weight) function, the outside diameter, and the constraint function versus x is shown in Figure 8.4. It is seen that for values of $x = 0 \sim 0.54$, $g(x)$ is negative (i.e., all solutions within this range are feasible). The optimal solution is at $x = 0.54$ because the weight in this case is the lowest of all feasible solutions.

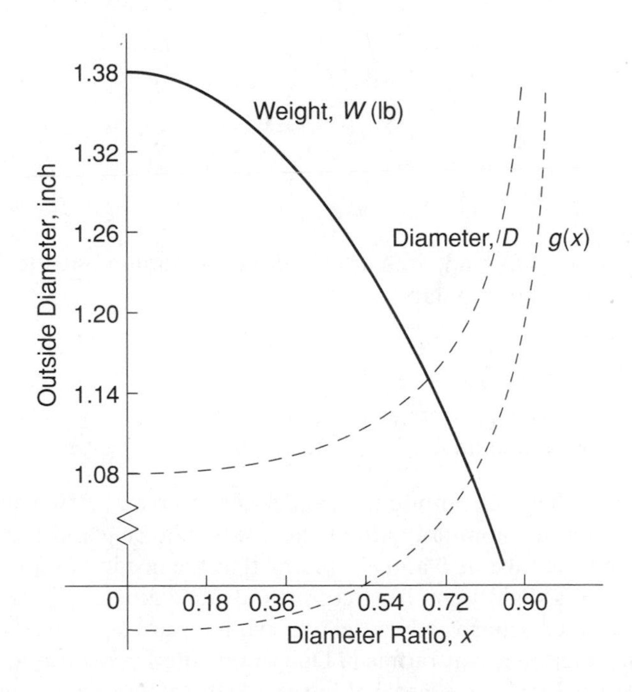

FIGURE 8.4
Weight versus diameter ratio

8.7 ▪ Unconstrained Optimization with Two Variables

An objective function of two design variables may be represented graphically in three-dimensional space, as shown in Figure 8.5. It is thus possible to obtain an optimal solution by the use of computer graphics. The optimal design is represented by the lowest point on the objective function surface.

The lowest point on the objective surface can be found by drawing its three-dimensional model. A more convenient way of searching for the optimal point, however, is to cut the objective function space with a number of parallel planes and plot the cross-sections. This approach requires less computation and computer memory. The method is illustrated in Example 8.4.

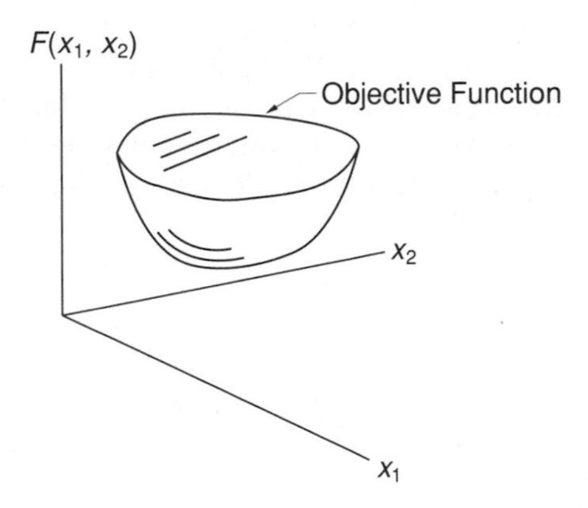

FIGURE 8.5
Objective function in three-dimensions

EXAMPLE ▪ 8.4

Design a compound circular cylinder of minimum outside diameter, given the following data:

Internal pressure	5,000 psi
Inside diameter	20 in.
Allowable tensile stress	25,000 psi
Young's modulus	3,000,000 psi

Assume that the compound cylinder is comprised of two circular tubes. Let the nominal radii of the inside tube be a and b, and those of the outside tube be b and c. Assume that the inside radius of the outer cylinder is slightly smaller than the outside diameter of the inside cylinder. Obviously, when the two cylinders are assembled, there is some interference at radius b. Due to the interference, compressive tangential stress is generated in the smaller cylinder and tensile tangential stress in the outer cylinder. The variation of the tangential stress due to interference is shown in dashed lines in Figure 8.6(a).

When the compound cylinder is subjected to an internal pressure, tensile tangential stress is generated in the tubes. The variation of this stress across the tube thickness is parabolic, as shown by the solid

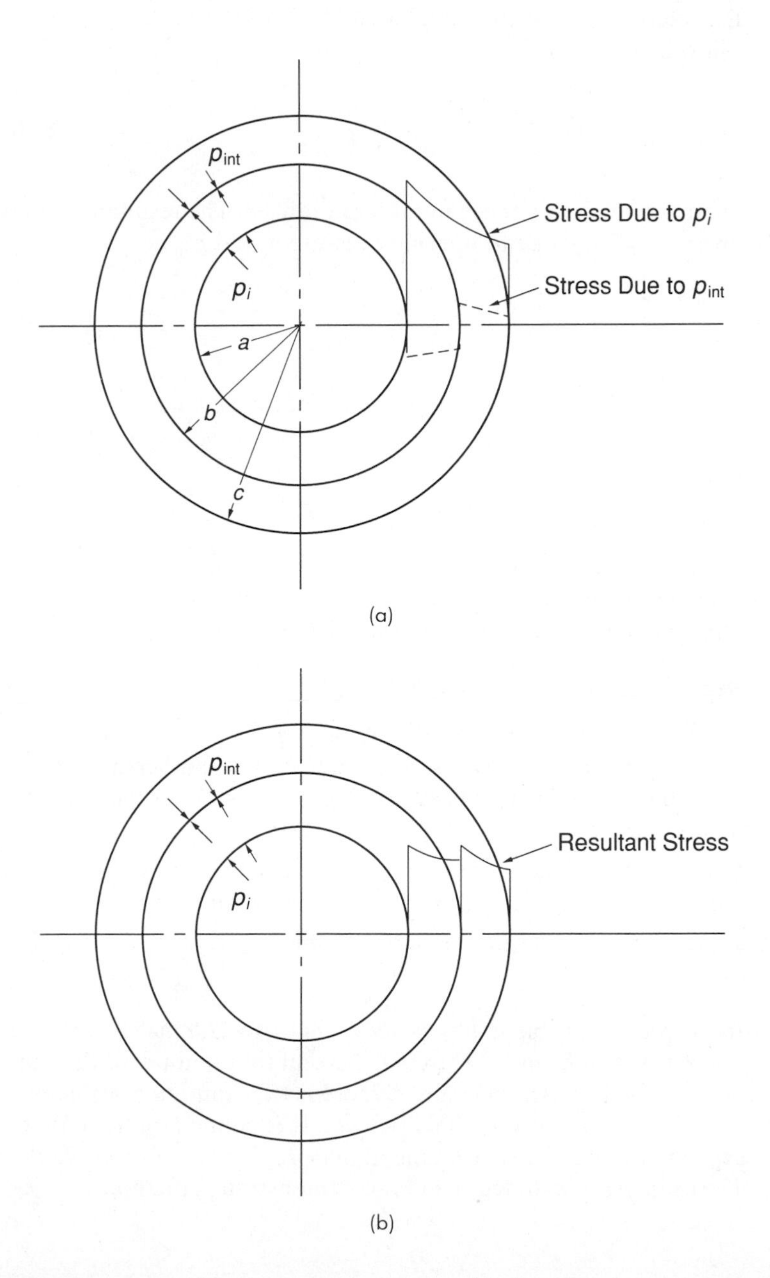

FIGURE 8.6
Stresses in a compound cylinder

curve in Figure 8.6(a). On comparing this curve with the curve of resultant stress shown in Figure 8.6(b), it appears that the stress distribution is more uniform in the latter case. The resultant stress is computed by adding the stresses due to the internal pressure and interference. In other words, by providing interference, it is possible to utilize the material of the cylinders more effectively.

The interference pressure generated at radius b is given by the following equation [1]:

$$p_{\text{int}} = \frac{E(c^2 - b^2)(b^2 - a^2)}{2b^3(c^2 - a^2)}\delta \tag{8.10}$$

where δ is the interference in radii at radius b. The resultant tensile stress at radius a due to internal pressure p_{i} and p_{int} is

$$\sigma_a = p_{\text{i}}\frac{c^2 + a^2}{c^2 - a^2} - p_{\text{int}}\frac{2b^2}{b^2 - a^2} \tag{8.11}$$

The resultant tensile stress at radius b (outside cylinder) is

$$\sigma_b = p_{\text{i}}\frac{a^2(1 + c^2/b^2)}{c^2 - a^2} + p_{\text{int}}\frac{c^2 + b^2}{c^2 - b^2} \tag{8.12}$$

For the optimal cylinder design, radius b and interference δ may be treated as the design variables. A computer program written for this problem finds the minimum in the following way:

Step 1 A value for b is selected.

Step 2 A value for δ is selected.

Step 3 An outer radius c is then found by trial and error, using equations (8.10) through (8.12), such that the stress at the critical points does not exceed the allowable value.

Step 4 Steps 2 and 3 are repeated until a sufficient number of points are obtained in order to plot curves, as shown in Figure 8.7.

Step 5 Another iteration is started from step 1, using a new value for b.

In the present problem, b is varied from 21 to 22.8 inches, and in each case δ is varied from 0.002 to 0.02 percent (of nominal b). The curves of δ versus c, shown in Figure 8.7, predict the minimum value of c at $b = 22.8$ inches and $\delta = 0.009$ percent. It is pointed out here that explicit formulas for solving the above problem are also available. The example is included mainly to demonstrate a method.

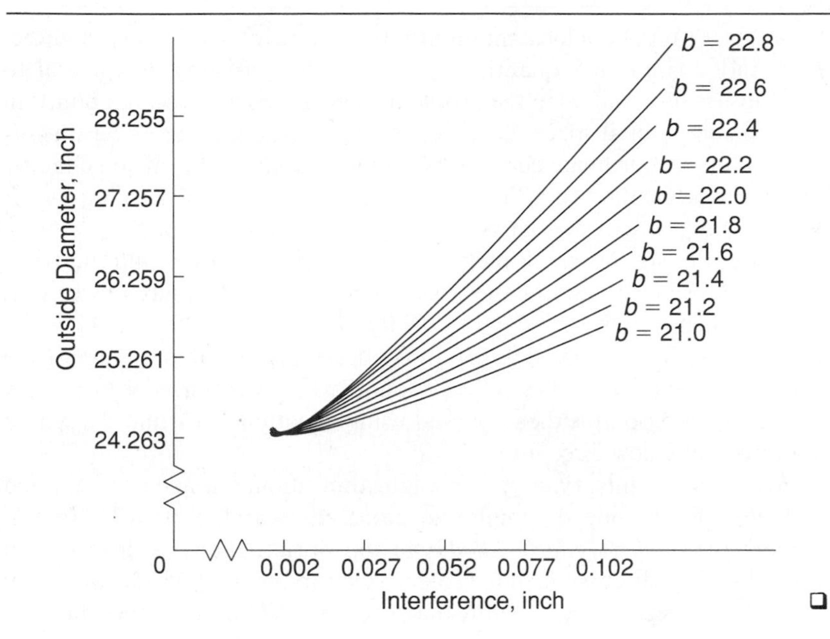

FIGURE 8.7 Outside diameter versus interference

8.8 ▪ Unconstrained Optimization with Multiple Design Variables

When there are more than two design variables in the objective function, graphical methods are no longer feasible. In such cases, mathematical means have to be used. Some of the popular methods of unconstrained optimization are discussed in the following six subsections.

❑ UNIVARIATE SEARCH METHOD

In this and all other methods described in this chapter, an arbitrary point—say, X_q—is first selected. A search is then performed for the local minimum (i.e., by moving along certain preferred directions in the design hyperspace, a point is found at which the objective function has the lowest possible value). Next, a search is started from this later point in a new direction and another minimum located. By proceeding in this manner, the global minimum is located in a number of attempts. Mathematically speaking, the search moves from point X_q to point X_{q+1} in an n-dimensional space according to the following scheme:

$$X_{q+1} = X_q + \alpha S_q \tag{8.13}$$

such that $F(X_{q+1})$ is a local minimum. In the above, S_q is the search direction, and α is a scalar quantity. S_q is a vector consisting of n (equal to the number of variables in the problem) elements as is the search point. In a three-dimensional space, $S_q = \{a \quad b \quad c\}^T$, meaning that along this direction, the x coordinate changes by a units and the y coordinate changes by b units, while the z coordinate changes by c units. $X_q = \{x_1 \quad x_2 \quad x_3\}^T$, where x_1, x_2, and x_3 are the x-, y-, and z-coordinates of point X_q.

In the univariate search method, the search progresses, alternatively, along the principal axes of the design space. Thus, the search directions assume such values as $S_0 = \{1 \quad 0 \quad 0 \quad \cdots \quad\}^T$, $S_1 = \{0 \quad 1 \quad 0 \quad \cdots\}^T$, etc., in turn. Once a search direction is selected, the local minimum of the objective function in that direction is found, which yields a value of α—say α^*. The search point is then updated using equation (8.13) and the search continued in a new direction.

An efficient univariate search algorithm should implement a good technique for finding the minimum along the search directions. In this context, the *Golden section method* is an appropriate choice. As is seen from the following discussion, the Golden section method yields the local minimum expeditiously, without the need for differentiating the objective function.

❑ GOLDEN SECTION METHOD

Let's say that the minimum of a function of single variable, $F(x) = x^3 - 3x^2 - 2x + 4$, is to be found. Obviously, the minimum of the function is at $x = 2.291$, as shown in Figure 8.8. To find the minimum of this

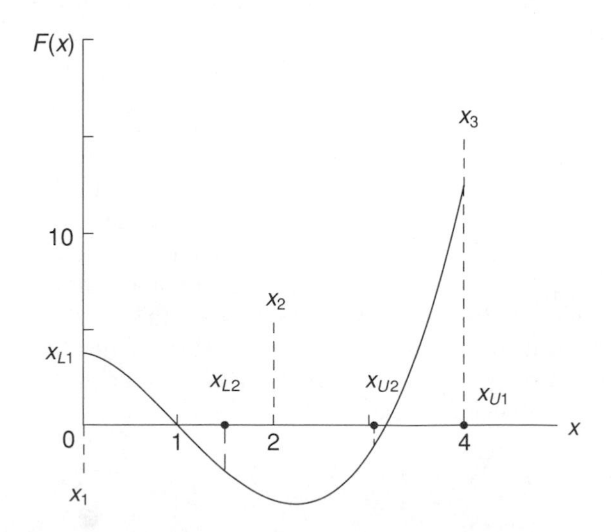

FIGURE 8.8
Golden section method

function by the Golden section method, the following scheme is adopted.

First, find the bounds on the minimum. Pick a value—say, $x_1 = 0$—for which $F(x) = 4$. Pick $x_2 = 2$, giving $F(x) = -4$. Since $F(x_2) < F(x_1)$, it is apparent that the search is moving toward the minimum; thus, x_1 may serve as a lower bound on the minimum. Let $x_{L1} = x_1$. Pick $x_3 = 4$, giving $F(x_3) = 12$. Since $F(x_3) > F(x_2)$, the minimum is toward the left of point x_3. Therefore, x_3 is an upper bound on the minimum. Let $x_{U_1} = x_3$. These two bounds are shown in Figure 8.8.

In the above example, rather coarse steps are used in selecting the bounds. Generally, a more cautious approach is recommended for functions that are known to have more than one minimum. Care is also needed not to mix up the bounds (i.e., the lower bound of the first minimum and the upper bound of the next minimum should not be paired together).

Next, refine the bounds until they are nearly equal. If the bounds are correctly determined, the function is guaranteed to be unimodal in between the bounds. In order to refine the bounds, select two points, x_1 and x_2, between x_{L1} and x_{U1}, preferably such that

$$x_1 = 0.62x_{L1} + 0.38x_{U1} \tag{8.14}$$

$$x_2 = 0.38x_{L1} + 0.62x_{U1} \tag{8.15}$$

or

$$x_1 = 1.52 \quad \text{giving } F(x_1) = -2.4594$$

$$x_2 = 2.48 \quad \text{giving } F(x_2) = -4.1582$$

Since $F(x_1) > F(x_2)$ and $x_2 > x_1$, x_1 is adopted as the new lower bound (i.e., $x_{L2} = x_1$). If $F(x_2) > F(x_1)$, then $x_{U2} = x_2$ is adopted. The new bounds on the minimum are also shown in Figure 8.8.

Two new search points are now selected. Calling these points x_1 and x_2 as before, let x_1 (new) $= x_2$ (old), and find x_2 (new) from equation (8.15). Thus, $x_1 = 2.48$, $x_2 = 3.056$, $F(x_1) = -4.1582$, and $F(x_2) = -1.5735$. Now, since $F(x_1) < F(x_2)$, obviously x_2 is the new upper bound, so make $x_{U_2} = x_2$. By repeating the above iterations, new bounds are continuously evaluated until a convergence is attained.

EXAMPLE ■ 8.5

Find the minimum of the function $F(X) = x_1^2 - 2x_1x_2 + 2x_2^2 - 3x_1 + 6$ by the univariate method.

Contours of the objective function for a few selected values of $F(X)$ are shown in Figure 8.9. Let's start the search from a point $X_0 = \{0 \quad 0\}^T$. The global minimum is obtained by repeating, a number of times, steps similar to steps 1 and 2 described below.

Step 1 Search in the direction $S_0 = \{1 \quad 0\}^T$. In this search direction, the variable x_2 remains constant. Let $x_2 = 0$ initially. The objective

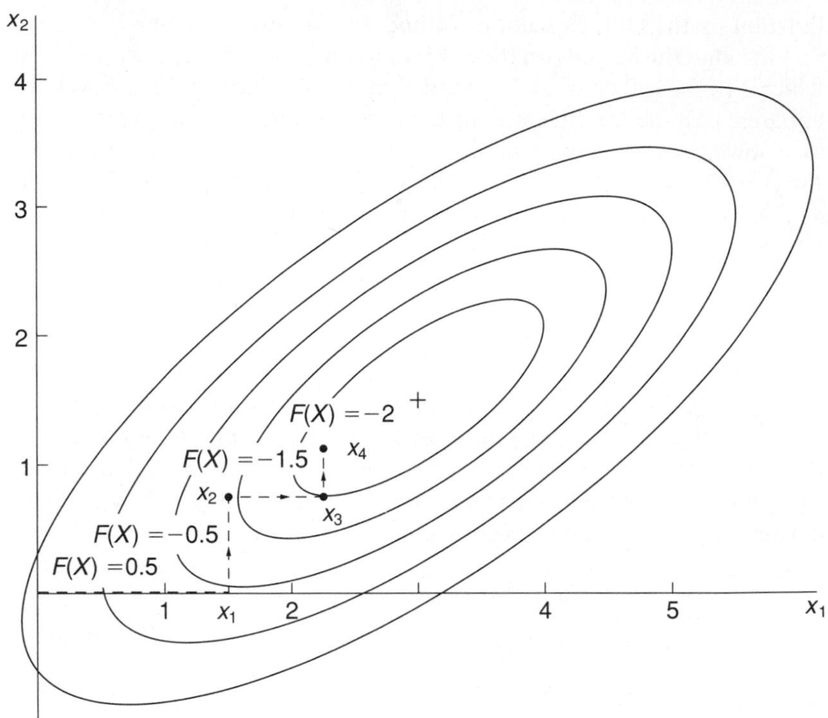

FIGURE 8.9 Contours of $F(X)$

function may therefore be modified to read $F(X) = x_1^2 - 3x_1 + 6$. At the local minimum, $\alpha^* = 1.5$ is found. So the new search point is $X_1 = \{0 \quad 0\}^T + 1.5\{1 \quad 0\}^T = \{1.5 \quad 0\}^T$.

Step 2 Select $S_1 = \{0 \quad 1\}^T$ as the new search direction. Modify the objective function to read $F(X) = 3.75 - 3n_2 + 2n_2^2$ by substituting $x_1 = 1.5$ as at the last minimum. The local minimum, in this direction, is found at $\alpha^* = 0.75$; therefore, $X_2 = \{1.5 \quad 0\}^T + 0.75\{0 \quad 1\}^T = \{1.5 \quad 0.75\}^T$.

Step 3 Let $S_2 = \{1 \quad 0\}^T$, yielding $\alpha^* = 0.75$ again, which moves the search point to $X_3 = \{1.5 \quad 0.75\}^T + 0.75\{1 \quad 0\}^T = \{2.25 \quad 0.75\}^T$.

Step 4 Let $S_3 = \{0 \quad 1\}^T$, yielding $\alpha^* = 0.375$ and $X_4 = \{2.25 \quad 1.125\}^T$.

The progress of the search in the above four steps is depicted in Figure 8.9 by dashed lines. The global minimum is situated at $\{3 \quad 1.5\}$. To check the optimality of the result, use equations (8.6) and (8.7). ❑

❑ POWELL'S METHOD

Powell [3] suggests a method for finding the unconstrained minimum of quadratic functions of several variables. This method is guaranteed to converge to the global minimum in a finite number of steps. It is a *zero-order method*, just as the univariate method described above. In all zero-order methods, the minimum is searched using the value of the function itself, not its derivatives.

It is observed that most objective functions may be approximated by quadratic functions in the vicinity of a minimum. Powell's method may therefore be applied to almost every type of objective function. Convergence takes place in less than n (equal to the number of variables) steps if the function is truly quadratic. If the initial starting point is in the close vicinity of the minimum, the convergence is quicker.

The reason Powell's method converges so rapidly is that the search is conducted in conjugate directions. Two directions, S_i and S_j, are said to be conjugate (or A-conjugate) if

$$S_i^T A S_j = 0 \quad (i \neq j) \tag{8.16}$$

where A is the coefficient matrix of the quadratic equation

$$F(X) = X^T A X + X^T B + C$$

To use Powell's method, follow these steps.

Step 1 Select a starting point—say, X_0—and minimize the function locally in n linearly independent directions—say, $S_1, S_2, \ldots, S_n$—to obtain a new starting point, $X_n = X_0 + \sum_{r=1}^{n} \alpha_r S_r$.

Step 2 Make $X_0 = X_n$, and minimize once again in all n directions as before, but start from X_0(new), and obtain X_n. Find a new search direction, $S_{n+1} = X_n - X_0$. It can be shown that S_{n+1} and S_n, used in the earlier step, are conjugate directions.

Step 3 Make $X_0 = X_n$, and renumber the search directions—S_1 (new) $= S_2$ (old), S_2 (new) $= S_3$ (old), . . . , S_n (new) $= S_{n+1}$ (i.e., drop old search direction S_1).

Step 4 Perform minimization along the new set of n directions, and obtain point X_n and search direction S_{n+1}. Renumber the search directions as in step 3 above.

EXAMPLE ▪ 8.6

Find the global minimum of the function given in Example 8.5 by Powell's method.

A good way to start Powell's method of search is first to minimize along the principal directions, as done below. Starting from

$X_0 = \{0 \quad 0\}^T$, after minimizing the first time along $S_1 = \{1 \quad 0\}^T$ and $S_2 = \{0 \quad 1\}^T$, $X_1 = \{1.5 \quad 0.75\}^T$ is obtained. Minimizing along the same directions once again, we get $X_2 = \{2.25 \quad 1.125\}^T$. A new search direction, $S_3 = X_2 - X_0 = \{0.75 \quad 0.375\}^T$, is next obtained. S_1 and S_3 are conjugate directions, as is verified in the next paragraph.

The function given in Example 8.5 may be expressed in matrix form:

$$X^T \begin{bmatrix} 1 & -1 \\ -1 & 2 \end{bmatrix} X + X^T \begin{bmatrix} -3 \\ 0 \end{bmatrix} + 6$$

On substituting for S_1 and S_3 in the first term of the equation, it is found that

$$\{0 \quad 1\} \begin{bmatrix} 1 & -1 \\ -1 & 2 \end{bmatrix} \begin{Bmatrix} 0.750 \\ 0.375 \end{Bmatrix} = 0$$

which proves that the directions S_1 and S_3 are indeed A-conjugate.

The search point at this stage is X_2, and the search directions are $S_1 = \{0 \quad 1\}^T$ and $S_2 = \{0.75 \quad 0.375\}^T$. The function may be minimized along S_1 using the Golden section method. For minimizing along S_2, the method described below is suggested.

Assuming that a function is quadratic, the minimum along any arbitrary search direction, S_q, is found by evaluating α^* from the following equation:

$$\alpha^* = \frac{1}{2} \frac{(\alpha_2^2 - \alpha_3^2)F_1 + (\alpha_3^2 - \alpha_1^2)F_2 + (\alpha_1^2 - \alpha_2^2)F_3}{(\alpha_2 - \alpha_3)F_1 + (\alpha_3 - \alpha_1)F_2 + (\alpha_1 - \alpha_2)F_3} \qquad (8.17)$$

α^* yields a minimum if the following is satisfied:

$$\frac{(\alpha_2 - \alpha_3)F_1 + (\alpha_3 - \alpha_1)F_2 + (\alpha_1 - \alpha_2)F_3}{(\alpha_2 - \alpha_3)(\alpha_3 - \alpha_1)(\alpha_1 - \alpha_2)} \leq 0 \qquad (8.18)$$

In the above, $F_1 = F(X_q + \alpha_1 S_r)$, $F_2 = F(X_q + \alpha_2 S_r)$, and $F_3 = F(X_q + \alpha_3 S_q)$.

Going back to Example 8.6, let's take $\alpha_1 = 0$, $\alpha_2 = 0.6$, and $\alpha_3 = 1$. Also, let $X_q = \{1.5 \quad 0.75\}$; then, $F_1 = -1.375$, $F_2 = -1.94875$, and $F_3 = -2.21875$ in $S_2 = \{0.75 \quad 0.375\}^T$ direction. From equations (8.17) and (8.18), $\alpha^* = 2$, which leads us to a new search point, $X_{q+1} = \{3 \quad 1.5\}^T$, which is the global minimum itself. ❑

In the algorithm described above, the search points are related to each other, after n number of searches, in the following manner:

$$X_{q+1} = X_q + \alpha_1^* S_1 + \alpha_2^* S_2 + \cdots + \alpha_n^* S_n$$

Conjugate direction S_{n+1} for the $(n+1)^{\text{th}}$ search is therefore given as

$$S_{n+1} = \alpha_1^* S_1 + \alpha_2^* S_2 + \cdots + \alpha_n^* S_n.$$

If $\alpha_1^* \approx 0$, then S_{n+1} does not contain any component of S_1. Since in the next iteration S_1 is dropped, it is obvious that in all searches, direction S_1 has no bearing at all. Because the search algorithm in such instances does not span the entire variable space, convergence occurs slowly or not at all. In order to obviate a problem like this, Powell suggests the following modified algorithm:

Step 1 Minimize $F(X)$ along the n search directions, and find $X_{q+1} = X_q + \sum_{j=1}^{n} \alpha_j^* S_j$ as before.

Step 2 Find an integer m, $1 \leq m \leq n$, such that $\Delta = F(X_{m-1}) - F(X_m)$ is a minimum.

Step 3 Define $F_1 = F(X_0)$, $F_2 = F(X_n)$, and $F_3 = F(2X_n - X_0)$.

Step 4 If $F_3 > F_1$ and/or $(F_1 - 2F_2 + F_3)(F_1 - F_2 - \Delta)^2 > 0.5\{\Delta(F_1 - F_2)\}^2$, use the old directions, and make $X_n = X_0$ for the next iteration. Otherwise, define $S_{n+1} = X_n - X_0$, and find α_{n+1}^* that minimizes the function in direction S_{n+1}. For the next iteration, use search directions $S_1, S_2, \ldots, S_{m-1}, S_{m+1}, S_{m+2}, \ldots, S_n, S_{n+1}$. For the starting point, use $X_0 = X_n + \alpha_{n+1}^* S_{n+1}$.

In iterative processes, a convergence tolerance is defined. In Powell's algorithm, tolerances δ_r, $r = 1, 2, \ldots, n$ for the variables $x_1, x_2, \ldots, x_n$ are set. If subsequent iterations produce changes in the variables that are less than the corresponding tolerances, the search is terminated. This straightforward approach, however, causes premature termination of the search in some instances. Powell suggests the following approach that leads to a true convergence:

Step 1 Apply the normal procedure until the change in every variable is reduced to $< 0.1\,\delta_r$. Let this search point be called a.

Step 2 Increase every variable to $x_r = x_r + 10\,\delta_r$, and use these to define X_0 for further search.

Step 3 Apply the normal search procedure again until the variations are reduced to $< 0.1\,\delta_r$ once again. Call this point b.

Step 4 Find the minimum of $F(X)$ on the line joining b and a. Call this point c.

Step 5 Assume ultimate convergence if components of the vectors $(a - b)$ and $(b - c)$ are all less than $0.1\,\delta_r$. Otherwise, follow step 6.

Step 6 Use $(a - c)$ as a new search direction in place of S_1, and repeat the iterations starting from step 1 above.

❑ FLETCHER AND REEVES' METHOD

Powell's method, discussed earlier, is useful in determining the minimum of functions for which the gradients are not easily available. The method is also more stable than most other search procedures; as a result, it tolerates significant inaccuracies arising from the rounding off in computer calculations. When the number of variables is large, the computer memory requirement in Powell's method becomes excessive. In such situations, the first-order method of Fletcher and Reeves [4] is found to be more efficient.

Fletcher and Reeves' method is one of conjugate gradients. It is quadratically convergent, just as Powell's method is. However, it requires more computations at every iteration step when compared with Powell's algorithm. The method is summarized below.

Step 1 Select a starting point, X_0.

Step 2 Compute the gradients, $G_0 = \nabla F(X_0)|_{X=X_0}$

Step 3 Select α_q^* to minimize $F(X_q + \alpha^* S_q)$ along S_q.

Step 4 Check for convergence. If there is convergence, stop; otherwise, continue to step 5.

Step 5 $X_{q+1} = X_q + \alpha^* S_q$

Step 6 $G_{q+1} = \nabla F(X)|_{X=X_{q+1}}$

$$\beta_q = |G_{q+1}|^2/|G_q|^2$$

$$S_{q+1} = -G_{q+1} + \beta_q S_q$$

Renumber the search directions, $S_q = S_{q+1}$, and the search point, $X_q = X_{q+1}$.

Step 7 Go to step 3.

The algorithm is so devised that the search directions are A-conjugate for a quadratic of the form $X^T A X + X^T B + C$. The search converges in n iterations or less if the function is truly quadratic. The method is, however, sensitive to the accuracy of computations. In some instances, where the function has eccentric contours, more than n iterations are required. A wise thing to do in such instances is to restart the iterations after a specified number of trials. One may choose to restart after, say, n iterations, in which case $S_{n+1} = -G_{n+1}$ is used in step 4.

Fletcher and Reeves' algorithm may also not converge in n iterations if the initial guess is bad (i.e., if it is far removed from the global minimum). In such instances, too, it is advisable to restart the iteration after n tries.

❑ SCALING OF THE VARIABLES

Often the objective function takes on a complex shape in engineering problems, mainly because of the difference in the order of magnitude of the variables. For example, if consistent units are used, there may be a difference of one order in the magnitudes of the radius of a shaft and its length. If proper scaling of the variables is carried out, the objective function becomes fairly easy to handle mathematically (i.e., it does not have pronounced eccentricity, sharp peaks, or deep valleys). In order to utilize full benefits from scaling, some fairly sophisticated logic has been developed [5]. Due to lack of space, this topic cannot be treated to any great length in this book. Described below is a simple method that is often applied to reap some advantages of scaling.

One method of scaling the variables is the *diagonal equalization method.* In this, the coefficients of the square terms in a given quadratic function are all made equal. Thus, a function $F(X) = 100x_1^2 + 2x_2^2 - 6x_1x_2$ is rewritten as $G(Y) = y_1^2 + y_2^2 - (3\sqrt{2}/10)y_1y_2$ by selecting $y_1 = 10x_1$ and $y_2 = \sqrt{2}x_2$.

❑ DAVIDON–FLETCHER–POWELL METHOD

This method is also known as the *variable metric method* [6, 7]. It is a first-order method requiring that the derivative of the function be evaluated at every stage of iteration. It is useful for objective functions having 50 to 200 variables.

The algorithm is summarized below.

Step 1 Assume a starting point, X_0, and a positive definite matrix, H_0 (an identity matrix serves the purpose well). Assume an initial search direction, $S_0 = -H_0 \nabla F(X_0)$.

Step 2 Minimize along S_q ($q = 0, 1, 2, \ldots$), and find $X_{q+1} = X_i + \alpha_q^* S_q$.

Step 3 Update the H matrix:

$$H_{q+1} = H_q + M_q + N_q$$

where

$$M_q = \alpha_q^* S_q S_q^T / S_q^T Y_q$$
$$N_q = -(H_q Y_q)(H_q Y_q^T)/(Y_q^T H_q Y_q)$$
$$Y_q = \nabla F(X_{q+1}) - \nabla F(X_q)$$

Step 4 Find a new search direction: $S_{q+1} = -H_{q+1} F(X_{q+1})$. If convergence is obtained, stop; otherwise, go to step 2.

8.9 ▪ Constrained Optimization: Indirect Methods

There are two different ways in which the constrained minimum is determined. One way is to incorporate the constraint functions into the objective function itself and form a *pseudo-objective function*. Optimization is then carried out using the methods described in Section 8.8 above. This approach gives rise to several indirect methods of constrained optimization. Another way is to keep the objective function and the constraint functions separate and to minimize using a *direct method*. In the present section, only the indirect methods are discussed. Direct methods are dealt with in Section 8.10.

Pseudo-objective functions are created by the use of *penalty functions* or *Lagrange multipliers*. Penalty functions are of two types: *exterior penalty functions* and *interior penalty functions*. In the exterior penalty function approach, the minimum is approached from outside the feasible domain, whereas in the internal penalty function method, the minimum is approached from inside the feasible domain. In next two subsections, use of these penalty functions is demonstrated.

❑ EXTERIOR PENALTY FUNCTION METHOD

The pseudo-objective function is formed, whether using the external or internal penalty function approach, in the following manner:

$$\phi(X, r_i) = F(X) + r_i P(X) \tag{8.19}$$

where $P(X)$ is a penalty function and r_i a scalar multiplier. If there are l inequality constraints and m equality constraints in a design problem, an exterior penalty function may assume the following form[8]:

$$P(X) = \sum_{j=1}^{l} [\max\{0, g_j(X)\}]^2 + \sum_{k=1}^{m} [h_k(X)]^2 \tag{8.20}$$

$P(X)$ given above imposes a penalty only if a constraint is violated. As a result, $\phi(X)$ is not likely to have a minimum too deep into the infeasible regions. The extent of constraint violation depends on the magnitude of r. Thus, in the feasible region $\phi(X) = F(X)$, the minimum of $\phi(X) \approx$ the constrained minimum of $F(X)$.

Since $P(X)$ contains square terms, it has first derivatives, and, thus, if $F(X)$ has first derivatives, then $\phi(X)$, too, has first derivatives. $P(X)$ may thus be used for minimization by the zero- or first-order methods described above. In the first iteration for a rapid convergence, r_i is kept small. The solution is refined subsequently by using larger values of the multiplier.

Larger values of the multiplier ensure that the minimum obtained is closer to the true minimum. A large value of r initially causes instability in the search algorithms, and, therefore, very small iterative steps are required, causing extremely slow convergence. Two examples are given below to illustrate the use of exterior penalty functions.

EXAMPLE ■ 8.7

$$\text{minimize: } F(X) = 3x^2 - 5x + 10$$

$$\text{subject to: } 1.5 \geq x \geq 1$$

A plot of $F(X)$ is shown in Figure 8.10. The unconstrained minimum of the function is at $x = 5/6 = 0.833$, giving $F(X) = 7.9166$. Two inequality constraint functions may be formed from the data given in the problem: $g_1(X) \equiv 1 - x \leq 0$ and $g_2(X) \equiv x - 1.5 \leq 0$. The pseudo-objective function is formed as follows:

$$\phi(X) = F(X) + r_q[\{\max(0, g_1(x))\}^2 + \{\max(0, g_2(x))\}]^2$$

Plots of $\phi(X)$ for $r_1 = 1$, $r_2 = 10$, and $r_3 = 100$ are superposed on the original curve in Figure 8.10. The minimum of $F(X)$ is the lowest point on the plot of $\phi(X)$ for a given r. As can be seen, closer and

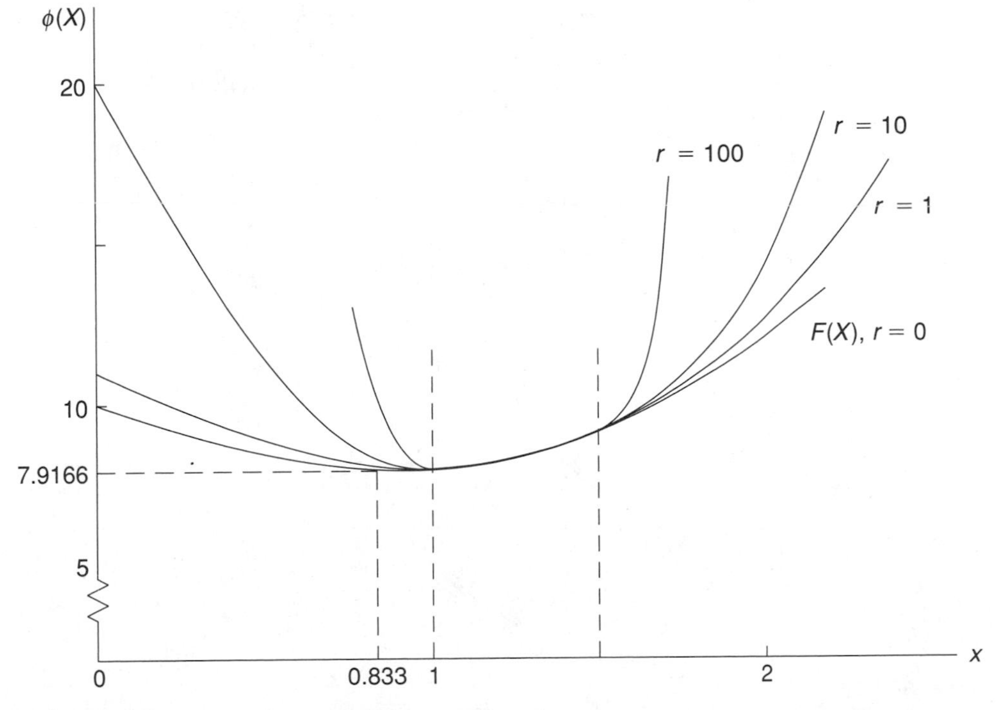

FIGURE 8.10 $F(X)$ and $\phi(X)$

closer estimates of the constrained minimum ($x = 1$, $F(X) = 8$) are obtained as the value of r is increased. At $r \to \infty$, the exact minimum is obtained by this method. ❑

It is noteworthy to realize here that for all finite values of r, the solution obtained is in the infeasible region, as is clearly demonstrated in Figure 8.10. In other words, the method never yields a feasible solution. This gives rise to some skepticism regarding the usefulness of the approach. The method is, nonetheless, good for obtaining a fairly close estimate of the minimum from which an experienced designer may obtain a feasible solution. Moreover, the approach is very simple to work with, causing few problems with instability in the algorithms.

EXAMPLE ■ 8.8

$$\text{minimize: } F(X) = x_1^2 - x_1x_2 + x_2^2 - x_1 + 3$$

$$\text{subject to: } x_1 \geq 1$$

$$x_1 - x_2 \leq 2$$

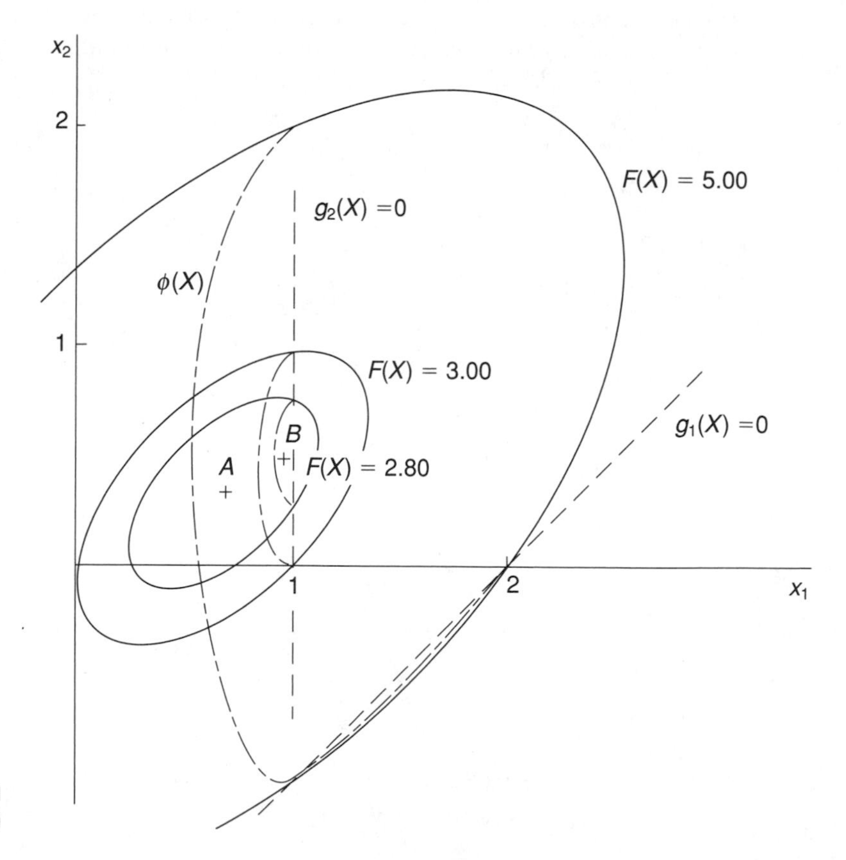

FIGURE 8.11
$F(X)$ and $\phi(X)$

Isoclines of the function for $F(X) = 2.8$, 3, and 5 are shown in solid lines in Figure 8.11. The unconstrained minimum, found mathematically, is at point $A(x_1 = 2/3, x_2 = 1/3)$.

Two inequality constraint functions are formed:

$$g_1(X) \equiv x_1 - x_2 - 2 \leq 0$$

$$g_2(X) \equiv 1 - x_1 \leq 0$$

$$\phi(X) = F(X) + r[\{\max(0, g_1(X))\}^2 + \{\max(0, g_2(X))\}^2]$$

Isoclines of $\phi(X)$ in dashed-dotted lines for $r = 10$ are superposed on the contours of $F(X)$ in Figure 8.11. The constrained minimum, found by the univariate method, is predicted at point B (0.98, 0.49), which is inside the infeasible region as expected. ❑

It is important to note the difference in the nature of the contours of $\phi(X)$. The contours show increased eccentricity and distortion as r is increased. The difference in the nature of the contours is more pronounced for higher values of r. This is the reason for slow convergence at higher values of r.

❑ INTERIOR PENALTY FUNCTION METHOD

Penalty functions may be so devised that convergence to the constrained minimum is achieved from inside the feasible region. Thus, even though the solution is not the correct minimum, it is at least feasible. A common type of internal penalty function is:

$$P(X) = -r_i \sum_{j=1}^{m} [1/g_j(X)] \qquad (8.21)$$

where r_i is a scalar multiplier in the ith cycle of iterations, the value of which is reduced in every cycle, and $g_j(X)$ are the inequality constraint functions.

EXAMPLE ■ 8.9

minimize: $F(X) = x_1^2 - x_1x_2 + x_2^2 - x_1 + 3$

subject to: $1 - x_1 \leq 0$

A psuedo-objective function is derived using equation (8.21):

$$\phi(X) = x_1^2 - x_1x_2 + x_2^2 - x_1 + 3 - r/(1 - x_1)$$

The contours of $\phi(X) = 5$ and $r = 1$, $r = 0.1$, and $r = 0.01$ are shown in Figure 8.12(a). It is observed from the figure that as r decreases

FIGURE 8.12
$\phi(X)$

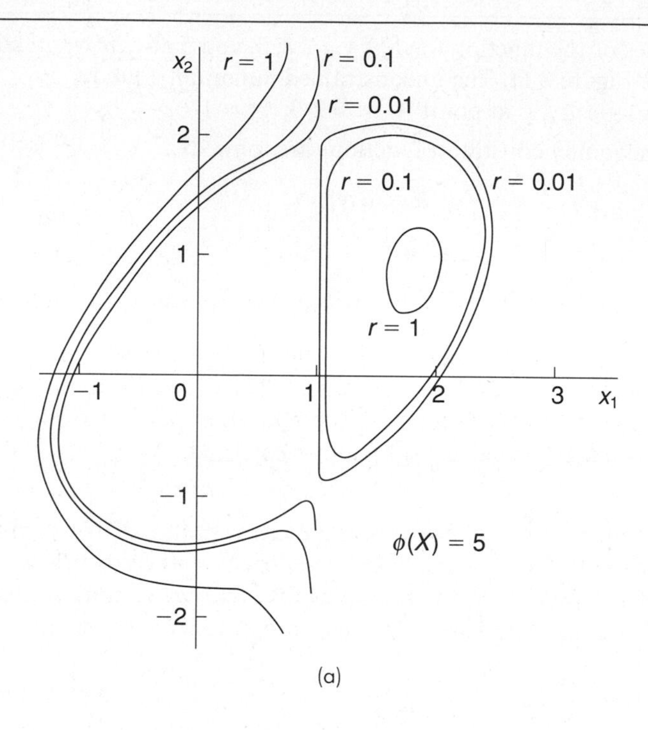

(a)

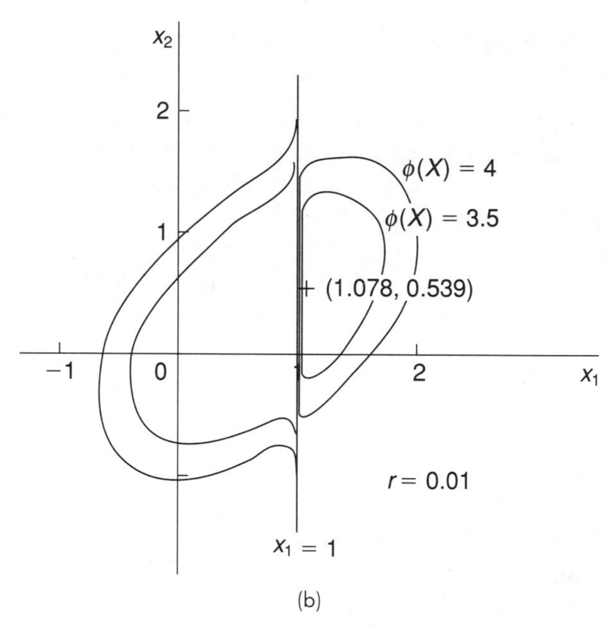

(b)

(for better solutions), the corners in the contours become sharper and the gradient larger. This is often the cause of instability in many situations at or near the minimum. Also noticeable is the fact that near the constraint boundaries, the objective function becomes discontinuous. This often leads to infeasible solutions.

A value of $r = 0.01$ is used for solving the above. The contours for $\phi(X) = 3.5$ and $\phi(X) = 4$ are shown in Figure 8.12(b). The minimum is found to be at (1.078, 0.539), which is in the feasible region. ❑

❑ EXTENDED INTERIOR PENALTY FUNCTION METHOD

This method yields feasible estimates of the minimum as in the interior penalty method. The penalty functions are, however, so formulated that the psuedo-objective function is continuous at the constraint boundaries.

An extended interior penalty function along the lines suggested by some authors [9, 10] is expressed as follows:

$$P(X) = \sum_{j=1}^{m} g'_j(X) \tag{8.22}$$

where $g'_j(X) = -1/g_j(X)$ if $g(X) \leq \varepsilon$; otherwise, $g'_j(x) = -\{2\varepsilon - g_j(X)\}/\varepsilon^2$. ε is a small negative number. It has been suggested [11] that

$$\varepsilon = -C(r)^a, \quad 1/3 \leq a \leq 1/2 \tag{8.23}$$

where C is a constant that is determined from the initial values of ε and r. For initial calculations, $-0.3 < \varepsilon < -0.1$ is a good start.

If, by some means, an idea of a feasible solution is available, then in all optimizations involving interior penalty functions, an estimate of the initial value of r is obtained as follows [12]:

$$r_{\text{int}} = |F(X_{\text{int}})|/|P(X_{\text{int}})| \tag{8.24}$$

Having obtained an initial value for r, values for a and ε are selected from the range indicated above. These values are then substituted into equation (8.23) above, and a value for C is determined. For all subsequent iterations, C may remain unchanged.

EXAMPLE ■ 8.10

Minimize the function given in Example 8.9 using an extended interior penalty function.

A conservative estimate of the variables for a feasible minimum is $x_1 = 1.1$ and $x_2 = 0.55$. An initial value for the multiplier r may be evaluated, based on the above estimates and equation (8.24), as follows:

$$r_{\text{int}} = [1.1^2 - (1.1 \times 0.55) + .55^2 - 1.1 + 3]/[1/(1 - 1.1)]$$
$$= 0.28$$

Assuming $\varepsilon = -0.2$ and $a = 0.5$, from equation (8.23)

$$-0.2 = -C(0.28)^{0.5}$$

giving $C \approx 0.38$.

The pseudo-objective function is then written as

$$\phi(X) = F(X) + rP(X)$$

where

$$P(X) = -1/(1 - x_1) \qquad \text{if } 1 - x_1 \leq \varepsilon$$
$$= -\{2\varepsilon - (1 - x_1)\}/\varepsilon^2 \quad \text{if } 1 - x_1 > \varepsilon$$
$$\varepsilon = -0.38(r)^{0.5}$$

The contour plots for $\phi(X) = 4$, $\phi(X) = 3.5$, and $\phi(X) = 3.2$ for $r =$ 0.001 are shown in Figure 8.13. As can be seen, there are no dis-

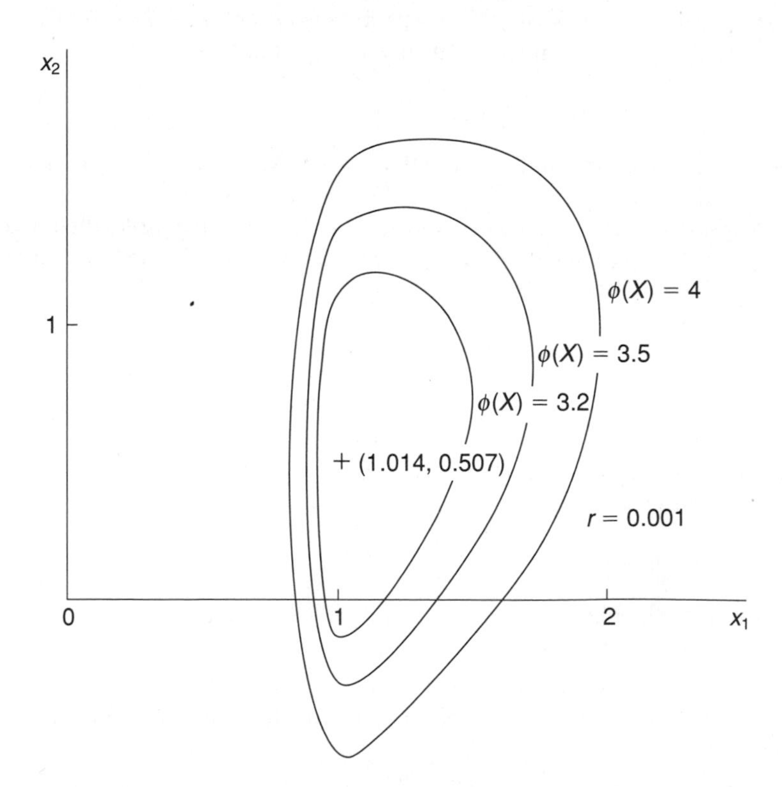

FIGURE 8.13
$\phi(X)$

continuities in the pseudo-objective function. The minimum predicted is at $x_1 = 1.014$ and $x_2 = 0.507$. ❑

❑ METHOD OF AUGMENTED LAGRANGIAN

In the penalty function methods discussed earlier, the solution depends on the choice of the multiplier r, and a precise optimal solution is impossible to find. These are major handicaps. Another inherent problem is that since no optimality condition is included in the algorithms, the search is often quite time consuming.

By the use of Lagrange functions, exact minimum subject to the given constraints is determined. Optimality criteria are embedded in the search algorithm so that faster convergence is achieved. The appropriate methods for optimization using Lagrangian (or, more precisely, the augmented Lagrangian) functions are discussed in the next few paragraphs.

Augmented Lagrangian Method for Equality Constraints ■ A Lagrangian is formed, based on the Kuhn-Tucker criterion of optimality, as follows:

$$L(X, \lambda) = F(X) + \sum_{j=1}^{m} \lambda_j h_j(X) \tag{8.25}$$

where λ_j are scalars known as Lagrangian multipliers and $h_j(X)$ are the equality constraints. The minimum of the Lagrangian above yields the constrained optimal solution. Thus, the problem here is to find a set of values for the Lagrangian multipliers that yields the minimum of $L(X, \lambda)$. It is seen that, at least theoretically, it is possible to determine the exact minimum in the above case. The situation is unlike that in the exterior penalty function method where r is infinite for the exact solution.

Although optimization may be carried out by minimizing the Lagrangian as stated above, problems are often encountered with nonconvex objective functions. In such cases, the Lagrangian contains saddle points that often make determination of the minimum difficult. Over the years, some augmented Lagrangian algorithms have been developed for which the convexity requirements are relaxed. One suggested way of forming an augmented Lagrangian is

$$A(X, \lambda, r_i) = L(X, \lambda) + \sum_{k=1}^{m} r_i [h_k(X)]^2 \tag{8.26}$$

The augmented Lagrangian may be minimized for finite values of the multiplier r (i.e., an exact constrained optimum may be found even when r is finite).

If values of the Lagrange multipliers are consistent with the optimal solution, then only one iteration is necessary to find that value of the vector X at which the augmented Langrangian has the minimum value. Usually, no closed-form solutions are available for the set of multipliers. These are found iteratively by methods such as those described below [8]:

Step 1 Assume $\lambda_j^q = 0$ for all j and a small value of r_q where q represents the iteration number.

Step 2 Minimize $A(X, \lambda, r_i)$, yielding $X^q_{\min}$.

Step 3 $\lambda_j^{q+1} = \lambda_j^q + 2r_q h_j(X^q_{\min})$ if $r_q < r_{\max}$ and $r_{q+1} = r_q + \Delta r$; otherwise, $r_{q+1} = r_q$. $r_{\max}$ is a user-defined maximum for r to accelerate convergence. Convergence still takes place if r is never modified.

Step 4 If $\lambda_j^{q+1} \approx \lambda_j^q$ for all j, convergence is assumed. It is recommended, however, before ending the search that variation in the values of the objective function also be checked.

Augmented Lagrangian Method for Inequality Constraints ▪ All inequality constraints may be expressed as equality constraints by the use of *slack variables*. Thus, the constraint function $g_j(X) \leq 0$ may be rewritten as $g_j(X) + Z_j^2 = 0$, where Z_j are slack variables.

The Langrangian function then is formed as follows:

$$L(X, \lambda, Z) = F(X) + \sum_{j=1}^{m} \lambda_j [g_j(X) + Z_j^2] \tag{8.27}$$

and the augmented Lagrangian as follows:

$$A(X, \lambda, Z, r) = L(X, \lambda, Z) + r_q \sum_{j=1}^{m} [g_j(X) + Z_j^2] \tag{8.28}$$

It has been shown [13] that the two functions given may also be expressed in a somewhat simpler form:

$$L(X, \lambda) = F(X) + \sum_{j=1}^{m} \lambda_j \psi_j \tag{8.29}$$

$$A(X, \lambda, r) = L(X, \lambda) + r_q \sum_{j=1}^{m} \psi_j^2 \tag{8.30}$$

where $\psi_j = \max[g_j(X), -\lambda_j/2r_i]$. The multipliers, λ_j, are modified during iterations by the following rule:

$$\lambda_j^{q+1} = \lambda_j^q + 2r_q \psi_j \tag{8.31}$$

EXAMPLE ▪ 8.11

minimize: $F(X) = x_1^2 - x_1 x_2 + x_2^2 - x_1 + 3$

subject to: $g(X) \equiv 1 - x_1 \leq 0$

$h(X) \equiv x_2 - 0.5 = 0$

The augmented Lagrangian for this problem, based on equations (8.26), (8.29), and (8.30) may be written as

$$A(X, \lambda, r) = F(X) + \lambda_1 \psi + r_q \psi^2 + \lambda_2 h(X) + r_q\{h(X)\}^2$$

$$\psi = \max[g(X), -\lambda_1/2r_q]$$

When $g(X) > -\lambda_1/2r_q$, then $\psi = g(X)$; assuming that this is true and that $r_q = 1$ in all iterations, the augmented Lagrangian may be written as

$$A(X, \lambda, r) = 2x_1^2 - x_1x_2 + 2x_2^2 - (3 + \lambda_1)x_1$$
$$- (1 - \lambda_2)x_2 + (4.25 + \lambda_1 - 0.5\lambda_2)$$

For stationary value of A

$$4x_1^* - x_2^* - (3 + \lambda_1) = 0 = \frac{\partial A}{\partial x_1}$$

$$-x_1^* + 4x_2^* - (1 - \lambda_2) = 0 = \frac{\partial A}{\partial x_2}$$

or $x_1^* = 4x_2^* - 1 + \lambda_2$ and $x_2^* = (7 + \lambda_1 - 4\lambda_2)/15$.

Iteration I: Assume $\lambda_1 = \lambda_2 = 0$ and a starting point of $x_1 = 0$ and $x_2 = 0$, in which case $\psi = g(X) = 1 - x_1$. Stationary values of A are $x_1^* = 0.866666$ and $x_2^* = 0.466666$, and

$$g(X^*) = 1 - x_1^* = 0.133333$$

$$h(X^*) = x_2^* - 0.5 = -0.033333$$

$$\psi = \max[0.133333, 0] = 0.133333$$

Iteration II: Using equation (8.31) with $r_q = 1$, $\psi_1 = -0.133333$, and $\psi_2 = -0.033333$, and

$$\lambda_1 = \lambda_1 + (2 \times 0.133333) = 0.266666$$

$$\lambda_2 = \lambda_2 + (2 \times -0.033333) = -0.066666$$

Since $g(X^*) > -\lambda_1/2$, $\psi = 1 - x_1$. New stationary values for A can be obtained by following the same procedure as above. The values are $x_1^* = 0.942222$ and $x_2 = 0.502222$, and

$$g(X^*) = 0.057778$$

$$h(X^*) = 0.002222$$

$$\psi = 0.05778$$

Iteration III: By following the same procedure as outlined above, one will get the following values:

$$\lambda_1 = 0.382222 \qquad \lambda_2 = -0.062222$$

$$x_1^* = 0.972741 \qquad x_2 = 0.508741$$

As is seen above, the solution is converging at $x_1 = 1$ and $x_2 = 0.5$, which represents the true minimum of the objective function. ❑

8.10 ▪ Constrained Optimization: Direct Method

Several methods are available in which the constraints are considered separately, rather than in combination with the objective function as with the penalty functions discussed above. Such methods are known as *direct methods.* In the direct method of optimization, the constraints are regarded as the limiting surfaces that divide the design space into feasible and infeasible regions. The constrained optimum is the lowest value of the objective function in the feasible space bounded by the constraint surfaces. The algorithm for the direct method involves the identification of a search path in a usable feasible region. This procedure requires complex mathematical manipulations, especially when a large number of design variables and constraints are involved. Detailed formulations for such algorithms are available in reference [12]. In the following section we will focus our attention on the working principle of a more contemporary technique, design optimization by the finite element method.

8.11 ▪ Use of the Finite Element Method in Optimization

The use of the finite element method for various types of analyses is discussed in Chapters 6 and 7. By means of this technique, it is possible to determine the effects (stress, strain, temperature distribution, etc.) of certain applied physical actions (force, pressure, heat, etc.) on geometrically complex systems (a roof truss, a pressure vessel, an aircraft wing, etc.). The analysis may be carried out even where the explicit mathematical relationships between the cause and the effects are not known. For example, we may be able to determine stress distribution in each member of a complex transmission tower even though we may not be able to derive mathematical equations for them. Likewise, in cases where it is not possible to obtain mathematical expressions for either the objective or the constraint functions, optimization may be carried out by means of the finite element method.

A basic requirement for using finite element analysis in design optimization is determination of the objective and constraint functions. One of the easiest ways of doing so may be along the lines discussed in Section 8.5 (i.e., by curve fitting). Suppose there are n known design variables that are expected to affect the optimum design. As mentioned in Section 8.8, most objective functions are often quadratic in nature at or around the optimal point. It is reasonable therefore to assume a trial objective function of the following type:

$$F(X) = a_0 + \sum_{i=1}^{n} (a_i x_i + b_i x_i^2) \tag{8.32}$$

Assuming different linearly independent combinations of the design variables, finite element analysis is performed a number of times. In each run of the program, the objective function is evaluated. From these results, the coefficients in equation (8.32) are evaluated. As can be seen, to evaluate $2n + 1$ coefficients in equation (8.32), a similar number of computer runs are required. This becomes an expensive proposition in problems involving a large number of design variables. All commercial optimization packages therefore include some algorithms that help reduce the number of runs.

The constraint functions are also evaluated in a manner similar to that used for the objective function. Once these functions are known, a global minimum is found by the techniques discussed earlier, and the corresponding values for the design variables are determined.

Results of the above optimization are often utilized to refine the values of the coefficients of equation (8.32). This may be achieved by eliminating data points far removed from the optimal point and reevaluating the objective function in the vicinity of the optimum. With the new set of coefficients, a new optimum may be determined and the process repeated until convergence is obtained.

The method described above outlines the broad principles by which the optimization algorithms based on the finite element method work. There are many variations in the actual implementation, and in most cases, an attempt is made to reduce the computational time and effort. The authors are aware of two commercially available optimization modules that may be of interest to the reader: Optisen (from SDRC) and the optimization module of ANSYS (from Swanson Analysis Systems, Inc.).

Examples 8.12 and 8.13 demonstrate the procedure for optimization using the finite elements method. Example 8.12 deals with a simple problem. Is is included so that fewer number of design variables and governing equations need to be considered. This example clearly demonstrates the principles of combining the classical design optimization theory and the finite element solution.

EXAMPLE ■ 8.12

Design a tension bar of circular cross-section, 30 inch long, to carry a load of 10,000 pounds. The weight of the bar should not exceed 2.5 pounds, and the longitudinal stress should be limited to 25,000 psi. The elongation in the bar must be a minimum, and the bar should have a uniform diameter throughout its length. Available material for the bar are steel and an aluminum alloy with the following properties:

modulus of elasticity: $E_{\text{steel}} = 30 \times 10^6$ psi $\quad E_{\text{al}} = 10 \times 10^6$ psi

specific weight: $\rho_{\text{steel}} = 0.282\ \text{lb/in}^3 \quad \rho_{\text{al}} = 0.10\ \text{lb/in}^3$

In order that the elongation in the bar be kept to its lowest value, the largest possible diameter should be used. Largest diameter means the heaviest possible bar. Apparently, under the given design criteria the target weight of the bar should be 2.5 pounds. Also, since the stress in the bar is not to exceed 25,000 psi, the diameter d (Figure 8.14) should not exceed $d_{\min}$ obtained from the following relationship:

$$25{,}000 = \frac{10{,}000}{\frac{\pi}{4} d_{\min}^2}, \text{ or } d_{\min} \simeq 0.75 \text{ in}$$

Intutively, steel would be the right material for the tension-bar because of its higher modulus of elasticity and thus its greater rigidity. But, the weight of a steel bar of 0.75 inch diameter and 30 inch length is around 4 pounds. To keep the weight down, a trade-off must to be made between the weight and rigidity of the bar. A composite bar consisting of steel and aluminum offers the mechanism for such compromise. In Figure 8.14, the length of the steel portion of the bar is l inch.

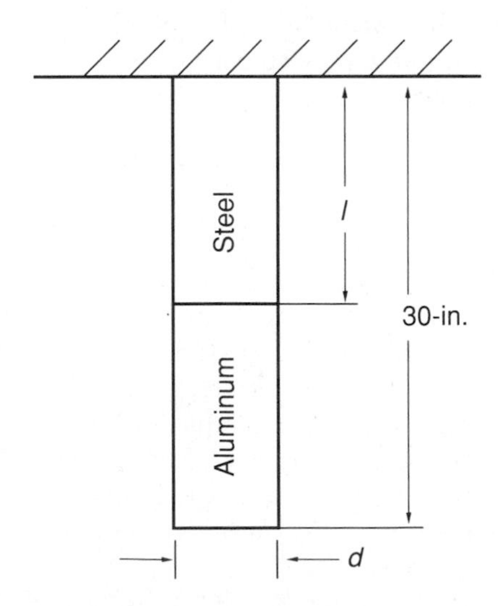

FIGURE 8.14
Tension bar

From the discussions above, it is appartent that the design variables in the problem are d and l. Given the geometry of the bar and the nature of loading (or the stress), elongation and weight can be calculated directly using exact mathematical relations. However, in the finite element approach, these quantities are calculated using numerical methods as proposed in equation (8.32). Thus, the weight of the bar is expressed as follows:

$$W(d, l) = a_0 + a_1 d + a_2 d^2 + a_3 l$$

and the elongation is

$$F(d, l) = b_0 + b_1 d + b_2 d^2 + b_3 l.$$

A second look at Figure 8.14 is enough to convince anyone that there is an order of magnitude difference between the values of d and l. This difference would cause serious errors in numerical evaluations. Therefore, it is necessary that the variables be scaled as mentioned earlier in Section 8.6. In the computations given below, the following scaling has been adopted:

$$x_1 = 2d \quad \text{and} \quad x_2 = 0.05l. \tag{8.33}$$

The revised weight and deflection equations now take the following form:

$$W(X) = A_0 + A_1 x_1 + A_2 x_1^2 + A_3 x_2 \tag{8.34}$$

$$F(X) = B_0 + B_1 x_1 + B_2 x_1^2 + B_3 x_2 \tag{8.35}$$

On repeating finite element analysis on the bar four times, with arbitrary sets of d and l values, the corresponding weights and elongations are obtained. These values are shown in Table 8.1. Two sets of four simultaneous equations are generated from these results using equations (8.34) and (8.35). The coefficients in the two equations

TABLE 8.1
Weight and elongation of tension bar (Applied load 10,000 lbs)

Run No.	Design Variables d, in	l, in	Weight lbs	Elongation inch
1.	0.75	20	2.933462	3.772562×10^{-2}
2.	0.80	16	2.971695	3.846244×10^{-2}
3.	0.85	12	2.941662	3.876992×10^{-2}
4.	0.90	10	3.066351	3.667768×10^{-2}

are evaluated and the equations rewritten as follows:

$$W(X) = -14.785937 + 15.42332x_1 - 3.4133x_1^2 + 2.22988x_2 \tag{8.36}$$

$$F(X) = 5.396772 + 3.45083x_1 - 2.1467x_1^2 - 1.97038x_2 \tag{8.37}$$

In equation (8.37) above, the function $F(X)$ represents elongations that are scaled-up 100 times. Thus, for $x_1 = 1.50$ and $x_2 = 1.0$, the value for $F(X)$ used in the corresponding simultaneous equation is 3.772562.

As decided earlier, the weight of the bar is to be fixed at 2.5 pounds. Combining this constraint with the weight function (8.36) obtained above, an equality constraint function may be derived thus:

$$2.5 = -14.785937 + 15.42332x_1 - 3.4133x_1^2 + 2.22988x_2$$

or

$$h(X) = -17.285937 + 15.42332x_1 - 3.4133x_1^2 + 2.22988x_2 = 0 \tag{8.38}$$

From stress consideration it has already been decided that the diameter of the bar may not be less than 0.75 inch. This constraint gives rise to the following constraint function:

$$g(X) = x_1 - 1.50 \leq 0 \tag{8.39}$$

The optimization problem may now be stated in the following way:

minimize: $F(X) = 5.396772 + 3.45083x_1 - 2.1467x_1^2 - 1.97038x_2$

subject to: $h(X) = -17.285937 + 15.42332x_1 - 3.4133x_1^2 + 2.22988x_2 = 0$

and: $g(X) = x_1 - 1.50 \leq 0.$

If the exterior penalty function approach is adopted for the optimization, then the psuedo-objective function may be expressed as follows:

$$\phi(X) = F(X) + P(X), \tag{8.40}$$

where

$$P(X) = r[\{h(X)\}^2 + \{\max(O, g(X))\}^2].$$

After several iterations, the following values for the variables are obtained consistent with the minimum elongation

$$x_1 = 1.50 \quad \text{and} \quad x_2 = 1.019199$$

The above values of the variables correspond to $d = 0.75$ inch and $l = 20.38$ inch. The corresponding values for the weight and deflection are 2.964 pound and 0.037 inch respectively.

Notice that the weight constraint in the solution has been violated. This is to be expected in all optimizations involving the exterior penalty functions. ❑

EXAMPLE ■ 8.13

A steel disk (6-inch inside diameter × 30-inch outside diameter) is shrink fitted over a 6-inch diameter shaft, as shown in Figure 8.15. The shaft rotates with an angular velocity of $\omega = 500$ rad/sec. Design a disk of minimum volume such that the tangential stress (σ_θ) in the material is limited to 30 kpsi and the enlargement (u_r) of the hole in the disk is limited to 0.003 inch over the 3-inch radius.

To fully understand the issues involved in the above problem, look at Figure 8.16, in which a disk of outside diameter d_o is shown mounted over a shaft of diameter d_i. Due to rotation of the shaft, centrifugal forces are introduced in the disk material. The stresses in a typical element of the disk are shown in Figure 8.16(b). Due to the combined effect of the radial stress (σ_r) and the tangential stress (σ_θ), the diameter of the disk tends to increase everywhere. In this problem, the increase in the radius of the hole must be limited to 0.003 inch, and the maximum value of σ_θ at any radius must be limited to 30 kpsi.

A solution to this problem has been worked out by Swanson and Marx [15]. The authors used the finite element program ANSYS and

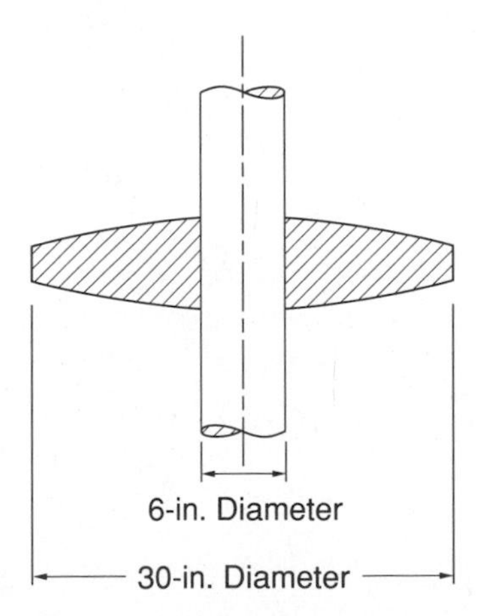

FIGURE 8.15
Shaft and disk assembly

FIGURE 8.16
Stresses in a rotating disk

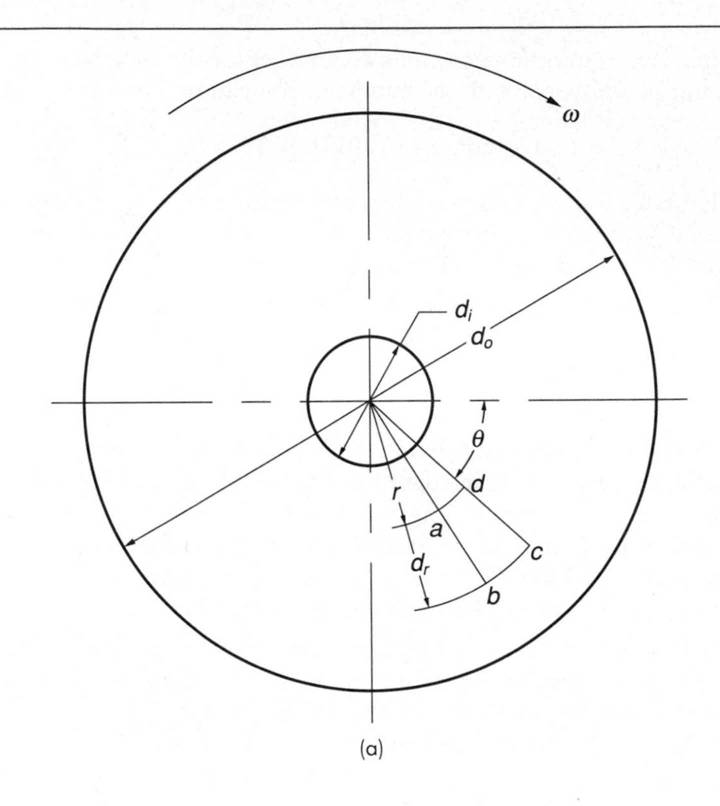

(a)

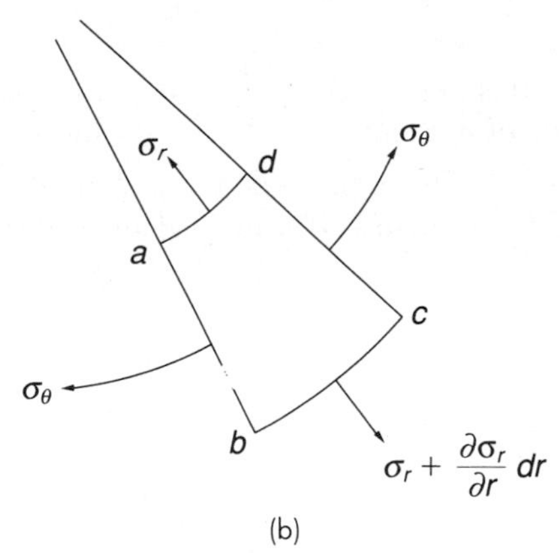

(b)

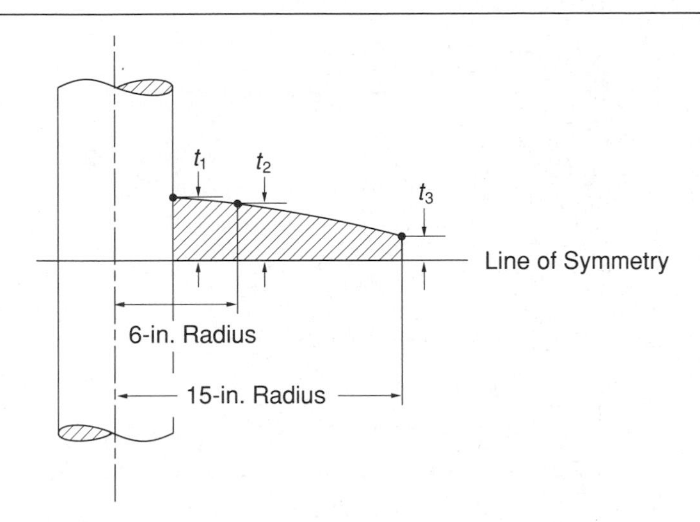

FIGURE 8.17
Symmetrical one-half section of disk

its optimization module. In view of the symmetry, only one-quarter of the section need be considered for analysis. The authors chose to model the cross-section using the axial height (t_1, t_2, and t_3) of the disk at radii of 3 inches, 6 inches, and 15 inches, respectively, as shown in Figure 8.17. The profile of the section was determined by fitting a quadratic spline through the three specified points. The optimal solution was obtained using the following algorithm:

Step 1 Define t_1, t_2, and t_3 (design variables).

Step 2 Generate the finite element mesh with triangular ring elements.

Step 3 Solve the finite element problem.

Step 4 Extract the maximum radial displacement at the hole.

Step 5 Extract the maximum tangential stress.

Step 6 Determine the total volume (objective function).

Step 7 Enter the optimization module with the values obtained at steps 4, 5, and 6.

Table 8.2 summarizes the results obtained by the authors in three different cases. The following constraints were applied in the three cases mentioned in Table 8.2:

Case 1 $t_1 > t_2 > t_3$ and $t_1 - t_2 > t_2 - t_3$
Case 2 Same as case 1, but $t_3 \geq 0.5$ inch
Case 3 $t_1 > t_2$

TABLE 8.2
Summary of results of optimization

Variables	Case 1	Case 2	Case 3
t_1 (in)	1.666	1.256	1.055
t_2 (in)	1.032	1.028	1.025
t_3 (in)	1.024	0.810	0.529
σ_θ (max.) (psi.)	28,774	29,474	27,381
Δr (in)	0.298E−2	0.298E−2	0.274E−2
Volume (in^3)	1430	1260	1060
Iterations	18	22	17

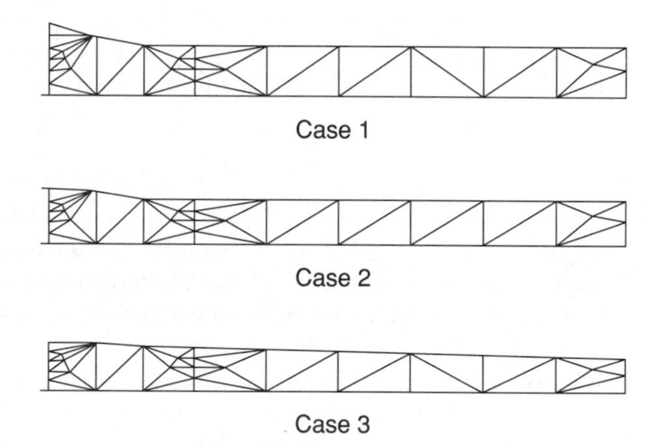

FIGURE 8.18
Optimum disk designs

The optimized half cross-sections for the three cases are shown in Figure 8.18. ❑

8.12 ▪ Summary

This chapter explains the concept of design optimization by means of the available numerical techniques, including the meaning of different terms used in optimization algorithms. Through actual design problems, the method of formulating objective and constraint functions is illustrated.

This chapter also presents possibilities of utilizing computer graphics in everyday engineering design problems. The chapter also describes several popularly known and used algorithms for constrained and unconstrained optimization methods. Numerical examples demonstrate the use of penalty functions and Lagrange multipliers in constrained optimization.

Finally, this chapter explains the method of using the finite element method in optimization. An illustrative example of this approach is presented using a commercial package.

References

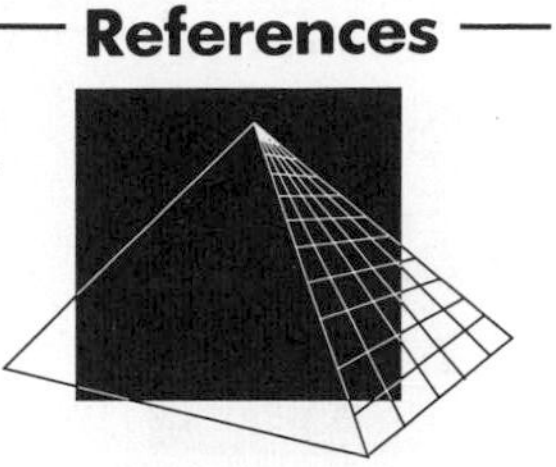

1. Shigley, J.E., and C.R. Mischke. *Mechanical Engineering Design.* New York: McGraw-Hill, 1989.
2. Hinkle, R.T., and I.E. Morse, Jr. "Design of Helical Springs for Minimum Weight, Volume and Length." *Journal of Engineering for Industry* 81, no. 7 (1959): 37–42.
3. Powell, M.J.D. "An Efficient Method for Finding the Minimum of a Function of Several Variables Without Calculating Derivatives." *Computer Journal* 7, no. 4 (1964): 155–62.
4. Fletcher, R., and C.M. Reeves. "Function Minimization by Conjugate Gradients." *Computer Journal* 7, no. 2 (1964): 149–54.
5. Balachandran, S. "Dynamic Scaling of Nonlinear Programs and Its Implementation Within Augmented Lagrangian Penalty Function Technique." Ph.D. diss., Virginia Polytechnic Institute and State University, 1984.
6. Davidon, W.C. *Variable metric method for minimization.* Argonne National Laboratory ANL-5590 Rev. Chicago: University of Chicago, 1959.
7. Fletcher, R., and M.J.D. Powell. "A Rapidly Convergent Descent Method for Minimization." *Computer Journal* 6, no. 2 (1963): 163–68.
8. Vanderplaats, G.N. *Numerical Optimization Techniques for Engineering Design.* New York: McGraw-Hill, 1984.
9. Kavlie, D., and J. Moe. "Automated Design of Frame Structures." *ASCE Journal of Structural Engineering* 97, no. ST1 (1971): 33–62.
10. Cassis, J.H., and L.A. Schmit. "On Implementation of the Extended Interior Penalty Function." *International Journal for Numerical Methods in Engineering* 10, no. 1 (1976): 3–23.
11. Haftka, R.T., and J.H. Starnes. "Application of a Quadratic Extended Interior Penalty Function for Structural Optimization." *AIAA Journal* 14, no. 6 (1976): 718–24.
12. Fox, R.L. *Optimization Methods for Engineering Design.* Reading, Mass.: Addison-Wesley, 1971.
13. Rockafellar, R.T. "The Multiplier Method of Hestenes and Powell Applied to Convex Programming." *Journal of Optimization Theory and Application* 12, no. 6 (1973): 555–62.
14. Vanderplaats, G.N., and F. Moses. "Structural Optimization by Method of Feasible Directions." *Computer and Structures* 3 (July 1973).
15. Swanson, J.A., and F.J. Marx. *Design Optimization Including Integrated Solid Modeling Using the Finite Element Program ANSYS*, National OEM Design '85 Show and Conference, Philadelphia, September 9–11, 1985.

Problems

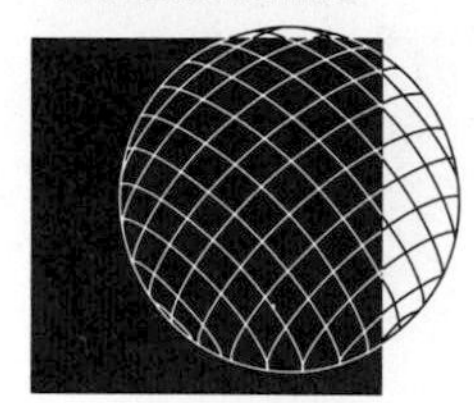

8.1 Design a helical compression spring of minimum weight from the following design data:

Deflection	1 in.
Load	100 lb.
G	11.5×10 psi
Allowable shear stress	40,000 psi

The spring is required to have plain ends.

8.2 Determine the optimal weight of the spring in Example 8.1, using the data given below:

Spring Index	Weight
6.10	0.7989
6.60	0.7900
7.40	0.7870
7.90	0.7909

8.3 Design a torsion bar of minimum weight from the following design data:

Torque	5,000 N.m
Minimum twist	0.08 rad
G	79 GPa
Allowable shear stress	190 MPa
Density	7,800 kg/m
Length	1 m

8.4 Design a compound cylinder of minimum outside diameter from the following data:

Inside radius	500 mm
Internal pressure	60 MPa
Allowable tensile stress	200 MPa
E	207 GPa

8.5 Design a hollow steel column of minimum weight, given the following:

Axial load	5,000 lb.
Column height	120 in.
E	30×10^6 psi
Poisson's ratio	0.3
Specific weight	0.28 lb./in^3
Allowable compressive stress	30,000 psi

The wall thickness should not be less than 0.1 inch, and the mean diameter of the tube should be between 5 and 6 inches.

A minimum factor of safety of 2.5 is required. (*Note:* Two possible modes of failure should be considered: compressive and buckling.)

8.6 Find the global minimum of $F(X) = 2x^3 - x^2 - x + 2$, using the Golden section method.

8.7 Find the global minimum of $F(X) = x_1^3 - 2x_1x_2 + 0.5x_2^2 - x_1 + 4$ by the univariate search method.

8.8 Find the global minimum of the function given in problem 8.7 by Powell's method.

8.9 Find the global minimum of the function given in problem 8.7 by Fletcher and Reeves' method.

8.10 Find the minimum of the function given in problem 8.7, using an exterior penalty function, if it is specified that $x > 0.5$.

8.11 Find the minimum of $F(X) = x_1^3 - x_2^2 - x_1x_2 - x_1 - x_2 + 2$ subject to $x_1 \geq 0.5$ and $x_2 = 1.0$. Use the interior penalty function method.

8.12 Solve problem 8.11, using an extended interior penalty function method.

8.13 Solve problem 8.11, using the augmented Lagrangian function method.

CHAPTER 9

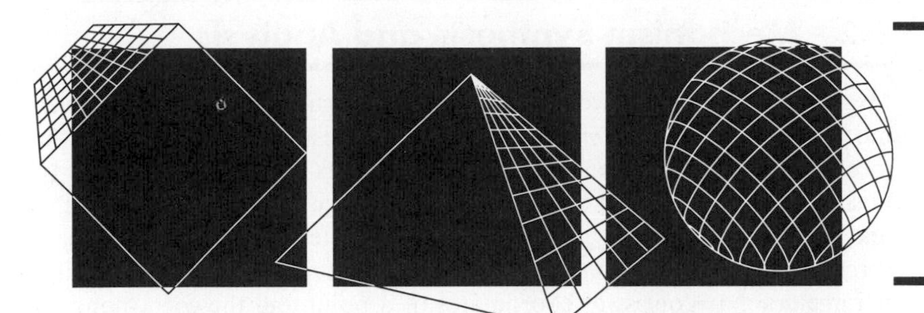

Linkage Synthesis on Microcomputers

9.1 Introduction

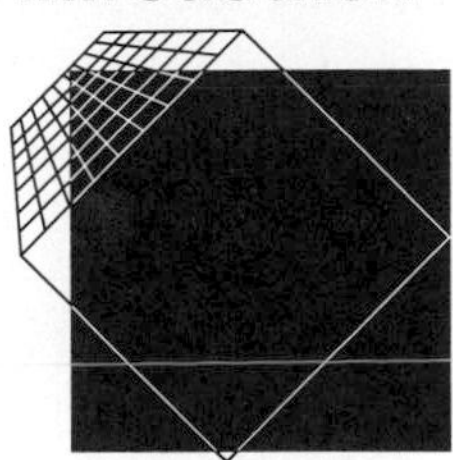

As trade competition from around the world grows, the industrialized nations will have to find means of reducing the expensive labor content in their manufactured goods. A higher level of automation is the only way of confronting cheap labor in other countries. One of the ways of achieving this is by replacing human workers with machines. A machine can replace a worker only if it is able to repeat the necessary movements with precision and in certain desired ways. In order to develop such special-purpose machinery, some relevant techniques need to be understood. These techniques are covered in an area known as *kinematic synthesis.* Kinematic synthesis has been somewhat of a neglected subject in many engineering programs, primarily due to lack of students' time. Computers, however, now make the teaching of kinematic synthesis much more expeditious. In this chapter, the reader is introduced to the basics of linkage synthesis, which may be further reinforced by classroom or laboratory demonstrations of suitable synthesis software.

9.2 ▪ Mechanism Synthesis and Analysis

Mechanism synthesis involves creation of new mechanisms from given motion parameters. *Motion parameters* are the design specifications for the mechanisms. In simple cases, motion parameters may consist of a set of position coordinates of a single point on the mechanism. For example, it may be specified that a point on the candidate mechanism must pass through points (x_0, y_0) and (x_1, y_1). In more complex cases, the motion parameters may consist of coordinates of a point and the corresponding angles subtended by one or two cranks from a reference direction. For example, in a three-position synthesis problem, the motion parameters may read as follows: $\{(x_0, y_0), \alpha_0, \gamma_0\}$, $\{(x_1, y_1), \alpha_1, \gamma_1\}$, and $\{(x_2, y_2), \alpha_2, \gamma_2\}$.

Mechansim analysis involves analyzing the kinematic and kinetic behavior of given mechanisms. Usually, the interacting forces between the members and the time-dependent positions, velocities, and accelerations of members and certain points are determined in mechanism analysis.

9.3 ▪ Mechanism and Linkage

A mechanism is a device that is used for transferring and suitably modifying an applied motion or force to perform a certain function at a different location. A mechanism may be comprised of gears, cams, belts, hydraulic or air cylinders, etc. The design of most of these components is fairly well documented [1, 2]. In the present chapter, the design of another class of mechanisms, consisting primarily of rigid links, is considered.

A linkage system consists of links of different shapes and sizes in which relative motion (rotary, turning, or sliding) is permitted through appropriate joints. Given below are some definitions that are relevant to the study of linkage systems.

- *Rigid body:* A rigid body is that which does not deform in any way when external loads are applied to it. The distance between any two points on the rigid body therefore remains constant.
- *Links:* Links are rigid members that collectively form a machine.
- *Joints:* Two or more links are connected to each other through joints. Joints allow relative motion between the links.
- *Binary links:* These are links with two flexible joints [Figure 9.1(a)].
- *Ternary links:* These are links with three joints [Figure 9.1(b)].
- *Quarternary links:* These are links with four joints [Figure 9.1(c)].
- *Connectors:* Joints between two links may be formed through the use of higher-pair connectors, lower-pair connectors, or wrapping connectors. Figure 9.2 shows some possible types of connectors. In each connector there are two working surfaces (elements). These together form a

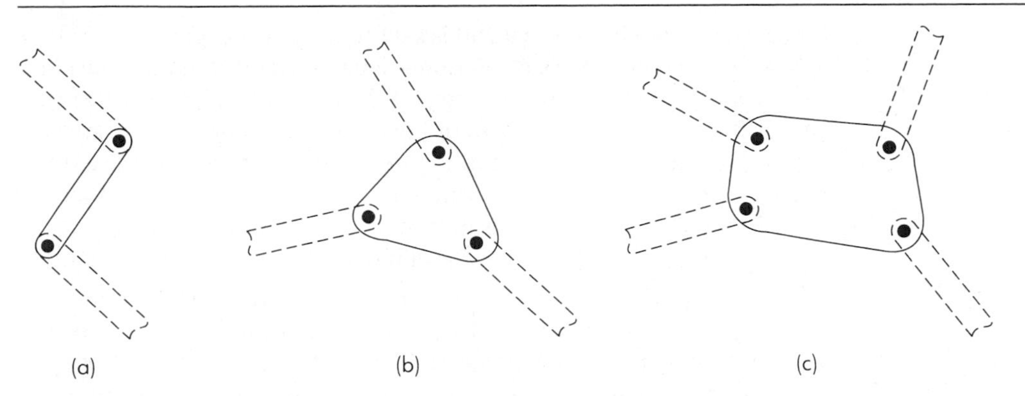

FIGURE 9.1 Links: (a) binary; (b) ternary; (c) quarternary

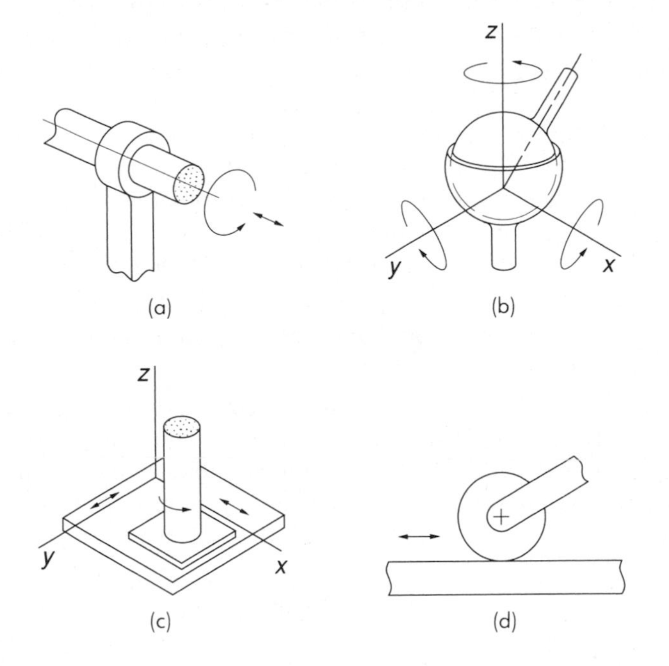

FIGURE 9.2 Different types of connectors

kinematic pair. If the mating elements in a pair take the form of two-dimensional areas or three-dimensional surfaces, then the pair is known as a *lower pair*. In the examples shown in Figure 9.2(a) through (c), the contacting surfaces form a lower pair. In Figure 9.2(a) and (b), one (hollow) element envelops the other (solid) element. Such pairs are referred to as *wrapping pairs* [3]. The connector shown in Figure 9.2(d) consists of a roller and a rigid surface. The two elements in this example contact along a straight line. A pair such as this, in which the nature of contact between the elements is either a straight line or a single point, is known as a *higher pair*. A *wrapping connector* consists of chain or belt elements.

The pair shown in Figure 9.2(a) is said to have two degrees of freedom because it allows two types of motion—axial movement and the turning of the shaft. The pairs shown in Figure 9.2(b) and (c) each have three degrees of freedom. The pair shown in Figure 9.2(d) has only one degree of freedom. In the example shown in Figure 9.2(a), by preventing the axial movement of the pin, a pin or revolute joint is obtained. On the other hand, if the turning of the pin is prevented (say, by replacing it with a square pin or by providing a key), a sliding or prismatic joint is obtained.

- *Kinematic chain:* A kinematic chain is formed by joining two or more links through connectors. These chains can be *open chains* or *closed chains.* In a closed chain, each link is connected to at least two other links. A chain that is not closed is said to be open. Four links forming a quadrangular chain is an example of a closed chain. An open chain consisting of two links and a single joint is known as a *dyad.*

- *Mechanism:* A mechanism is formed by fixing one of the links of a kinematic chain, provided that relative motion between the links is still possible and only a unique geometrical configuration is obtainable for every crank position. A mechanism is referred to as closed or open depending on whether it is formed from a closed or an open chain.

Mechanisms can be either *planar* (two-dimensional) or *spatial* (three-dimensional). In planar mechanisms, all links move in parallel planes. In spatial mechanisms, the links move in nonparallel planes.

- *Inversion:* A distinct mechanism may be formed by fixing, in turn, each link of a given kinematic chain. These mechanisms are known as (kinematic) inversions of the chain.

- *Position variables:* When a linkage system is in motion, certain angles and lengths keep changing continuously. Those angles and/or lengths that do change are called position variables of the system. The geometrical configuration of the system at any moment in time can be obtained if the exact values of the position variables are known.

- *Degree of freedom (DOF) of a linkage system:* DOF represents the minimum number of position variables that should be specified to completely define the geometry of a linkage system. In the systems shown in Figure 9.3(a) and (b), if the angular position of one of the cranks (O_aA or O_bB) is known, the system can be geometrically defined. The systems therefore have a single degree of freedom. The system shown in Figure 9.4(a) has two degrees of freedom.

- *Grubler's equation:* The DOF in a planar mechanism consisting of j_1 number of pin joints and N number of links is given by Grubler's equation:

$$\text{DOF} = 3(N - 1) - 2j_1 \tag{9.1}$$

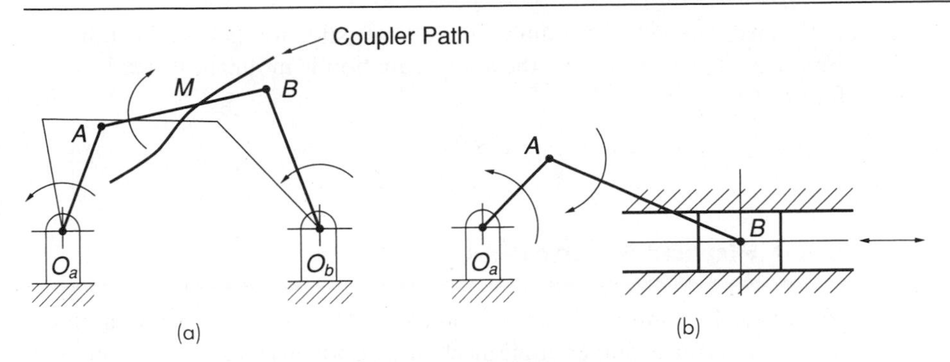

FIGURE 9.3 (a) four-bar and (b) slider-crank mechanisms

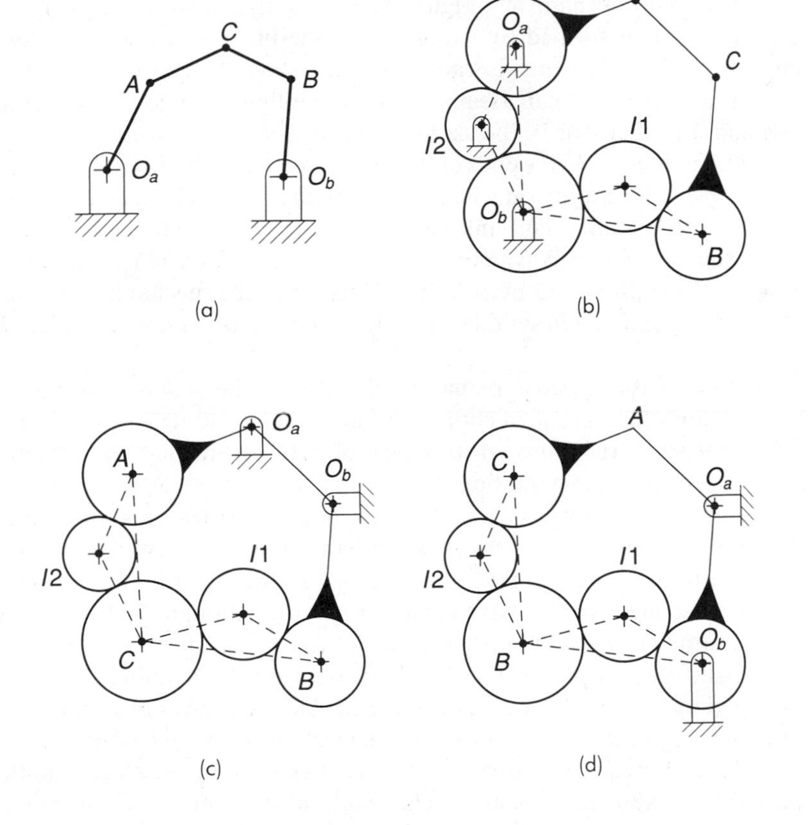

FIGURE 9.4 (a) a five-link chain; (b)–(d) geared five-bar mechanisms

If the mechanism has j_2 number of roller or slide joints [as shown in Figure 9.3(b)] as well, then the above equation is modified to read as follows [4]:

$$DOF = 3(N - 1) - 2j_1 - j_2 \tag{9.2}$$

9.4 ▪ Planar Mechanisms

Planar mechanisms may vary in complexity and design depending on the duty they perform. Some simple mechanisms, however, are versatile enough to perform many functions on their own. These basic mechanisms are described in the next few paragraphs below. Most other special-purpose and somewhat more complex mechanisms may be formed by combining several of these mechanisms.

A four-bar mechanism [Figure 9.3(a)] is the most popular planar mechanism. It is formed by fixing one of the links of a four-link closed chain. The link fixed thus is known as the *frame*. Once a frame has been established, the chain can assume only one configuration for a given crank position. In other words, the chain yields a viable mechanism.

Links connected to the ends of the frame are known as either the *crank* or the *follower*, depending on whether it is driving the mechanism or is being driven by the crank. The link connecting the crank and the follower is known as the *coupler*. Any point on the coupler is known as a *coupler point*. Usually the path traced by coupler points when the mechanism moves is the design parameter in synthesis problems. This path is known as a *coupler curve*.

Another basic type of planar mechanism is the *slider-crank* mechanism. It contains a crank, a coupler, a frame, and a sliding (prismatic) pair [Figure 9.3(b)]. It is through this type of mechanism that reciprocating engines produce rotary motion.

The two mechanisms described above have many features in common. As a matter of fact, the slider-crank mechanism may be viewed as a special case of a four-bar mechanism in which the follower is of infinite length.

Four-bar mechanisms have certain limitations though. Not many motion parameters can be satisfied with these; for example a maximum of only eight motion parameters (angles turned by the members, positions occupied by the coupler, etc.) can be specified in a synthesis problem. When motion requirements are more complex, mechanisms with a larger number of links are used. In geared five-bar mechanisms, up to eleven motion parameters, and in six-bar mechanisms, a maximum of 14 motion parameters can be specified. Some discussion of geared five-bar and six-bar mechanisms follows in the next two paragraphs.

A planar mechanism with five bars can be formed if proper constraints are present. In Figure 9.4(a), we have a kinematic chain with five links. Link

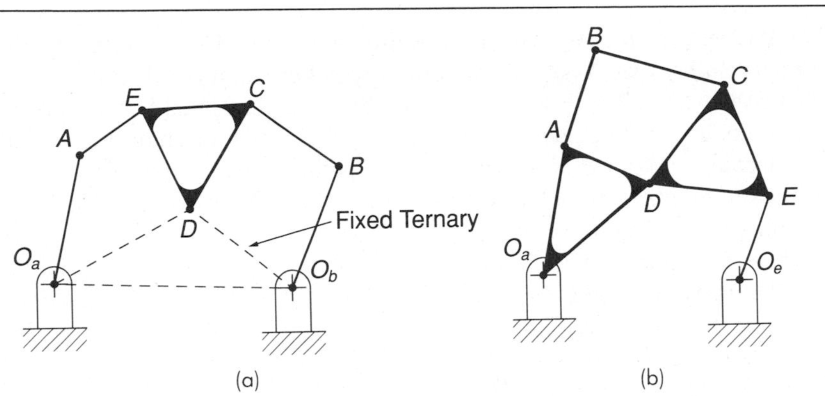

FIGURE 9.5 (a) WATT2 mechanism; (b) WATT1 mechanism

O_aO_b is fixed, and therefore, it acts as the frame. Fixing one link does not produce a useful mechanism in this case, as happens in the case of four-bar mechanisms. The reason for this can easily be grasped by considering the motion of the linkage system shown in Figure 9.4(a). Links O_aA and O_bB can be turned independently of each other (i.e., there is no uniqueness in the system). In order to introduce uniqueness, gears are attached to the chain, as shown in Figure 9.4(b). In the case of links O_aA and BC, no relative motion is allowed between the links and the gears attached to them. Idlers $I1$ and $I2$ are attached to the ternary link O_bB and to the frame (O_aO_b), respectively. The device is now a mechanism and is known as a *geared five-bar* mechanism. Figure 9.4(c) and (d) illustrates inversions of the original chain.

Still more stringent motion requirements can be met by six-bar mechanisms. Figure 9.5 shows two types of six-bar mechanisms (each consisting of two tenary and four binary links) known as WATT2 and WATT1, respectively. WATT mechanisms are formed when the ternary links

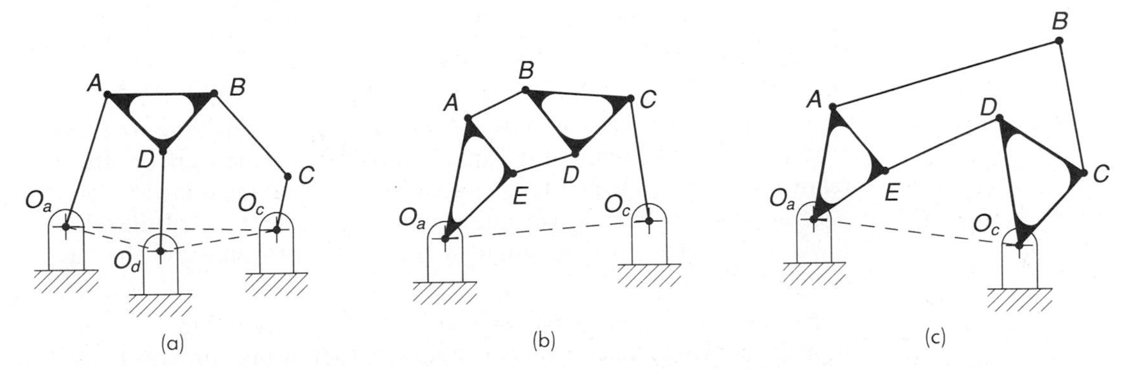

FIGURE 9.6 (a) Stephenson 3 mechanism; (b) Stephenson 2 mechanism; (c) Stephenson 1 mechanism

are connected together with a pin joint such as D. Fixing either of the ternary links forms WATT2. Fixing any other binary link results in a WATT1 mechanism. Figure 9.6 shows Stephenson 3, Stephenson 2, and Stephenson 1 mechanisms. These differ from WATT mechanisms in that the ternary links are never connected together. Fixing any ternary link results in Stephenson 3. Fixing any link in the dyad, such as (CO_cO_a) gives Stephenson 2 [Figure 9.6(b)]. Fixing either of the remaining binary links results in Stephenson 1 [Figure 9.6(c)]. Notice that several of the six-bar mechanisms discussed above, in fact, are two four-bar mechanisms combined in certain ways.

9.5 ▪ Linkage Synthesis

As mentioned earlier, mechanism synthesis is associated with the creation of new mechanisms. A new mechanism may be needed because in certain machinery or gadgets, points, links, or planes are required to move along a given path or to maintain a certain orientation, velocity, and acceleration. Although a large number of different synthesis problems arise in engineering operations, most of the commonly occurring problems can be put into a few general categories, some of which are listed below [5]:

1. *Path generation,* in which the only requirement is to guide a point—say, a tool tip P–along a given path [Figure 9.7(a)].

2. *Path coordination,* in which the position (l_i, m_i), (l_j, m_j), etc., occupied by point P [Figure 9.7(b)] traveling along a given path is required to be related to the angular rotation ϕ_{ij} of a crank. Such problems arise in assembling and packaging plants. In a somewhat more complicated situation, coordination with two cranks may be required.

3. *Coplanar motion synthesis,* in which a plane moving along a given path may be required to maintain a certain orientation as it passes through a given path [Figure 9.7(c)]. Examples of coplanar motion synthesis may arise in profile machining operations where the cutting tool is required to maintain a given angular relationship with the workpiece.

4. *Angular coordination of two or three cranks,* in which the cranks move in such a fashion that the angular positions of these cranks are related to each other. This type of problem is encountered in *function generation*—for example, when the angular movement (θ_{ij}) of the follower must be equal to the square of the angular movement of the crank, ϕ_{ij} (i.e., $\theta = \phi^2$).

5. *Coplanar motion coordination,* in which it is required that the motion of a moving plane be coordinated with that of one (or more) cranks [Figure 9.7(d)].

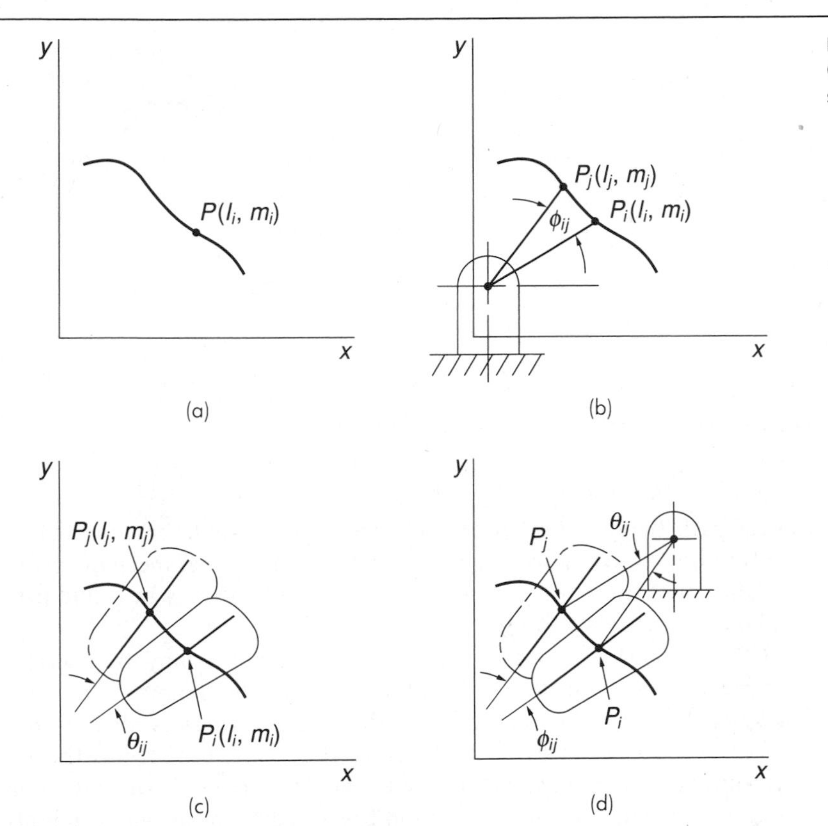

FIGURE 9.7 Categorization of synthesis problems

9.6 ▪ Precision Points and Their Selection

Let's look at Figure 9.7(a) again. In order to move the point P from its present position, it has to be attached to some linkage system. Suppose a four-bar mechanism is chosen, as shown in Figure 9.8(a). It is not difficult to perceive that it would generally be impossible to find a mechanism that would carry P along the desired curve without ever deviating from it. In mechanism synthesis, we strive therefore to find a mechanism that takes P to at least a few selected points on the given path. The actual path traced by P is different from the desired path. The two paths intersect each other at, say, two points—P_1 and P_2, as shown in Figure 9.8(a). Points P_1 and P_2 in the above case are referred to as *precision points* (or *positions*).

In mechanism synthesis, a path is usually given. For a good design, the precision points are preselected such that the difference between the actual path and the desired path is as small as possible. Having determined the precision points, synthesis is carried on with respect to these points only. No consideration whatsoever is given to the actual and the desired paths. The synthesis process is thus a discretized solution to a continuous problem. Depending on the number of precision points selected on the

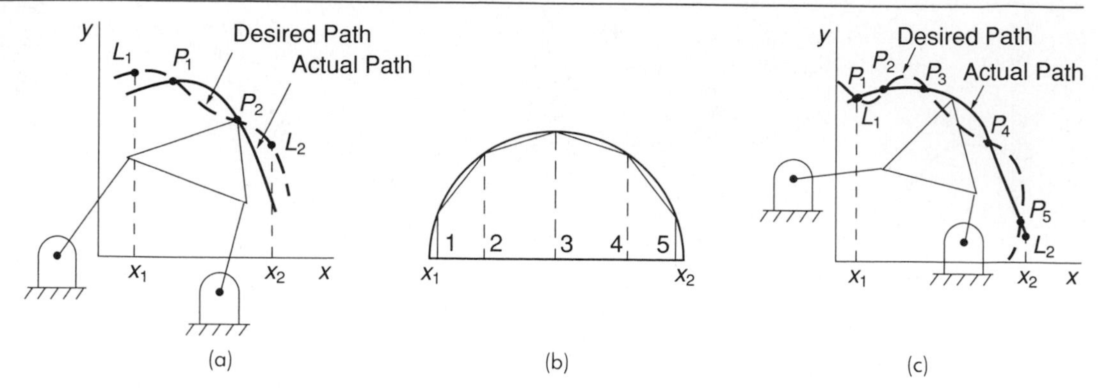

FIGURE 9.8 Precision points and Chebyshev spacing

desired path, the synthesis process is known as two-point, three-point, four-point (and so on) synthesis. It is shown later that a four-bar mechanism for a a maximum of five finitely separated precision points may be found using available analytical methods.

Before embarking on the process of synthesis, it may be useful to describe a popular method of obtaining precision points—known as *Chebyshev spacing.* The method identifies precision points for minimum error in cases where the given function (e.g., the function describing the path of a point or the relationship between the angles of the links) is polynomial in nature. It is valid only if the function generated by the mechanism is also polynomial and is of degree less than that of the original function [3].

Suppose that the path shown in Figures 9.7(a) and 9.8(a) can be expressed in terms of the Cartesian coordinates in the following form:

$$y = f(x) \tag{9.3}$$

It can be shown that in order to reduce to a minimum the deviation of the actual path from the desired path, the precision points should be obtained by Chebyshev spacing, mentioned above. Geometrically, Chebyshev spacing is obtained from the coordinates of the vertices of a polygon of $2n$ sides inscribed within a circle whose diameter is equal to the range of displacement, travel, rotation, etc., in the synthesis problem. The integer number n mentioned above is the number of accuracy points desired. Thus, in Figure 9.8(a), if it is desired that P travel along the desired path between positions L_1 and L_2, then the range of travel for the Chebyshev spacing scheme is taken as $x_2 - x_1$. Let's say that we require five accuracy points between the two extreme positions. We therefore draw a ten-sided polygon within a circle (diameter $= x_2 - x_1$) such that the x axis intersects the polygon symmetrically [Figure 9.8(b)]. The x coordinates of the accuracy points are then found by dropping perpendiculars down to the x axis from the corners of the polygon. The precision points are marked 1, 2, . . . , 5 in the diagram. The corresponding y coordinates of the precision points are found from equation (9.3) or graphically from the path itself [Figure 9.8(c)].

EXAMPLE ■ 9.1

Using a four-bar mechanism, generate a path between $x = 1.5$ and $x = 3.5$, defined by equation $y = 1.2x^{0.3}$. Select precision points such that the deviation of the generated path from the above curve is minimum for a (i) two-point, and (ii) three-point synthesis.

In order to find the precision positions based on Chebyshev's spacing scheme, mark the points $x = 1.5$ and $x = 3.5$ on the x axis as shown in Figure 9.9. Draw a semi-circle of radius $= (3.5 - 1.5)/2 = 1$ with its center at $x = (1.5 + 3.5)/2 = 2.5$.

Step 1 For a two-point synthesis, draw a half square (i.e., half of a four-sided polygon) inside the semi-circle as shown in Figure 9.9(a). The precision points (P_1 and P_2) are found to be located at $x_1 = 1.7929$ and $x_2 = 3.2071$ with a corresponding $y_1 = 1.2(1.7929)^{0.3} = 1.4297$ and $y_2 = 1.2(3.2071)^{0.3} = 1.7022$. The curve generated by the candidate mechanism will thus pass through the points P_1 and P_2 marked on Figure 9.10(a).

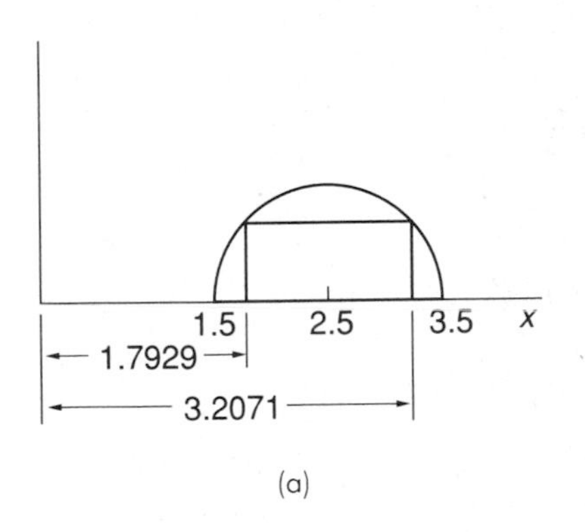

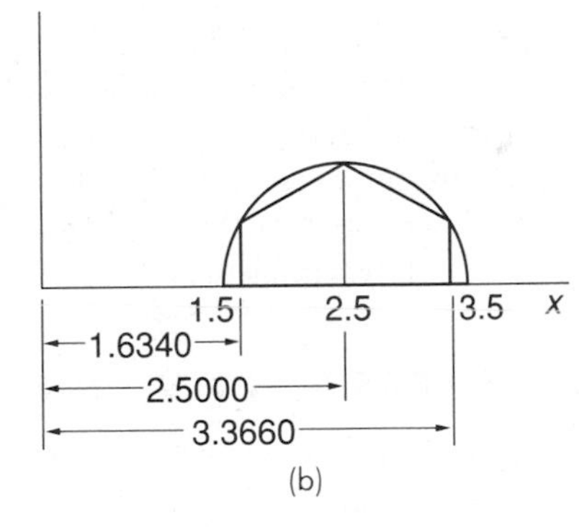

FIGURE 9.9 Selection of precision points for (a) two-position synthesis, and (b) three-position synthesis

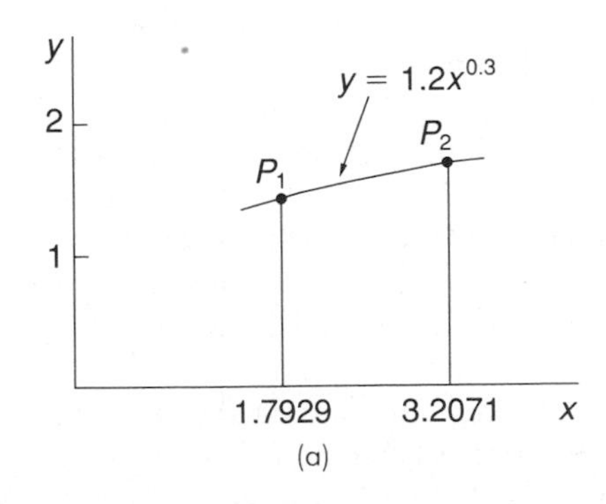

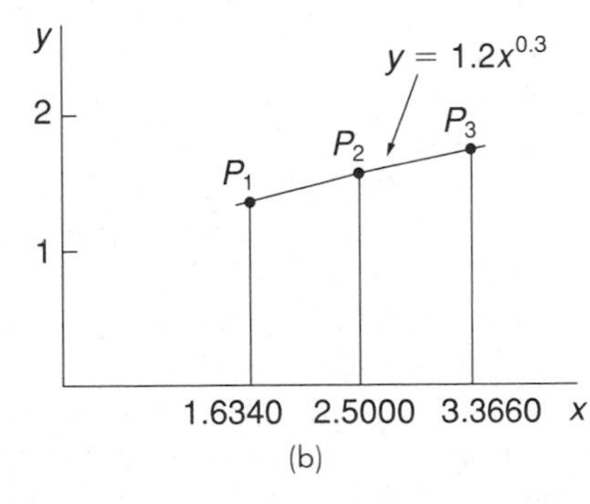

FIGURE 9.10 Precision points for (a) two-position synthesis, and (b) three-position synthesis

Step 2 For a three-point synthesis, draw a half hexagon (i.e., half of a six-sided polygon) inside the semi-circle as shown in Figure 9.9(b). The precision points are found to be located at $x_1 = 1.6340$, $x_2 = 2.500$ and $x_3 = 3.3660$ with a corresponding $y_1 = 1.2(1.6340)^{0.3} = 1.3905$, $y_2 = 1.2(2.5000)^{0.3} = 1.5796$ and $y_3 = 1.2(3.3660)^{0.3} = 1.7271$. The curve generated by the candidate mechanism will thus pass through the points marked P_1, P_2, and P_3 in Figure 9.10(b).

It should be pointed out that the above results can also be easily derived mathematically. ❑

9.7 ▪ Geometric Method of Mechanism Synthesis

There are graphic or geometric methods for solving synthesis problems, which, though tedious, can solve many of the synthesis problems. The trouble with the geometric methods is that they require a long time to complete and the accuracy of the solutions is often questionable. Analytic solutions are also available in many cases and should be used in preference to geometric methods. But the analytic solutions tend to become mathematically too involved in somewhat complex problems. In many cases, it may not be feasible for many of us to develop the appropriate solution. If such is the case, one is left with no option but to use geometric methods. It is imperative therefore that the reader have some insight into geometric synthesis techniques as well. For a detailed discussion on the geometric methods, one might refer to references [3] and [4]. Given below is a very brief description of these methods. Our main thrust is toward algebraic analysis of synthesis problems.

❑ TWO-POSITION SYNTHESIS OF A COUPLER

Assume that a coupler has to be moved from position AB to $A'B'$ by a four-bar mechanism. Let's also assume that ends A and B of the coupler lie on the curves described by the functions $y_a = f(x_a)$ and $y_b = g(x_b)$, respectively. It can be seen that any crank, such as O_aA, linking the coupler at point A [Figure 9.11(a)] to any point on the perpendicular bisector $O_aO'_a$ of the line joining AA', will move point A to point A'. An infinite number of such cranks can be found. In an exactly similar manner, an infinite number of separate cranks can be found for moving the end B to the two specified positions. Thus, if a four-bar mechanism is to be formed by taking any one crank from each of these two families of cranks, we can obtain a total of $\infty \times \infty = \infty^2$ distinct four-bar mechanisms—a large number, indeed. This is not the end of the story though.

We arrived at the above number by assuming that points A and B had to be connected directly to the cranks. It may not always be necessary. Assume

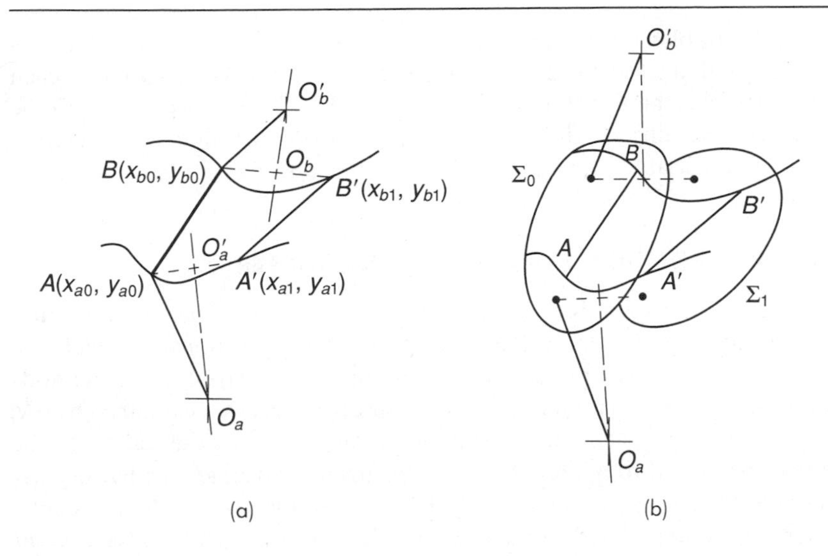

FIGURE 9.11
Two-position synthesis

that the coupler is attached to a moving plane (Σ) [Figure 9.11(b)]. Cranks can now be attached to any point on the moving plane in order to move the coupler exactly as before. Corresponding to each point on the plane we can obtain an infinite number of cranks. There is a total of ∞^2 points on the plane (∞ in the x direction and ∞ in the y direction); thus, a total number of ∞^3 cranks can be found corresponding to each of the two paths. By combining any two cranks from these, a total of $\infty^3(\infty^3 - 1)/2 \approx \infty^6$ distinct four-bar mechanisms can be found.

Although any power of infinity is infinity again, it is preferable that the number of options be expressed as powers of infinity. The information thus conveyed to the reader is more precise because the reader gets to know the exact number of free choices available in any particular solution.

To explain this concept further, we might review the above problem. The two-position synthesis problem imposes two constraints on the mechanism. The problem specifies (1) that end A of the coupler must pass through point A' whose x coordinate is x_{a1} and (2) that end B must pass through point B' whose x coordinate is x_{b1}. Since y coordinates of these points are functions of the x coordinates, the corresponding values (i.e., y_{a1} and y_{b1}) can be determined from the equation of the paths. The initial positions (x_{a0}, y_{a0}) and (x_{b0}, y_{b0}) do not impose any constraints on the linkage system. The mechanism can be moved in the fashion of a rigid body such that ends A and B rest at these positions to start with. Now, in a four-bar mechanism, there are eight design variables—namely, the x and y coordinates of the four pin joints. By altering any one of these coordinates while keeping the others fixed, a new mechanism can be found. In mechanism synthesis, for every constraint imposed by the problem, one of these design variables is predetermined. If the mechanism has got more design variables (say, q) than the number of constraints (say, s), then $q - s$

design variables are left undefined (free). Every one of the free variables can be assigned arbitrary values by the designer. Thus, corresponding to each free variable a family of an infinite number of solutions is possible. In the problem explained earlier, there are only two constraints and therefore $8 - 2$ free variables, which results in ∞^6 mechanisms.

❑ THREE-POSITION SYNTHESIS OF A COUPLER

Figure 9.12 depicts a three-position synthesis problem. The initial position of the coupler is marked AB. Two subsequent positions to which the coupler must move are marked as $A'B'$ and $A''B''$. The graphic methods for obtaining four-bar mechanisms for the above case are explained below.

In its simplest form, the synthesis consists of finding the centers of the circles passing through point $AA'A''$ and $BB'B''$. A viable four-bar mechanism is then obtained by joining the ends of the coupler to the corresponding centers. Thus, O_aABO_b is the solution mechanism. In the present problem, there are four constraints. It follows therefore that we should have four free variables and ∞^4 solutions. Figure 9.12 shows one solution. In order to find the other feasible solutions, the coupler should be attached to a moving plane and synthesis carried out in the manner explained for the two-position synthesis.

The trouble with the approach described above is that although the mechanism so designed takes on the geometric configurations desired, there is no guarantee that it is able to move from one position to another in the correct sequence and without locking. Alternatively, if it does move as desired, there may be instances of acute transmission angle, requiring large

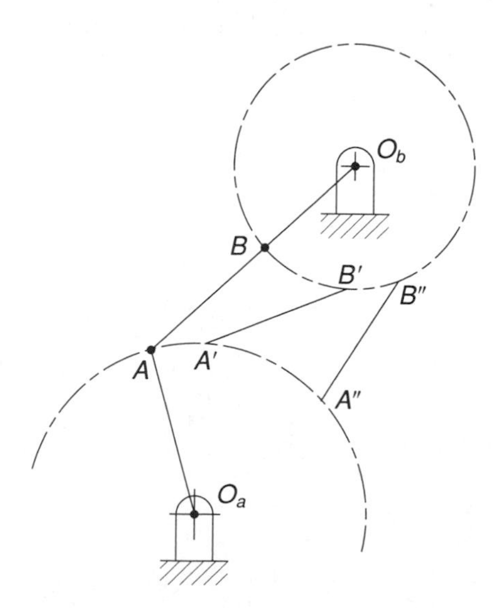

FIGURE 9.12
Three-position synthesis

driving torques. For a successful design, it is important therefore that there be options for trying different solutions. And that is why it is necessary to express the number of options more precisely, as is being done here.

❑ FOUR-POSITION SYNTHESIS OF A COUPLER

In the two synthesis problems discussed above, our main objective has been finding the centers or center points for circles that pass through either two or three points. This is basically high school geometry. Things suddenly become tough when we try to have more than three accuracy points. As a matter of fact, it may not be possible at all to draw a circle through any arbitrary set of four or five points. The possible cranks for more than three accuracy points are determined by an indirect method described below.

Burmester has shown that there exist on a moving lamina (plane) an infinite number of fixed points (circle points) that lie on circles of a fixed plane for four distinct positions of the lamina. As the lamina moves, these points travel along circular paths with respect to the fixed plane. A collection of the circle points is known as the *circle point curve.* Similarly, the collection of the centers of the circular paths of the circle points is known as the *center point curve.* The geometric method of four-position synthesis essentially involves determining these curves. In four-position synthesis, as in other cases, the coupler is assumed to be embedded onto a moving Σ plane (lamina), and, therefore, the motion of the coupler becomes identical to that of the Σ plane. With this improvisation, it is possible to construct geometrically the circle point curve attached to the Σ plane and a center point curve in a fixed plane.

Typical examples of circle point and center point curves are given in Figure 9.13. There is one-to-one correspondence between the two curves

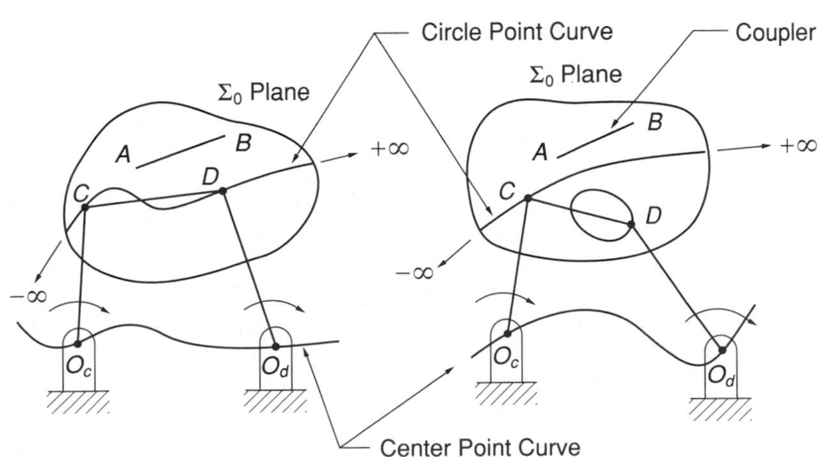

FIGURE 9.13
Circle point and center point curves

[i.e., for every point on the circle point curve, there is a (unique) corresponding point on the center point curve]. The implication of these two curves is that if a crank, such as O_cC, is obtained by joining a circle point (C) and the corresponding center point (O_c), the crank would carry the Σ plane to the other three specified positions (besides Σ_0 where it originally rests) as it turns about a pivot attached to the center point O_c. By taking any other point—say, D—on the circle point curve, a second crank (or follower) O_dD can be found. O_cCDO_d is a viable four-bar mechanism for the positions of the coupler defined above. The center point and circle point curves are cubic in nature.

Since there are an infinite number of points on the circle point curve, it is possible to obtain an infinite number of separate cranks. The total number of distinct four-bar mechanisms possible for four-position syntheses is $\infty(\infty - 1)/2 = \infty^2$. Constructional details of the curves may be found in Reference [3].

❑ FIVE-POSITION SYNTHESIS

A five-position synthesis problem is solved as two separate four-position synthesis problems. Let the positions taken by the coupler be as shown in Figure 9.14. We can consider positions 0, 1, 2, and 3 in the first instance and obtain the circle point and center point curves shown in Figure 9.14. We can consider positions 0, 1, 2, and 4 next and obtain the corresponding curves marked by dashed lines in Figure 9.14. Obviously, the points of intersection of the two circle point curves are relevant to all five positions of the coupler. In a similar manner, the points of intersection of the two center point

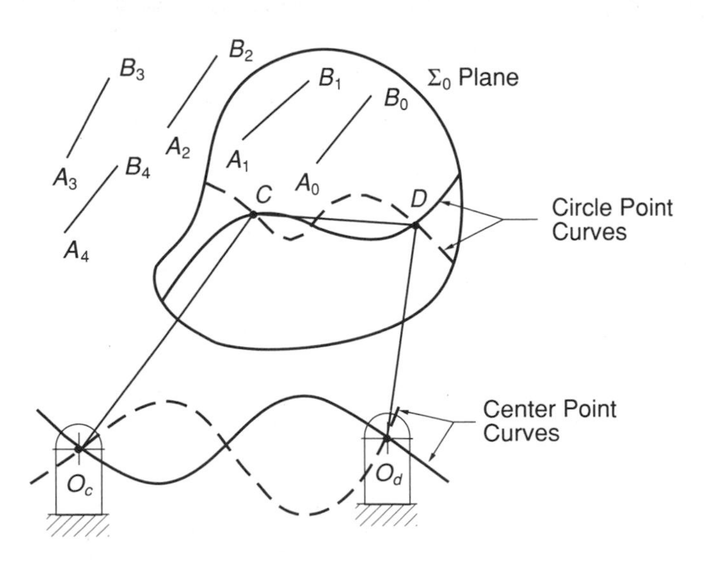

FIGURE 9.14
Five-position synthesis

curves are relevant to all five positions of the coupler. There is one-to-one correspondence between the circle point and center point curves, as explained before; therefore, for each of the points of intersection between the two center point curves, there is a corresponding point of intersection between the two circle point curves. If you take one of the intersection (circle) points—say, C—and join it to the corresponding center point, O_c (Figure 9.14), you have a viable crank. Take a second intersection (circle) point, D, and find its corresponding center point, O_d. You now have a four-bar mechanism O_cCDO_d that might carry the coupler AB to all five specified positions.

9.8 ▪ Analytic Method of Linkage Synthesis

The method described in this section uses complex numbers to express polar coordinates. Similar equations may also be derived using Cartesian coordinates.

First, let's learn the use of complex numbers in linkage synthesis. Figure 9.15 shows a coupler that is situated such that the coupler point M is at the location (l, m) and the coupler makes an angle α from the horizontal axis. Corner A of the coupler is pin-jointed to a link AO_a. If the links are treated as vectors, it can be shown that the link $OO_a = he^{i\theta}$, link $AO_a = re^{i\Gamma}$, etc. in which $i = \sqrt{-1}$. Considering the loop formed by the links OO_a, O_aA, AM, and MO as a vector polygon, a loop closure equation may be written as follows:

$$te^{i\beta} = he^{i\theta} + re^{i\Gamma} + se^{i(\alpha+\gamma)} \tag{9.4}$$

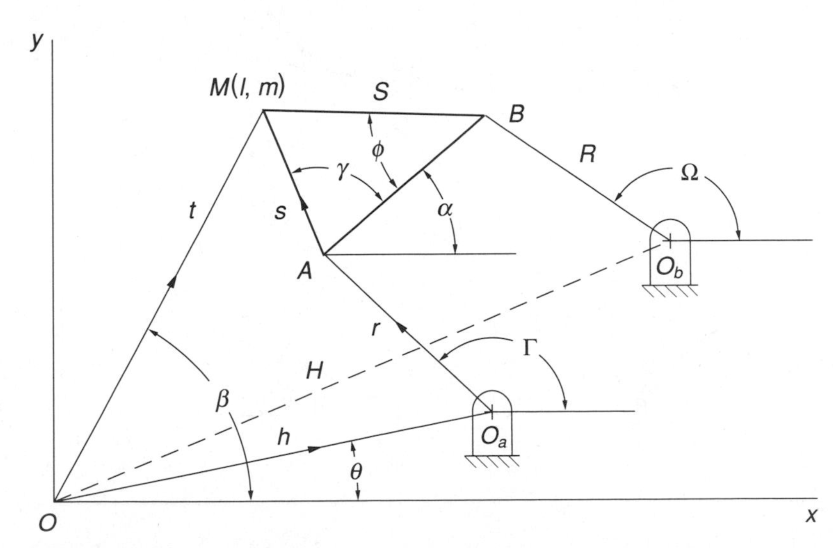

FIGURE 9.15
Links as vectors

Notice also that

$$te^{i\beta} = l + im \tag{9.5}$$

❑ TWO-POSITION SYNTHESIS

Now, let's assume that we are required to find a four-bar mechanism such that it carries a given point through two positions (l_0, m_0) and (l_1, m_1), while the corresponding coupler angles are α_0 and α_1, respectively. The above are the motion parameters for the candidate mechanism. The designer should find a suitable coupler and the center points for the two cranks. The problem is shown graphically in Figure 9.16.

In the above problem, let MAB be a candidate coupler and O_a a center point of the mechanism. The coupler and the crank O_aA appear in the attitudes shown in Figure 9.16, corresponding to the motion parameters given in the problem. As done above, we may now write the loop closure equation for the two positions of the coupler in the following form:

$$t_0e^{i\beta_0} = he^{i\theta} + re^{i\Gamma_0} + se^{i(\alpha_0+\gamma)}$$

$$t_1e^{i\beta_1} = he^{i\theta} + re^{i\Gamma_1} + se^{i(\alpha_1+\gamma)}$$

Subtracting the second equation from the first:

$$t_1e^{i\beta_1} - t_0e^{i\beta_0} = r(e^{i\Gamma_1} - e^{i\Gamma_0}) + se^{i\gamma}(e^{i\alpha_1} - e^{i\alpha_0}) \tag{9.6}$$

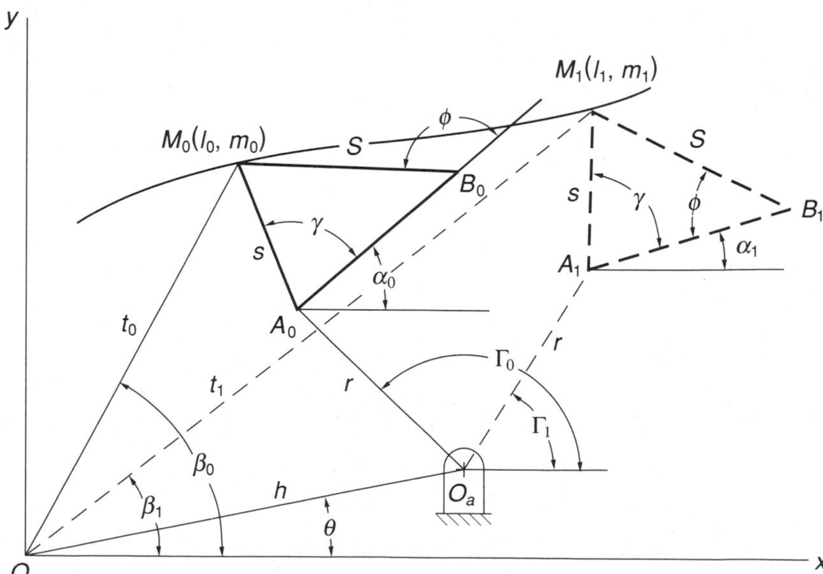

FIGURE 9.16
Two-position synthesis

Equation (9.6) yields two separate equations when the real and imaginary parts are separated. The left-hand side of the equations contains known quantities [i.e., (l_0, m_0) and (l_1, m_1)]. The right-hand side of the equations contain unknown quantities $(r, s, \gamma, \Gamma_0, \Gamma_1)$. Obviously, there are three free variables. Arbitrary values can be assigned to these variables to solve the equations. Since each of the free variables can be assigned an infinite number of different values, it is possible to find ∞^3 different combinations of the crank O_a and the circle point A. By changing the values of s or γ, the location of the circle point can be changed. By assigning different values for Γ_0 and Γ_1, the location of the center point O_a can be altered.

In order to complete the design of the four-bar mechanism, a similar process can be repeated with the other joint (B, Figure 9.15) of the coupler and the corresponding center point—say, O_b. By preassigning values for S and ϕ, a different location for the circle point (B) can be selected. By altering the values of the angles (say) Ω_0 and Ω_1 made by crank BO_b from the horizontal, the position of the center point O_b can be altered.

Upon considering all the possible feasible four-bar mechanisms, it is obvious from the above that a total of ∞^6 different option are available Recall that a similar conclusion is arrived at in Section 9.6.

❑ THREE-POSITION SYNTHESIS

If a four-bar mechanism is required to carry a coupler through three separate positions—say, 0, 1, and 2—two equations of the following type are obtained in a manner similar to that used above:

$$t_1 e^{i\beta_1} - t_0 e^{i\beta_0} = r(e^{i\Gamma_1} - e^{i\Gamma_0}) + s e^{i\gamma}(e^{i\alpha_1} - e^{i\alpha_0}) \tag{9.7}$$

$$t_2 e^{i\beta_2} - t_0 e^{i\beta_0} = r(e^{i\Gamma_2} - e^{i\Gamma_0}) + s e^{i\gamma}(e^{i\alpha_2} - e^{i\alpha_0}) \tag{9.8}$$

In equations (9.7) and (9.8) above, there are six unknowns—namely, r, s, γ, Γ_0, Γ_1, and Γ_2—and four equations. Obviously, there are two free variables. Thus, by assigning different values for, say, s and γ, ∞^2 different locations of the circle point A and the center point O_a can be found. Considering the four-bar mechanism as a whole, a total of ∞^4 different mechanisms can be found. The conclusion here is similar to that drawn in Section 9.6.

EXAMPLE ■ 9.2

Design a four-bar mechanism so that a coupler point passes through the points $P_0(4.50, 6.00)$, $P_1(7.65, 6.90)$, and $P_2(8.92, 5.83)$. The corresponding coupler angles measured from the horizontal are 5°, −3°, and −3°.

In a three position synthesis, two of the variables in equations (9.7) and (9.8) should be predefined. For one solution, assume the length of the coupler arm $MA\text{-}s = 2.8$ and the coupler angle $MAB\text{-}\gamma = 25°$ as in Figure 9.17. Therefore, the known set of data reads as follows:

$$l_0 = 4.50, l_1 = 7.65, l_2 = 8.92$$

$$m_0 = 6.00, m_1 = 6.90, m_2 = 5.83$$

$$\alpha_0 = 5°, \alpha_1 = -3°, \alpha_2 = -3°$$

$$\gamma = 25°, s = 2.800$$

In order to find the length (r) of the crank AO_a, the above values are substituted in equations (9.7) and (9.8). On separating the real and imaginary terms the following equations are obtained:

$$7.65 - 4.50 = r(\cos\Gamma_1 - \cos\Gamma_0) + 2.8(\cos 22° - \cos 30°)$$

$$6.90 - 6.00 = r(\sin\Gamma_1 - \sin\Gamma_0) + 2.8(\sin 22° - \sin 30°)$$

$$8.92 - 4.50 = r(\cos\Gamma_2 - \cos\Gamma_0) + 2.8(\cos 22° - \cos 30°)$$

$$5.83 - 6.00 = r(\sin\Gamma_2 - \sin\Gamma_0) + 2.8(\sin 22° - \sin 30°)$$

Note the use of equation (9.5) above to replace $t_0 \cos\beta_0 = l_0$ and $t_0 \sin\beta_0 = m_0$, etc. On solving the above equations simultaneously, the unknown variables are found to have the following values:

$$\Gamma_0 = 155°, \Gamma_1 = 72°, \Gamma_2 = 31° \text{ and } r = 2.4$$

On substituting the above values in equation (9.4) the x and y coordinates of the center point O_a are found:

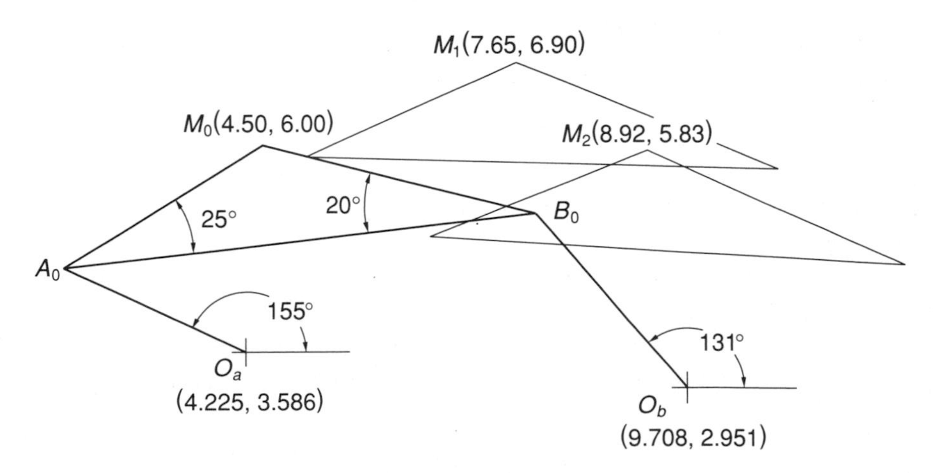

FIGURE 9.17 A four-bar mechanism solution

$$\text{h}\cos\theta = 4.50 - 2.4\cos 155^\circ - 2.8\cos 30^\circ = 4.225$$

$$\text{h}\sin\theta = 6.00 - 2.4\sin 155^\circ - 2.8\sin 30^\circ = 3.586$$

For finding the crank length BO_b-R assume length MB-$S = 3.452$ and angle $MBA = 20^\circ$ (i.e. $\phi = 160^\circ$). On substituting the values in equations (9.7) and (9.8), the following set of equations is obtained:

$$7.65 - 4.50 = R(\cos\Omega_1 - \cos\Omega_0) + 3.452(\cos 157^\circ - \cos 165^\circ)$$

$$6.90 - 6.00 = R(\sin\Omega_1 - \sin\Omega_0) + 3.452(\sin 157^\circ - \sin 165^\circ)$$

$$8.92 - 4.50 = R(\cos\Omega_2 - \cos\Omega_0) + 3.452(\cos 157^\circ - \cos 165^\circ)$$

$$5.83 - 6.00 = R(\sin\Omega_2 - \sin\Omega_0) + 3.452(\sin 157^\circ - \sin 165^\circ)$$

Solution of the above simultaneous equations yields the following:

$$\Omega_0 = 131^\circ, \Omega_1 = 67^\circ, \Omega_2 = 32^\circ, \text{and } R = 2.856$$

The x and y coordinates of the center point O_b are obtained from equation (9.4) as follows:

$$x \text{ coordinate} = 4.50 - 2.856\cos 131^\circ - 3.452\cos 165^\circ = 9.708$$

$$y \text{ coordinate} = 6.00 - 2.856\sin 131^\circ - 3.452\sin 165^\circ = 2.951$$

The solution of simultaneous equations obtained above often requires tedious algebraic manipulations. However, some computer programs are available for quick answers [4]. ❑

❑ FOUR- AND FIVE-POSITION SYNTHESIS

If four separate positions—say, 0, 1, 2, and 3—of a coupler are defined in a synthesis problem, the following three complex equations can be derived:

$$t_1 e^{i\beta_1} - t_0 e^{i\beta_0} = r(e^{i\Gamma_1} - e^{i\Gamma_0}) + se^{i\gamma}(e^{i\alpha_1} - e^{i\alpha_0}) \quad (9.9)$$

$$t_2 e^{i\beta_2} - t_0 e^{i\beta_0} = r(e^{i\Gamma_2} - e^{i\Gamma_0}) + se^{i\gamma}(e^{i\alpha_2} - e^{i\alpha_0}) \quad (9.10)$$

$$t_3 e^{i\beta} - t_0 e^{i\beta_0} = r(e^{i\Gamma_3} - e^{i\Gamma_0}) + se^{i\gamma}(e^{i\alpha_3} - e^{i\alpha_0}) \quad (9.11)$$

In the above, there are six equations and seven unknowns—r, s, γ, Γ_0, Γ_1, Γ_2, and Γ_3. One variable is free. It is possible to select s as the free variable. By selecting all possible values (numbering ∞) of s, the circle point curve can be defined. Corresponding values of r and Γ define the center point curve.

When five separate positions—say, 0, 1, 2, 3, and 4—are given, a second set of equations in addition to equations (9.9) through (9.11), as given below, can also be derived:

$$t_1e^{i\beta_1} - t_0e^{i\beta_0} = r(e^{i\Gamma_1} - e^{i\Gamma_0}) + se^{i\gamma}(e^{i\alpha_1} - e^{i\alpha_0}) \tag{9.12}$$

$$t_2e^{i\beta_2} - t_0e^{i\beta_0} = r(e^{i\Gamma_2} - e^{i\Gamma_0}) + se^{i\gamma}(e^{i\alpha_2} - e^{i\alpha_0}) \tag{9.13}$$

$$t_4e^{i\beta} - t_0e^{i\beta_0} = r(e^{i\Gamma_4} - e^{i\Gamma_0}) + se^{i\gamma}(e^{i\alpha_4} - e^{i\alpha_0}) \tag{9.14}$$

Equations (9.12) through (9.14) also yield a pair of center and circle point curves. The intersections of the two center point curves yield the center points for the five-position synthesis. The intersections of the circle point curves obtained from equations (9.9) through (9.14) yield the circle points.

As mentioned in Section 9.6, the circle point and center point curves are cubic in nature. Two cubic curves can have a total of nine intersection points. Two of the intersection points are at infinity, and three are situated at poles and often are not suitable for use. Thus, the number of feasible circle points is only four. By selecting any two circle points at a time, a different four-bar mechanism can be designed. The total number of options in a five-position synthesis problem is therefore only six.

Several computer programs, such as LINKS [7], and LINCAGES [7], are available for four- and five-position synthesis. It is therefore possible for students interested in linkage synthesis to use such software for quick solutions, rather than spending long hours solving the complex equations given above.

9.9 ▪ Cognate Four-Bar Mechanisms

A theorem developed by mathematicians Roberts and Chebyshev states that for a given four-bar mechanism, there are two other four-bar mechanisms (*cognates* or *path cognates*) that trace identical coupler paths. This property of four-bar mechanisms has been widely used in synthesis. It was mentioned before that in some instances a four-bar mechanism had to be discarded because the links were too large or its geometry unwieldy. If confronted with such a problem, the designer has the option of looking at the other two mechanisms (path cognates) to see if these have any better features. This, however, is an obvious and somewhat trivial use of the Roberts-Chebyshev theorem. More effective use of the theorem can be made in a variety of synthesis problems, as explained in this section. First, we discuss the methods for geometrically constructing the cognates.

❑ GEOMETRIC CONSTRUCTION OF COGNATE FOUR-BAR MECHANISMS

In Figure 9.18, O_aABO_b is the given four-bar mechanism, and M is a point on the coupler whose path is shown. We wish to obtain the cognates for the

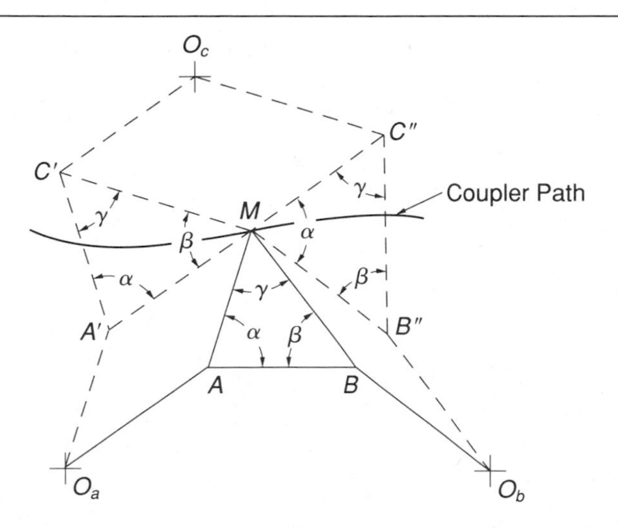

FIGURE 9.18
Roberts-Chebyshev construction

given mechanism such that all have point M in common and in all cases the path traced by point M is identical. Draw parallelograms O_aAMA' and O_bBMB'' on the two cranks of the mechanism. Next, on the sides MA' and MB'', construct triangles similar to the coupler triangle such that the angles around point M are in the same cyclic order as within the coupler. Thus, in the example shown in Figure 9.18, the angles occur in the sequence (α, β, γ) as within the coupler ABM. Another parallelogram is then constructed with MC' and MC'' as its adjacent sides. Call this parallelogram $MC''O_cC'$. We now have the two cognates designated as $O_aA'C'O_c$ and $O_bB''C''O_c$. O_c is another pivot point, just like O_a and O_b. The proof of this construction is quite simple and can be found in reference [3].

It is interesting to compare the nature of the angular movements of the various cranks in the three mechanisms discussed above. Looking at Figure 9.18, one can see that the angular movements of the crank O_aA and the coupler $MA'C'$ would be exactly the same at all times because crank O_aA remains parallel to the side MA' of the coupler. To elucidate by a numerical example, the above statement implies that if the crank O_aA rotates, say, 10° clockwise, the coupler $MA'C'$ will also rotate 10° in the same direction. Again, since the side MC' remains parallel to the crank O_cC'' at all times, the crank O_cC'' will have the same angular movements as the coupler $MA'C'$. Thus, cranks O_aA and O_cC'' will have similar angular movements. It can also be shown that cranks O_cC' and O_bB undergo exactly the same angular motions. These features of path cognates are found useful in many synthesis situations, some of which are considered later.

The construction illustrated in Figure 9.18 becomes cumbersome when the links are in some awkward position. A much simplified construction suggested by Cayley may be used beneficially if the primary concern of the designer is to obtain accurate measurements of the links in the cognate

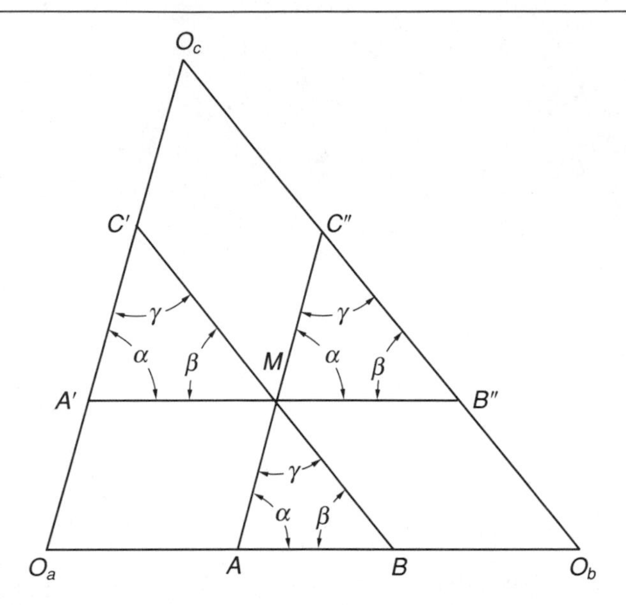

FIGURE 9.19
Cayley's construction

mechanisms. In Figure 9.19, Cayley's construction for the mechanism shown in Figure 9.18 is given. The original four-bar mechanism is stretched out such that the crank and the follower become collinear. Lines AM and BM are extended as shown. Line $O_aA'C'O_c$ is drawn parallel to line AMC''. Line $O_bB''C''O_c$ is drawn parallel to line BMC'. Line $A'MB''$ is drawn parallel to line O_aABO_b. The cognate mechanisms are given as $O_aA'C'O_c$ and $O_bB''C''O_c$.

The path cognates can also be found analytically, using the same principles as for the geometric methods described above. The analytic approach is liable to yield more accurate results. It may also be computerized, as in the program PATHS [6], for quick results.

9.10 ▪ Miscellaneous Synthesis Problems

The problems dealt with in Sections 9.6 and 9.7 are of the coplanar motion synthesis type. In this section, problems of other types are dealt with. An explicit method for function generation is given. It may, however, be pointed out that by combining the method of coplanar motion synthesis with the Roberts-Chebyshev construction, it may be possible to solve any synthesis problem.

❑ FUNCTION GENERATION

As mentioned earlier, in function generation, the angular rotations of different links in a mechanism must be related in some specified manner.

Assume that a four-bar mechanism is required in which the follower rotates through $\Delta\phi_1$ and $\Delta\phi_2$ as the driving crank rotates through $\Delta\theta_1$ and $\Delta\theta_2$. The above represents a three-position synthesis problem.

Direct Method ■ Let's assume the initial crank and follower angles to be θ_0 and ϕ_0 (Figure 9.20). Let $\theta_0 + \Delta\theta_1 = \theta_1, \theta_0 + \Delta\theta_2 = \theta_2, \phi_0 + \Delta\phi_1 = \phi_1$, and $\phi_0 + \Delta\phi_2 = \phi_2$. The corresponding coupler angles are γ_0, γ_1, and γ_2, respectively. Figure 9.20 shows a candidate mechanism. The links in the mechanism have the following lengths: the crank, r units; the follower, R units; and the coupler, C units. By writing loop equations as demonstrated in Section 9.7 and separating the real and imaginary parts, the following sets of equations can be obtained:

$$r\cos\theta_k + C\cos\gamma_k - R\cos\phi_k - B = 0 \tag{9.15}$$

$$r\sin\theta_k + C\sin\gamma_k - R\sin\phi_k = 0 \tag{9.16}$$

where $k = 0, 1, 2$, or

$$C\cos\gamma_k = R\cos\phi_k + B - r\cos\theta_k$$

$$C\sin\gamma_k = R\sin\phi_k - r\sin\theta_k$$

On squaring and adding the above equations, the Freudenstein equation [8] is obtained, as given below:

$$\cos(\theta_k - \phi_k) = K_1\cos\phi_k - K_2\cos\theta_k + K_3 \tag{9.17}$$

where $K_1 = B/r$, $K_2 = B/R$, and $K_3 = (B^2 + r^2 - C^2 + R^2)/(2rR)$.

Equation (9.17) contains four unknowns. Three separate equations can be generated corresponding to $k = 0$, $k = 1$, and $k = 2$. Usually, a value for the frame length (B) is assumed and other values determined by solving the equations simultaneously.

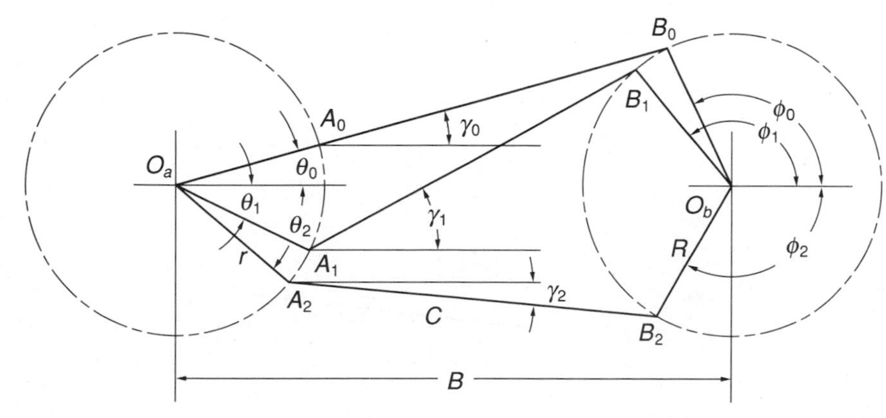

FIGURE 9.20
Function generation: direct method

EXAMPLE ■ 9.3

Design a four-bar mechanism in which the crank angles and the corresponding follower angles are as given below:

$$\theta_0 = 30°, \theta_1 = 75° \text{ and } \theta_2 = 96°$$
$$\phi_0 = 30°, \phi_1 = 62° \text{ and } \phi_2 = 77°$$

Assume the base $B = 5$ in.

From the data given in the problem, three equations can be generated using Freudenstein equation (9.17).

$$\cos(30° - 30°) = \cos 30° K_1 - \cos 30° K_2 + K_3$$
$$\cos(75° - 62°) = \cos 62° K_1 - \cos 75° K_2 + K_3$$
$$\cos(96° - 77°) = \cos 77° K_1 - \cos 96° K_2 + K_3$$

Applying Cramer's rule, the following equations are obtained:

$$\frac{K_1}{\begin{vmatrix} \cos 0° & -\cos 30° & 1 \\ \cos 13° & -\cos 75° & 1 \\ \cos 19° & -\cos 96° & 1 \end{vmatrix}} = \frac{K_2}{\begin{vmatrix} \cos 30° & \cos 0° & 1 \\ \cos 62° & \cos 13° & 1 \\ \cos 77° & \cos 19° & 1 \end{vmatrix}}$$

$$= \frac{K_3}{\begin{vmatrix} \cos 30° & -\cos 30° & \cos 0° \\ \cos 62° & -\cos 75° & \cos 13° \\ \cos 77° & -\cos 96° & \cos 19° \end{vmatrix}}$$

$$= \frac{1}{\begin{vmatrix} \cos 30° & -\cos 30° & 1 \\ \cos 62° & -\cos 75° & 1 \\ \cos 77° & -\cos 96° & 1 \end{vmatrix}}$$

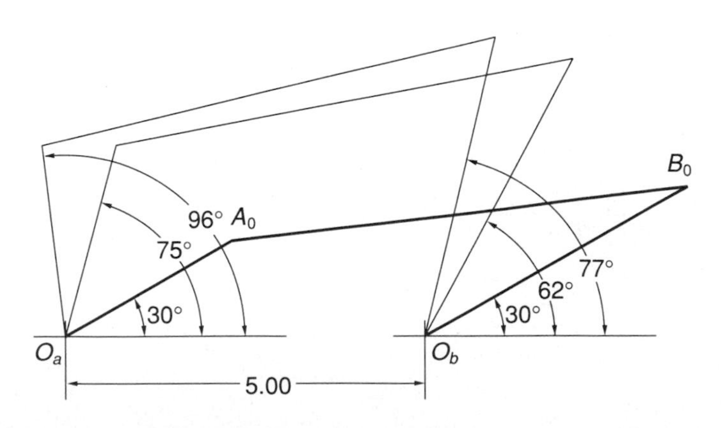

FIGURE 9.21
Four-bar mechanism

The above yield values for K_1, K_2, and K_3 are: $K_1 = 1.87$, $K_2 = 1.179$, and $K_3 = 0.402$. On further solving for the value of the linkage lengths, the following results are obtained: $r = 2.67$ in., $R = 4.24$ in., and $C = 6.40$ in. The mechanism is shown in Figure 9.21. ❑

Indirect Method ■ Function generation problems can also be solved in a somewhat indirect method, as described below.

Assume a value for the crank length, and draw the three crank positions (O_aA_0, O_aA_1, and O_aA_2) shown in Figure 9.22. Assume a coupler length and draw it congruent with the crank such that the coupler makes angles ϕ_0, ϕ_1, and ϕ_2 with the horizontal. At this stage, the problem resembles the coplanar motion synthesis discussed in Section 9.6. A four-bar mechanism can be found by locating the center points at O_a and O_b. O_aABO_b is not the solution mechanism, however. This is so because the coupler has the motion characteristic required of the follower. To correct the above deficiency, a path cognate of the above mechanism should be used for the solution.

The mechanism is shown redrawn in Figure 9.23. A triangular coupler of arbitrary shape (MAB) is constructed. Path cognates are then obtained by the Roberts-Chebyshev method. Notice that in mechanism $O_cC''B''O_b$, crank O_bB'' has motion characteristics similar to coupler MAB and crank O_cC'' has motion characteristics similar to crank O_aA. It is thus the desired function generator.

It may be worthwhile to note here that the indirect approach can handle a set of two additional motion specifications, consistent with our ability

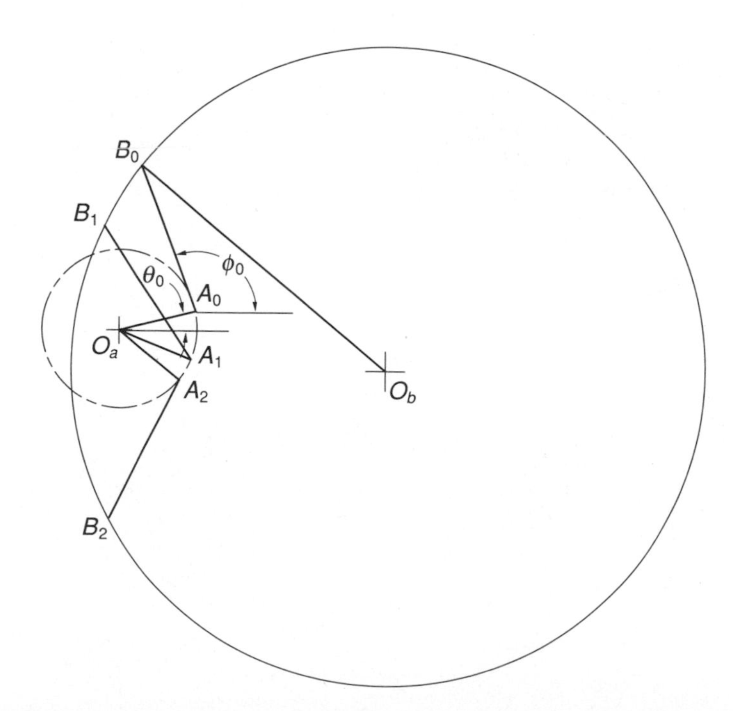

FIGURE 9.22
Function generation: indirect method

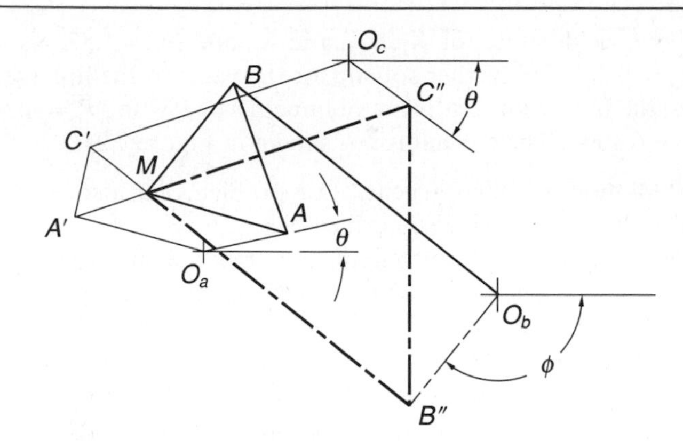

FIGURE 9.23
Function generator

to solve up to five-position synthesis. Assume that it is also desired in the above problem that for $\Delta\theta_3$ angular rotation of the crank, the follower rotates through $\Delta\phi_3$. Let $\theta_0+\Delta\theta_3=\theta_3$ and $\phi_0+\Delta\phi_3=\phi_3$. The function generation can then be represented as four-position synthesis, as shown in Figure 9.24. Corresponding to the positions $A_0B_0, \ldots, A_3B_3$ of the coupler, a crank such as CO_c can be found. In the resulting four-bar mechanism, O_aACO_c, the coupler rotates through angles $\Delta\phi$ corresponding to the angle $\Delta\theta$ rotation of the crank. Once again, a path cognate can be found in which the angular rotations are in the right cranks.

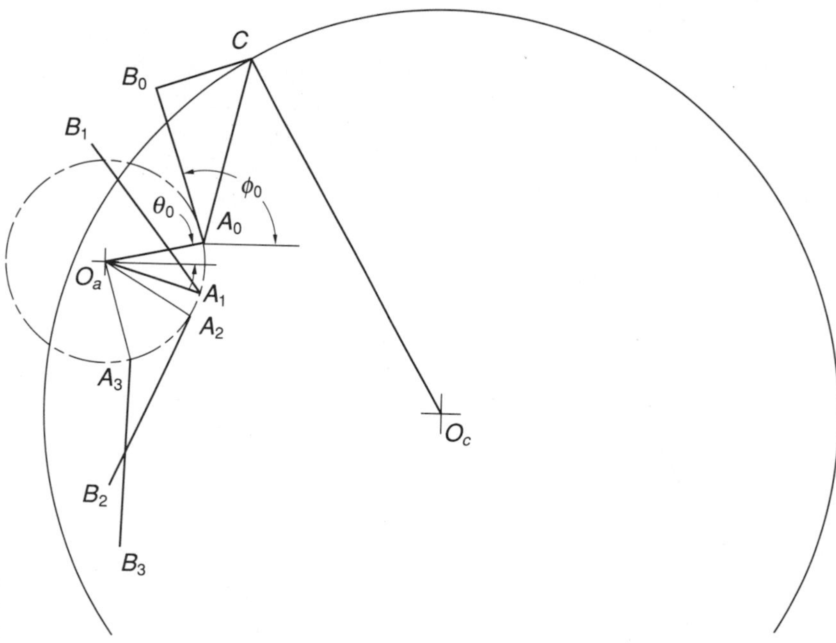

FIGURE 9.24
Function generation: four-position synthesis

❑ COPLANAR MOTION COORDINATION

Assume that the motion of a moving plane is to be coordinated with that of a crank, as shown in Figure 9.7(d). This problem can be solved using the methods described in Section 9.6 and 9.7. The problem requires several manipulations. One possible way of obtaining a solution is described below.

To keep the explanation simple, let's carry on a three-position synthesis. In step 1, assume a coupler of any size and complete a coplanar motion synthesis, as shown in Figure 9.25. Notice that the solution does not take into account the angular motion requirement of the crank. In step 2, solve a similar problem, but assign to the coupler (Figure 9.26) the angular motion of the crank. Obtain path cognates of mechanism $O_c CDO_d$, as shown

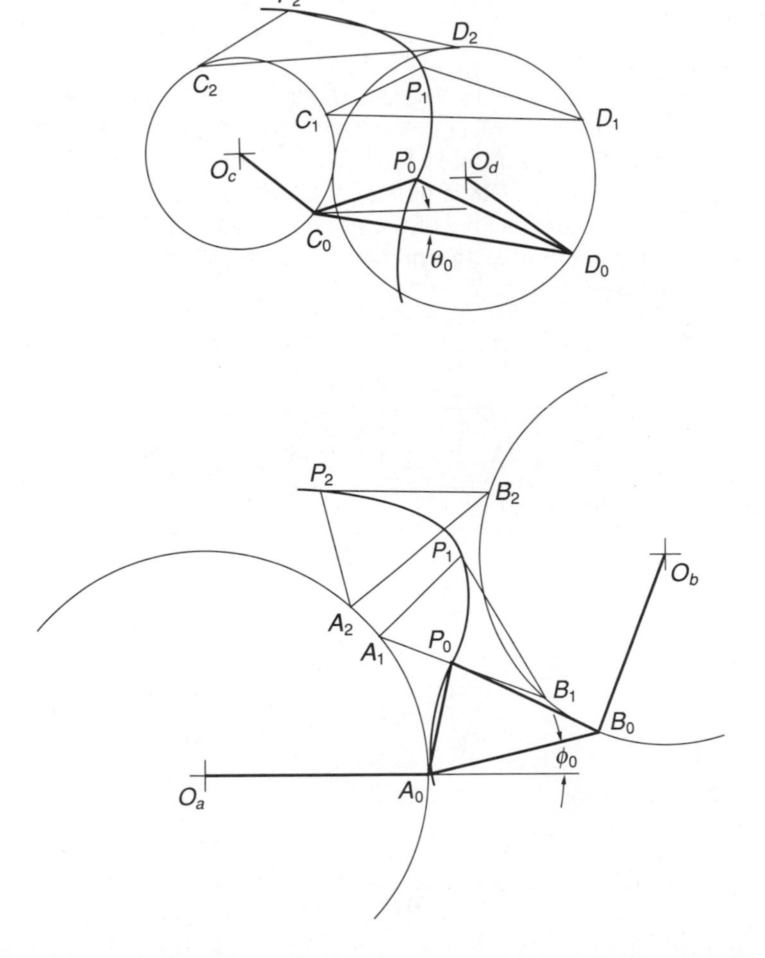

FIGURE 9.25 Coplanar motion coordination: step 1

FIGURE 9.26 Coplanar motion coordination: step 2

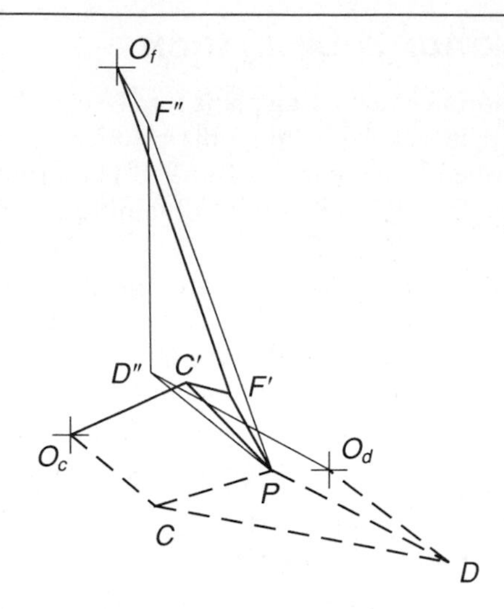

FIGURE 9.27 Coplanar motion coordination: Step 3

in Figure 9.27. In the four-bar system $O_fF'C'O_c$, crank O_cC' has angular motions similar to those of coupler CDP. Thus, if system $O_fF'C'O_c$ and system O_aABO_b (Figure 9.25) are connected at coupler point P, the complex system shown in Figure 9.28 is obtained. Crank O_fF' does not serve any purpose and may cause locking of the system. It can therefore be removed. Thus, the final solution to the problem is a six-bar mechanism of the Stephenson 3 type.

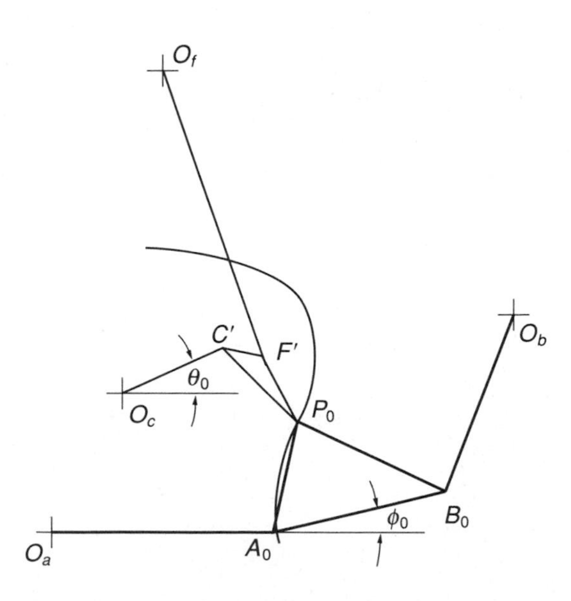

FIGURE 9.28 Coplanar motion coordination: a completed solution

9.11 ▪ Summary

This chapter explains the difference between mechanism synthesis and analysis and discusses the reasons for including synthesis in this book. Definitions of commonly used terms in the fields of synthesis and analysis are provided. The chapter also presents Grubler's equation for determining the degree of freedom in a linkage system and describes popular types of mechanisms using linkage systems.

Common types of linkage synthesis problems are categorized, and graphical and analytical methods for carrying out direct synthesis are given in some detail. Techniques are also given for the indirect approach to synthesis, in which the Roberts-Chebyshev construction is often used.

References

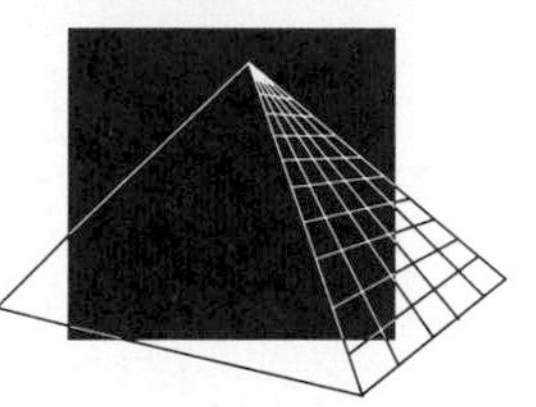

1. Paul, Burton. *Kinematics and Dynamics of Planar Machinery.* Englewood Cliffs, N.J.: Prentice-Hall, 1979.
2. Shigley, J.E., and J.J. Uicker, Jr. *Theory of Machines and Mechanisms.* New York: McGraw-Hill, 1980.
3. Hartenburg, R.S., and J. Denavit. *Kinematic Synthesis of Linkages.* New York: McGraw-Hill, 1964.
4. Sandor, G. N., and A.G. Erdman. *Advanced Mechanism Design: Analysis and Synthesis.* Vol. 2. Englewood Cliffs, N.J.: Prentice-Hall, 1984.
5. Tesar, D. 'Lecture Notes.' University of Florida.
6. Sinha, D.K., and T.R. Hsu. *Advanced Machine Design by Microcomputers.* Duxbury, Mass.: Kern International, 1989.
7. LINCAGES, copyright 1979, University of Minnesota.
8. Freudenstein, F. "Approximate Synthesis of Four-Bar Linkages." Transaction of the American Society of Mechanical Engineers, *Journal of Mechanical Engineering for Industry* 77, no. 6 (1955): 583.

Problems

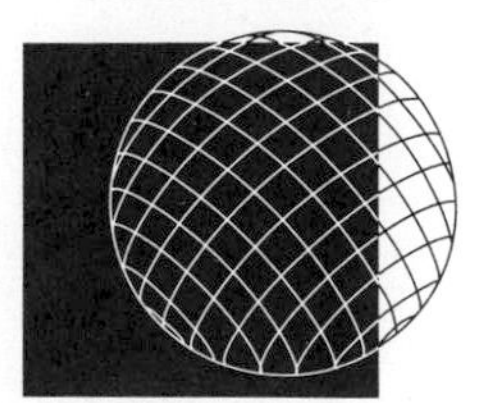

9.1 Design a four-bar mechanism in which the position of a coupler point is defined by the coordinates $(x, x^2 - 2x + 4)$ for the interval $x = 1$ to $x = 2$.

9.2 In the following figure, two positions of the blade of a crop-shear are shown. Design a suitable four-bar mechanism for operating the shear.

9.3 In the following figure, the arrangement for pouring metal into a mold is shown. Design a four-bar mechanism consistent with the three positions of the tundish shown.

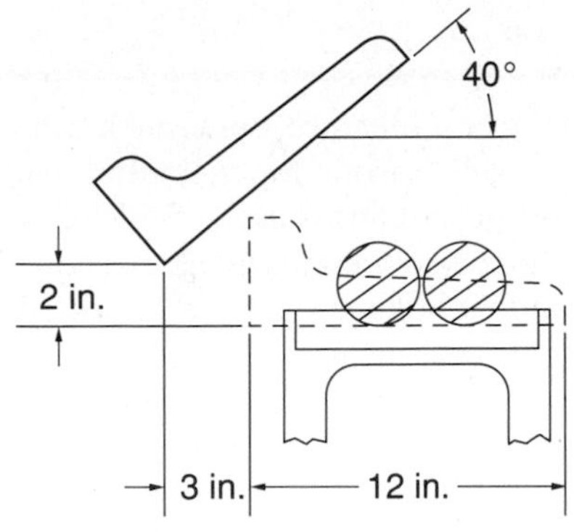

Figure for Problem 9.2

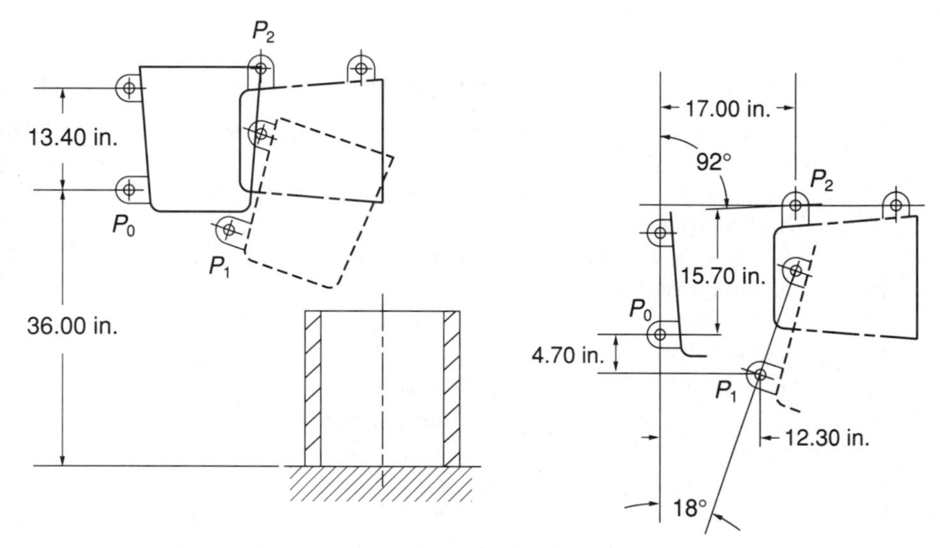

Figure for Problem 9.3

9.4 For the four-bar mechanism shown below, find the path cognates.

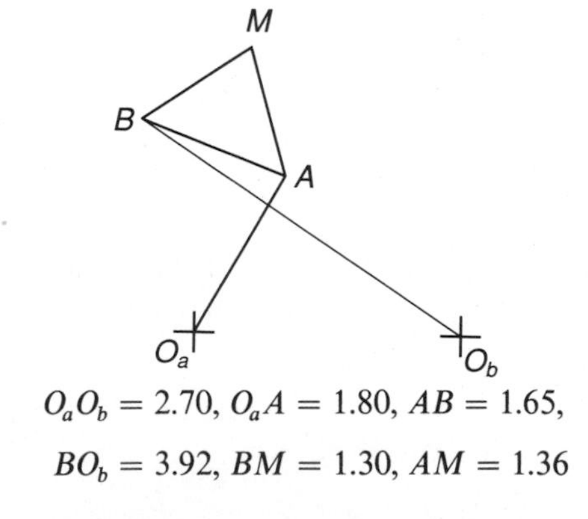

$O_aO_b = 2.70$, $O_aA = 1.80$, $AB = 1.65$,

$BO_b = 3.92$, $BM = 1.30$, $AM = 1.36$

9.5 In the mechanism shown below, the angular motion of the coupler is determined by the variable γ and the angular motion of the follower by the variable ϕ. Find a mechanism in which the motion of the crank is given by γ and that of the follower by ϕ, as before.

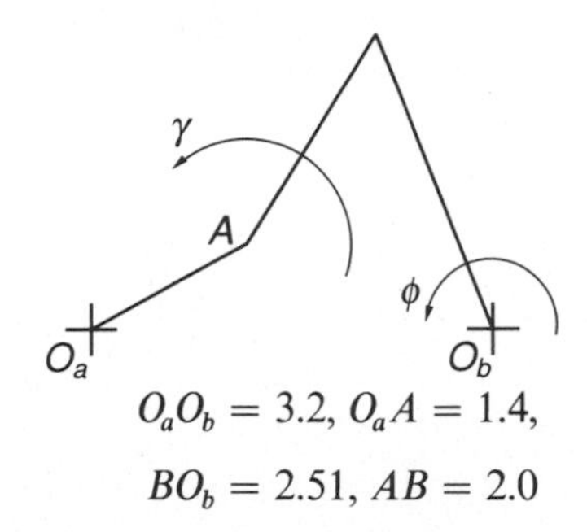

$O_aO_b = 3.2$, $O_aA = 1.4$,

$BO_b = 2.51$, $AB = 2.0$

9.6 Find a four-bar function generator for the following motion parameters: $\Delta\theta_0 = 36°$, $\Delta\theta_1 = 51°$, $\Delta\phi_0 = 17°$, and $\Delta\phi_1 = 134°$.

9.7 Redesign the four-bar mechanism for the shear shown in problem 9.2 such that the driving crank rotates 240° during the cutting stroke of the shear.

9.8 For the metal-pouring device shown in problem 9.3, it is desired that the driving crank rotate through 77° and 99°, respectively, for the corresponding movement of the tundish from position 0 to 1 and from position 1 to 2. Design a six-bar operating mechanism for the drive.

CHAPTER 10

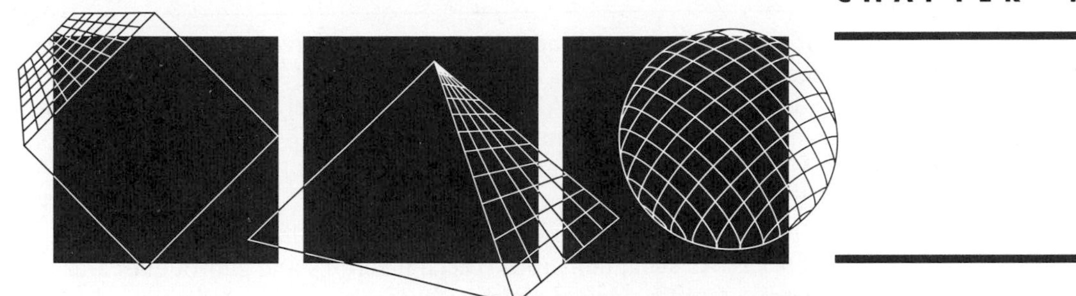

Integrated CAD Systems

10.1 Introduction

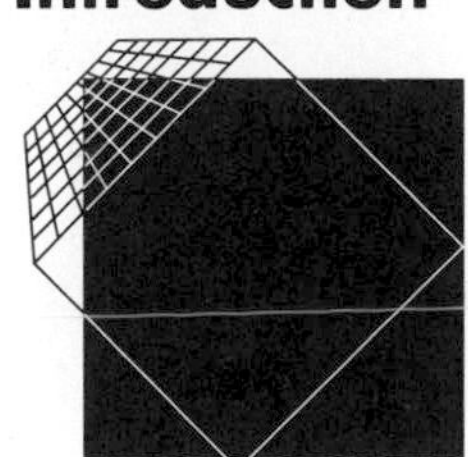

We have thus far devoted almost all our efforts to studying the principles and mathematical formulations of computer graphics and design analyses, for they are the essential ingredients of modern CAD systems. It is now the time for us to see how these elements can be integrated into a single package to provide the user with a design tool that is both effective and efficient. It is hoped that by virtue of this exercise, the reader will acquire a good understanding of how commercial CAD packages work.

Most commercial CAD software packages, commonly known as *turnkey systems*, are created by integrating major components that include geometric modeling, drafting, and finite element analysis. Among many such systems in the marketplace are CALMA (General Electric Company), SDRC I-DEAS (Structural Dynamics Research Corporation), CADAM/CAEDS/CATIA (IBM Corporation), APPLICON (Schlumberger Company), and MEDUSA (Prime Computer). This list is by no means complete, as new products are made available all the time. A survey of some of these commercial systems operating on mainframe computers and minicomputers is available in some recent publications (e.g., see references [1] and [2]). The astounding advances in microcomputer technology in the last decade have made microcomputer-based commercial CAD systems, such as AutoCAD, CADKEY, and Versacad, enormously popular in industry. A comprehensive study on the relative merits of these systems is available in reference [3].

Integrated turnkey CAD systems are extremely powerful design tools as a result of the many unique capabilities already mentioned in Section 1.3. Some of the more significant advantages to industrial users are described in the two subsequent paragraphs.

Most turnkey systems are developed to handle a variety of design problems. As such, numerous special functions and subroutines are integrated into a single package, ready for practical applications. In most cases, no further in-house development or system modifications need be carried out by the user. Different users of the same system can share both experiences and program libraries.

Good documentation and the turnkey nature of these systems make it possible for users with little knowledge or specialized training in software programming to make effective use of these systems. Another great advantage of using turnkey systems is that all future upgrading of the system and software maintenance are the vendor's responsibility.

Coupled with the advantages of turnkey CAD systems, as mentioned above, are a number of disadvantages. One major disadvantage is that they are expensive in terms of both capital and maintenance costs. The cost of acquiring many of these systems requiring mainframe computers used to be in the millions of dollars. The price of the required hardware, however, has been drastically reduced in recent years, as most turnkey CAD software can now be operated by powerful work stations that can be purchased for less than $10,000. Unfortunately, many sophisticated systems are still beyond the reach of small-sized industry, as the annual costs for software and hardware maintenance can range from $50,000 to $250,000. Another shortcoming is that, unlike many other sophisticated engineering tools, most turnkey CAD systems can only be used as "black boxes" due to their proprietary nature. This inaccessibility to the system hinders their maximum possible effective use by industry.

Integrated CAD systems can perform sophisticated design analysis that one could not have imagined possible just a decade ago. We will present the reader with two design case studies to illustrate the effectiveness of integrated CAD systems, one using the MICROCAD system for microcomputers and the other using the ANSYS code. The ANSYS program was originally developed for finite element analysis, as described in Section 7.7. It has been extended to handle solid geometric modeling and has incorporated design optimization in the analysis. It has thus become an effective CAD tool for solving highly complex problems.

10.2 ▪ Overview of Integrated CAD Systems

As we pointed out in Chapter 1, a unified definition of CAD is far from firmly established. One ongoing argument concerns the pertinent elements that constitute a CAD system. In an early article, Krouse [4] suggested that CAD should involve four major components: geometric modeling, analysis, kinematics, and animation. Yet if one agrees with the general engineering design procedure outlined in Section 1.6 and Figures 1.1 and 1.6, the above definition obviously can only cover a small portion of the entire design practice. From the authors' observations, major components included in most turnkey CAD systems can be categorized as (1) geometric modeling, (2) design analysis, (3) engineering drafting, and (4) data management, storage, and transfer.

❑ GEOMETRIC MODELING

The primary function of geometric modeling is to define an object or a structure from the point of view of its geometric properties [5]. Geometric properties—such as a structure's size; configuration, and surface conditions; both initial and deformed conditions; and geometric relations with other mating components—are essential input for the subsequent design analysis and the design of an automated production process. For example, accurate geometric information is critical to the production of control tapes for numerically controlled (NC) machines, the determination of optimum machining procedure, and the final assembly of various components, as illustrated in Plates 1, 2, 3, 4, 7, and 8.

Typically, a geometric model of a solid object can be created by the user through outlines drawn on the monitor screen, using either a mouse or light pen, or a digitizer tablet connected to the host computer. The outline, or profile, so sketched can be displayed on the monitor screen. The user can then select appropriate algorithms by invoking the function options specified in the menu. Three common types of models can be obtained: two-dimensional, three-dimensional, and 2.5-dimensional models.

A two-dimensional model is constructed with such graphic primitives as dots, lines, arcs, and curves, as described in Chapter 5. This type of model can describe the geometry of planar structures. It is inexpensive to produce and hence suitable for industries involved in the production of components of simple geometries (e.g., the sheet metal industry). However, it is inadequate for the description of objects of more complex geometries.

A three-dimensional model, on the other hand, can provide the user not only with a full description of the exterior geometry of a solid, but also with the detailed geometric information of its interior. Typical three-dimensional models produced by advanced turnkey systems are depicted in

Plates 1, 2, 3, 7, and 8. Simple three-dimensional models can be in the form of wireframes with or without hidden lines removed, as described in Chapter 4. Spectacular solid models, as shown in the aforementioned plates, can also be obtained by using some very sophisticated algorithms. At the present time, complex solid models can be produced by turnkey CAD systems using a CPU with sufficient memory, as well as a monitor screen with high resolution. Although solid models are of great visual value to the user and have unique advantages over wireframe models for visualizing objects of complex geometries, substantial cost differences make the wireframe model with hidden lines removed more popular for most engineering applications.

The 2.5-dimensional model fits in between the two- and three-dimensional models. It is basically a two-dimensional model, but with the capability of producing additional views in planes other than the viewing plane on which the profile is created by the user. These views normally include three orthographic projections, as described in Chapter 4. Extra perspective views at selected projection angles and viewing distances, as described in that chapter, can also be made available. The latter views can be made by either line drawings or wireframes. Both hardware and software requirements for the production of 2.5-dimensional models are substantially less than those for the full three-dimensional model. Yet the 2.5-dimensional model can retain almost all the necessary geometric information for the purpose of design. Good-quality 2.5-dimensional models can be produced easily on microcomputers [6]. The reader will find that this type of model is used in the MICROCAD system described in Section 10.5. A comprehensive detailed description of the above three types of models can be found in reference [7].

❑ DESIGN ANALYSIS

The use of analysis in CAD has been described in Section 1.7. A viable CAD system must have strong analytic capability. Most turnkey systems contain sophisticated graphic and drafting software, which can be used in conjunction with one or more commercial finite element analysis programs, such as ANSYS and SDRC I-DEAS. Some also include analytic programs to handle design optimization. The basic requirements for the analysis package in an integrated CAD system are as follows:

1. It must have graphic input and output capabilities. In finite element analysis, it should be possible to generate meshes through a digitizer board, a mouse, or a light pen. The output of the analysis can be displayed graphically, as shown in Plates 9, 10, 13, 15, and 16. The ability to animate the configurations of a moving, or deformable solid, as demonstrated in Plates 5 and 6, is another desirable feature.

2. The manual effort required for data input must be kept to a minimum. Again, in the case of finite element analysis, features like automatic mesh generation are essential.

3. Efficient data transfer and storage are desirable. It was only recently that internal automatic data transmission between various modules in a CAD system (e.g., between the geometric modeling module and the finite element analysis module) became a reality. It is highly desirable that the configuration of the object and other pertinent design parameters can be stored, retrieved, and transferred efficiently within the integrated system itself.

❑ ENGINEERING DRAFTING

Drafting is essentially the documentation of the end results of a design process. In a fully integrated CAD system, all finalized results from the geometric modeling and analysis are expected to be transformed into engineering drawings. Flexible editorial options are desirable so the user can explicitly illustrate the design product. Many experts predict that future drafting systems will be paperless, with computer monitors displaying drawings only as required.

❑ DATA MANAGEMENT, STORAGE, AND TRANSFER

From the description of the above three major components in an integrated CAD system, one may readily see that efficient management of pertinent design data and object configurations is an extremely important function in the system. Various methods for the management of large amounts of information have been presented in Chapter 5. The subject of solid geometric data management, however, has not yet reached a desirable maturity. While it is beyond the intended level of this textbook to present an in-depth description of this topic, extensive treatment of this subject is available in reference [5].

Few commercial CAD systems in the marketplace have included all the aforementioned four functions in a single package. Integration of these functions, however, can be achieved by means of *interpreters*, as indicated in Figure 10.1. Some major systems operating on mainframe computers or sophisticated work stations have been interfaced with manufacturing. Some have even incorporated optimization in their design analysis. The major elements and the logical relationships among these elements are illustrated in Figure 10.1.

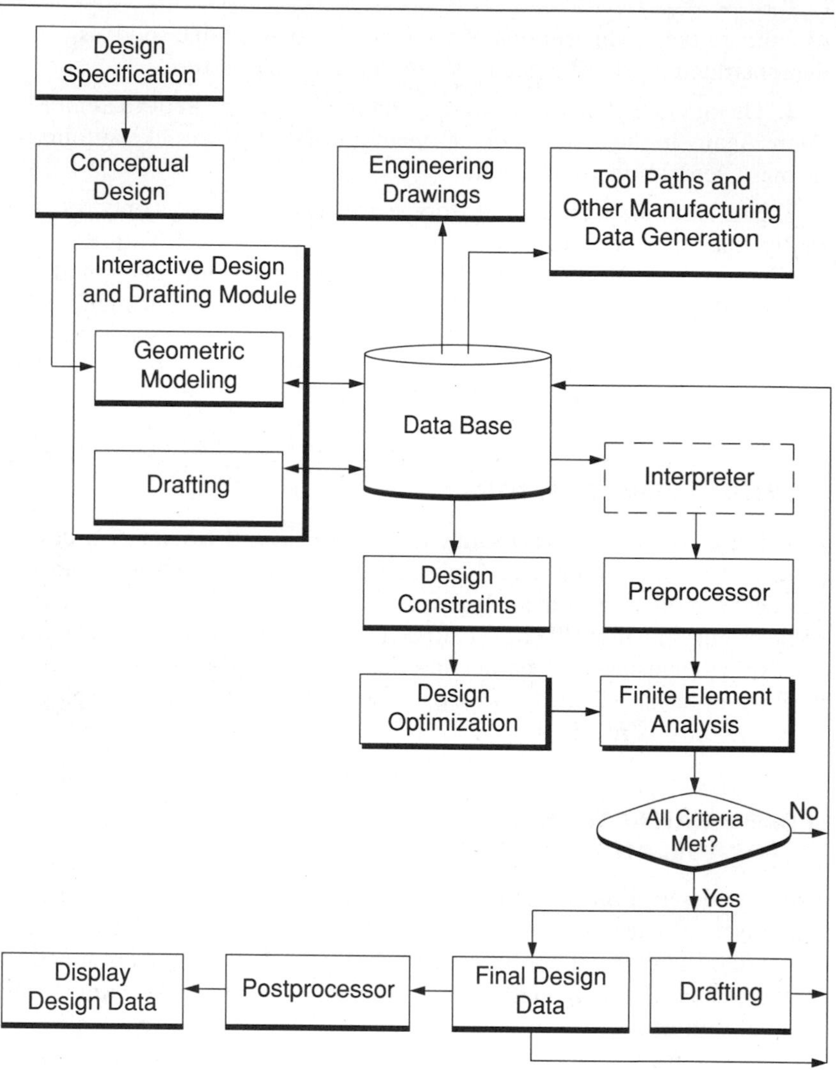

FIGURE 10.1 Major functional elements in an integrated CAD system

10.3 ▪ Integrated CAD Systems

Few commercial turnkey CAD systems encompass all four major components described in the foregoing section. Major systems operating on mainframe computers or work stations, such as CALMA/SDRC I-DEAS and ANSYS, can handle integrated CAD functions, as described in Figure 10.1, through interpreters. Most microcomputer-based CAD systems, such as AutoCAD and CADKEY, are constructed primarily to perform only geometric modeling and drafting. Their performance is effectively limited to what is included in the interactive design and drafting modules and the database, as illustrated in Figure 10.1.

In addition to the difference in scope between major CAD systems and those microcomputer-based systems, other major differences are as follows:

1. There is obviously a significant difference in capital costs between the two systems. Figures 10.2 and 10.3 illustrate the hardware required for both types of CAD systems. This difference is diminishing rapidly with the recently developed multiuser, multitasking work stations. Such work stations are often perceived as a "bridge" between minicomputers (Figure 10.2) and personal computers (PCs) (Figure 10.3) from the point of view of cost and computing power. Those now being built are coming closer to PCs in terms of size and price but have the power of a mainframe computer. Most major CAD systems can now be handled by work stations. Following are some of the unique features of work stations:

- They have sufficient processing power to carry out all normal functions required for CAD (e.g., the creation and modification of geometric models, finite element mesh generation, and editing).
- Most work stations have twin screens or a single screen with a

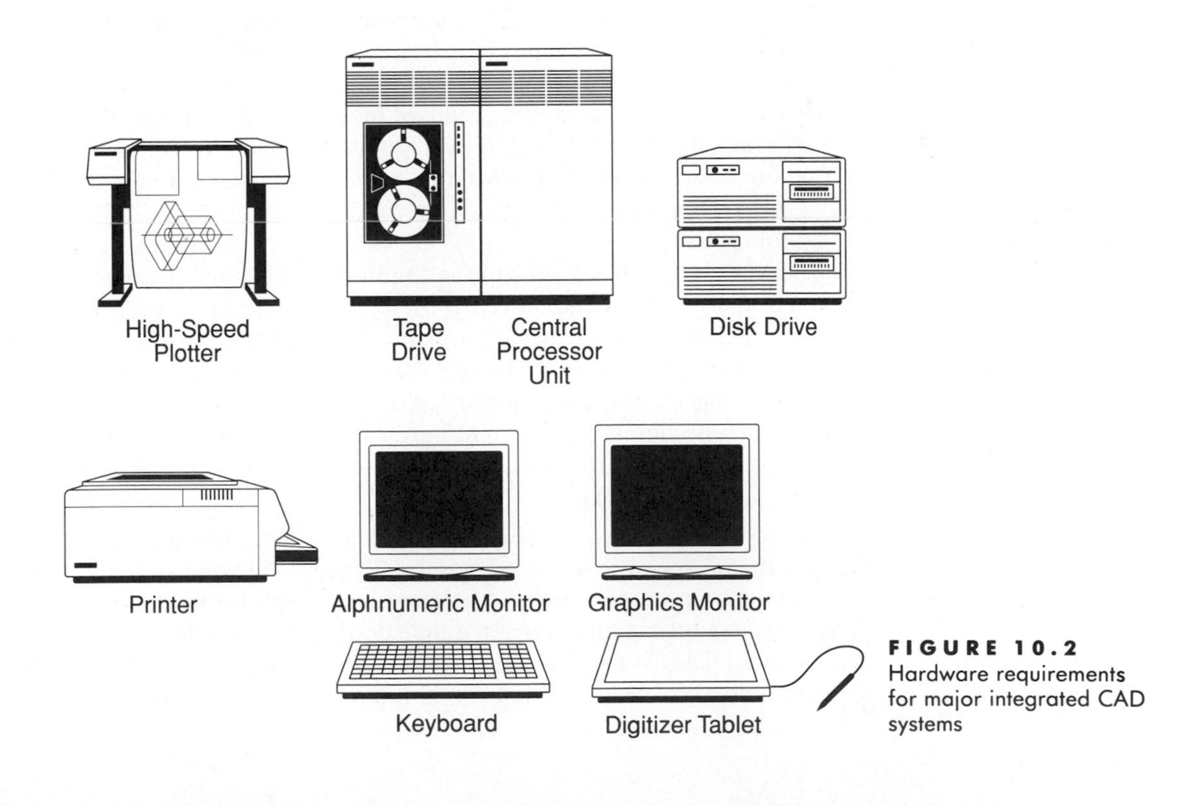

FIGURE 10.2
Hardware requirements for major integrated CAD systems

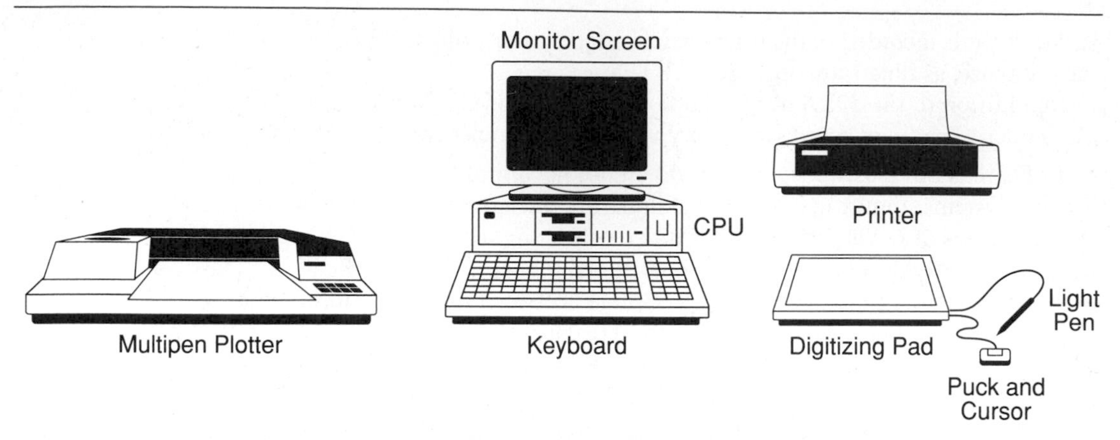

FIGURE 10.3 Hardware requirements for microcomputer-based CAD systems

window manager for effective multiple-window displays of the results from different programs executed simultaneously.

- They can be used as a stand-alone computer or as a server in a network system.
- New work stations accept natural forms of communication (e.g., voice input or input by pattern recognition, including handwritten characters).
- They are constructed using reduced instruction set computer (RISC) technology, which is much more efficient in terms of computational speed and memory requirement than the complex instruction set computer (CISC) technology used by many other computers.

Major CAD systems, being versatile and multifunctional, require much larger storage memories and faster CPUs. Typically, these systems utilize dual monitors or windows, one for alphanumeric I/O and the other for graphic I/O provided by work stations or minicomputers, as shown in Figure 10.2. The microcomputer-based systems, on the other hand, require relatively simple and low-cost hardware. As one can visualize from Figure 10.3, the entire system can be readily fit into the trunk space in a compact automobile.

2. Major systems are much more amenable to multiuser situations than are the microcomputer-based systems. Because the latter systems are designed primarily for single users, they are suitable for small-sized design offices working on relatively simple design problems. Multi-user-oriented CAD systems are important to major industries, such as the automobile and aerospace industries, as described in Section 5.8.

Microcomputer-based systems thus have limited applications in such industries.

3. Some major systems have been integrated with computer-aided manufacturing (CAM). By referring to Figure 10.1, the reader can see how the CALMA system, for example, can provide the user with optimum tool paths to machine a solid component once its geometry is finalized. Other manufacturing data, such as control tapes for NC machines, can be produced by this CAD system. Microcomputer-based CAD systems have limited capabilities in this area.

4. Major systems are obviously more versatile and flexible, with many more options and functional commands. For example, there are 281 commands available to the user with a CALMA system, in comparison to 60 commands with a CADKEY system. Many microcomputer-based systems can produce engineering drawings limited to A and B sizes ($8\frac{1}{2}''$ × 11″ and 11″ × 17″, respectively), whereas major systems can produce drawings up to E size (34″ × 44″).

The above two types of commercial CAD systems are compared in Table 10.1. Despite the obvious limitations of microcomputer-based CAD systems, they have gained considerable acceptance by small and medium-sized industries because of the value they provide in return for the moderate investment necessary to acquire such systems. The rapid advancement of microcomputer technology, as we are witnessing now, will undoubtedly close the gap between these two systems in the conceivable future.

TABLE 10.1 Comparison of commercial CAD systems

Features	Major Systems	Microcomputer-Based Systems
Degree of integration	Close to full integration, as described in Figure 10.1	Limited to geometric modeling and drafting
Options	Versatile, with large numbers of functional commands	Limited, with options
Number of users	Designed for multiple users	Intended for single users
CAD/CAM interface	Most systems provide such interface	Little direct interface available
Colors of graphics, and drawing sizes	As many as 4,000 colors are available for geometric models, with drawings up to E size	Limited to 16 colors, with small drawing sizes; large drawings can be produced at slow speeds
User friendliness	Users require special training	Easy to use; user friendly
Cost of software	High	Low

10.4 ▪ Major Functions of Commercial CAD Systems

We have shown in the foregoing section that the capabilities of commercial CAD systems can vary significantly from one system to another. However, most commercial CAD systems can perform the major functions discussed in this section. These functions can be classified into six distinct groups: (1) editing/filing, (2) geometric construction, (3) drafting, (4) design analysis, (5) display control, and (6) error messages and help. Commands for invoking these functions are available either from the template attached to the digitizer tablet or from the menu appearing on the monitor screen.

❑ EDITING/FILING

In addition to the usual editing and filing of documents and design data, a CAD system should have the capability to edit geometric models, as well as to record, delete, or store the complete design file, including geometric information. Geometric editing may involve such functions as moving, duplicating, mirroring, and verifying.

❑ GEOMETRIC CONSTRUCTION

The procedure normally begins with setting the viewport. A viewport is the portion of the graphic display screen in which the geometric model is to be viewed. There are 2-dimensional, 2.5-dimensional, and 3-dimensional models available to the user, as described in Section 10.2. In the case of three-dimensional models, most CAD systems offer both wireframe and solid models. The basic elements for constructing geometric models are points, lines, arcs, and curves. The principles of constructing with these basic elements have been presented in Chapters 3 and 5. A variety of basic geometries can be constructed by combining these basic elements. These basic geometries can also be conveniently called up from the built-in modules. Many systems offer the following basic geometries and features:

- Polygons
- Arc/circle tangents
- Lines of chosen weight in various formats, such as solid, dashed, center, and phantom (double center), and in parallel or tangent to a selected curve or arc
- Rectangle corners with chosen width/height
- Trimming of lines, arcs, or circles
- Translation or duplication of an enclosed geometry
- Mirroring for geometries that can be constructed by means of repeating part of the geometry in mirror images, such as gear teeth

- Duplication for identical geometries that appear in several places in the viewport (e.g., bolts and nuts in a machine component)
- Scaling
- Surface creation in the form of B-surfaces, ruled surfaces, surfaces of revolution, cylinder surfaces, bonded surfaces, etc.
- Clipping and pasting

❑ DRAFTING

Most commercial CAD systems can offer two-dimensional drawings with three-dimensional viewing (i.e., views from the top, front, back, bottom, right, or left, plus either an isometric or perspective view) of the solid. These views can either be recalled from an existing file produced by the geometric construction or be created by following procedures used in the geometric construction. Drafting function commands should include labeling, dimensioning, windows, symbols, and schematics. An intelligent drafting function can also offer crosshatching with the chosen weight and format of lines, as described in the foregoing paragraph. Commonly used graphic elements, such as fillets and chamfers, are available to the user.

❑ DESIGN ANALYSIS

Finite element analysis usually forms the basis for design analysis in most CAD systems. Interface between design analysis and geometric modeling is usually available. Some systems also offer analysis on design optimization in conjunction with finite element analysis. Objective functions used in some commercial systems involve weight and selected overall dimensions. There are multiple choices of types of analyses as well. For example, the ANSYS program can handle a variety of analyses, as described in Section 7.7. The SDRC I-DEAS program is capable of solving both static and dynamic (modal) problems, as well as linkage and mechanisms analysis. Most of these systems offer a variety of elements to be used in the analytic model. Choices are also available for the loading options and boundary conditions. Automatic mesh generation is a common option built into most commercial packages. Users are also given the choice of the types of postprocesses for their analytic results. Output of the analysis in the forms shown in Plates 1 through 17 is made possible by most commercial CAD systems.

❑ DISPLAY CONTROL

Commands involved in display control usually include the following options:

- Selection of views: any one or all orthographic views, plus either one isometric or one perspective view of the object

- Isometric or perspective views in rotation
- Windowing and zooming of views with selected magnifications for detailed geometric information

❑ ERROR MESSAGES AND HELP

In addition to printed user's manuals, commercial systems have made available comprehensive help menus on the monitor screen. Users may diagnose errors in the design process without having to refer to the printed manuals.

10.5 ▪ The MICROCAD System

As mentioned in Section 10.3 and also in Table 10.1, most microcomputer-based commercial CAD systems offer excellent computer-aided geometric modeling and drafting, but virtually no system offers design analysis. Complete mechanical engineering design thus cannot be carried out by using these systems alone. Integrated CAD is, however, still possible by using a noncommercial MICROCAD system.* The hardware configuration and software requirements for this system are specified in Appendix IV.

MICROCAD is an integrated CAD system operational on microcomputers with low-cost peripherals that include a dot-matrix printer, a multipen plotter, and a digitizer pad. This system was developed by the author (TRH) and his students. In many ways, it can be regarded as a miniature version of a major commercial CAD system with limited capabilities. However, it possesses all four major functions described in Section 10.2 for integrated CAD systems. It is thus possible to demonstrate the principle of integrated CAD systems by using this simplified system.

❑ GENERAL DESCRIPTION

MICROCAD is intended for use in the mechanical engineering design of two-dimensional plane structures. Major software components included in this system are (1) GEOMOD for geometric models, (2) ELAS for elastic stress analysis by the finite element method, and (3) DRAFT for two-dimensional engineering drafting.

A 2.5-dimensional geometric model can be produced by entering the outline of the object involving line segments, arcs, and curves specified by four-point spline functions, as described in Chapters 3 and 5. An oblique view formulated by the three-dimensional projection at selected viewing

---

* The reader should not be confused by the fact that there is a commercial system under the name of microCADD available in the marketplace.

angles and distances is available in addition to the three orthographic views. Interactive manipulation and modifications of the geometry of the model can be readily carried out by following appropriate instructions appearing on the monitor screen. Once finalized, the geometric data can be transferred to a central database following the procedure described in Section 5.5.

If so desired, stress analysis on the object of the finalized configuration can be performed next by using the finite element method described in Chapter 7. The profile of the model can be retrieved and displayed on the monitor screen once again. A finite element mesh can be marked directly on the profile of the model by using a digitizer pad and a moving puck. In order to facilitate demarcation of the mesh, a function menu is displayed on the monitor screen. Functions can be invoked by moving the puck over the digitizer board so that the cross hairs rest on the function command menu appearing on the screen, and then pressing an appropriate button on the puck. Upon completion of the stress analysis, graphic output on the deformed geometries, color-coded stress contours, and critical zones is made available to the designer. Satisfaction of specified design criteria can be readily determined by visual inspection.

Engineering drawings of the finalized design configurations can be produced at the end of the design process. Additional drafting by solid, dashed, and center lines can be carried out by the user. Unwanted lines between two selected points can be removed from the screen.

Automatic dimensioning is available for linear and angular distance. Labels with selected letter sizes and positions can be applied anywhere within the specified region.

❑ PROGRAM STRUCTURES

As we have discussed in Section 10.2, geometric modeling, finite element analysis, and engineering drafting constitute the skeleton of most integrated turnkey CAD systems, including the MICROCAD system. The linkage among these three components is provided by the central database, as illustrated in Figure 10.4. The two-dimensional geometry of an object is created first by a profile generator in the GEOMOD module. The main functions of the profile generator are to assemble the descriptions of lines, curves, and arcs entered by the user and to formulate the outline of the object. The profile generator also allows the user to modify and prettify the input outline. Once the profile of the object is completely described, a description of the components that form this profile (e.g., the coordinates of the terminal points for line segments, the coordinates of the center and the starting and ending points of an arc, the coordinates of spline points for a curve) is then transferred and stored in the central database. Visual display routines are available, which make it possible to retrieve these data from the central database and display them pictorially in four ways, including three orthographic (i.e., front, top, and side) views and one oblique view. The

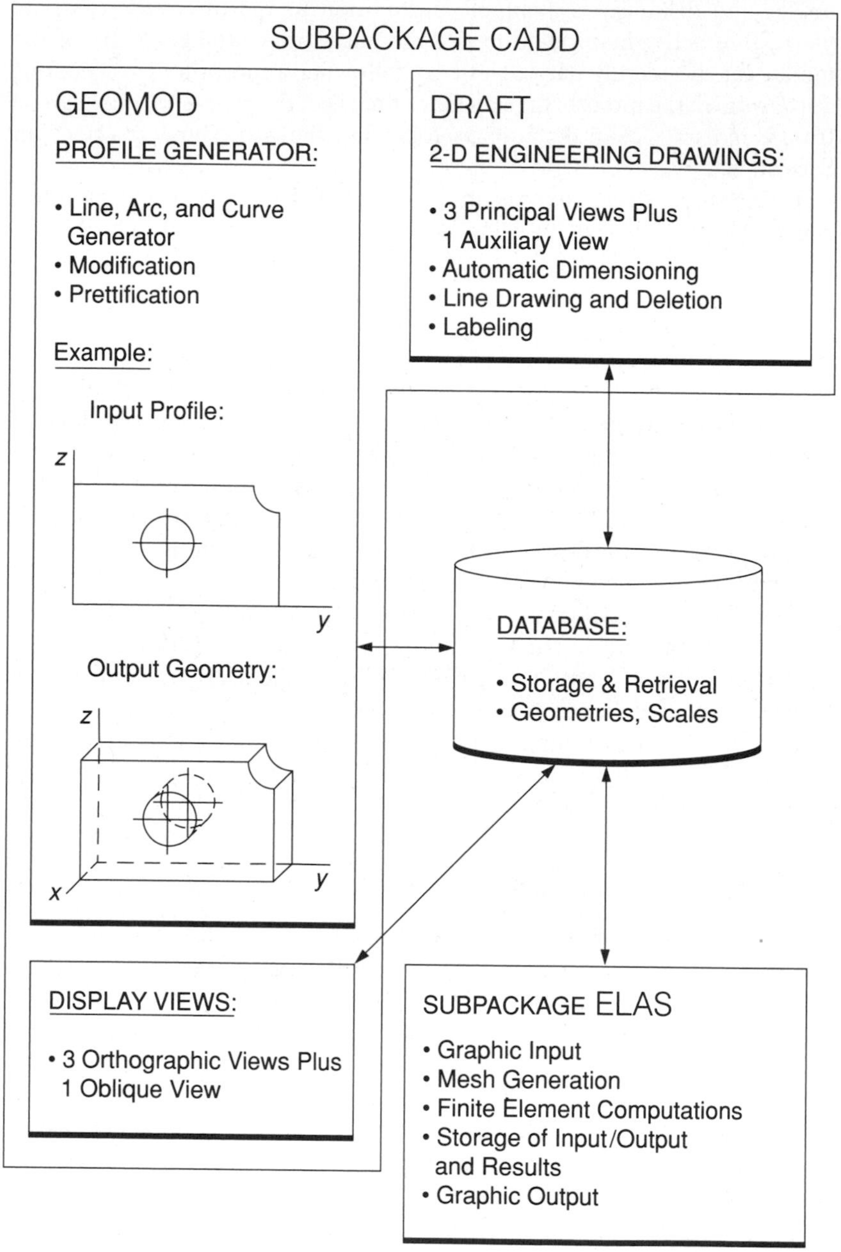

FIGURE 10.4 Program structure of MICROCAD

SUBPACKAGE CADD

GEOMOD

PROFILE GENERATOR:

- Line, Arc, and Curve Generator
- Modification
- Prettification

Example:

Input Profile:

z
y

Output Geometry:

z
x
y

DRAFT

2-D ENGINEERING DRAWINGS:

- 3 Principal Views Plus 1 Auxiliary View
- Automatic Dimensioning
- Line Drawing and Deletion
- Labeling

DATABASE:

- Storage & Retrieval
- Geometries, Scales

DISPLAY VIEWS:

- 3 Orthographic Views Plus 1 Oblique View

SUBPACKAGE ELAS

- Graphic Input
- Mesh Generation
- Finite Element Computations
- Storage of Input/Output and Results
- Graphic Output

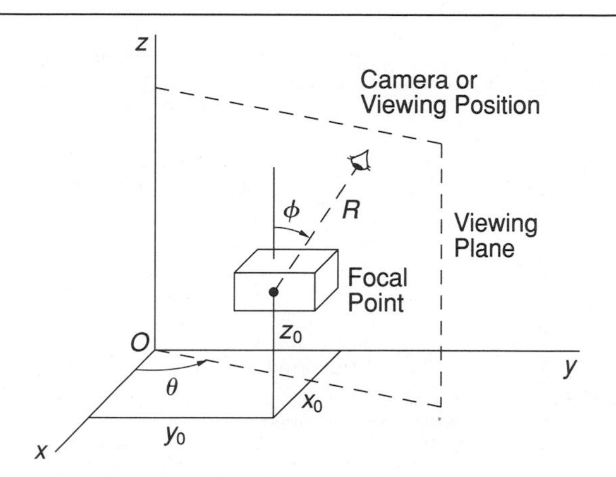

FIGURE 10.5
Definition of viewing position for oblique views of an object

latter can be produced by specifying the six required values [8] that define the position of the "camera" or "viewer's eye" in a spherical coordinate system and the position of the focal point in a rectangular coordinate system. These six values are illustrated in Figure 10.5.

The oblique view provides the user with a visual appreciation of the object geometry from various viewing angles and positions. A typical graphic display of a planar object created by MICROCAD on an IBM PC is shown in Figure 10.6. The object profile stored in the central database can

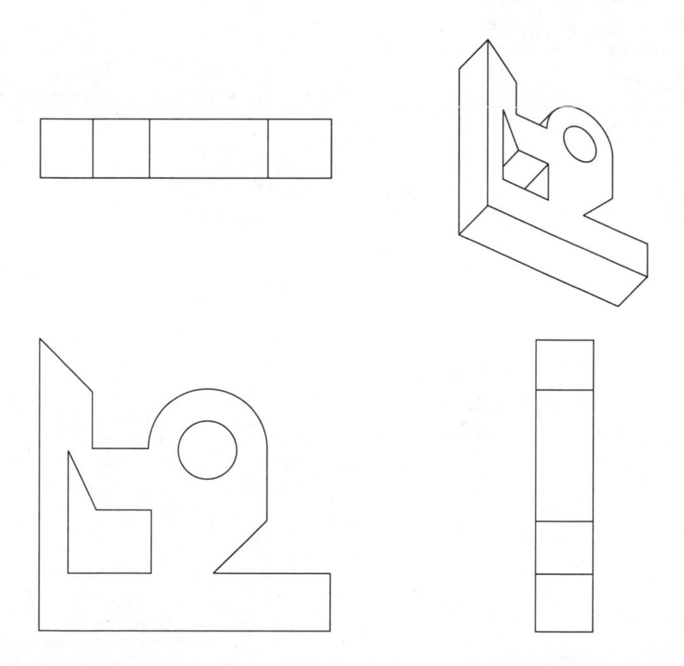

FIGURE 10.6
Geometric modeling of a planar structure by the MICROCAD system

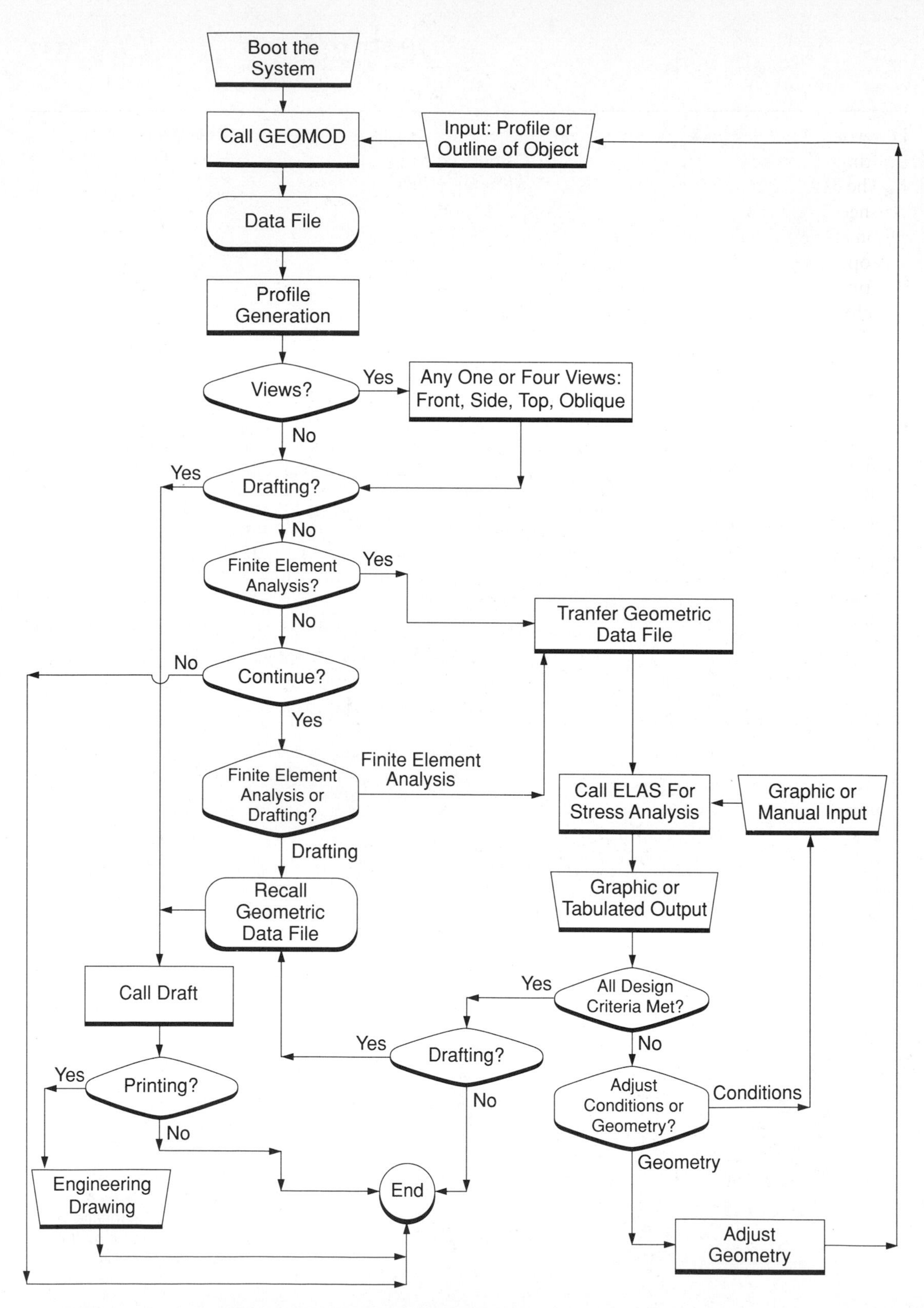

FIGURE 10.7 Flowchart of the MICROCAD system

be retrieved either for the finite element stress analysis or for engineering drafting.

The DRAFT module in the MICROCAD system is used to produce engineering drawings in three orthographic views and one oblique view. Upon invoking the DRAFT command, a menu that contains several drafting options is displayed along with these four views. Any one of those drafting functions can be prompted by appropriately placing the cursor attached to the digitizer puck. The functions mentioned above include the creation and deletion of existing solid, dashed, and center lines; both linear and radial dimensioning; and labeling. A typical drawing produced by this routine will be presented in the next section.

The profile retrieved from the central database can be used for graphic discretization in finite element stress analysis. The software program used for finite element analysis is the ELAS code, the formulations of which are presented in reference [9]. Specifications of the nodes and elements can be entered by moving the cursor over the sketch of the discretized finite element model attached to the digitizer pad. The results of the stress analysis can be displayed graphically, as described in Chapter 6 and shown in Plates 9, 13, 15, and 16. The flowchart for the MICROCAD system is shown in Figure 10.7.

❑ CASE STUDY USING MICROCAD

MICROCAD is constructed to be interactive. As such, the user can operate the system by simply responding to the instructions and questions prompted on the monitor screen without having to refer to the written manual.

A design case study will be presented to illustrate the use of this system with a microcomputer and other peripherals, as illustrated in Figure 10.3. This case involves the integrated design of a gear tooth. The process will begin with geometric modeling, and the geometry created in the geometric modeling will be automatically transferred for the subsequent finite element stress analysis and engineering drafting.

The geometry and dimensions of the gear tooth are shown in Figure 10.8. The tooth is made of steel. Following is pertinent information required for the design:

Material Properties ■

Young's modulus:	30×10^6 psi (206,896 MPa)
Shear modulus of elasticity:	12×10^6 psi (82,760 MPa)
Poisson's ratio:	0.25
Yield strength:	75,000 psi (517 MPa)

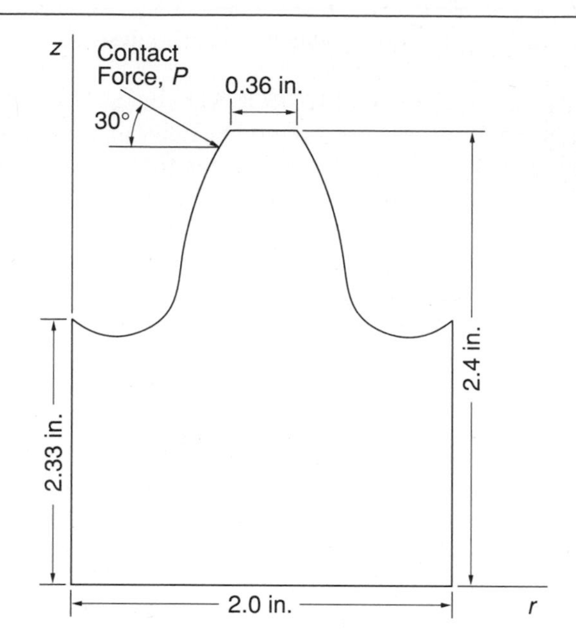

FIGURE 10.8 Geometry and dimensions of a gear tooth

Design Conditions ▪ Allowable stresses (refer to Figure 10.8 for the coordinate systems) are

$$\sigma_{rr} = \sigma_{zz} = \sigma_{rz} = 10{,}000 \text{ psi (69 MPa)}$$

Allowable maximum displacements are

$$\delta_{\max} = 0.01 \text{ inch (0.25 mm)}$$

A contact force, $P = 3{,}000$ lb_f(13,344 N) was applied to the tooth, as shown in Figure 10.8.

Geometric Modeling ▪ The input profile of the gear tooth required in Figure 10.7 was carried out by entering the coordinates of the terminal and data points (knots) along the perimeter of the tooth, as indicated in Figure 10.9. The sequence of input was as follows: three line segments between *A* and *B*, one curve with six knots between *B* and *C*, one line between *C* and *D*, and 1 curve with five knots between *D* and *A*.

The coordinates of all these points were entered from the digitizer pad. The input profile generated the three orthographic views and the one oblique view displayed in Plate 14.

Finite Element Stress Analysis ▪ As shown in Figure 10.10, a finite element idealization with 105 nodes and 90 elements was used. Users may enter nodal coordinates and element descriptions through either the keyboard or the digitizer pad. The latter option provides great speed for

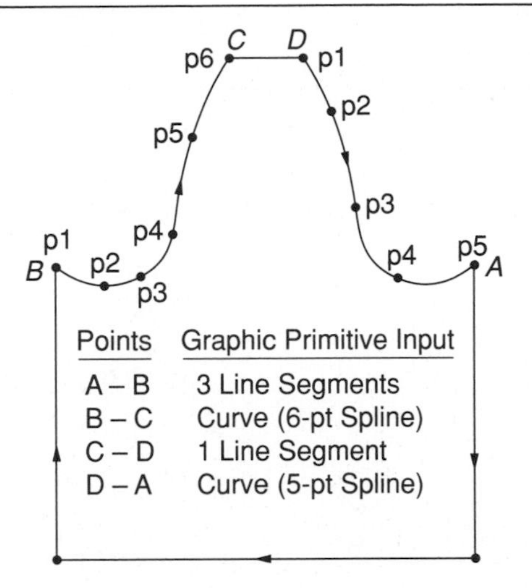

FIGURE 10.9
Input profile of a gear tooth

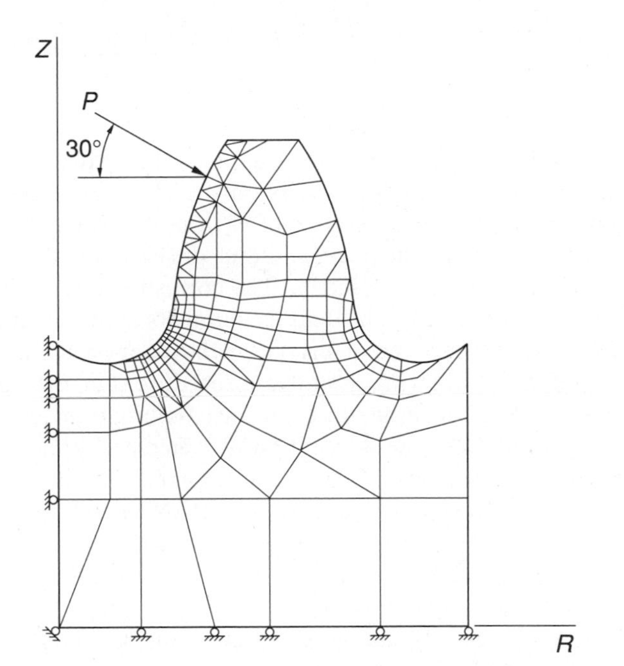

FIGURE 10.10
Finite element idealization of a gear tooth

models involving large numbers of nodes and elements. This process began with the retrieval of the gear tooth profile on the monitor screen, along with a function menu, as shown in Figure 10.11. A hard copy of the display was obtained, and the desired discretization pattern (or finite element mesh) was drawn on this copy by the user. The pattern was then pasted on the digitizer pad. Discretization of the tooth then became the digitization of the pattern.

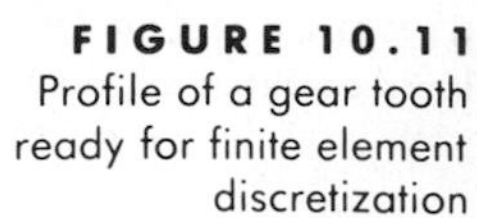

FIGURE 10.11 Profile of a gear tooth ready for finite element discretization

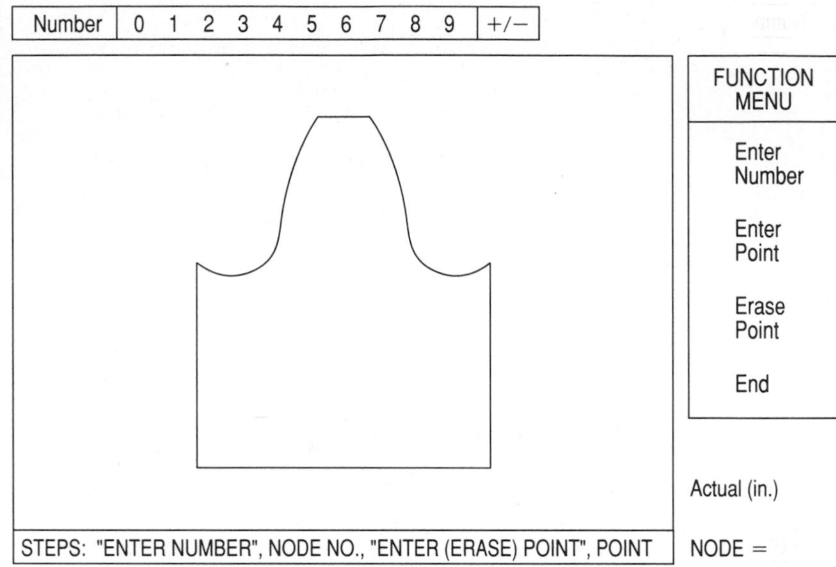

In order to do this, the puck was used over the pattern, which synchronized the motion of the crosshairs on the monitor screen, to input the location and thus the coordinates of each node. Nodal numbers can also be entered by moving the puck, and thus the crosshairs, to the appropriate part of the menu on the screen. The corresponding numeric value for the node was then entered through the numbers provided at the top of Figure 10.11. Prescribed force or displacement components at any node could be entered by following a similar procedure, but with a different menu and a set of codes designating various combinations of forces and displacement conditions. The description of elements by their respective node numbers was carried out in a similar manner, with the menu at the bottom of Fig. 10.11 being replaced by ELEMENTS–STEP 1: "ENTER NUMBER" to be followed by another set of instructions of "NODE I, J, K, L." The complete discretized model for the gear tooth as it appeared on the monitor screen is shown in Figure 10.12.

MICROCAD offers the following graphic display functions at the conclusion of the finite element stress analysis:

1. Undeformed mesh
2. Deformed mesh
3. Superposition of 1 and 2
4. Stress contour plots
5. Zoom

The user may choose to view one or all of the above graphic outputs. Figure 10.13 shows the third option: the superposition of the deformed and

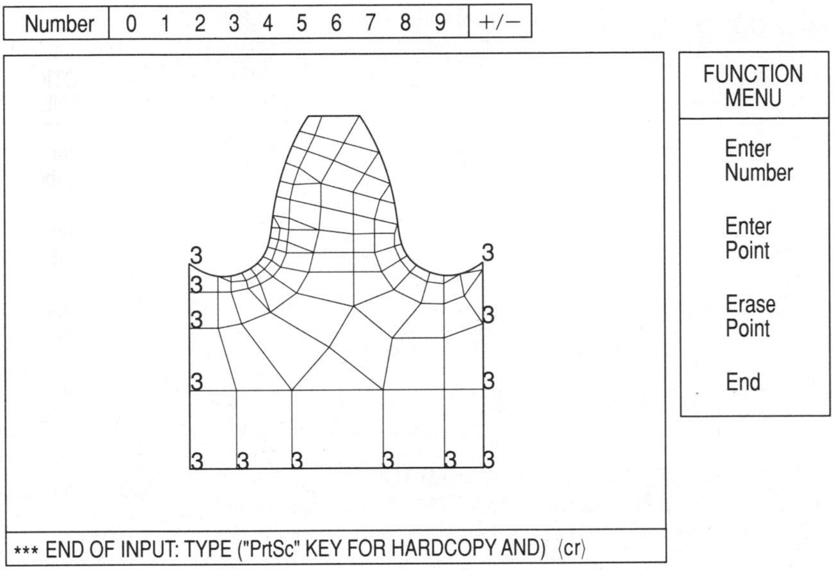

Number	0	1	2	3	4	5	6	7	8	9	+/−

FUNCTION MENU

- Enter Number
- Enter Point
- Erase Point
- End

*** END OF INPUT: TYPE ("PrtSc" KEY FOR HARDCOPY AND) ⟨cr⟩

FIGURE 10.12
A gear tooth profile after finite element discretization

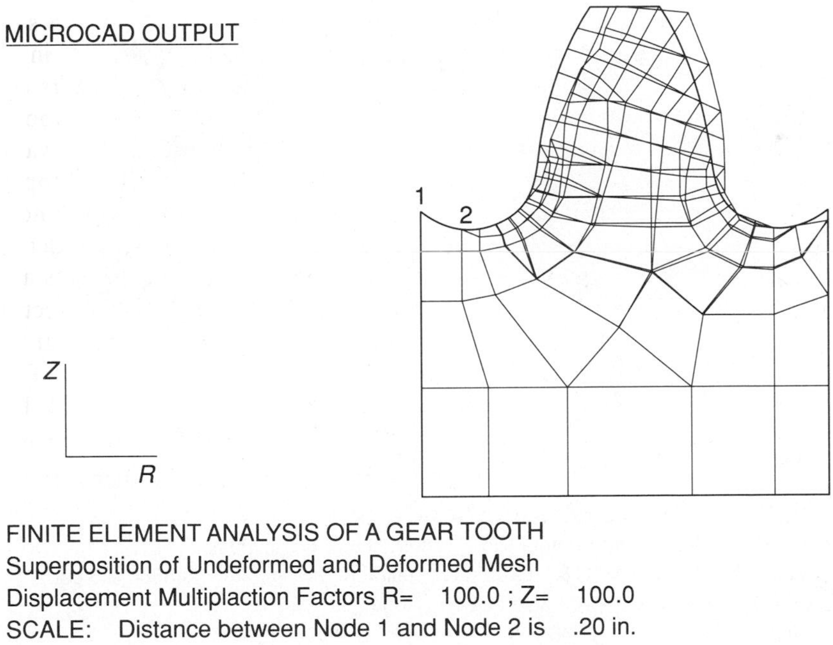

MICROCAD OUTPUT

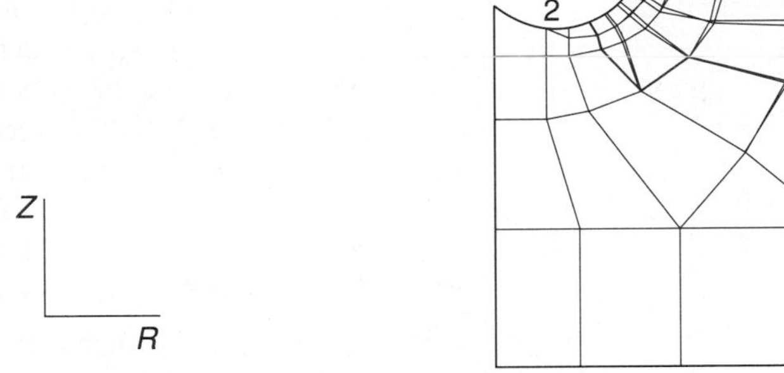

FINITE ELEMENT ANALYSIS OF A GEAR TOOTH
Superposition of Undeformed and Deformed Mesh
Displacement Multiplaction Factors R= 100.0 ; Z= 100.0
SCALE: Distance between Node 1 and Node 2 is .20 in.

FIGURE 10.13
Superposition of deformed and undeformed finite element meshes in a gear tooth

FIGURE 10.14 Stress contours in a gear tooth: (a) stress in R direction; (b) stress in Z direction; (c) shear stress

MICROCAD OUTPUT

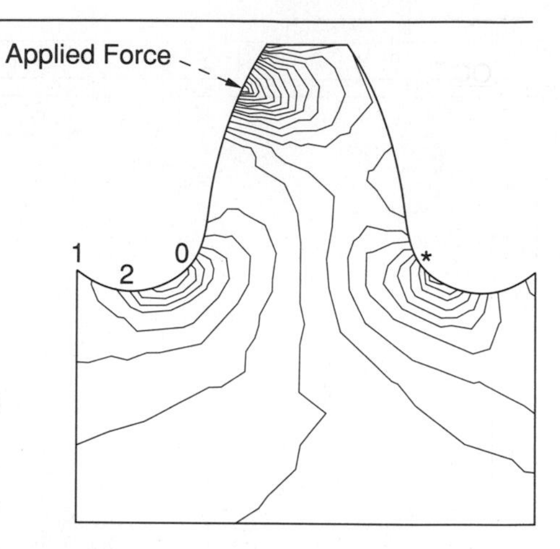

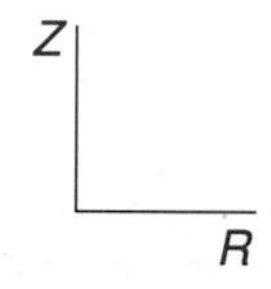

FINITE ELEMENT ANALYSIS OF A GEAR TOOTH
R Stress Component Contour Plot (1500.00 psi. increments)
Maximum Stress (0) = .12133E+05 psi. ; Minimum Stress (*) = −.20943E+05 psi.
SCALE: Distance between Node 1 and Node 2 is .20 in.
Stress Colour Code: Negative : Positive : Over Allowable

(a)

MICROCAD OUTPUT

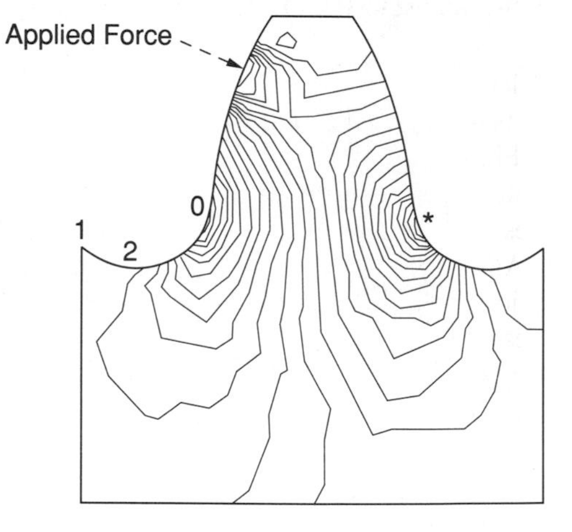

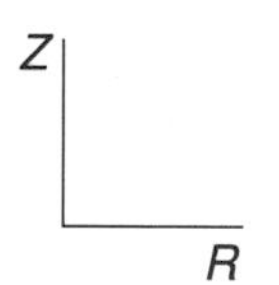

FINITE ELEMENT ANALYSIS OF A GEAR TOOTH
Z Stress Component Contour Plot (1500.00 psi. increments)
Maximum Stress (0) = .17323E+05 psi. ; Minimum Stress (*) =−.24886E+05 psi.
SCALE: Distance between Node 1 and Node 2 is .20 in.
Stress Colour Code: Negative : Positive : Over Allowable

(b)

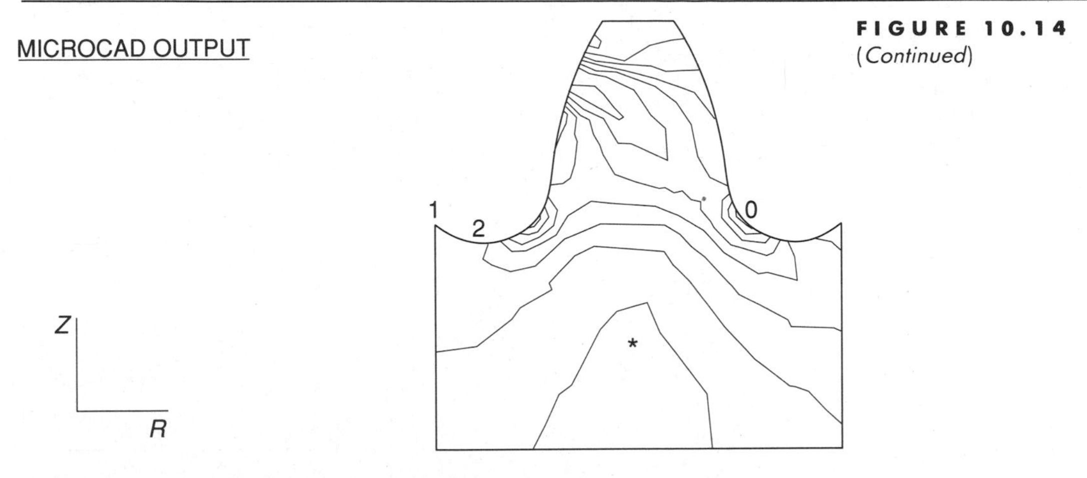

FIGURE 10.14 (*Continued*)

FINITE ELEMENT ANALYSIS OF A GEAR TOOTH
RZ Stress Component Contour Plot (1500.00 psi. increments)
Maximum Stress (0) = .11845E+05 psi. ; Minimum Stress (*) = −.54382E+03 psi.
SCALE: Distance between Node 1 and Node 2 is .20 in.
Stress Colour Code: Negative : Positive : Over Allowable

(c)

undeformed finite element meshes. The stress contours for stress components in the *R* and *Z* directions, as well as for the shear stress, are displayed in Figure 10.14. Figure 10.15 shows the zoomed view of the *Z* stress contours. The zooming option allows the user a better view of the detailed stress distributions in critical regions in the structure.

All graphic output from MICROCAD has color codes to designate various conditions, as shown in Plate 15 for deformations and Plate 16 for stress contours. The regions with a stress level exceeding the specified design allowable values are shown in red, which makes the designer aware of the criticality of the design with a single glance.

Engineering Drafting ▪ Once all design criteria are satisfied from finite element analysis and the geometry of the gear tooth is finalized, the next logical step for the user is to produce an engineering drawing for the purpose of production. The DRAFT component of MICROCAD was developed to produce two-dimensional engineering drawings.

The DRAFT command can be invoked by entering the DRAFT command from the main menu. The four views of the object, such as those shown in Plate 14, will be displayed immediately on the monitor screen, along with a menu, as shown in Figure 10.16. Solid, dashed, and center lines can be superimposed on these views at any selected place by prompting the LINES function and then specifying the starting and ending points of

FIGURE 10.15 Detail of stress distribution in a gear tooth using zooming option

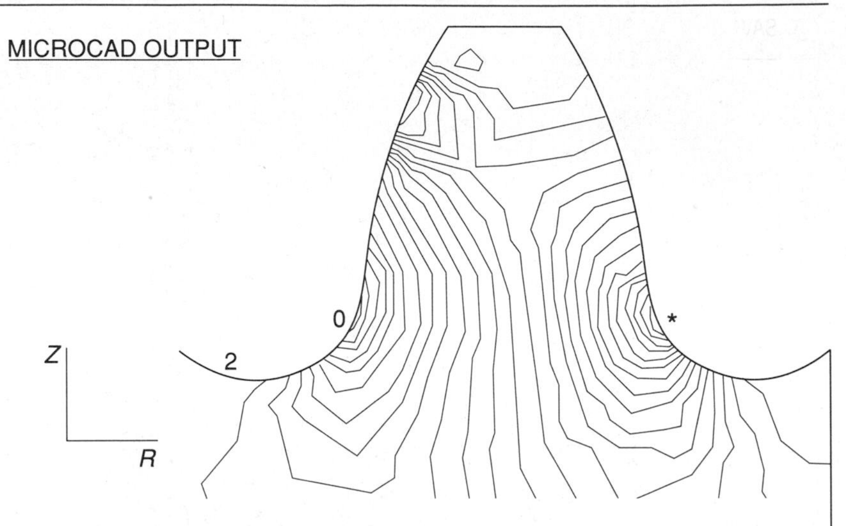

FINITE ELEMENT ANALYSIS OF A GEAR TOOTH
Z Stress Component Contour Plot (1500.00 psi. increments)
Maximum Stress (0) = .17323E+05 psi. ; Minimum Stress (*) = –.24886E+05 psi.
Upper Right Corner with Zoom Facter of 1.60
Stress Colour Code: Negative : Positive : Over Allowable

the lines by moving the crosshairs (located over the END function in Figure 10.16) to appropriate positions. Existing lines can be removed in a similar manner.

Linear dimensioning between any pair of points can be carried out by the DIMEN function. The terminal points are identified by positioning the crosshair over them. The distance between these two points is automatically computed by the program and shown where the user desires. The radial dimensions can be shown by following the same procedure except that a pair of points must be specified. In this case, these points are the location of the center of the curve and a point on the curve. Again, distances are automatically computed by the program. The LABEL function can be used to label various parts of the structural components and to specify the title of the drawing at the desired locations. The LABEL function provides most capabilities that common word-processing software does.

The reader will find that the dimensions of the gear tooth created by the MICROCAD system (Figure 10.16) are inconsistent with the actual dimensions shown in Figure 10.8. Corresponding dimensions in these two figures are close, but not identical. These discrepancies are due to the fact that the dimensions given in Figure 10.16 are produced by the program, which depends on how accurately the user can position the crosshair over the desired terminal points. Consequently, minor errors can be introduced by relatively unskilled operators.

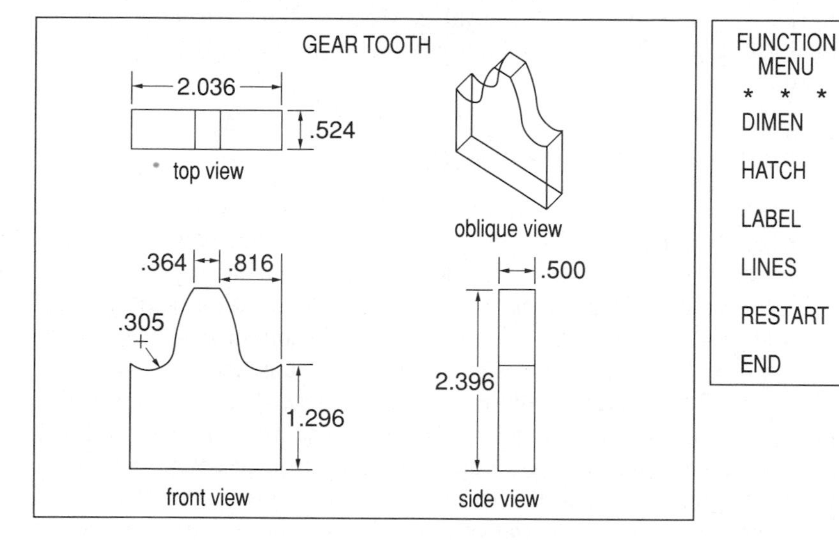

FIGURE 10.16 Computer-aided drafting produced by the MICROCAD system

MICROCAD also offers partial shading of drawings. Plate 17 illustrates the enhanced image filling of a larger gear tooth.

10.6 ▪ Design Case Study Using the ANSYS Program

The ANSYS code described in Section 7.7 was used to optimize the design of a rocker arm for an overhead valve engine manufactured by the Onan Corporation, Minneapolis, Minnesota. As illustrated in Figure 10.17, the

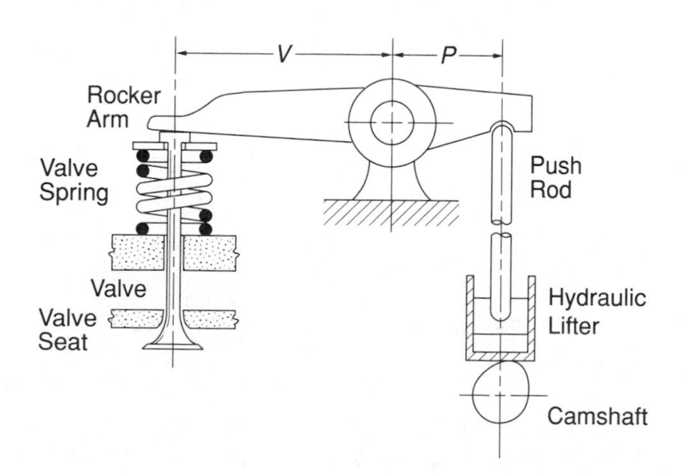

FIGURE 10.17 Schematic of a valve train

rocker arm rotates, transmitting the push rod's axial motion to the valve and thus opening and closing the valve. Much of the information for this case illustration is based on an internal company report [10].

❑ DESCRIPTION OF THE PROBLEM

"A valve train designer must strike a balance between engine performance and valve train life. To maximize engine air flow, hence power output, it is desirable to have instantaneous valve openings and closings" [11]. Due to reduced valve train life, these rapid valve events will not be acceptable if they strongly excite valve train resonances. Resonant behavior can be minimized by increasing the fundamental valve train resonant frequency.

Resonant frequencies are the natural frequencies of a structure, which can be determined by modal analysis involving the structure's stiffness and mass characteristics. In this case study, the structure is the entire valve train. The modal analysis of the valve train has been presented in reference [11]. The stiffness and mass properties of the rocker arm have a strong influence on the natural frequencies of the valve train. The design case is thus related to the optimization of the rocker arm geometry that will maximize the natural frequencies of the valve train. This is obtained by maximizing the stiffness-to-rotary-inertia ratio of the rocker arm.

An initial rocker arm design was obtained as a result of the conceptual design process. The geometry of the rocker arm is illustrated in Figure 10.18. Due to geometric constraints, the center plane of the arm was skewed from the rotational axis. Two V/P ratios of 1.51 and 1.65 were analyzed, with the parameters V and P designated in Figure 10.17. Both cast aluminum and steel were considered for the rocker arm.

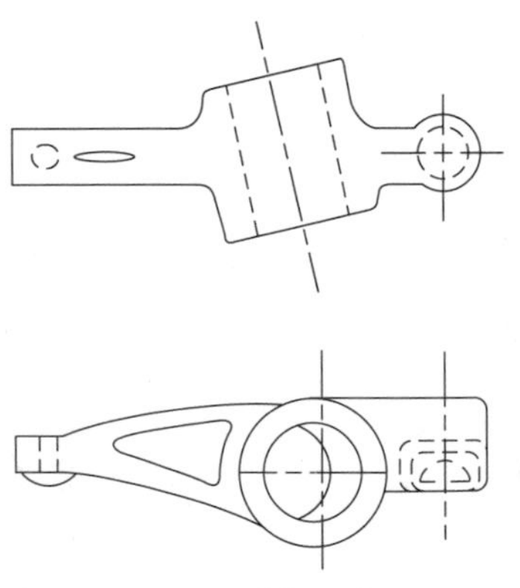

FIGURE 10.18
Schematic of rocker arm

Source: Courtesy of Onan Corp.

❑ FINITE ELEMENT ANALYSIS

The ANSYS program was used to generate a solid model of the initial rocker arm. To simplify the analysis, the skewness was neglected in the finite element analysis. This simplification resulted in a plane of symmetry through the center of the arm. Thirty-two parameters were used to describe the rocker arm geometry. The line segments, areas, and volumes used in a variation of the initial rocker arm are illustrated in Plate 11. The finite element mesh used for the rocker arm consisted of 625 brick- and wedge-shaped three-dimensional isoparametric elements. The mesh used for the initial rocker arm is illustrated in Plate 12.

Figure 10.19 shows the applied loading and boundary conditions used in the analysis. A 100-pound vertical force was distributed to four nodes on the valve side. Vertical translation of four nodes on the push-rod side was assumed to represent the push-rod reaction, with the radial reaction on the hub assumed to be distributed over a certain portion of the hub.

The superimposed deformed and undeformed shapes of the initial rockker arm are shown in Figure 10.20. The arm stiffness was computed within the ANSYS postprocessing routine from the applied loading and deformed geometry. Mass properties of the arm are determined from the material density and the solid model volume blocks. The arm stiffness and inertia

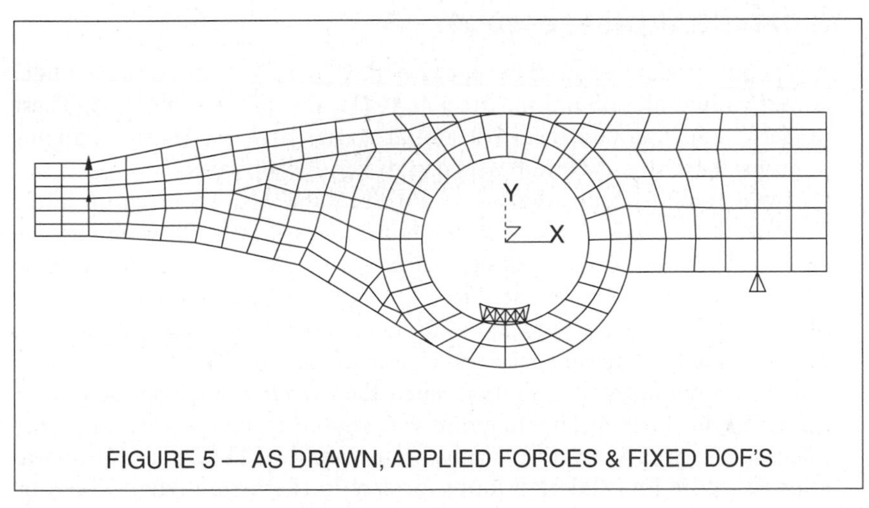

FIGURE 10.19 Force and boundary conditions in the finite element analysis
Source: Courtesy of Onan Corp.

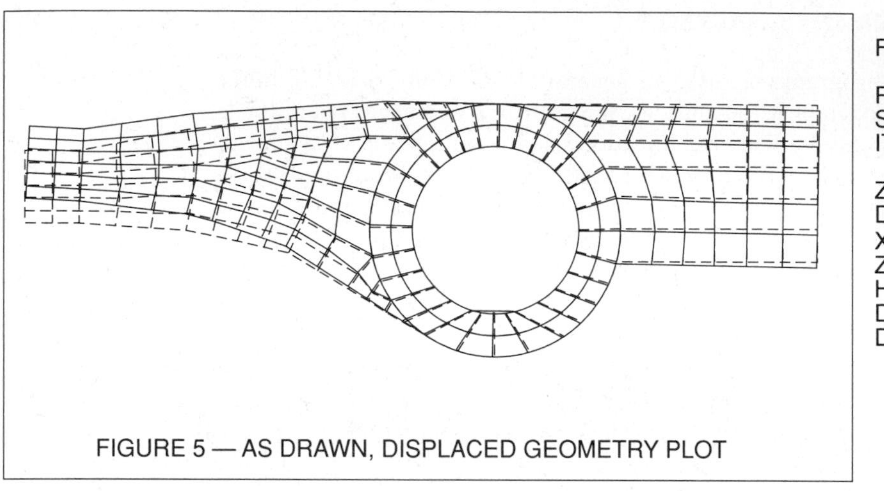

FIGURE 10.20 Superimposed deformed and original shapes of a rocker arm
Source: Courtesy of Onan Corp.

were used with the other valve train component stiffnesses and mass properties to compute the valve train natural frequency.

❑ OPTIMIZATION PROCEDURE

A separate model was used for each of the four rocker arm designs under consideration: aluminum and steel at V/P ratios of 1.51 and 1.65. These models were obtained from the initial rocker arm model by changing appropriate geometry parameters and material properties.

The design attributes that are changed by the ANSYS program in the optimization process are called *design variables*. Of the 29 geometry parameters, 13 were used as design variables in the optimization process. These 13 variables are indicated in Figure 10.21. Upper and lower bounds of these variables were specified as design constraints. These bounds were required due to physical space and clearance requirements.

ANSYS minimizes a quantity called the *objective function* during the optimization process. This function was chosen to be the natural period (reciprocal of frequency) of the valve train assembly. The optimization was allowed to run for 50 design loops for each rocker arm design. The total CPU time required to optimize each design was approximately nine hours on a DEC VAX Model 3600.

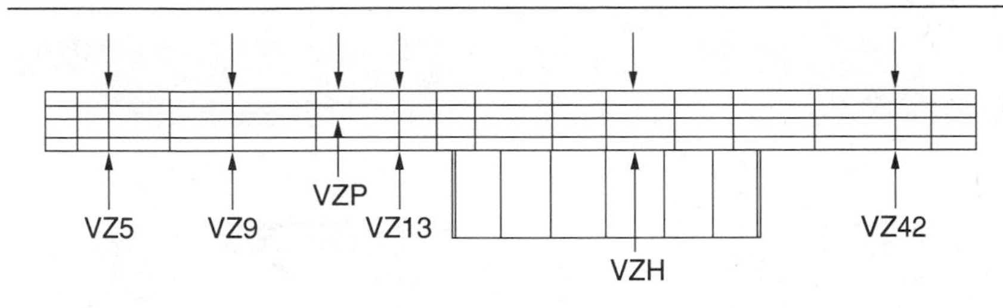

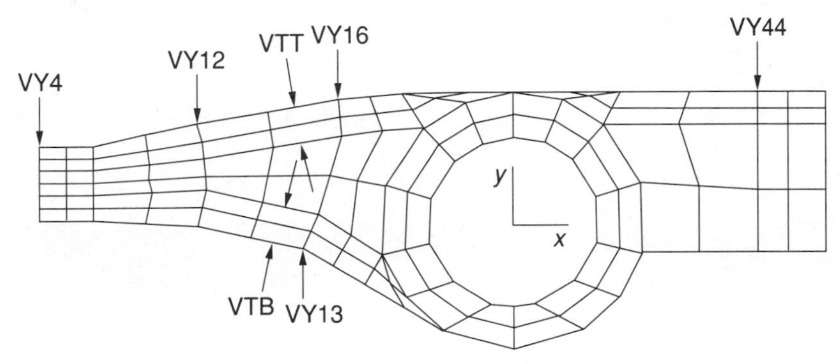

FIGURE 10.21 Design variables in an optimization process

Source: Courtesy of Onan Corp.

❑ OPTIMIZATION RESULTS

The four optimized rocker arms are shown in Figure 10.22. The plots show the volumes used in the solid model representation of the arms. As a result of design optimization, a number of the design variables reached their upper or lower bound. Note that the aluminum designs maximized the variable widths to a greater extent than did the steel designs.

The fundamental natural frequency of the valve train increased by 13 and 14 percent for the aluminum rocker arms with respective V/P ratios of 1.65 and 1.51. Further improvements in natural frequency were possible with steel arms, with an 18 percent increase at both V/P ratios.

The Von Mises equivalent stresses for the initial rocker arm and the aluminum arm with a 1.51 V/P ratios are expressed graphically in Plate 13. Note that different color scales were used for the two contour displays.

❑ INTERPRETATION OF RESULTS

The results obtained from the design analysis were considered satisfactory for the following reasons:

1. The optimized rocker arm geometries have increased the valve train's natural frequency. With an increased natural frequency, it

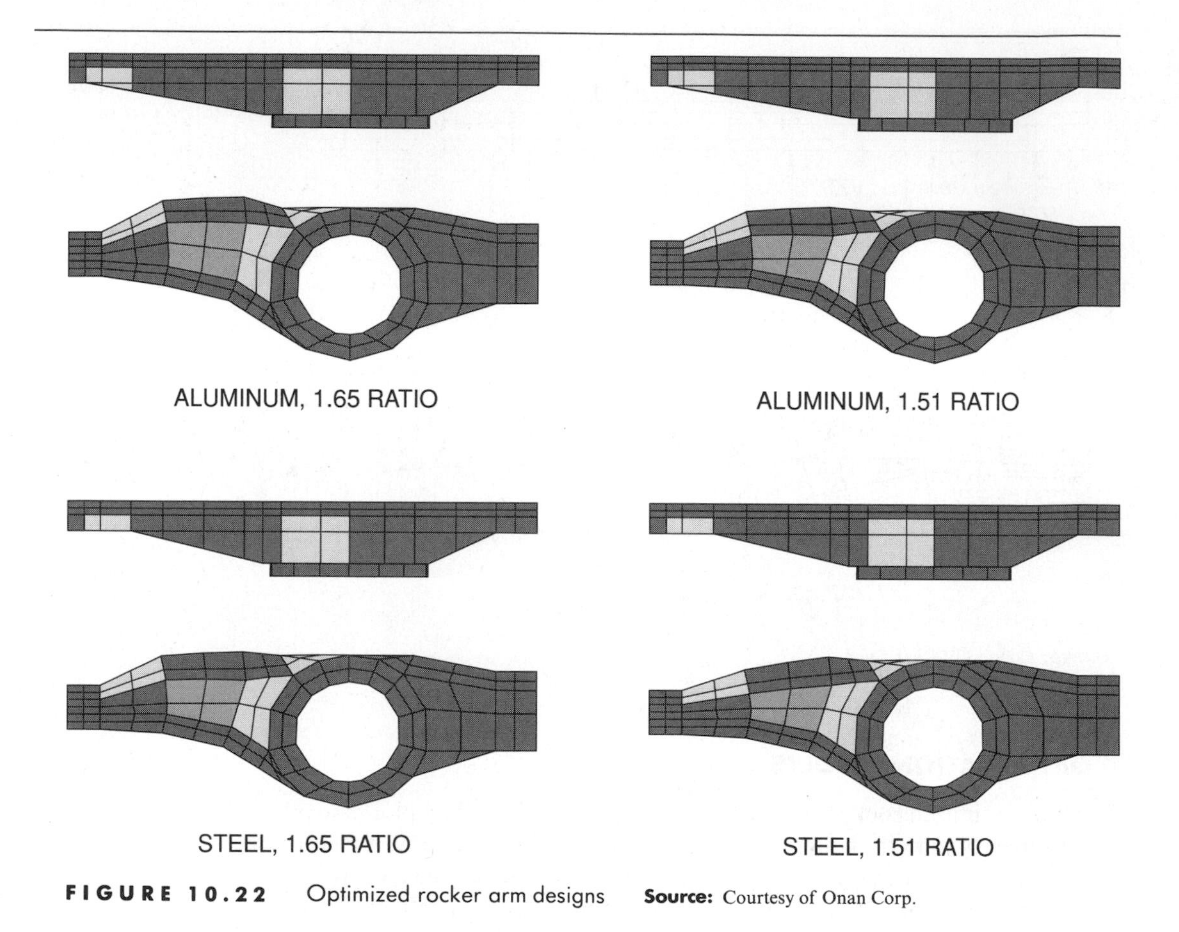

FIGURE 10.22 Optimized rocker arm designs **Source:** Courtesy of Onan Corp.

is expected that the dynamic loading will decrease, thus increasing the valve train life.

2. The maximum equivalent stresses in all the optimized geometries were below the fatigue strengths for both aluminum and steel designs. No structural fatigue problems are expected with the optimized rocker arm geometries.

10.7 ▪ Expert Systems for CAD

Many features unique to CAD, as opposed to the traditional design approach, have been presented in Chapter 1. In addition to the obvious improvement in the quality of the design, another principal advantage of

CAD is the increased productivity of the designer [12]. CAD can help the designer to conceptualize the product and thus reduce the time the designer needs to synthesize, analyze, and document the design. What modern CAD systems can accomplish today is well beyond what the designers could ever imagine just a decade ago.

As computer hardware and CAD software continue to advance at an astounding pace, designers are expected to handle more and more complex problems involving huge numbers of design variables and constraints. Designers will require a broad range of special knowledge in order to handle these problems. Expert systems that are developed to assist designers in synthesizing these design variables and to offer best optimum solutions to problems have thus become the most promising and desirable enhancements to the already powerful CAD systems.

Expert systems can be viewed as a subset of artificial intelligence. They are constructed to deal with specific problems—much the way human experts do. An expert system established for gear design can only handle problems associated with gear design. Likewise, an expert system for hydraulic transmission systems can only be expected to handle the problems related to the hydraulic transmission. Revenues from the sale of expert systems have increased dramatically from $35 million in 1986 to $275 million in 1989. Forecasts indicate this revenue will double in 1990 and rise to a staggering $900 million in 1991.

An expert system consists of four major components, as described in reference [13]: (1) acquisition module, (2) knowledge base, (3) inference engine, and (4) explanatory interface. The relationships of these four components are illustrated in Figure 10.23.

The *knowledge base* in an expert system stores information on the subject domain, which contains all experts' rules of judgment in a way that allows the inference engine to deduce from it. Most knowledge bases are built on rules in the IF–THEN format. This format specifies conditions and offers the optimum solutions corresponding most closely to these conditions.

An *inference engine* consists of search and reasoning procedures that enable an expert system to offer optimum solutions with justifications. In a

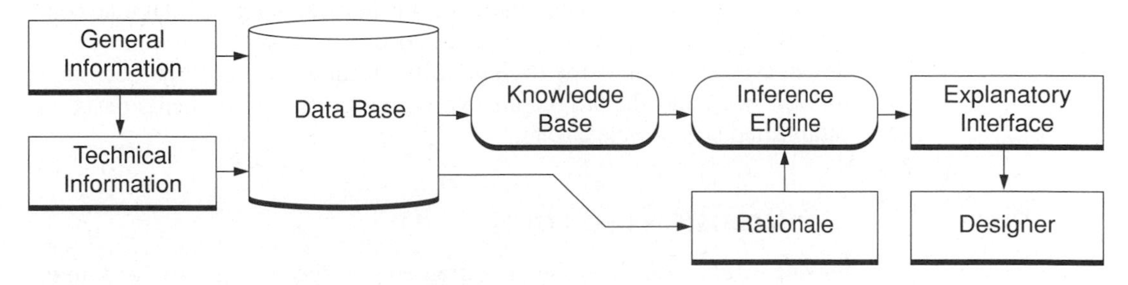

FIGURE 10.23 Expert systems for CAD

rule-based system, it involves matching the IF conditions to the facts and thus coming up with a conclusion that can lead to a recommendation or diagnosis of the problem.

Included in the rules for a design knowledge base should be the following major items.

❑ DESIGN SPECIFICATION

A design specification essentially is a list of requirements for the specific design endeavor. It thus constitutes the rules that the designer must follow. According to Pahl and Beitz [14], a design specification should clearly specify the requirements in the following areas:

- Geometry and dimensions of the product or parts
- Kinematics in the types of motion or other dynamic aspects
- Forces and loads applied to the product or parts
- Energy that relates to the output efficiency, required input energy, pressure, temperature, heating or cooling, etc.
- Material, both types and requirements
- Operational and environmental safety
- Human comfort and human-machine compatibility
- Production within factory limitations, preferred production methods, requirements in quality and tolerances
- Quality assurance, as required
- Operations requirements for noise level, wear, life expectancy, etc.
- Maximum permissible cost to produce the product or parts
- Schedules in production and delivery

One may readily translate the above requirements into the IF part of the IF–THEN rules in a knowledge base. For an example, one rule may well be this: IF the part is a rectangular block, THEN the length is ____, the width is ____, and the height is ____.

In some simpler cases involving the mechanical design of machine components, simple rules can be established on the basis of commonly used design criteria, such as the design limits on the maximum tolerances and interferences, and the fatigue life expectancy of the components based on established stress-cycle curves.

❑ CONCLUSIONS TO RULES

Having established the design requirements as specified above, the knowledge base also requires input regarding possible conclusions corresponding to specific requirements. Following is an example of how an expert system can contribute to a small portion in the process of designing a beam structure. Some of the rules with corresponding conclusions may be stated as

Rule 1: IF (material is ____), THEN (allowable stress is ____).

Rule 2: IF (the section modulus of the beam is ____ and the maximum bending moment is ____), THEN (the maximum bending stress is ____).

Rule 3: IF the maximum bending stress is greater than or equal to the allowable stress, THEN the design is unsatisfactory.

Rule 1 relates to the material specification. Information required to complete this rule can be found from the design specification, which should be available from the design database. The completion of Rule 2 requires computations of the section modulus, Z, based on the cross-section of the beam, and the maximum bending moment, M, in the beam subject to the loading and end conditions. The maximum bending stress, of course, can be determined by the relation of M/Z, which is used to reach the conclusion as stated in Rule 3.

The above illustration of the use of the Expert system demonstrates that the construction of a knowledge-based expert system requires input from existing information stored in the database as well as from analytic means. It is also common to use empirical rules in establishing rules in the knowledge base. For example, a common practice in pressure vessel design is to allow extra material in the structure when those parts of the vessel are expected to be in contact with corrosive contents. Rules such as that presented below may be present in an expert system for pressure vessel design:

Rule: IF (the part is in contact with corrosive material), THEN (a corrosion allowance of ____ should be added to the *diameter of the shell*, or *the thickness of flat faces*).

An *inference engine* in an expert system is used to derive plausible reasoning for whatever recommendation the expert system offers to the designer. The rationale for such recommendations may be established on the basis of optimization procedures, as presented in Chapter 8 and also as illustrated in the design case study in the foregoing section. A value assessment scheme for various design output, proposed by Pahl and Beitz [14], may be used as a basis of such recommendations. This scheme assigns merit points on a scale of 0 to 10 for various conclusions drawn from a design process. A typical scheme is presented in Table 10.2.

Obviously, the use of the above scheme will require that the conclusions in Table 10.2 be included in the rules in the knowledge base. The inference engine would synthesize the merit points associated with various design schemes and options and offer its recommendations based on the total points scored.

TABLE 10.2 Value scale for a design process

Conclusion	Points	Conclusion	Points
Absolutely useless	0	Good with few drawbacks	6
Very inadequate	1	Good	7
Weak	2	Very Good	8
Tolerable	3	Exceeds requirement	9
Adequate	4	Ideal	10
Satisfactory	5		

The recommendations offered by a design expert system will have to be interpreted and evaluated by the designer. This endeavor and the implementation of these recommendations by the expert system constitute the major task of the *explanatory interface* component in Figure 10.23.

10.8 ▪ CAD/CAM Interface

The linkage between mechanical design and manufacturing through a common database has been illustrated in Figure 1.1. We have also demonstrated in Chapter 5 that many of the design data are also necessary for the manufacturing and production of the product. The interface of CAD and CAM databases has been described in Section 5.9 and Figure 5.10.

In this section, we will take a look at the fabrication aspect of the CAD/CAM interface. We will show how the geometric models—in particular, the solid models—produced by a CAD process can provide sufficient surface information to drive the computer numerically controlled (CNC) machine tools (e.g., milling machines, lathes, and drill presses) that make the part.

According to Groover [15], a numerical control system consists of the following three basic components: (1) the program of instructions, (2) the machine control unit (MCU), and (3) the process equipment (PE). The program of instructions gives detailed commands to the PE, with the MCU serving as the translator that converts the program commands into appropriate signals that prompt the electromechanical movements of the cutting tools.

Figure 10.24 illustrates a three-axis milling machine in which the motion of the working table on the *xy* plane is controlled by the stepping motors M_x and M_y, whereas the motion of the tool on the *z* axis is controlled by the third stepping motor, M_z. This device may be regarded as the process equipment in a CNC system. The motion of the stepping motors is

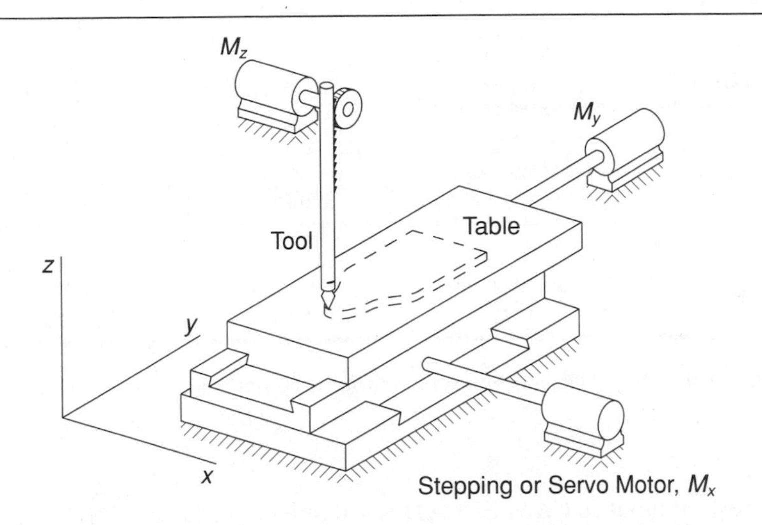

FIGURE 10.24
Schematic of a three-axis milling machine

triggered by a pulse train in an open control-loop. Each pulse drives the motor by a fraction of one revolution. Alternately, one can replace these stepping motors with servo motors for closed control-loop CNC operations. In this case, the motion of the DC servo motors will be controlled by encoded signals received from position sensors attached to the machine tool table. Whatever the signals that prompt the motions of the control motors, they are correlated to the coordinates of the tool that cuts the material. Apparently, the ideal situation is to have the coordinates of the tool match the exact profile of the part to be cut. For example, the profile of a gear tooth in Figure 10.24 becomes the desirable tool path. As profiles of two-dimensional objects can be created by CAD and stored in a design database, as described in Chapter 5, the coordinates of any point on a two-dimensional profile can be defined by the curve fitting technique illustrated in Chapter 3. The tool path for cutting the gear tooth by a CNC milling machine, as shown in Figure 10.24, can thus be determined by the geometry of the gear tooth produced in the CAD process.

Tool paths for cutting parts of more complex geometries can be determined by following principles similar to those used for the two-dimensional cases. Obviously, the coordinates of the tool along the tool path will have to be interpolated from the surface topography of the object, as illustrated in Figure 10.25. There, the solid model produced by CAD provides the necessary surface information for the determination of the coordinates of the points on the surface. These coordinates are used to generate the CNC codes or automatically programmed tooling (APT) statements that determine the tool path for the CNC operation.

We have thus far demonstrated how the profile and surface information of a solid created from CAD can be used to determine the tool path. Another

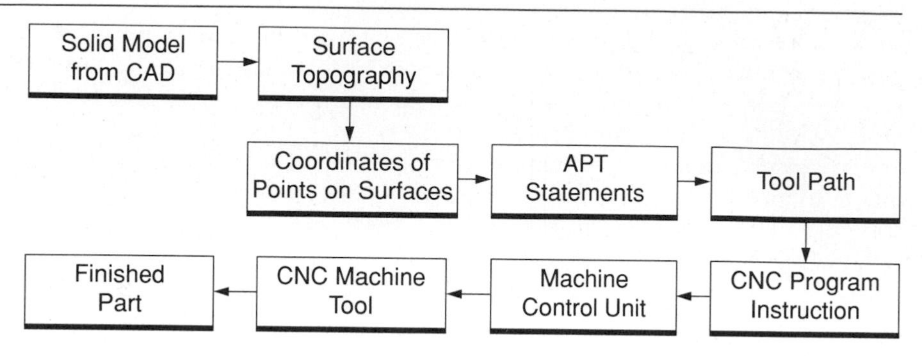

FIGURE 10.25 CNC machine tool path for solid parts generated by CAD

important input to CAM by CAD is the rate of cutting by CNC machine tools. There is obviously an optimum cutting rate for any CNC operation; for example, from the cost-saving point of view, one would desire a fast rate. However, a fast cutting process will result in shorter tool life and also unsatisfactory surface conditions. Breakage of either the workpiece or the tool can be expected if the feeding rate and tool speed become excessively high. Optimum cutting rates for various materials using CNC machine tools can be ideally determined by a knowledge-based, or even an expert, system, as described in the foregoing section. One plausible approach to developing such a system is to use the finite element method to assess the structural integrity of both the workpiece and the tool under various contacting forces between these two solids. The results of such an analysis may very well be used to establish some rules in the knowledge base, which is capable of recommending the optimum cutting speed and feed rate for certain specific CNC operations to the machine operators.

10.9 ▪ Summary

An overview of integrated CAD systems was presented in this chapter. An integrated CAD system should have the capability of performing four major functions of geometric modeling; design analysis; drafting; and data storage and transfer. Commands for performing these functions by most commercial CAD systems were also described. Traditionally, integrated CAD required expensive mainframe computers and peripherals. However, this situation has changed in recent years, and now a number of these systems can be performed by microcomputers. In spite of their limited capabilities, microcomputer-based CAD systems are suitable for small and medium sized industry because of their moderate maintenance and acquisition costs. Design cases using MICROCAD, a microcomputer-based integrated CAD system, and the ANSYS program have been presented. The latter case relates to the optimum modal analysis of rocker arms for an overhead valve engine.

Two special topics on "Expert systems for CAD" and "CAD/CAM interface" have been presented. These two subjects are intended to provide the reader with the directions of immediate extension of the current CAD technology. As indicated in Section 10.7, the strong demand for expert systems by the industry has been demonstrated by the rapid expansion of market demands for such systems in recent years. Such systems in CAD will further enhance engineers' capability in dealing with much more complex problems.

The importance of integrating CAD and CAM has been emphasized in Chapters 1, 5 and 10. The CAD/CAM interface presented in this chapter intends to offer a fundamental concept of how geometry models created in a CAD practice may be transmitted to a CNC machine for the production of an object. Details of the linkages between CAD and CAM can be found in special books and monographs.

References

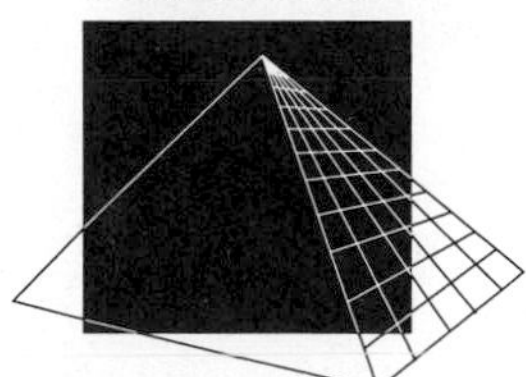

1. Allan, J.J., III, ed. *A Survey of CAD CAM Systems*. 3d ed. Dallas: Productivity International, 1983.
2. O'Connell, C., J.K. Sheriak, A. Browne and S.V. Johnson, eds., *Directory of Microcomputer Software for Mechanical Engineering Design*. 1985 ed. New York: Marcel Dekker, 1985.
3. Gile, J.R. "Computer-Aided Design." *Infoworld*, July 17, 1989, pp. 62–72.
4. Krouse, J.K., "CAD/CAM—Bridging the Gap from Design to Production." *Machine Design*, June 12, 1980, pp. 117–25.
5. Gardan, Y., and M. Lucas. *Interactive Graphics in CAD*. New York: UNIPUB, 1984. Chapter 3.
6. Lichten, L. "Computer-Aided Design Applications on Microcomputer." *IEEE Computer Graphics and Animation* Vol. 4, No. 10 (October 1984):25–28.
7. Krouse, J.K. "Geometric Models for CAD/CAM." *Machine Design*, July 24, 1980, pp. 99–106.
8. Myers, R.E. *Microcomputer Graphics*. Reading, Mass.: Addison-Wesley, 1982.
9. Hsu, T.R., and D.K. Sinha. *Finite Element Analysis by Microcomputers*. Duxbury, Mass.: Kern International, 1988.
10. Gilberg, M. *A Series Rocker Arm Finite Element Analysis, Shaft*

Design. Rep. no. 64 62-89-0001. Minneapolis, Minn.: Onan Corp., 1989.
11. Seidlitz, S. "Valve Train Dynamics—A Computer Study." SAE paper 89 0620. 1989.
12. Groover, M.P., and E.W. Zimmers, Jr. *CAD/CAM: Computer-Aided Design and Manufacturing*. Englewood Cliffs, N.J.: Prentice-Hall, 1984.
13. Richard Forsyth, ed. *Expert Systems, Principles and Case Studies.* 2d ed. London: Chapman and Hall Computing, 1989.
14. Pahl, G., and W. Beitz. "Engineering Design, A Systematic Approach." edited by K. Wallace. Berlin: Springer-Verlag, 1988.
15. Groover, M.P. *Automation, Production Systems, and Computer Integrated Manufacturing*. Englewood Cliffs, N.J.: Prentice-Hall, 1980.

Problems

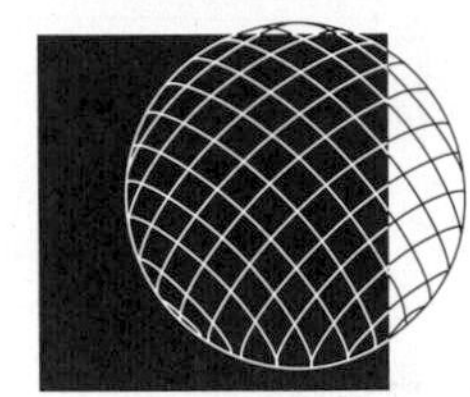

10.1 Use the commands available in a major commercial CAD system to produce the profile of a gear with six teeth.

10.2 How would you modify the flowchart of the MICROCAD program in Figure 10.7 to incorporate the design optimization theories.

10.3 Sketch a cube viewed with two different focal distances as defined in Figure 10.5 at a ratio of 10 to 1.

10.4 How would the above cube look with a negative focal distance?

10.5 Prepare the instructions to input the profile of a handle with a geometry shown below to MICROCAD through a keyboard? (*Hint:* Follow the example in Section 10.5)

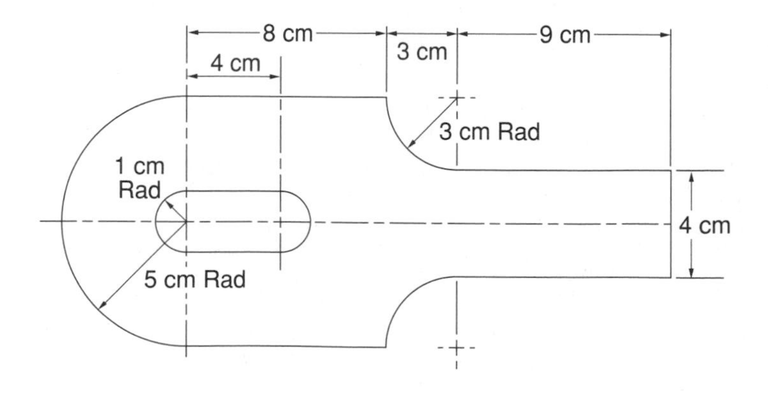

10.6 Prepare IF–THEN statements for problem 1.11 on the design analysis of a simple beam structure. Present your statements in a logical sequence.

10.7 How would you incorporate the IF–THEN statements for problem 10.6 into a knowledge-based design analysis of beam structures?

10.8 As described in Section 10.8, a CNC milling machine works by moving the workpiece attached to the machine table against the cutting tool. The movement of the table in the x, y and z directions is controlled by servo motors, as illustrated in Figure 10.24. Assume that each of the servo motors takes 200 steps to complete a revolution, and that each of the steps produces a 0.5 mm linear motion of the table in the respective directions. (One should realize that the table can move in both directions with counterclockwise rotation of the motor shaft for advancing and clockwise rotation for retracting). Find the percentage of errors in the cuts (shown below) by the machine for

(a) A straight edge.
(b) A curved edge.
(c) A circular edge.

(*Hint:* You will first have to set the origin of the coordinates x and y at a convenient location on the table.)

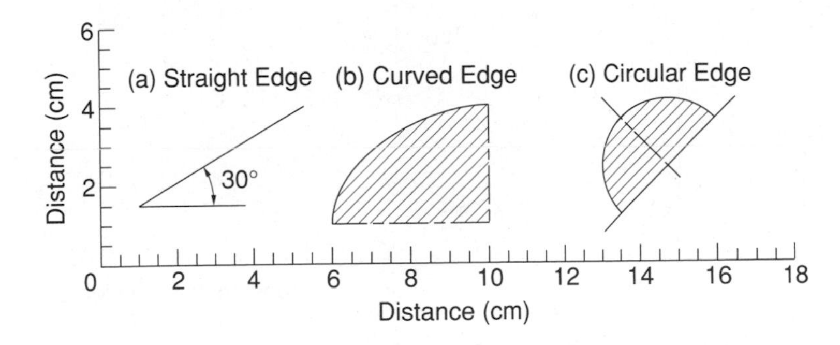

10.9 The cutting instructions for a CNC machine are normally written in the C language [15]. The principle in preparing such instructions is to trigger the servo motors to turn the shafts in the desirable directions and in steps. The controlled rotations of the shafts can drive the workpiece along the desirable paths, as described in Section 10.8. Prepare a list of such a sequence of events by filling in a table like the following for the production of a metal plate of the shape illustrated below. The designations of the motors are also illustrated below. Use the configurations of the servo motors as described in problem 10.8, and assume that the width of the cutting tool is the same as the thickness of the plate.

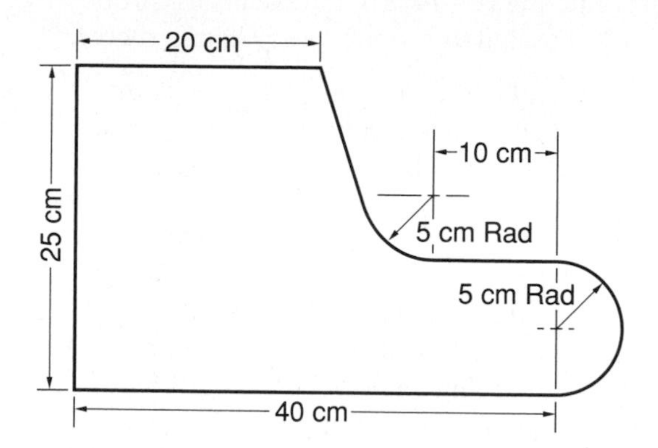

Allow 5 mm Material to be Removed

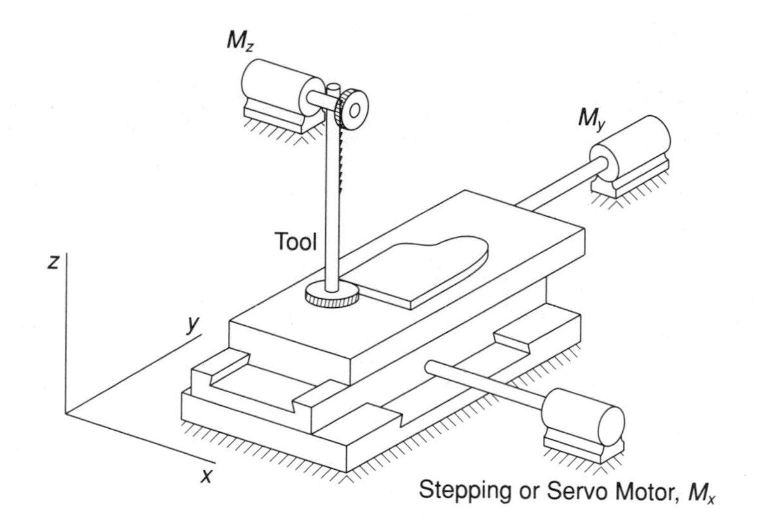

Instructions to Servo Motors

Steps Required	No. of Revolutions	Motor M_x	Motor M_y	Both M_x and M_y

APPENDIX I

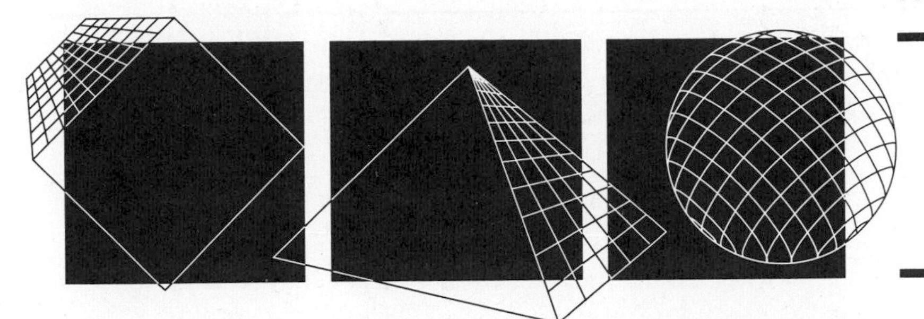

Review of Matrix Operations

AI.1 ▪ Definitions

If the rows and columns of a matrix $[A]$ are interchanged, the new matrix resulting from this operation is known as the *transpose*, here $[A]^T$, of the first matrix. Thus, if a matrix has the following form:

$$[A] = \begin{bmatrix} a_{11} & a_{12} & \cdots & a_{1n} \\ \vdots & \vdots & & \vdots \\ a_{m1} & a_{m2} & \cdots & a_{mn} \end{bmatrix}$$

then its transposed form becomes

$$[A]^T = \begin{bmatrix} a_{11} & a_{21} & \cdots & a_{m1} \\ \vdots & \vdots & & \vdots \\ a_{1n} & a_{2n} & \cdots & a_{mn} \end{bmatrix} \tag{AI.1}$$

A square matrix is known as a *symmetric matrix* if the elements on either side of the diagonal are identical. For instance, the matrix $[A]$ is said to be a symmetric matrix if it takes the following form:

$$[A] = \begin{bmatrix} 2 & 3 & 4 & 5 \\ 3 & 1 & 2 & 8 \\ 4 & 2 & 6 & -1 \\ 5 & 8 & -1 & 2 \end{bmatrix} \tag{AI.2}$$

A symmetric matrix is equal to its transpose (i.e., $[A] = [A]^T$).

In the finite element and other methods, one often comes across matrices that contain several zero elements. These matrices are known as *sparse matrices*. For example,

$$[A] = \begin{bmatrix} 1 & 3 & 0 & 0 & 0 \\ 3 & 2 & 1 & 0 & 0 \\ 0 & 1 & 1 & 2 & 0 \\ 0 & 0 & 2 & 2 & -1 \\ 0 & 0 & 0 & -1 & 1 \end{bmatrix} \tag{AI.3}$$

where $[A]$ is also a *banded matrix* because all the nonzero elements are clustered around the main diagonal. The bandwidth of the above matrix is three, as there are three elements in each row except the first and last rows.

If each element in the columns of a square matrix is equal in magnitude and opposite in sign to the corresponding number in its rows, or vice versa, such a matrix is known as a *skew-symmetric matrix*. For example,

$$a_{12} = -a_{21}, a_{13} = -a_{31}, \ldots$$

$$[A] = -[A]^T$$

All elements on the main diagonal of a skew-symmetric matrix are zero (i.e., $a_{11} = a_{22} = \cdots = a_{nn} = 0$).

It is often convenient to express a symmetric matrix as a product of two simpler matrices, known as *diagonal* and *triangular matrices*, such as $[A] = [L][D][L]^T$. The following are examples of diagonal matrices:

$$[D] = \begin{bmatrix} 2 & 0 & 0 & \cdots & 0 & 0 \\ 0 & -8 & 0 & \cdots & 0 & 0 \\ \vdots & \vdots & \vdots & & \vdots & \vdots \\ 0 & 0 & 0 & \cdots & 5 & 0 \\ 0 & 0 & 0 & \cdots & 0 & 3 \end{bmatrix} \tag{AI.4}$$

$$[S] = \begin{bmatrix} 2 & 0 & 0 & \cdots & 0 & 0 \\ 0 & 2 & 0 & \cdots & 0 & 0 \\ \vdots & \vdots & \vdots & & \vdots & \vdots \\ 0 & 0 & 0 & \cdots & 2 & 0 \\ 0 & 0 & 0 & \cdots & 0 & 2 \end{bmatrix} \tag{AI.5}$$

$$[I] = \begin{bmatrix} 1 & 0 & 0 & \cdots & 0 \\ 0 & 1 & 0 & \cdots & 0 \\ 0 & 0 & 1 & \cdots & 0 \\ \vdots & \vdots & \vdots & & \vdots \\ 0 & 0 & 0 & \cdots & 1 \end{bmatrix} \tag{AI.6}$$

The $[S]$ matrix, which has identical diagonal elements, is known as a *scalar matrix*, whereas $[I]$, in which all the diagonal elements are unity, is called the *unit* or *identity matrix*.

A *triangular matrix* is one in which all nonzero elements are on either the upper or the lower side of the main diagonal. Thus,

$$[L] = \begin{bmatrix} 1 & 0 & 0 \\ -1 & 1 & 0 \\ 0 & 1 & -2 \end{bmatrix} \tag{AI.7}$$

is a lower triangular matrix, and

$$[U] = \begin{bmatrix} 1 & -2 & 0 \\ 0 & 1 & 1 \\ 0 & 0 & 2 \end{bmatrix} \tag{AI.8}$$

is an upper triangular matrix.

A *null matrix* is one in which all individual elements are zeros; that is,

$$[N] = \begin{bmatrix} 0 & 0 & 0 & 0 & 0 \\ 0 & 0 & 0 & 0 & 0 \\ 0 & 0 & 0 & 0 & 0 \\ 0 & 0 & 0 & 0 & 0 \end{bmatrix} \tag{AI.9}$$

AI.2 ▪ Matrix Algebra

❑ ADDITION AND SUBTRACTION

Addition and subtraction of two matrices is possible only if all the participating matrices have the same number of rows and columns. The sum or the difference of the matrices $[A]$, $[B]$, $[C]$, ... is given as follows:

$$[A] \pm [B] \pm [C] \pm \cdots$$

$$= \begin{bmatrix} a_{11} \pm b_{11} \pm c_{11} \pm \cdots & a_{12} \pm b_{12} \pm c_{12} \pm \cdots & a_{1n} \pm b_{1n} \pm c_{1n} \pm \cdots \\ a_{21} \pm b_{21} \pm c_{21} \pm \cdots & a_{22} \pm b_{22} \pm c_{22} \pm \cdots & a_{2n} \pm b_{2n} \pm c_{2n} \pm \cdots \\ \vdots & \vdots & \vdots \\ a_{m1} \pm b_{m1} \pm c_{m1} \pm \cdots & a_{m2} \pm b_{m2} \pm c_{m2} \pm \cdots & a_{mn} \pm b_{mn} \pm c_{mn} \pm \cdots \end{bmatrix} \tag{AI.10}$$

From the above definition, it should be easy to derive the commutative and associative laws, which can be expressed as

$$[A] + [B] = [B] + [A]$$
$$[A] + ([B] + [C]) = ([A] + [B]) + [C] \tag{AI.11}$$

❑ MULTIPLICATION BY A SCALAR

The multiplication of a matrix by a scalar, such as α, is achieved by multiplying all the individual elements by α; that is,

$$\alpha[A] = \begin{bmatrix} \alpha a_{11} & \alpha a_{12} & \cdots & \alpha a_{1n} \\ \alpha a_{21} & \alpha a_{22} & \cdots & \alpha a_{2n} \\ \vdots & \vdots & & \vdots \\ \alpha a_{m1} & \alpha a_{m2} & \cdots & \alpha a_{mn} \end{bmatrix} \tag{AI.12}$$

❑ MATRIX MULTIPLICATION

Let us consider multiplication of two matrices, $[A]$ and $[B]$, resulting in a third matrix, $[C]$; that is,

$$[A][B] = [C] \tag{AI.13}$$

In the above, $[A]$ is referred to as the *premultiplier* and $[B]$ the *postmultiplier*. Beginners can best carry out the operation of multiplication by drawing a diagram, as will be explained in the following paragraph.

Place the premultiplier matrix $[A]$ in the top left-hand corner of the diagram and the postmultiplier matrix $[B]$ in the bottom right-hand corner of the diagram. Draw horizontal lines through each row of $[A]$ and vertical lines through each column of $[B]$. The points of intersection of these lines represent the positions of the elements of the matrix $[C]$ in the upper right-hand corner of the diagram. Any element of $[C]$ may now be regarded as a product of the row of $[A]$ and the column of $[B]$ meeting at that particular position. This product is obtained by summing the individual products of each element of the row and the corresponding element of the column in question. Thus, for example, if we wish to work out the element c_{11}, we should have

$$c_{11} = a_{11}b_{11} + a_{12}b_{21} + \cdots + a_{1n}b_{n1} \tag{AI.14}$$

where $[A]$ is an $m \times n$ matrix and $[B]$ is an $n \times k$ matrix.

From equation (AI.14), it should be quite evident that the number of elements in the rows of $[A]$ must be the same as the number of elements in the columns of [B]. The [C] matrix would be an $m \times k$ matrix. Now, the reader can easily appreciate how the group of equation (2.1) was represented by the product of matrices [A] and $\{x\}$ given in equation (2.3) and equation (2.4) in Chapter 2.

We would like to urge the reader to verify that

$$[A][B] \neq [B][A] \quad \text{(AI.15)}$$

and also that, except in the case of row and column vectors and square matrices, the fact that [A][B] exists does not imply that [B][A] exists, too.

One may also express equation (AI.14) in an indexical form as

$$c_{ij} = a_{ir}b_{rj} \qquad (r = 1 \text{ to } n)$$

where n is the number of columns in the premultiplier or the number of rows in the postmultiplier.

An algorithm written in FORTRAN IV for the matrix multiplication is presented below.

Caution: For variables with asterisk signs (i.e., "*"), the actual *numerical* values have to be in place before using the program.

COMPUTER PROGRAM
For matrix multiplication

```
c  THIS PROGRAM MULTIPLIES TWO MATRICES:
c  MATRIX A(MA,NAMB) AND MATRIX B (NAMB, NB)
c  THE PRODUCT MATRIX IS C (MA,NB)
   DIMENSION A(MA*,NAMB*), B(NAMB*,NB*), C(MA*,NB*)
   READ (5,1) MA,NAMB, NB
 1 FORMAT (3I6)
   READ (5,2)((A(I,K), K=1,NAMB), I=1, MA)
 2 FORMAT (XEXX)*
   READ (5,3)((B(K,J), J=1,NB), K=1,NAMB)
 3 FORMAT (XEXX)*
   DO 10 I=1, MA
   DO 20 J=1, NB
   WJ=0.0
   DO 30 K=1, NAMB
30 WJ=WJ + A(I,K)*B(K,J)
20 W(J) = WJ
   DO10 K=1,NB
10 C(I,K) = W(K)
   WRITE (6,4)((C(I,K), K=1,NB), I=1,MA)
 4 FORMAT (1H1,/,5X,10E15.2,/)
   STOP
```

AI.3 ▪ Determinants

❑ DEFINITIONS

A *determinant* is a single number obtained from arrays of numbers contained in a square matrix when operated on according to a specific scheme.

The numerical value of the determinant $|A|$ of a second-order matrix $[A]$ is given by the following expression:

$$|A| = \begin{vmatrix} a_{11} & a_{12} \\ a_{21} & a_{22} \end{vmatrix} = a_{11}a_{22} - a_{21}a_{12} \tag{AI.16}$$

Proceeding in a similar manner as above, the determinant of a third-order matrix is given by

$$|B| = \begin{vmatrix} b_{11} & b_{12} & b_{13} \\ b_{21} & b_{22} & b_{23} \\ b_{31} & b_{32} & b_{33} \end{vmatrix} = b_{11}\begin{vmatrix} b_{22} & b_{23} \\ b_{32} & b_{33} \end{vmatrix} - b_{21}\begin{vmatrix} b_{12} & b_{13} \\ b_{32} & b_{33} \end{vmatrix} + b_{31}\begin{vmatrix} b_{12} & b_{13} \\ b_{22} & b_{23} \end{vmatrix}$$

or

$$|B| = b_{11}B_{11} - b_{21}B_{21} + b_{31}B_{31}$$

where B_{11}, etc., are known as the minors of the respective elements. The submatrices are obtained by eliminating the row and column of the matrix to which the element in question belongs. Hence, the submatrix $[B_{11}]$ is obtained by dropping the first row and first column of matrix $[B]$. Now, the general expression for the determinant of a matrix of size $n \times n$ can be easily derived as follows:

$$|C| = \begin{vmatrix} c_{11} & c_{12} & \cdots & c_{1n} \\ c_{21} & c_{22} & \cdots & c_{2n} \\ \vdots & \vdots & & \vdots \\ c_{n1} & c_{n2} & \cdots & c_{nn} \end{vmatrix}$$
$$= c_{11}|C_{11}| - c_{21}|C_{21}| + c_{31}|C_{31}| + \cdots + (-1)^{n+1}c_{n1}|C_{n1}| \tag{AI.17}$$

❑ PROPERTIES OF DETERMINANTS

1. The value of a determinant does not change if rows are written as columns and columns as rows as long as they are written in the same order (i.e., $|A| = |A^T|$).

2. The value of a determinant is not changed if to every element of any row or column is added a constant multiple of the corresponding element of any other row or column. For example,

$$\begin{vmatrix} 1 & 1 \\ 1 & 2 \end{vmatrix} = \begin{vmatrix} 2 & 1 \\ 3 & 2 \end{vmatrix} = \begin{vmatrix} 1 & 2 \\ 1 & 3 \end{vmatrix} = \begin{vmatrix} 2 & 3 \\ 1 & 2 \end{vmatrix} = \begin{vmatrix} 1 & 1 \\ 2 & 3 \end{vmatrix}$$

3. If all the elements of a row or a column are multiplied by the same constant, α, the value of the determinant is multipled by α.

4. If all the elements in a row or a column are zero, the value of the determinant is zero.

5. If the corresponding elements of any two rows or columns are proportional, the value of the determinant is zero. For example,

$$\begin{vmatrix} 1 & 2 \\ 3 & 6 \end{vmatrix} = 0 \qquad \begin{vmatrix} 2 & 4 \\ 3 & 6 \end{vmatrix} = 0$$

6. The sign of the determinant changes once corresponding to every single interchange between the columns and the rows. For example,

$$\begin{vmatrix} 1 & 1 \\ 3 & 6 \end{vmatrix} = +3 \qquad \begin{vmatrix} 3 & 6 \\ 1 & 1 \end{vmatrix} = -3 \qquad \begin{vmatrix} 6 & 3 \\ 1 & 1 \end{vmatrix} = +3$$

7. If $[C] = [A][B]$, then $|C| = |A||B|$.

AI.4 ▪ Adjoint and Inverse Matrices

From equation (AI.17), it may be possible to write the cofactor for any element c_{ij} in the square matrix $[C]$ as

$$\text{Cofactor } c_{ij} = (-1)^{i+j}|C_{ij}| \tag{AI.18}$$

where $|C_{ij}|$ is the minor obtained by eliminating the ith row and the jth column from the matrix $[C]$.

If every element of the matrix $[C]$ is replaced by its cofactor and the resulting matrix transposed, the *adjoint matrix* $[C^+]$ is obtained:

$$[C^+] = \begin{bmatrix} C_{11} & C_{21} & C_{31} & \cdots & C_{n1} \\ C_{12} & C_{22} & C_{32} & \cdots & C_{n2} \\ \vdots & \vdots & \vdots & & \vdots \\ C_{1n} & C_{2n} & C_{3n} & \cdots & C_{nn} \end{bmatrix} \tag{AI.19}$$

The *inverse matrix* $[C]^{-1}$ of the square matrix $[C]$ is obtained by dividing the adjoint matrix $[C^+]$ by the determinant $|C|$; that is,

$$[C]^{-1} = \frac{[C^+]}{|C|} \tag{AI.20}$$

It can be shown easily that

$$[C][C]^{-1} = [I] \tag{AI.21}$$

It may be necessary to mention here a few important properties of the inverse matrices:

1. If the determinant $|A| = 0$, the inverse of $[A]$ does not exist; therefore, such a matrix is known as a *singular* matrix.
2. The inverse of the product of matrices is the same as the product of the inverse individual matrices taken in reverse order; that is,

$$([A][B][C])^{-1} = [C]^{-1}[B]^{-1}[A]^{-1} \tag{AI.22}$$

3. The inverse of the inverse matrix is the matrix itself; that is,

$$[A] = ([A]^{-1})^{-1} \tag{AI.23}$$

4. The inverse of the transpose is the transpose of the inverse; that is,

$$([A]^T)^{-1} = ([A]^{-1})^T \tag{AI.24}$$

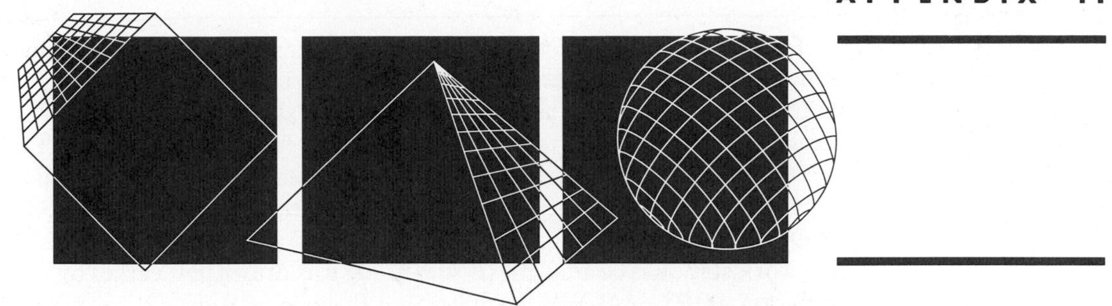

The Graphical Kernal System (GKS)

AII.1 ▪ GKS Work Station

The concept of graphics work stations has been formalized in GKS. A number of functions have been provided with the help of which the user can synthesize a work station that conceptually resembles the one at which he or she is working. A work station has been defined as that which is capable of graphic input and/or output. A work station capable only of input is categorized as an *input work station*. Similarly, there are *output* and *outin work stations* as well. At any work station, only a single output device is envisaged, but there may be multiple input devices. A single computer may be connected to a number of work stations.

A program must identify the work station(s) it is going to use. The function OPEN WORKSTATION (WS,CID,TYPE) allows the user to select work station WS. The parameter CID specifies the channel number that should be used to communicate with that work station. The parameter TYPE specifies the type of station. A rather extensive list of work station types is provided in GKS. By specifying a number for TYPE, one can determine whether the work station is the input/output or outin type. The TYPE number also contains information regarding the types of logical input devices (see Section AII.5) supported at the station, the type of display used, and the resolution of the display surface. The FORTRAN 77 binding for the above function is CALL GOPWK (WS,CID,TYPE). If several work stations are used simultaneously, each is identified separately.

Having identified the work stations, the user activates them. Output may be directed to only the active work stations. The function ACTIVATE WORKSTATION (WS) is used to activate work station WS. The user may deactivate selected work stations by using the function DEACTIVATE WORKSTATION (WS). If it is necessary to clear the previous graphics display before a new image is drawn at a work station, the function CLEAR WORKSTATION (WS,IOPT) is used. If IOPT = CONDITIONALLY, the work station is cleared only if something was drawn at it earlier by the program. IOPT = ALWAYS implies that the work station will be cleared before every drawing. At the end of a graphics session, all active work stations are deactivated and closed. For closing, the function CLOSE WORKSTATION (WS) is used. The FORTRAN 77 bindings for the above functions are as follows:

GKS function	FORTRAN 77 binding
ACTIVATE WORKSTATION	GACWK
DEACTIVATE WORKSTATION	GDAWK
CLEAR WORKSTATION	GCLRWK
CLOSE WORKSTATION	GCLWK

Work station window to viewport mapping functions are provided in GKS, thus affording added flexibility to the user in presenting images. As explained in Chapter 4, through the use of normalization transformations, the details from a window are mapped onto the unit square defined by the normalized lengths of the display surface. Through the use of the work station window to viewport mapping functions, one can control the contents of the display at various active work stations. There are two such fuctions available: (1) SET WORKSTATION WINDOW (WS,XWMIN, XWMAX,YWMIN,YWMAX), which is used for transmitting the image to work station WS within a rectangular window whose corner points have normalized device coordinates (NDC) of (XWMIN,YWMIN) and (XWMAX,YWMAX); and (2) SET WORKSTATION VIEWPORT (WS,XVMIN,XVMAX,YVMIN,YVMAX), which is used to specify a viewport on the display device at work station WS. XVMIN,XVMAX, . . . are the normalized device coordinates of the corners of the viewport. The FORTRAN 77 names of the above functions are, respectively, GSWKW and GSWKVW.

To elucidate the use of the above functions, let's look at the image in Figure AII.1. The image contains a pond, a few birds, and the sun within the unit square. Let's say that the following work station to viewport mapping

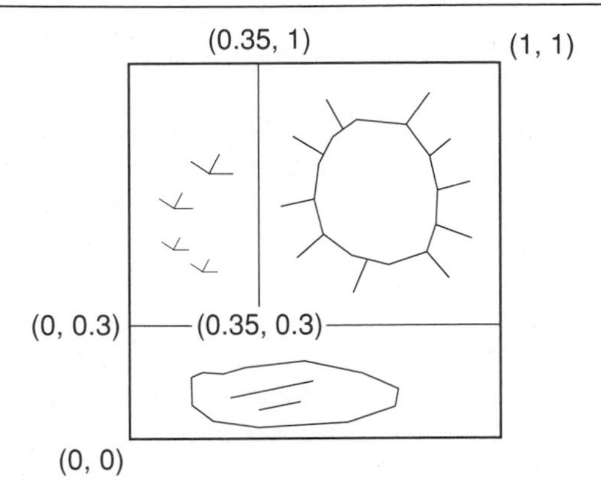

FIGURE AII.1
A picture composed within the unit square

functions are used:

```
SET WORKSTATION WINDOW (1,0.,1.,0.,0.3)
SET WORKSTATION VIEWPORT (1,XVMIN1,XVMAX1,YVMIN1,YVMAX1)
SET WORKSTATION WINDOW (2,0.,0.35.,0.3,1.)
SET WORKSTATION VIEWPORT (2,XVMIN2,XVMAX2,YVMIN2,YVMAX2)
SET WORKSTATION WINDOW (3,0.35,1.,0.3,1.)
SET WORKSTATION VIEWPORT (3,XVMIN3,XVMAX3,YVMIN3,YVMAX3)
```

As a result of the above statements, the image produced at work station 1 will contain only the image of the pond, work station 2 will contain the image of the birds, and so on. GKS lets the operator choose the mode in which the images at different output devices are updated. For example, if the output is to be directed to an off-line plotter at a remote site, it is better to complete the entire image and then transmit the information through the network in one step. On the other hand, if the user is constructing an image interactively using a digitizer board, it is imperative that the changes be displayed instantaneously. With such diverse needs vis-à-vis the timing of the graphics display, the following *deferral modes* are provided:

ASAP, in which the visual change is displayed As Soon As Possible

BNIG, in which the visual change is displayed Before the Next Interaction Globally (i.e., the change will be effected once an input has been requested from any of the active work stations)

BNIL, in which the visual change is displayed Before the Next Interaction Locally (i.e., the display will be updated as soon as an input is requested from that particular work station)

ASTI, in which the visual change is allowed to take place At Some Time

It may happen that in order to affect the visual changes discussed in the above paragraph, the entire image will have to be regenerated. Such instances may occur if, say, the user wishes to change the color of the display while an image is being created. The user can let the regeneration take place as usual or suppress it by specifying IRM = ALLOWED or SUPPRESSED in the function SET DEFERRAL STATE (WS,DEFM,IRM). The parameter DEFM can be set equal to ASAP, BNIG, BNIL, or ASTI, as discussed above. The FORTRAN 77 binding given to the above function is GSDFS.

AII.2 ▪ Output Attributes

In Section 3.11, a single index number for the attributes of the GKS primitives is used. These numbers represent bundled attributes. Bundled attributes are work station dependent and are defined by GKS representation functions. SET POLYLINE REPRESENTATION (WS,I, TYPE,WIDTH,COL) sets the attributes of polylines at work station WS and labels the set of attributes as *I*. In Section 3.11, the statement SET POLYLINE INDEX (N) implies I = N. Attribute TYPE defines the type of line. TYPE = 1 represents continuous lines, TYPE = 2 represents broken (hidden) lines, TYPE = 3 represents dashed-dotted (center) lines, etc. The attribute WIDTH is used to specify the width of the lines. If WIDTH = 2 is specified, the lines drawn at the device are twice the width of the normal lines. COL is the color index. The color indices are often predetermined at the work stations; additionally, new colors may be created. New colors are defined by using the function SET COLOUR REPRESENTATION (WS,COL,RED,GREEN,BLUE). COL = 0 specifies the background color in raster display. COL = 1, 2, . . . corresponds to the indices for the display colors of the lines. By specifying 0–1 values for the parameters RED, GREEN, and BLUE, the intensities for the corresponding colors in the color mix are determined. Thus, if green lines are to be drawn against a white background on a raster display at work station 3, color index 2, as defined below, should be used:

```
SET COLOUR REPRESENTATION (3,0,1,1,1)    Set background color
SET COLOUR REPRESENTATION (3,2,0,1,0)    Set green color for lines
SET POLYLINE INDEX (2)                   Select green color for lines
```

The FORTRAN 77 bindings for the first two functions are, respectively, GSPLR and GSCR.

As opposed to the bundled attributes, it is also possible to use unbundled or individually specified attributes. Individually specified attributes are independent of the work stations and thus apply uniformly to all work stations. The GKS functions for specifying the attributes individually, along with the FORTRAN 77 bindings, are given below.

GKS function	FORTRAN 77 binding
SET LINETYPE (TYPE)	CALL GSLN(TYPE)
SET LINEWIDTH (WIDTH)	CALL GSLWC(WIDTH)
SET POLYLINE COLOUR INDEX (COL)	CALL GSPLCI(COL)

SET POLYMARKER REPRESENTATION (WS,I,TYPE,SIZE,COL) is used for specifying bundled attributes for a polymarker. WS, I, and COL have the same meaning as for the line type representation. TYPE specifies the type of the marker, such as dots, crosses, etc. SIZE specifies the size of the marker relative to the standard size at the work station. These attributes may also be specified individually. The names of the various functions relating to polymarker representation are given below.

GKS function	FORTRAN 77 binding
SET POLYMARKER REPRESENTATION (WS,I,TYPE,SIZE,COL)	CALL GSPMR(WS,I,TYPE,SIZE,COL)
SET MARKER TYPE (TYPE)	CALL GSMK(TYPE)
SET MARKER SIZE SCALE FACTOR (SIZE)	CALL GSMKSC(SIZE)
SET POLYMARKER COLOUR INDEX (COL)	CALL GSPMCI(COL)

The parameter I in the function SET FILL AREA INDEX (I) used earlier stands for bundled attributes for the fill area primitive specified by the representation function SET FILL AREA REPRESENTATION (WS,I,INTS,STYLE,COL). The parameter WS specifies the work station and I the fill area index. INTS specifies the interior style, which may be hollow, solid, patterned, or hatched. If the interior style *hollow* is chosen,

FIGURE AII.2
A fill area pattern

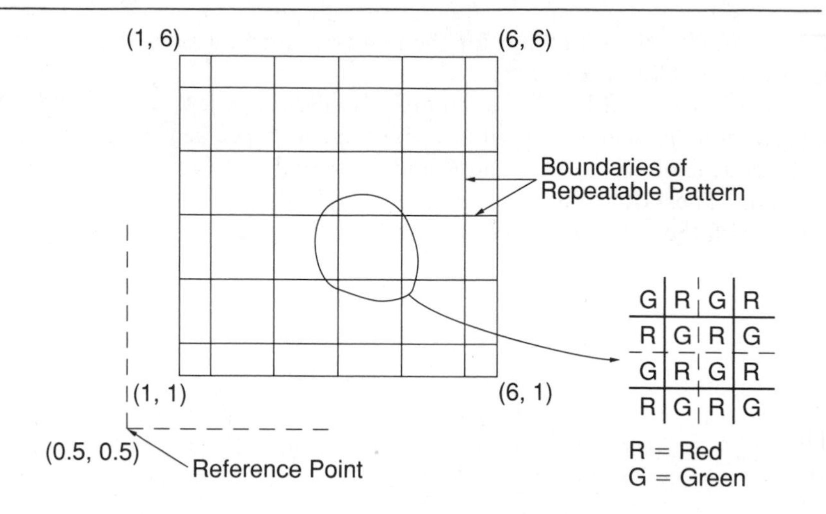

nothing is drawn inside the loop enclosed by the primitive FILL AREA. If the interior style *solid* is chosen, the enclosed area is painted solid with the color COL. If the interior style *pattern* is chosen, the area is filled with a checkered pattern described by the index STYLE. The style of hatching is also determined by the index STYLE. Given below is the method for determining styles of patterns and hatching.

The fill area patterns consist of rectangular boxes divided into a number of cells. These cells can have different colors. The pattern is placed side by side within the area and repeated as many times as necessary. The size and location of the pattern are determined by the functions SET PATTERN SIZE (DX,DY) and SET PATTERN REFERENCE POINT (RX,RY), where DX determines the length of the pattern in the x direction and DY determines the height of the pattern in the y direction. Coordinates (RX,RY) determine the location of a reference point from where pattern drawing starts. As mentioned earlier, the pattern can be subdivided into small cells and the color of each cell specified separately. These attributes are specified in a work station–dependent way by the function SET PATTERN REPRESENTATION (WS,STYLE,Nox,Noy,COLA). In the above function, WS specifies the work station, and STYLE is the fill area style index mentioned above. Nox and Noy specify the number of cells in the x and y directions that the pattern contains. COLA (Nox,Noy) contains information on the color of each of the cells.

Let's assume that we wish to fill a rectangular area of size 5×5 with a pattern of size 1×1 such that the pattern consists of four cells, two green and two red. The following program produces the pattern shown in Figure AII.2:

```
REAL X(4),Y(4)
DATA X/1,6,6,1/
DATA Y/1,1,6,6/
SET FILL AREA REPRESENTATION (1,3,PATTERN,2,0)
SET PATTERN SIZE (1,1)
SET PATTERN REFERENCE POINT (0.5,0.5)
SET PATTERN REPRESENTATION (1,2,2,2,COLA)
SET FILL AREA INDEX (1)
FILL AREA
```

The array COLA would have the size 2 × 2 in the above case and contain, say, numbers 1 and 2 corresponding to the colors green and red at work station 1. The FORTRAN 77 bindings for the functions described above are as follows:

GKS function	FORTRAN 77 binding
SET FILL AREA REPRESENTATION	GSFAR
SET PATTERN SIZE	GSPA
SET PATTERN REFERENCE POINT	GSPARF
SET PATTERN REPRESENTATION	GSPAR

There are more attributes to a text primitive than those described in Section 3.12. These remaining attributes are the text font and precision, the character expansion factor, character spacing, and the text color index. These attributes are embodied in a text index (TXI) defined in the function SET TEXT REPRESENTATION (WS,TXI,TFONT,TPREC,CEXF, CSPAC,TCOL). A particular text font is selected at work station WS by specifying a value for TFONT. Producing text takes more computer time than, say, drawing straight lines. Keeping this in mind, options are provided in GKS whereby the user can specify different levels of accuracy for producing text. The accuracy (TPREC) levels available for the text are as follows:

- STRING: This is the least accurate approach to producing text and hence the quickest. STRING precision is comparable to draft-quality printing on a dot-matrix printer. In the STRING precision, not all the text attributes specified by the user need be implemented. The only requirement laid down in GKS is that the position of the first character in the string must be correctly aligned with the position specified by the user. The character up vector, text path, text alignment, and character spacing requirements can be violated. The character height and expansion factors need be only approximately correct. Such latitude has been provided in STRING precision so that the characters can be rapidly produced using a character generator hardware device.

- CHARACTER: CHARACTER precision is more stringent than STRING precision and therefore a bit slower to produce. The character up vector, character expansion factor, spacing of characters, character height, etc., are honored as far as possible at the particular work station. Attributes relating to the font or the exact form of the individual characters may be ignored. In CHARACTER (or CHAR) precision, the character output may be compared with the near-letter-quality printing on a dot-matrix printer—it takes longer to print than does draft printing, but the printing is of better quality.
- STROKE: The best-quality character generation is obtained with STROKE precision. In STROKE precision, all text attributes are strictly adhered to. The characters are generated using POLYLINE or FILL AREA primitives. The characters produced in STROKE precision can be compared to those coming out of a painter's brush where the outline of the characters is drawn first and then the inside painted with strokes of the brush. Needless to say, in STROKE precision, the text takes much longer to produce.

Some other text attributes are as follows:

- Character expansion factor (CEXF): This factor determines the width of the characters. A character expansion factor of 1.6 specifies that the width of the characters will be 1.6 times the normal width consistent with the height.
- Character spacing (CSPAC): Normally, the characters are drawn with zero spacing in between them, which means that the characters may be touching each other or be closer to each other than envisaged by the font designers. If need be, the characters may be spaced farther apart by specifying character spacing. A character spacing of 0.2 implies that an extra gap equal to $\frac{1}{5}$ of the character height will be introduced in between adjacent characters.
- Text color index (TCOL): This determines the color of the characters.

AII.3 ▪ Window, Viewport, and Clipping Functions

In GKS, the lower left corner of the display surface is designated as point (0, 0) and the upper right corner as point (1, 1) in an NDC system. A viewport is defined by the function SET VIEWPORT (N,Xmin,Xmax, Ymin,Ymax) where Xmin, Xmax, . . . define the limits of the viewport in terms of the NDC. A window in the user's world coordinate system is defined by the function SET WINDOW (N,XMIN,XMAX,YMIN, YMAX), where the parameters XMIN, XMAX, etc., have meanings similar to those for the viewport. The FORTRAN 77 bindings for the two functions are, respectively, CALL GSVP (N,Xmin,Xmax,Ymin,Ymax) and CALL GSWN (N,XMIN,XMAX,YMIN,YMAX). Parameter N acts as labels for a window or a viewport. A user may predefine a number of windows and viewports in the beginning of a program and then select a particular set, using SET NORMALIZATION TRANSFORMATION, for a certain output primitive. Let's assume that two sets of windows and viewports are predefined as follows:

```
SET WINDOW (1,10.,100.,15.,55.)
SET VIEWPORT (1,.1,.9,.1,.9)
SET WINDOW (2,5.,60.,7.,80.)
SET VIEWPORT (2,.5,1.,.5,1.)
```

Now suppose that a graphic output from subroutine PROFIT is to be plotted using the second window and viewport combination. The following set of GKS functions should be used:

```
SET NORMALIZATION TRANSFORMATION (2)
CALL PROFIT
```

CALL GSELNT (N) is the FORTRAN 77 binding for the GKS function SET NORMALIZATION TRANSFORMATION (N).

With regard to the clipping of primitives, GKS provides two options—clip and noclip. These options are invoked using the function SET CLIPPING INDICATOR (ICLIP). The FORTRAN 77 binding for the above function is CALL GSCLIP(ICLIP). If ICLIP = CLIP, portions of the image outside the viewport are clipped. If ICLIP = NOCLIP, everything is drawn.

AII.4 ▪ Segments

The concept of segments has been introduced in GKS in order to enable programmers to store different parts of a graphic image independently. This enables one to manipulate some parts of an image, keeping the rest unaltered. Thus, for example, the image of Yogi Bear can be stored in one segment and the forest around him in a separate segment. By redrawing and erasing the Yogi image at different spots, Yogi can be shown to be roving the forest. The concept of segments is very useful for graphic animation.

A segment is created in three steps, as follows:

```
CREATE SEGMENT (I)
YOGI
CLOSE SEGMENT
```

where YOGI is the subroutine for creating the image of Yogi Bear.

Segments created by the above process are stored at all work stations active at the time. Thus, in the following example, segment (5) is stored at work station 1, while segment (6) is stored at work stations 1 and 2:

```
OPEN WORKSTATION (1,1,4)
OPEN WORKSTATION (2,2,4)

ACTIVATE WORKSTATION (1)

CREATE SEGMENT (5)
YOGI
CLOSE SEGMENT

ACTIVATE WORKSTATION (2)

CREATE SEGMENT (6)
FOREST
CLOSE SEGMENT
```

The above method of storing segments at work stations is known as work station–dependent segment storage (WDSS). Momentarily, work station–independent segment storage (WISS) will also be discussed.

A segment is deleted entirely from the system (i.e., from all work stations active at the time) when the function DELETE SEGMENT (I) is used. Segment (I) is deleted only from work station WS if the function DELETE SEGMENT FROM WORKSTATION (WS, I) is invoked. The FORTRAN 77 bindings for the segment functions described are as follows:

GKS function	FORTRAN 77 binding
CREATE SEGMENT	GCRSG
CLOSE SEGMENT	GCLSG
DELETE SEGMENT	GDSG
DELETE SEGMENT FROM WORKSTATION	GDSGWK

Images in a segment may be transformed in the manner explained in Sections 3.6 and 3.7. Segment transformation is achieved through the function SET SEGMENT TRANSFORMATION (ITRAN,MATRIX), the FORTRAN 77 binding for which is CALL GSSGT(ITRAN,MATRIX). The parameter MATRIX is the transformation matrix. The matrix may be evaluated for a single transformation or a set of different types of transformations (translation, rotation, or scaling) performed in a given sequence by invoking the function EVALUATE TRANSFORMATION MATRIX (X,Y,DX,DY,R,SX,SY,SWITCH,MATRIX) or for a concatenation of transformations by invoking the function ACCUMULATE TRANSFORMATION MATRIX (MATIN,X,Y,DX,DY, R,SX,SY,SWITCH,MATOUT). The FORTRAN 77 bindings for the above two functions, respectively, are CALL GEVTM(X,Y,DX,DY, R,SX,SY,SWITCH,MATRIX) and CALL GACTM(MATIN,X,Y,DX,DY, R,SX,SY,SWITCH,MATOUT).

In the above functions, (X,Y) are the coordinates of a fixed point. DX and DY specify a translation or shift vector. R is the angle of rotation in radians. SX and SY represent scale factors that are applied in the x and y directions, respectively. The parameter SWITCH can be specified as WC (world coordinates), in which case X, Y, DX, and DY are assumed to belong to the world coordinate system. If SWITCH = NDC, X, Y, DX, and DY are given in normalized device coordinates. The transformation matrix computed by the function is returned in the parameter MATRIX. The transformation matrix so formed is passed on to the function ACCUMULATE TRANSFORMATION MATRIX as MATIN along with the information on subsequent transformation required. The necessary matrix multiplication is carried out in the function, and a new matrix (MATOUT) is returned.

Suppose a square, as shown in Figure AII.3, has a corner at the point A (1, 1). It is desired that the square be moved such that the corner A moves to the location (2, 2.5) (i.e., the square is moved a distance of 1 unit in the x

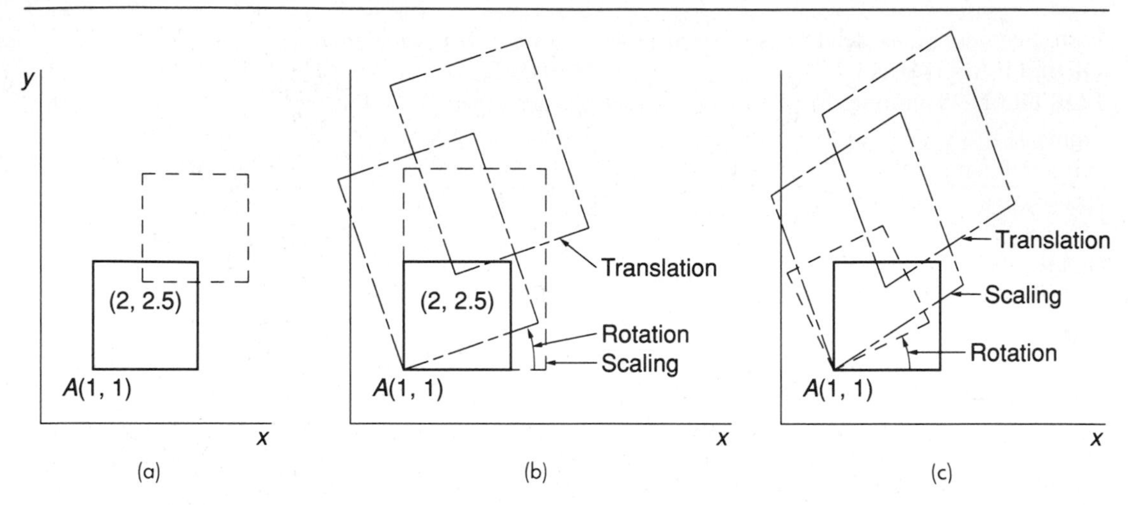

FIGURE AII.3 Square: (a) in its original and translated positions; (b) scaled, rotated, and translated; (c) rotated, scaled, and translated

direction and 1.5 units in the y direction). If no rotation or scaling is required, the call to the function EVALUATE TRANSFORMATION MATRIX contains the following: EVALUATE TRANSFORMATION MATRIX (1,1,1,1.5,0,1,1,WC,MATRIX).

The transformed image is now drawn as shown in dashed lines in Figure AII.3(a). More than one transformation can also be carried out by a single call to the above function. The following call evaluates the transformation matrix required to change the image in Figure AII.3(a) to that shown in Figure AII.3(b): EVALUATE TRANSFORMATION MATRIX (1,1,1,1.5,.4,1.5,2,WC,MATRIX).

In the preceding transformation, the square is first enlarged 1.5 times in the x direction and 2 times in the y direction; then it is rotated 0.4 radian in a counterclockwise direction and moved by 1 unit in the x direction and 1.5 units in the y direction. However, if it is desired that the rotation be carried out first, to be followed by scaling and then by shifting, the following statements are required:

```
EVALUATE TRANSFORMATION MATRIX (1,1,0,0,.4,1,1,WC,MATRIX)
ACCUMULATE TRANSFORMATION MATRIX (MATIN,1,1,0,0,0,2,2,WC,MATOUT)
ACCUMULATE TRANSFORMATION MATRIX (MATIN,1,1,1,1.5,0,1,1,WC,MATOUT)
```

The above functions cause the square to appear as shown in Figure AII.3(c).

In work station–dependent segment storage (WDSS), described above, segments are created at all active work stations simultaneously. However, this arrangement is unsatisfactory at times. If, say, one of the active work stations is a plotter, it is desirable that the plotting start only when the final version of the image is completed. In order for that to happen, it is necessary that the timing of segment transfer be in the hands of the user. WDSS offers no such mechanism; WISS, however, offers the necessary control of segment transfer to the user.

It is possible to provide WISS at a central work station implementing GKS. Segments are created/manipulated or deleted at the central work station exactly as they would be at any other work station. Segment manipulations particular to WISS are carried out through the following functions:

- ASSOCIATE SEGMENT WITH WORKSTATION (WS,ID): Work station WS is kept inactive while segment ID is created. Once ID is completed and closed, it may be associated with WS using the above function. The effect of the association function is exactly as though work station WS were active while segment ID was being created. The association function cannot be invoked while a segment is open.
- COPY SEGMENT TO WORKSTATION (WS,ID): This performs a function similar to ASSOCIATE SEGMENT WITH WORKSTATION, except that segment ID is not stored at work station WS.
- INSERT SEGMENT (ID,MAT): Unlike the ASSOCIATE and COPY functions described above, this function can be invoked while a segment (not segment ID) is open. Segment ID is stored in WISS. As a result of the segment transfer, ID becomes part of the open segment at all active work stations. Thus, segment ID becomes a user-defined primitive. Transformations and clipping parameters defined in segment ID are ignored. Instead, these are governed by the definitions contained in the open segment into which ID is inserted.

The FORTRAN 77 bindings of the above WISS functions are GASGWK GCSGWK, and GINSG.

The use of WISS functions is demonstrated below as a possible alternative to the cam and follower animation program described in Chapter 4. Suppose that we must create two separate segments containing the images of a cam and a follower and then combine the two segments to produce the image shown in Figure AII.4(a). The GKS program for the creation

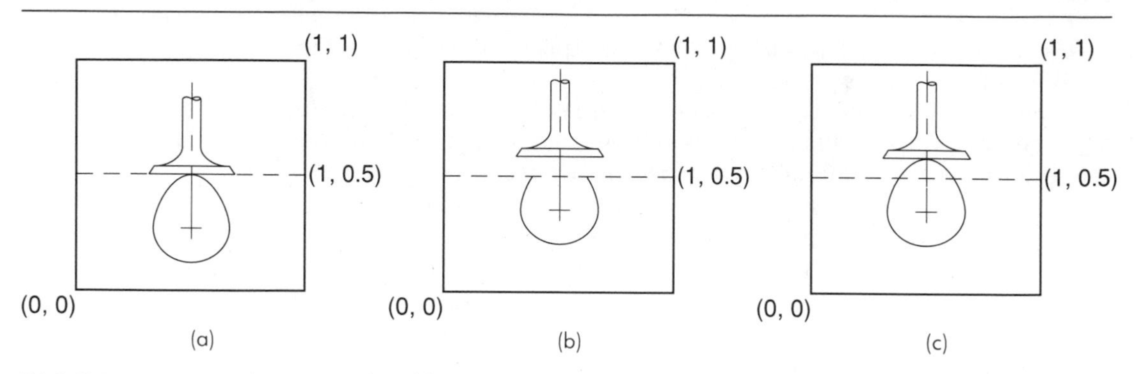

FIGURE AII.4 Images produced by various WISS functions

of the segments reads thus:

```
OPEN WORKSTATION (WIS,1,11)                    Open WISS work station
OPEN WORKSTATION (5,2,8)                       Open work station 8

SET WINDOW (1,0.,50.,0.,50.)
SET VIEWPORT(1,0.,1.,0.,.5)
SET WINDOW (2,0.,50.,50.,100.)
SET VIEWPORT(2,0.,1.,.5,1.)

ACTIVATE WORKSTATION (WIS)                     Store CAM in segment 1
CREATE SEGMENT (1)                             at the WISS station
SELECT NORMALIZATION TRANSFORMATION(1)
CAM
CLOSE SEGMENT

CREATE SEGMENT (2)                             Store FOLLOWER in
SELECT NORMALIZATION (2)                       segment 2 at the
FOLLOWER                                       WISS station
CLOSE SEGMENT
```

Now, if the following ASSOCIATE/COPY functions are invoked, the image created at work station 5 will look exactly like Figure AII.4(a):

```
ASSOCIATE SEGMENT TO WORKSTATION (5,1)
ASSOCIATE SEGMENT TO WORKSTATION (5,2)
COPY SEGMENT TO WORKSTATION (5,1)
COPY SEGMENT TO WORKSTATION (5,2)
```

In order to use the INSERT function, however, the following statements are required:

```
ACTIVATE WORKSTATION (5)
EVALUATE TRANSFORMATION MATRIX(0,0,0,0,0,1,1,WC,MAT)

CREATE SEGMENT (3)
INSERT SEGMENT (1,MAT)
INSERT SEGMENT (2,MAT)
CLOSE SEGMENT
```

The transformation matrix above leaves the images unaltered since no scaling, rotation, or translation is envisaged. Segments 1 and 2 are combined to form segment 3 at work station 5 and the WISS work station. The image produced in this case is also like Figure AII.4(a).

Suppose, now, that we want the image altered somewhat as it is displayed at work station 5. A translation of ten units in the y direction is required. The segments will have to be recreated before the ASSOCIATE/COPY functions can be invoked:

```
SET CLIPPING INDICATOR (CLIP)
EVALUATE TRANSFORMATION MATRIX (0,0,0,10,0,1,1,MAT)

ACTIVATE WORKSTATION (WIS)
CREATE SEGMENT(1)
SET SEGMENT TRANSFORMATION (1,MAT)
SELECT NORMALIZATION TRANSFORMATION(1)
CAM
CLOSE SEGMENT
CREATE SEGMENT (2)
SET SEGMENT TRANSFORMATION (2,MAT)
SELECT NORMALIZATION TRANSFORMATION (2)
FOLLOWER
CLOSE SEGMENT
```

By looking at Figure AII.4(a), one can easily guess that if the cam is moved up, some parts of its details will go out of the viewport specified by normalization transformation 1. In the ASSOCIATE/COPY functions, the clipping rectangles of the individual segments are retained. The image produced by these transfers would therefore have part of the cam missing, as shown in Figure AII.4(b).

Using the INSERT function for the same type of manipulation as discussed in the preceding paragraph, one would write the program thus:

```
SET CLIPPING INDICATOR (CLIP)
EVALUATE TRANSFORMATION MATRIX (0,0,0,10,0,1,1,MAT)
ACTIVATE WORKSTATION (5)
SET VIEWPORT(3,0.,1.,0.,1.)

CREATE SEGMENT (3)
SELECT NORMALIZATION TRANSFORMATION (3)
INSERT SEGMENT (1,MAT)
INSERT SEGMENT (2,MAT)
CLOSE SEGMENT
```

The image produced due to the above transfer will look like Figure AII.4(c), in which the cam and the follower have been moved up as desired, but no part of the cam is missing. In the INSERT function, the clipping rectangles of the individual segments are discarded, and clipping is governed by the viewport specified for the new segment.

AII.5 ▪ Graphic Input Devices in GKS

In GKS, provision has been made to allow graphic data to be input through six different logical types of input devices. All the commercially available input devices can be categorized as one of the following types:

1. *Locator* is a generic name given to devices such as cross hairs and thumbwheels on a storage tube. As is well known, these devices can be used to input positions of points [i.e., (x, y) coordinates].
2. *Pick* devices are those that identify an object displayed on the computer screen. A light pen is a pick device.
3. *Choice* is a device that makes it possible for the user to select from a given set of options—a button box or a screen menu, for example.
4. *Valuator* is a device that can be used to input a certain value. Thus, a keyboard may act as a valuator if it is used to enter the height of a point whose coordinates (x, y) are entered through a locator. A potentiometer is also a valuator.

5. *String* is used for entering a string of characters, as through the keyboard.
6. *Stroke* is a device used for entering a sequence of positions (x, y), as with a digitizer board.

❑ USING THE GRAPHIC INPUT DEVICES

Data are entered through the input devices in REQUEST (or SAMPLE or EVENT) mode. There is a separate function for REQUEST input involving different types of devices. These functions are explained below.

REQUEST LOCATOR (WS, DV, ST, N, X, Y) ▪ This function is used with locator devices. It returns the values of N and (X,Y). (X,Y) are the world coordinates of the position at which the locator is pointed. As is explained in the discussion of window and viewport transformations, one could use several such transformations in a single graphic image. The function determines the particular transformation relationship that is used in the viewport in question and uses the information to transform the device coordinates into world coordinates. The parameter WS refers to the particular graphics work station at which the locator is situated. DV specifies which particular device (of maybe several) at said work station is to be used. ST is a status indicator. If the locator is not able to provide the requested data, ST = 0 (or NONE) is returned. If the data are successfully obtained, ST = 1 (or OK) is returned. The FORTRAN 77 binding for the function is CALL GRQLC(WS,DV,ST,N,X,Y).

REQUEST VALUATOR (WS, DV, ST, VAL) ▪ This function performs a REQUEST on valuator device DV at work station WS. ST is the status indicator as before. VAL is the value returned. The FORTRAN 77 binding for the function is CALL GRQVL(WS,DV,ST,VAL).

REQUEST STROKE (WS, DV, NMAX, ST, N, NP, X, Y) ▪ This function returns to the application program a sequence of positions associated with coordinates X and Y. Thus, X and Y are arrays of numbers. NP is the number of points requested. NMAX designates the maximum value of NP. Other parameters have their usual meanings, as explained earlier. The FORTRAN 77 binding for the function is CALL GRQSK(WS,DV, NMAX,ST,N,NP,X,Y).

REQUEST PICK (WS, DV, ST, SEGN, PCID) ▪ This function returns the identifying number (SEGN) of the segment in which the object at which the pick device is pointed is situated. In some complicated images, there may be a large number of objects. Thus, if the pick device is pointed at any one of those objects, it will return the same value for SEGN. This may not be a satisfactory enough response for the application program. The function therefore allows each object within a segment to be numbered separately. PCID refers to the number of the particular object within segment SEGN.

The GKS function for allocating numbers to the objects is SET PICK IDENTIFIER (I). The following listing illustrates the use of the function:

```
CREATE SEGMENT (N)

DO 1 I = 1,10
SET PICK IDENTIFIER (I)
OBJECT (I)
1 CONTINUE
  CLOSE SEGMENT
```

where OBJECT is the subroutine for drawing the objects. OBJECT(1) is thus given the identification number 1, OBJECT(2) the number 2, etc., in SEGMENT number N. The FORTRAN 77 bindings for the two functions are CALL GRQPK(WS,DV,ST,SEGN,PCID) and CALL GSPCID (I).

REQUEST CHOICE (WS, DV, ST, CH) ▪ This function returns the choice number that has been invoked by the user either by pressing a button or by pointing a light pen on a menu. The FORTRAN 77 binding for the function is CALL GRQCH(WS,DV,ST,CH).

REQUEST STRING (WS, DV, NMAX, ST, NR, STR) ▪ This function returns the character string STR consisting of NR number of characters. NMAX represents the maximum number of characters that the string is allowed to contain. The FORTRAN 77 binding for the function is CALL GRQSTS(WS,DV,NMAX,ST,NR,STR).

AII.6 ▪ Modes of Operation of Input Devices

GKS provides three different modes in which the logical input devices may operate. The REQUEST mode has already been discussed in the preceding section. It is the default mode of operation of all input devices. When an application program REQUESTs an input, its own execution is stopped, and the logical device named by the calling statement becomes active. Once the data have been returned to the program, it becomes active again. Thus, in the REQUEST mode, either the logical device or the application program is active; they are never active at the same time.

An alternate mode of operation of the logical devices is the SAMPLE mode. In this mode, the application program and the input device are active at the same time. The input process works in the background while the execution of the application program continues. The program picks up the current input data when it requires them. The mode selection is carried out in the function SET LOCATOR [or any other device] MODE (WS,DV,SAMPLE,EC). If EC = ECHO is specified, an echo will be produced, showing the data that have been returned to the application program. EC = NOECHO produces no echo. The data are actually obtained by calling such functions as SAMPLE LOCATOR [or any other device] (WS,DV,N,X,Y). The FORTRAN 77 names for the SAMPLE input functions are as follows:

GKS function	FORTRAN 77 binding
SET LOCATOR MODE	GSLCM
SET STROKE MODE	GSSKM
SET VALUATOR MODE	GSVLM
SET CHOICE MODE	GSCHM
SET PICK MODE	GSPKM
SAMPLE LOCATOR	GSMLC
SAMPLE STROKE	GSMSK
SAMPLE VALUATOR	GSMVL
SAMPLE CHOICE	GSMCH
SAMPLE PICK	GSMPC

A third mode of operation is the EVENT mode. As in the SAMPLE mode, here again the input device(s) and the application program are active at the same time. The program can set several input devices in the EVENT mode. When an operator triggers any of these devices, data are entered into an event queue. The program reads each of the entries in the event queue in sequence and acts on them accordingly. There is a single event queue for all devices, but each entry into the queue contains information regarding the device from which the particular data came.

A function named AWAIT EVENT (TOUT,WS,DVC,DV) is provided for facilitating user interaction in the EVENT mode. The function checks the event queue. If the event queue is empty at the time AWAIT EVENT is invoked, the program waits for a time equal to TOUT for an event to occur. If no event is generated during this time, DVC = NONE is returned. If, on the other hand, the event queue has a few entries in it, the function AWAIT EVENT returns the work station identifier (WS), the input device class (DVC = LOCATOR, etc.), and the input device number of the first (the earliest) entry into the queue. This information is then used by the GET function to return data to the application program.

The various EVENT input functions and their FORTRAN 77 bindings are listed below.

GKS function	FORTRAN 77 binding
AWAIT EVENT	CALL GAWE(TOUT,WS,DVC,DV)
GET LOCATOR	CALL GGTLC(N,X,Y)
GET STROKE	CALL GGTSK(NMAX,N,NP,X,Y)
GET VALUATOR	CALL GGTVL(VAL)
GET CHOICE	CALL GGTCH(CH)
GET PICK	CALL GGTPC(ST,SEGN,PCID)
GET STRING	CALL GGTST(NR,STR)
FLUSH DEVICE EVENTS	CALL GFLDE(WS,DVC,DV)

The function FLUSH DEVICE EVENT is used to remove entries from the event queue. If invoked, it will remove from work station WS all entries that were introduced into the queue through device DV of class DVC.

A small program is given below to demonstrate the role of different EVENT input functions. The program has the capability to draw two sets of lines in different line styles interactively.

```
01      SET WINDOW (1,XMIN,XMAX,YMIN,YMAX)
        SET VIEWPORT (1,0.,1.,0.,1.)
        SET LOCATOR MODE (WS,DV,EVENT,ECHO)
        SET CHOICE MODE (WS,DV,EVENT,ECHO)
05      NPT1 = 0
        NPT2 = 0

      1 CONTINUE

        AWAIT EVENT (30,WS,DVC,DV)
        IF (DVC.EQ.NONE) GOTO 3
10      IF (DVC.EQ.LOCATOR) GOTO 2
        GET CHOICE (CH)
        GOTO 1
      2 GET LOCATOR (NT,X,Y)
        IF (CH.EQ.1) THEN
15              NPT1 = NPT1 + 1
                X1(NPT1) = X
                Y1(NPT1) = Y
        ELSEIF (CH.EQ.2) THEN
                NPT2= NPT2 + 1
```

```
20          X2(NPT2) = X
            Y2(NPT2) = Y
       ENDIF
     3 SET POLYLINE INDEX (1)
       POLYLINE (NPT1,X1,Y1)
25     SET POLYLINE INDEX (2)
       POLYLINE (NPT2,X2,Y2)
```

In line 8 above, a timeout of 30 seconds is set. Thus, if no data are entered in an empty queue for 30 seconds, DVC = NONE will be returned. The program envisages that two types of lines—say, continuous and broken—will be drawn. The user can enter data at random. Before entering the coordinates of a point through a locator device—say, a digitizer puck—the user should, however, first indicate whether the point belongs to the continuous line or the broken line. The user may differentiate the two types of data through the use of a choice device—say, a menu on the digitizer board. Depending on whether a point belongs to the solid line or the broken one, the application program stores the coordinates in arrays (X1,Y1) and (X2,Y2), respectively, as shown in lines 14–22.

Once the user has entered all the point data, he or she should do nothing further. After waiting for 30 seconds at line 8, the program will transfer control to line 23. Two types of lines will then be drawn.

AII.7 ▪ Initialization of Input Devices

A reference value for the input devices should first be determined before they are used to input data. The initial values are assigned using a GKS INITIALIZE function of the following type:

```
INITIALIZE xx(WS,DV,initial val,PE,XMIN,XMAX,YMIN,YMAX,IMAX,IA)
```

In the above function, WS is the work station number, and DV is the identification number for the device of the logical type xx. XMIN,XMAX, etc. define a rectangular box in the device coordinates where echos should appear. Parameter PE determines the prompt and echo types. PE = 1 can be used in all cases, which implies that the user has selected the device-dependent prompt and echo types. There are several prompt and echo types that may be generated by GKS. For complete information on these, the reader is advised to check the original documents mentioned at the end of Chapter 3. IA(MAX) is the data array that the device is required to return to the application program.

Depending on the type of input device in question, the nature of the input data differs. Given below are the exact names of the initialization functions and the associated parameters:

```
INITIALIZE LOCATOR (WS,DV,NRN,X,Y,PE,XMIN,YMIN,XMAX,YMAX,IMAX,IA)
```

where NRN is the normalization transformation number and X,Y are the coordinates of the initial locator position in world coordinates.

```
INITIALIZE STROKE (WS,DV,NRN,N,X,Y,PE,XMIN,YMIN,XMAX,YMAX,IMAX,IA)
```

where N is the number of points in the initial stroke and X(N),Y(N) are the coordinates of points in the initial stroke.

```
INITIALIZE VALUATOR (WS,DV,IVAL,PE,XMIN,YMIN,XMAX,YMAX,IMAX,IA)
```

where IVAL is the initial value.

```
INITIALIZE CHOICE (WS,DV,ICN,PE,XMIN,YMIN,XMAX,YMAX,IMAX,IA)
```

where ICN is the initial choice.

```
INITIALIZE PICK (WS,DV,IS,ISN,IPI,PE,XMIN,YMIN,XMAX,YMAX,IMAX,IA)
```

where IS is the initial status, ISN is the initial segment name, and IPI is the initial pick identifier.

```
INITIALIZE STRING (WS,DV,IS,PE,XMIN,YMIN,XMAX,YMAX,IMAX,IA)
```

where IS is the initial string.

The FORTRAN 77 bindings for the above functions are, respectively, GINLC, GINSK, GINVL, GINCH, GINPC, and GINST.

AII.8 ▪ GKS Metafile

We have discussed methods of storing segments at different work stations. It must, however, be remembered that segment storage is temporary in nature. Once the GKS system is closed down, the segments at all work stations are lost. A permanent way of storing segments is through the use of the metafile functions. GKS defines two types of metafiles: an input metafile and an output metafile. Metafiles are treated exactly like a work station; thus, a metafile is activated or deactivated using the corresponding functions for work stations. The installation manager decides which particular work stations to designate as metafiles.

An *output metafile* is one to which the output is directed for permanent storage. Thus, in the example given below, the subroutine YOGI will be stored in the output metafile (WS = 1):

```
OPEN WORKSTATION (1,1,10)
ACTIVATE WORKSTATION (1)
YOGI
DEACTIVATE WORKSTATION (1)
CLOSE WORKSTATION (1)
```

GKS also provides a function for storing nongraphic data in metafiles. This function is WRITE ITEM TO GKSM (WS,TYPE,L,REC). The parameter TYPE refers to the type of data, L to the length of the data record, and REC to the data record itself.

Application programs can obtain data from an *input metafile* using the following functions:

- GET ITEM TYPE FROM GKSM (WS,TYPE,LENGTH), which returns the type (TYPE) and length (LENGTH) of the current data record from the metafile.
- READ ITEM FROM GKSM (WS,MAXL,LENGTH,RECORD), which reads the current data record in the array RECORD. MAXL is specified by the user, suggesting the maximum possible length of the data record. LENGTH is the actual length of the record returned.
- INTERPRET ITEM (TYPE,LENGTH,RECORD), which follows the READ ITEM FROM GKSM function in order to interpret the data returned.

AII.9 ▪ Miscellaneous GKS Functions

Before any GKS function is invoked, it is necessary to initialize GKS by the following function:

```
OPEN GKS (EF)
```

where EF is the name of the file to which error messages should be directed. Similarly, the very last statement at the end of a GKS session should be CLOSE GKS.

We have seen before that there are two ways of defining attributes of the primitives: bundled and individual. The function SET ASPECT SOURCE FLAGS (LIST) allows users to specify the method for specifying attributes of the graphics primitives. The parameter LIST contains the information (aspect source flags) regarding the mode in which each of the attributes (aspect) will be determined. The aspect source flags (ASF = BUNDLED/UNBUNDLED) should follow the sequence given on the next page.

```
SET LINE TYPE
SET LINEWIDTH SCALE FACTOR
SET POLYLINE COLOR INDEX
SET MARKER TYPE
SET MARKER SIZE SCALE FACTOR
SET POLYMARKER COLOR INDEX
SET TEXT FONT AND PRECISION
SET CHARACTER EXPANSION FACTOR
SET CHARACTER SPACING
SET TEXT COLOR INDEX
SET FILL AREA INTERIOR STYLE
SET FILL AREA STYLE INDEX
SET FILL AREA COLOR INDEX
```

If all the attributes are to be bundled, the function is invoked thus:

```
SET ASPECT SOURCE FLAGS(BUNDLED,BUNDLED,...,BUNDLED)
```

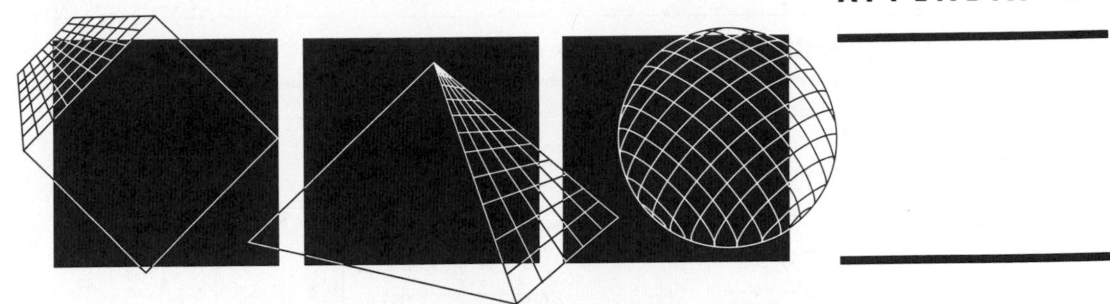

PROFILE — A Program for a Two-Dimensional Graphic Database

AIII.1 ▪ Symbols

Represents a terminal point, either the beginning or the end of a program or a subroutine

Indicates an input or an output operation

Denotes a processing operation

A yes/no decision is made here

An off-page connector indicates that the flowchart is continued on another page

AIII.2 ▪ Summary of Flowchart

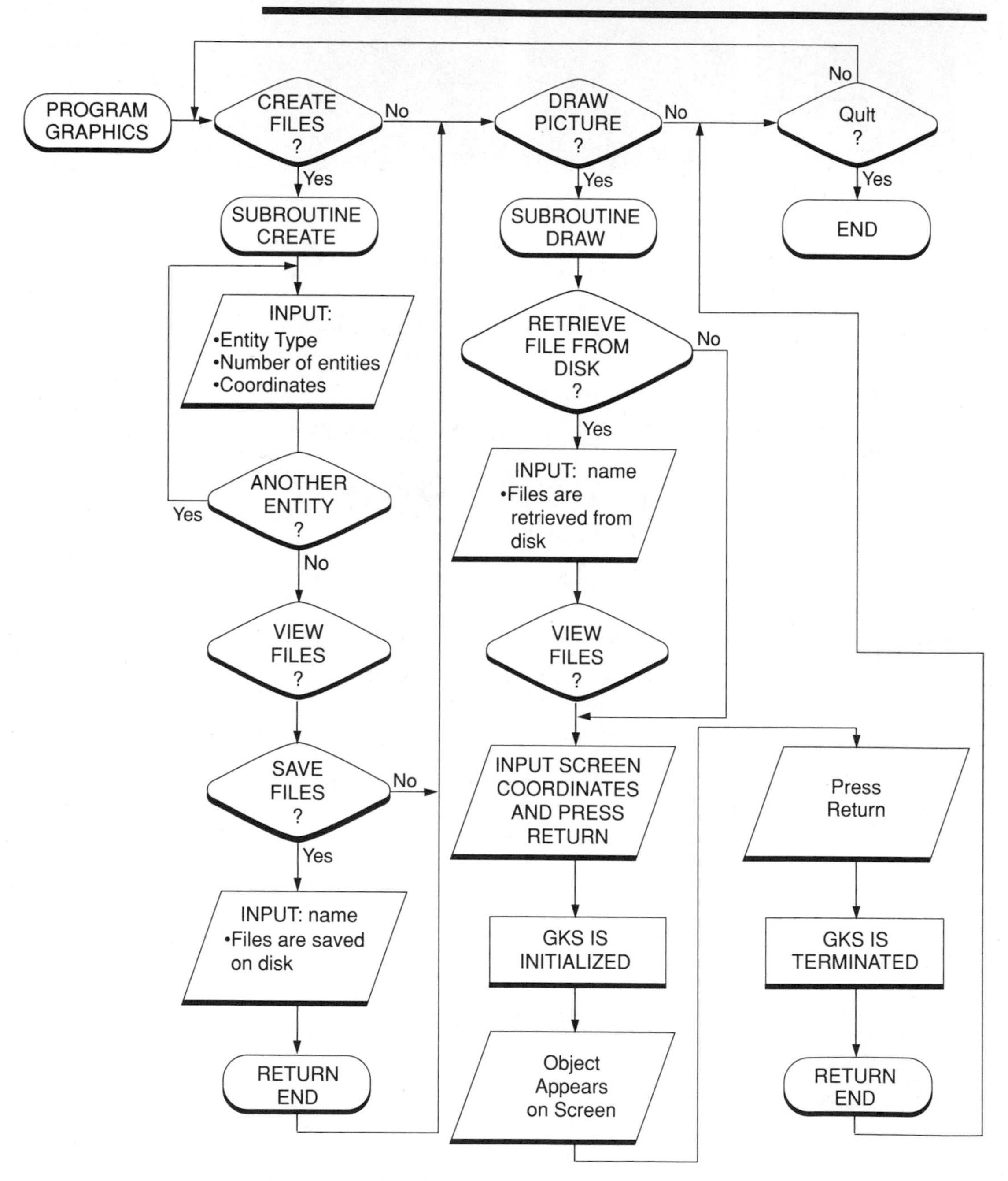

AIII.3 ▪ Flowcharts of Key Subroutines

MAIN PROGRAM

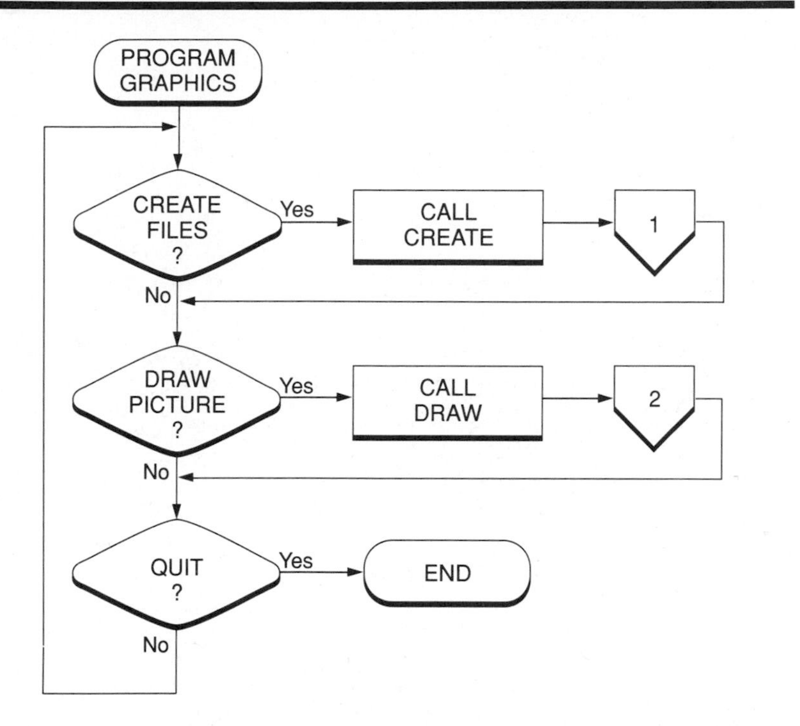

1

SUBROUTINE CREATE

INITIALIZE VARIABLES E12=E13 =E14=1, TOTENT=0

PRINT INPUT INSTRUCTIONS

5

TOTENT = TOTENT + 1

4

INPUT ENTITY TYPE, NUMBER OF ENTITIES

ENTITY IS A LINE ?

Yes

STORE ENTITY IDENTIFIER E12

INPUT COORDINATES OF LINE ENDS x_1, y_1, x_2, y_2

No

ENTITY IS AN ARC ?

Yes

STORE ENTITY IDENTIFIER E13

INPUT COORDINATES OF CENTER OF CURVATURE AND ARC ENDS $x_0, y_0, x_1, y_1, x_2, y_2$

No

ENTITY IS A CURVE ?

Yes

STORE ENTITY IDENTIFIER E14

INPUT NUMBER OF POINTS ON CURVE: MIN 3, MAX 20

INPUT COORDINATES OF POINTS $x_1, y_1, \ldots, x_n, y_n$

No

3

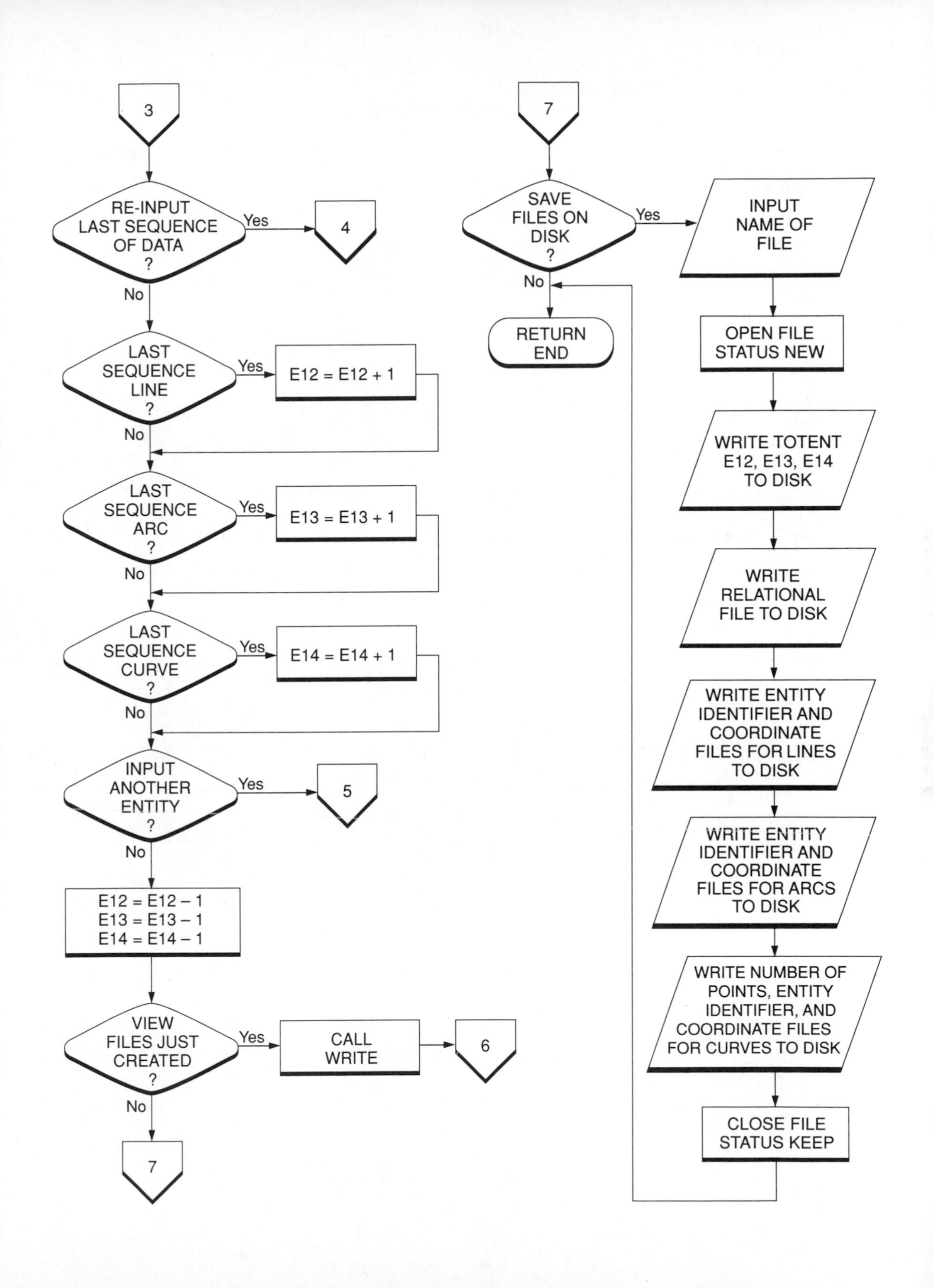
3
RE-INPUT LAST SEQUENCE OF DATA ?
Yes
4
No
LAST SEQUENCE LINE ?
Yes
E12 = E12 + 1
No
LAST SEQUENCE ARC ?
Yes
E13 = E13 + 1
No
LAST SEQUENCE CURVE ?
Yes
E14 = E14 + 1
No
INPUT ANOTHER ENTITY ?
Yes
5
No
E12 = E12 – 1
E13 = E13 – 1
E14 = E14 – 1
VIEW FILES JUST CREATED ?
Yes
CALL WRITE
6
No
7
7
SAVE FILES ON DISK ?
Yes
INPUT NAME OF FILE
No
RETURN END
OPEN FILE STATUS NEW
WRITE TOTENT E12, E13, E14 TO DISK
WRITE RELATIONAL FILE TO DISK
WRITE ENTITY IDENTIFIER AND COORDINATE FILES FOR LINES TO DISK
WRITE ENTITY IDENTIFIER AND COORDINATE FILES FOR ARCS TO DISK
WRITE NUMBER OF POINTS, ENTITY IDENTIFIER, AND COORDINATE FILES FOR CURVES TO DISK
CLOSE FILE STATUS KEEP

SUBROUTINE DRAW

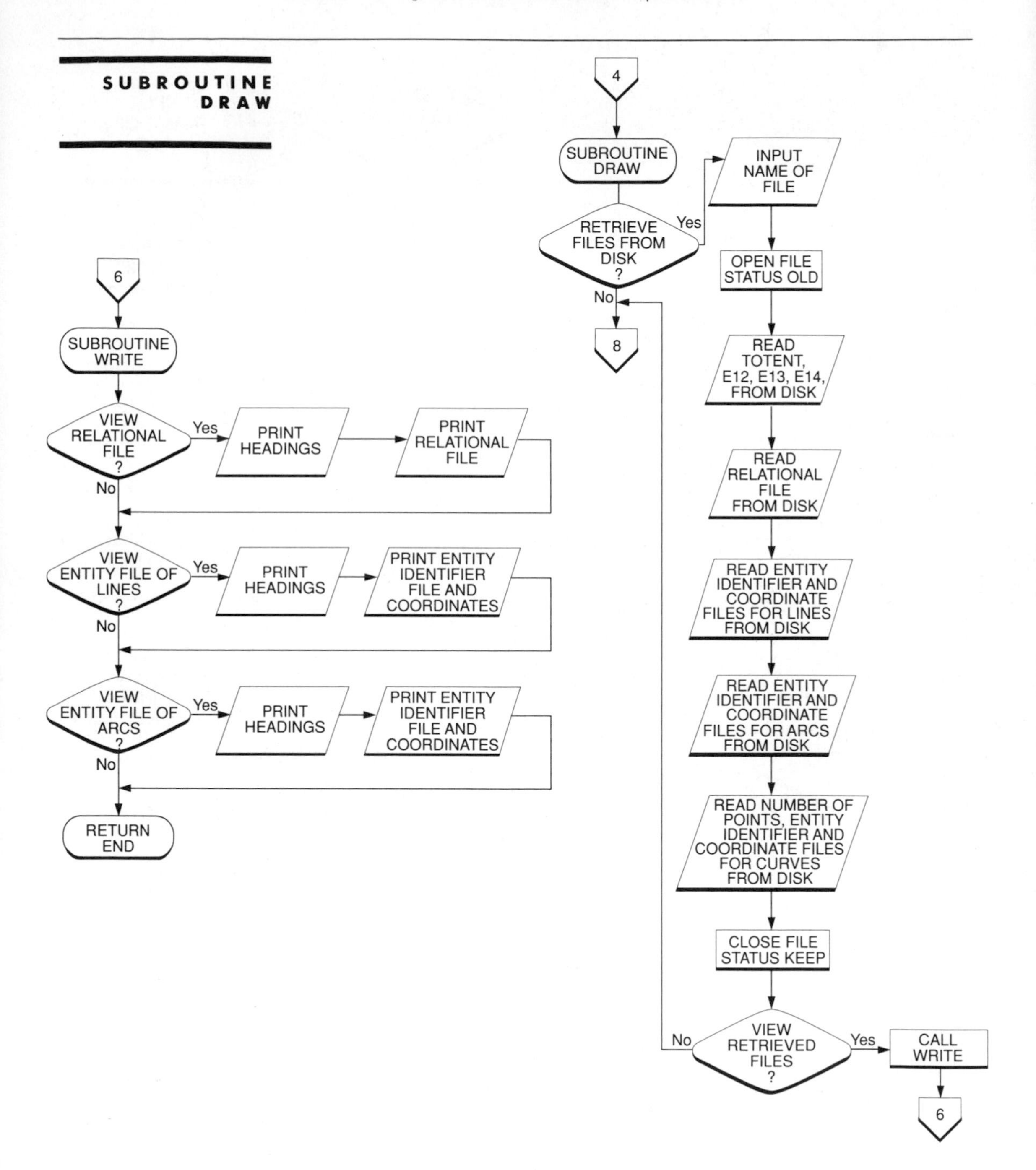

SUBROUTINE ARCS AND CURVE

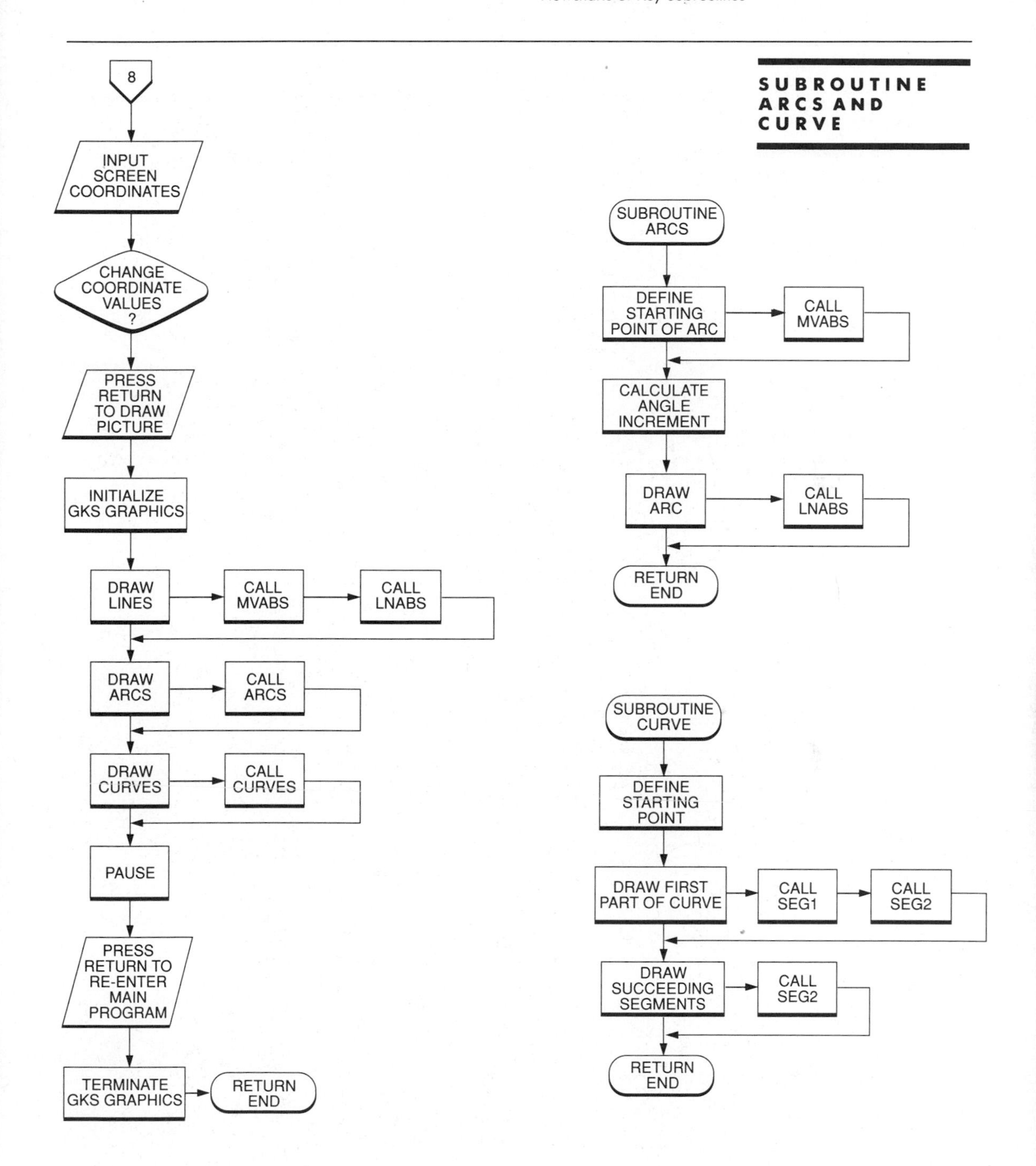

AIII.4 ▪ Program Listing

PROGRAM PROFILE

```
C This program is used to create a database required
C to generate a two-dimensional geometric object
C
C
C EI2 - entity identifier for entity type 2
C EI3 - entity identifier for entity type 3
C EI4 - entity identifier for entity type 4
C RELATE - relational file
C X0, Y0, X1, Y1, X2, Y2 - coordinates for arcs, circles, lines, curves
C EFT2L - entity file type 2, lines - coordinates
C EFT3A - entity file type 3, arcs - coordinates
C EFT4C - entity file type 4, curves - coordinates
C EIFL - entity identifier file for lines
C EIFA - entity identifier file for arcs
C EIFC - entity identifier file for curves
C TOTENT - total number of entities to be input
C
C-----------------------MAIN PROGRAM-----------------------
C
      INTEGER EI2, EI3, EI4, RELATE(100,3), CHOICE, TOTENT,
     *        EIFL(100), EIFA(100), EIFC(100), NUMPTS(20)
      REAL EFT2L(100,4), EFT3A(100,6), EFT4C(100,40)
      CHARACTER FNAME*25
      WRITE (*,'(1x,a,a)') CHAR(27), '[2J'
C
5     WRITE (*,50)
50    FORMAT (/ 'Create entity file? (Y=1/N=2) '\)
      READ (*,*,ERR=5) CHOICE
C
      IF ( CHOICE .EQ. 1 ) THEN
         CALL CREATE (RELATE,EIFL,EIFA,EIFC,EFT2L,EFT3A,EFT4C,
     *                EI2,EI3,EI4,FNAME,TOTENT,NUMPTS)
      END IF
C
6     WRITE (*,51)
51    FORMAT (/ 'Draw picture? (Y=1/N=2) '\)
      READ (*,*,ERR=6) CHOICE
C
      IF ( CHOICE .EQ. 1 ) THEN
         CALL DRAW (RELATE,EIFL,EIFA,EIFC,EFT2L,EFT3A,EFT4C,
     *              EI2,EI3,EI4,FNAME,TOTENT,NUMPTS)
      END IF
```

PROGRAM PROFILE
(Continued)

```
C
7     WRITE (*,52)
52    FORMAT (/ 'Quit? (Y=1/N=2) '\)
      READ (*,*,ERR=7) CHOICE
C
      IF ( CHOICE .EQ. 2 ) THEN
         GOTO 5
      END IF
C
      STOP
C
      END
C
C--------------------SUBROUTINE CREATE--------------------
C
      SUBROUTINE CREATE (RELATE,EIFL,EIFA,EIFC,EFT2L,EFT3A,EFT4C,
     *                   EI2,EI3,EI4,FNAME,TOTENT,NUMPTS)
C
      INTEGER EI2, EI3, EI4, RELATE(100,3), CHOICE, C2, C3, C4,
     * TOTENT, EIFL(100), EIFA(100), EIFC(100), NUMPTS(20)
      REAL EFT2L(100,4), EFT3A(100,6), EFT4C(100,40)
      CHARACTER FNAME*25
C
      EI2 = 1
      EI3 = 1
      EI4 = 1
      TOTENT = 0
C
      WRITE (*,130)
      WRITE (*,131)
      WRITE (*,132)
130   FORMAT (//'To create a file, the following data is required as pro

     *mpted for each sequence:')
131   FORMAT ('entity type:  2 - lines, 3 - arcs (including circles), 4
     *- curves;')
132   FORMAT ('number of entities; pertaining coordinates.'/)
      WRITE (*,*) 'Note:  curves must be functions, ie. one value X per
     *value Y.'
      WRITE (*,*) 'The more points entered, the better quality the curve

     * will be.'
      WRITE (*,*) 'Note:  arcs are drawn counter clockwise.'
```

(continues)

PROGRAM PROFILE *(Continued)*

```
C
8     TOTENT = TOTENT + 1
C
28    WRITE (*,55) TOTENT
11    WRITE (*,60)
55    FORMAT (//'Sequence  '\,I2)
60    FORMAT (/'Input entity type '\)
      READ (*,*,ERR=11) RELATE(TOTENT,1)
C
12    WRITE (*,57)
57    FORMAT ('Input number of entities '\)
      READ (*,*,ERR=12) RELATE(TOTENT,2)
C
C Write to relational and entity files for lines
C
      IF ( RELATE(TOTENT,1) .EQ. 2 ) THEN
         RELATE(TOTENT,3) = EI2
C
13       WRITE (*,*) 'Input coordinates of line ends'
         WRITE (*,*) 'X1, Y1, X2, Y2'
         DO 30 C2 = 1,4
            READ (*,*,ERR=13) EFT2L(EI2,C2)
   30    CONTINUE
         EIFL(EI2) = EI2
      END IF
C
C Write to relational and entity files for arcs
C
      IF ( RELATE(TOTENT,1) .EQ. 3 ) THEN
         RELATE(TOTENT,3) = EI3
C
14       WRITE (*,*) 'Input coordinates of center, ends of arc'
         WRITE (*,*) 'X0, Y0, X1, Y1, X2, Y2'
         DO 40 C3 = 1,6
            READ (*,*,ERR=14) EFT3A(EI3,C3)
   40    CONTINUE
         EIFA(EI3) = EI3
      END IF
C
C Write to relational and entity files for curves
C
      IF ( RELATE(TOTENT,1) .EQ. 4 ) THEN
         RELATE(TOTENT,3) = EI4
C
33       WRITE (*,140)
140      FORMAT (/'Number of points on curve (min 3, max 20) ? '\)
         READ (*,41,ERR=33) NUMPTS(EI4)
41       FORMAT (I2)
```

PROGRAM PROFILE *(Continued)*

```
C
         WRITE (*,*) 'Input coordinates of consecutive points'
         WRITE (*,*) 'from one end of the curve to the other'
34       WRITE (*,*) 'X1,Y1, X2,Y2, X3,Y3, . . ., Xn,Yn'
C
         C4 = NUMPTS(EI4) * 2
         DO 43 I = 1,C4
            READ (*,*,ERR=34) EFT4C(EI4,I)
   43    CONTINUE
         EIFC(EI4) = EI4
      END IF
C
C
26    WRITE (*,27)
27    FORMAT ('Re-input last sequence of data? (Y=1/N=2) '\)
      READ (*,*,ERR=26) CHOICE
      IF (CHOICE .EQ. 1) THEN
         GO TO 28
      END IF
C
      IF ( RELATE(TOTENT,1) .EQ. 2 ) THEN
         EI2 = EI2 + 1
      END IF
      IF ( RELATE(TOTENT,1) .EQ. 3 ) THEN
         EI3 = EI3 + 1
      END IF
      IF ( RELATE(TOTENT,1) .EQ. 4 ) THEN
         EI4 = EI4 + 1
      END IF
C
9     WRITE (*,29)
29    FORMAT ('Input another entity? (Y=1/N=2) '\)
      READ (*,*,ERR=9) CHOICE
C
      IF (CHOICE .EQ. 1) THEN
         GO TO 8
      END IF
C
      EI2 = EI2 - 1
      EI3 = EI3 - 1
      EI4 = EI4 - 1
C
C View files just created
C
24    WRITE (*,102)
102   FORMAT (//'View files just created? (Y=1/N=2) '\)
      READ (*,*,ERR=24) CHOICE
```

(continues)

PROGRAM PROFILE *(Continued)*

```
C
      IF (CHOICE .EQ. 1) THEN
         CALL WRITE (RELATE,EIFL,EFT2L,EIFA,EFT3A,TOTENT,EI2,EI3)
      END IF
C
C Save files just created on disk
C
19    WRITE (*,400)
400   FORMAT (// 'Save file? (Y=1/N=2) '\)
      READ (*,*,ERR=19) CHOICE
C
      IF ( CHOICE .EQ. 1 ) THEN
C
21       WRITE (*,410)
410      FORMAT (// 'Name of file? (type *(120)* in front of name) '\)
         READ (*,415,ERR=21) FNAME
415      FORMAT (A25)
C
         OPEN (1,FILE = FNAME, STATUS = 'NEW')
         WRITE (1,'(415)') TOTENT, EI2, EI3, EI4
C
         DO 70 I = 1,TOTENT
            WRITE (1,'(315)') (RELATE(I,J),J=1,3)
   70    CONTINUE
C
         DO 75 I = 1,EI2
            WRITE (1,'(I5,4E14.7)') EIFL(I),(EFT2L(I,J),J=1,4)
   75    CONTINUE
C
         DO 80 I = 1,EI3
            WRITE (1,'(I5,6E14.7)') EIFA(I),(EFT3A(I,J),J=1,6)
   80    CONTINUE
C
         DO 85 I = 1,EI4
            WRITE (1,*) NUMPTS(I),EIFC(I),(EFT4C(I,J),J=1,NUMPTS(I)*2)
   85    CONTINUE
C
         CLOSE (1, STATUS = 'KEEP')
C
      END IF
C
      RETURN
C
      END
```

PROGRAM PROFILE
(Continued)

```
C
C-----------------------SUBROUTINE WRITE-----------------------
C
      SUBROUTINE WRITE(RELATE,EIFL,EFT2L,EIFA,EFT3A,TOTENT,EI2,EI3)
      REAL EFT2L(100,4),EFT3A(100,6)
      INTEGER RELATE(100,3),R,EIFL(100),EIFA(100),TOTENT,EI2,EI3
C
C Print the relational file
C
18    WRITE (*,105)
105   FORMAT (// 'View relational file? (Y=1/N=2) '\)
      READ (*,*,ERR=18) CHOICE
C
      IF (CHOICE .EQ. 1) THEN
C
         WRITE (*,100)
         WRITE (*,110)
100      FORMAT (/ 'Entity Type',5X,'1 Entities',6X,
     *   'Entity Identifier')
110      FORMAT ('-----------',5X,'----------',6X,'-----------------')
C
         DO 20 R = 1,TOTENT
            WRITE (*,120) (RELATE(R,I),I=1,3)
120         FORMAT (5X,I1,15X,I1,19X,I1)
   20    CONTINUE
C
      END IF
C
C
C Print the entity file of lines
C
16    WRITE (*,205)
205   FORMAT (// 'View entity file of lines? (Y=1/N=2) '\)
      READ (*,*,ERR=16) CHOICE
C
      IF (CHOICE .EQ. 1) THEN
C
         WRITE (*,200)
         WRITE (*,210)
200      FORMAT (/ 'Identifier',5X,'X1',5X,'Y1',5X,'X2',5X,'Y2')
210      FORMAT ('----------     --     --     --     --')
C
         DO 15 R = 1,EI2
            WRITE (*,220) EIFL(R),(EFT2L(R,I),I=1,4)
```

(*continues*)

PROGRAM PROFILE *(Continued)*

```
220         FORMAT (4X,I3,7X,F4.1,3X,F4.1,3X,F4.1,3X,F4.1)
   15    CONTINUE
C
      END IF
C
C
C Print the entity file of arcs
C
17    WRITE (*,305)
305   FORMAT (// 'View entity file of arcs? (Y=1/N=2) '\)
      READ (*,*,ERR=17) CHOICE
C
      IF (CHOICE .EQ. 1) THEN
C
         WRITE (*,300)
         WRITE (*,310)
300      FORMAT (/ 'Identifier',5X,'X0',5X,'Y0',5X,'X1',5X,'Y1',5X,
     *     'X2',5X,'Y2')

310      FORMAT ('----------    ----   ----   ----   ----   --',
     *     '--   ----')
C
         DO 25 R = 1,EI3
            WRITE (*,320) EIFA(R),(EFT3A(R,I),I=1,6)
320         FORMAT (4X,I3,7X,F4.1,3X,F4.1,3X,F4.1,3X,F4.1,3X,F4.1,
     *        3X,F4.1)
   25    CONTINUE
C
      END IF
C
      RETURN
C
      END
C
C----------------------SUBROUTINE DRAW---------------------------
C
      SUBROUTINE DRAW (RELATE,EIFL,EIFA,EIFC,EFT2L,EFT3A,EFT4C,
     *                 EI2,EI3,EI4,FNAME,TOTENT,NUMPTS)
C
      INTEGER RELATE(100,3), EIFL(100), EIFA(100), EIFC(100), CHOICE,
     *        EI2,EI3,EI4,TOTENT,ERRFID,WKTYPE,WKID,CONNID,NUMPTS(20)
      REAL EFT2L(100,4), EFT3A(100,6), EFT4C(100,40)
      CHARACTER FNAME*25
C
22    WRITE (*,420)
420   FORMAT (// 'Retrieve file from disk? (Y=1/N=2) '\)
      READ (*,*,ERR=22) CHOICE
```

PROGRAM PROFILE (*Continued*)

```
C
      IF ( CHOICE .EQ. 1 ) THEN
C
23          WRITE (*,430)
430         FORMAT (// 'Name of file? (type ''(120)'' in front of name)')
            READ (*,435,ERR=23) FNAME
435         FORMAT (A25)
C
            OPEN (1,FILE = FNAME, STATUS = 'OLD')
            READ (1,'(4I5)') TOTENT, EI2, EI3, EI4
C
            DO 70 I = 1,TOTENT
               READ (1,'(3I5)') (RELATE(I,J),J=1,3)
   70       CONTINUE
C
            DO 75 I = 1,EI2
               READ (1,'(I5,4E14.7)') EIFL(I),(EFT2L(I,J),J=1,4)
   75       CONTINUE
C
            DO 80 I = 1,EI3
               READ (1,'(I5,6E14.7)') EIFA(I),(EFT3A(I,J),J=1,6)
   80       CONTINUE
C
            DO 85 I = 1,EI4
               READ (1,*) NUMPTS(I),EIFC(I),(EFT4C(I,J),J=1,NUMPTS(I)*2)
   85       CONTINUE
C
            CLOSE (1, STATUS = 'KEEP')
C
C View files retrieved from disk
C
26          WRITE (*,440)
440         FORMAT (//,'View files retrieved from disk? (Y=1/N=2) '\)
            READ (*,*,ERR=26) CHOICE
C
            IF (CHOICE .EQ. 1) THEN
               CALL WRITE(RELATE,EIFL,EFT2L,EIFA,EFT3A,TOTENT,EI2,EI3)
            END IF
C
      END IF
```

(*continues*)

PROGRAM PROFILE
(Continued)

```
C
C———GRAPHICS———
C
C Define maximum and minimum coordinate values for drawing picture
C
      WRITE (*,340)
      WRITE (*,345)
340   FORMAT (//'For the picture to cover the entire monitor screen,')
345   FORMAT ('define the maximum and minimum X and Y coordinate values'
     *)
31    WRITE (*,347)
347   FORMAT ('Xmin, Xmax, Ymin, Ymax')
      READ (*,*,ERR=31) WNXMIN, WNXMAX, WNYMIN, WNYMAX
C
32    WRITE (*,348)
348   FORMAT ('Change coordinate values? (Y=1/N=2) '\)
      READ (*,*,ERR=32) CHOICE
      IF (CHOICE .EQ. 1) THEN
         GO TO 31
      END IF
C
      WRITE (*,350)
350   FORMAT(//)
C
C Draw picture
C
      PAUSE 'Press return to draw picture, and again to re-enter program
     *'
C
      ERRFID=0
      WKID=1
      CONNID=0
      VPXMIN = 0.0
      VPXMAX = 0.8
      VPYMIN = 0.0
      VPYMAX = 1.0
C
C Initialize GKS
      CALL GOPKS (ERRFID)
      CALL GICGA6 (WKTYPE)
      CALL GOPWK (WKID, CONNID, WKTYPE)
      CALL GACWK (WKID)
      CALL GSCR (WKID, 1, 1.0, 0.0, 0.0)
      IDTRAN = 1
      CALL GSVP (IDTRAN, VPXMIN, VPXMAX, VPYMIN, VPYMAX)
      CALL GSWN (IDTRAN, WNXMIN, WNXMAX, WNYMIN, WNYMAX)
      CALL GSELNT (IDTRAN)
C
      CALL GSPLCI (1)
```

PROGRAM PROFILE
(Continued)

```
C
C Draw lines
C
      DO 48 I = 1,EI2
         CALL MVABS (EFT2L(I,1),EFT2L(I,2))
         CALL LNABS (EFT2L(I,3),EFT2L(I,4))
   48 CONTINUE
C
C Draw arcs and circles
C
      DO 49 I = 1,EI3
         CALL ARCS (EFT3A(I,1),EFT3A(I,2),EFT3A(I,3),EFT3A(I,4),
     *              EFT3A(I,5),EFT3A(I,6))
   49 CONTINUE
C
C Draw curves
C
      DO 47 N = 1,EI4
         CALL CURVE (EFT4C,NUMPTS,D,NSEG,N)
   47 CONTINUE
C
      pause ' '
C
C Terminate GKS
C
      CALL GDAWK (WKID)
      CALL GCLWK (WKID)
      CALL GCLKS ()
      WRITE (*,'(1X,A,A)') CHAR(27),'(=2h'
C
      RETURN
C
      END
C
C----------------------GRAPHICS SUBROUTINES----------------------
C
      SUBROUTINE MVABS (X, Y)
      COMMON /UOMCOM/ X1, Y1
      X1 = X
      Y1 = Y
      RETURN
      END
```

(continues)

PROGRAM PROFILE *(Continued)*

```
C
      SUBROUTINE LNABS (X,Y)
      REAL XA(2), YA(2)
      COMMON /UOMCOM/ X1, Y1
      XA(1) = X1
      YA(1) = Y1
      XA(2) = X
      YA(2) = Y
      CALL GPL (2, XA, YA)
      X1 = XA(2)
      Y1 = YA(2)
      RETURN
      END
C
C--------------------------SUBROUTINE ARCS--------------------------
C
      SUBROUTINE ARCS (X0,Y0,X1,Y1,X2,Y2)
      REAL X, Y, XNEW, YNEW, X0, Y0, X1, Y1, X2, Y2, C, S, DELT
      REAL A, B, R, ALPHA, THETA
      INTEGER I, NSEG
C
C X0, Y0 - location of center of curvature
C NSEG - number of segments in whole arc
C X, Y, XNEW, YNEW - coordinates of points on the arc
C C, S - storage for trig functions computed just once
C DELT - incremental angle for each segment
C THETA - total angle of arc
C
C Define Starting point of arc
C
      X = X1
      Y = Y1
      CALL MVABS(X,Y)
C
C Define the incremental angle and constants
C
      IF (X1  .EQ. X2 .AND. Y1 .EQ. Y2) THEN
         THETA = 2.0 * 3.141596
      ELSE
         A = SQRT((X2-X1)**2 + (Y2-Y1)**2)
         B = A/2
         R = SQRT((X2-X0)**2 + (Y2-Y0)**2)
         ALPHA = ASIN(B/R)
         THETA = 2*ALPHA
      END IF
      NSEG = 16
      DELT = THETA/FLOAT(NSEG)
      C = COS(DELT)
      S = SIN(DELT)
```

PROGRAM PROFILE
(Continued)

```
C
C Draw arc ccw
C
      DO 10 I=1,NSEG
         XNEW = X0 + (X-X0)*C - (Y-Y0)*S
         YNEW = Y0 + (X-X0)*S + (Y-Y0)*C
         CALL LNABS (XNEW,YNEW)
         X = XNEW
         Y = YNEW
   10 CONTINUE
      RETURN
      END
C
C-------------------------SUBROUTINE CURVE---------------------------
C
       SUBROUTINE CURVE (EFT4C,NUMPTS,D,NSEG,N)
       REAL X,Y,X0,Y0,X1,Y1,X2,Y2,D,INTRVL,EFT4C(100,40)
       INTEGER NSEG,C4,NUMPTS(20),EI4,R
C
C X, Y - coordinates of calculated points on the curve
C X0, Y0, X2, Y2 - coordinates of ends of curve segments
C NSEG - number of segments in curve
C
C Define Starting point of curve
C
      NSEG=80
      X = EFT4C(N,1)
      Y   EFT4C(N,2)
      CALL MVABS(X,Y)
C
C Draw first curve consisting of segment 1 and segment 2
C
      X0 = EFT4C(N,1)
      Y0 = EFT4C(N,2)
      X1 = EFT4C(N,3)
      Y1 = EFT4C(N,4)
      X2 = EFT4C(N,5)
      Y2 = EFT4C(N,6)
C
      D = 3/(X2-X0) * ((Y2-Y1)/(X2-X1) + (Y0-Y1)/(X1-X0))
C
      INTRVL = (X1 - X0) / FLOAT(NSEG)
C
      CALL SEG1 (NSEG,X,Y,X0,Y0,X1,Y1,INTRVL,D)
      CALL SEG2 (NSEG,X,Y,X1,Y1,X2,Y2,INTRVL,D)
C
C Draw succeeding segments
```

(*continues*)

PROGRAM PROFILE *(Continued)*

```
C
      DO 25 I=3,NUMPTS(N)*2-5,2
         X0 = EFT4C(N,I)
         Y0 = EFT4C(N,I+1)
         X1 = EFT4C(N,I+2)
         Y1 = EFT4C(N,I+3)
         X2 = EFT4C(N,I+4)
         Y2 = EFT4C(N,I+5)
         D = 3/(X2-X0) * ((Y2-Y1)/(X2-X1) + (Y0-Y1)/(X1-X0))
         INTRVL = (X1 - X0) / FLOAT(NSEG)
C
         CALL SEG2 (NSEG,X,Y,X1,Y1,X2,Y2,INTRVL,D)
C
   25 CONTINUE
C
      RETURN
      END
C
C————————————————————————SEGMENT SUBROUTINES——————————————————————
C
      SUBROUTINE SEG1 (NSEG,X,Y,X0,Y0,X1,Y1,INTRVL,D)
      REAL X,Y,X0,Y0,X1,Y1,INTRVL,D
      INTEGER NSEG
C
      DO 10 I=0,NSEG
         X = X0 + I * INTRVL
         Y = D/(6*(X1-X0)) * (X-X0)**3 + Y0/(X1-X0) * (X1-X)
     *        + (Y1/(X1-X0) - D*(X1-X0)/6) * (X-X0)
        CALL LNABS (X,Y)
10    CONTINUE
C
      RETURN
      END
C
      SUBROUTINE SEG2 (NSEG,X,Y,X1,Y1,X2,Y2,INTRVL,D)
      REAL X,Y,X1,Y1,X2,Y2,INTRVL,D
      INTEGER NSEG
C
      INTRVL = (X2 - X1) / FLOAT(NSEG)
      DO 20 I=0,NSEG
         X = X1 + I * INTRVL
         Y = D/(6*(X2-X1)) * (X2-X)**3 + Y2/(X2-X1) * (X-X1)
     *        + (Y1/(X2-X1) - D*(X2-X1)/6) * (X2-X)
         CALL LNABS (X,Y)
20    CONTINUE
C
      RETURN
C
      END
```

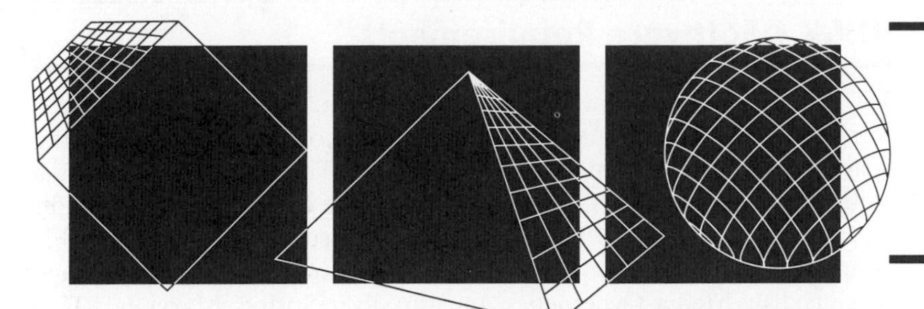

Computer Configuration of the MICROCAD System

AIV.1 Introduction

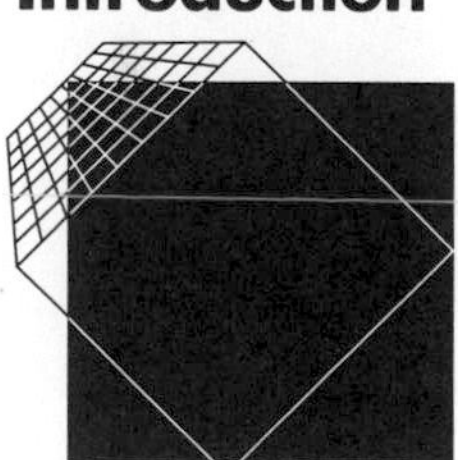

The MICROCAD system is designed to work like a miniature version of commercial CAD systems with integrated finite element stress analysis. The general features of this system are described in Section 10.5, with a flowchart presented in Figure 10.7. This system is designed to be interactive with the user. The user can operate the system by simply responding to the prompts that appear on the monitor screen. There are three principal modules involved in the system: the geometric model for two-dimensional plane geometries with extruded views, the drafting module for the similar geometries, and the finite element stress analysis. The last module can be used as a stand-alone program with prompts available in the system. It handles three basic geometries: three-dimensional axisymmetric, two-dimensional plane stress (usually for thin planar structures), and two-dimensional plane strain. Up to six different materials with their property variations up to five temperatures can be accommodated. The finite element nodal coordinates and element description can be input either manually through the keyboard or by a digitizer tablet, as described in Chapter 10. The output of nodal displacement, the element stresses and strains, and the nodal stresses can be in tabulated form or as graphic representations.

AIV.2 ▪ Software Requirements

The MICROCAD system has evolved from the early stage of microcomputer development. As such, the software components used for the system may not be as contemporary as desired. The bulk of the system is written in the FORTRAN 77 language using the Microsoft (MS) version 3.3 FORTRAN compiler. The operating system software is MS DOS 3.3. Graphic functions are supported by the HALO Graphics Library package developed by Media Cybernetics, Inc., in Silver Spring, Maryland. This graphic package, however, can be replaced by the Graphical Kernal System (GKS), as described in Appendix II. The user should be prepared to make some revisions to the FORTRAN programs in order to use GKS.

AIV.3 ▪ Hardware Requirements

A *minimum* computer hardware configuration to run the system would consist of

- An IBM PC with two double-sided (360K) floppy disk drives, an IBM color/graphics adapter, a serial port (COM1), a parallel printer port (PRN), and at least 320K of RAM.
- An EPSON printer with Graftex or an IBM dot-matrix printer attached to the printer port.
- A digitizing tablet with an active area of about 11″ × 11″ attached to the serial port; one of the following models should work: a GTCO model 7 or model 5, or a Summa Graphics Bit Pad One.

The following information may prove useful when configuring the digitizing tablet and serial interface.

The serial port should be configured with the DOS command called MODE. Use the following form just before running the programs:

```
MODE COM1: 9600,N,8,1
```

that is, 9,600 baud, no parity, 8 data bits, and 1 stop bit.

The digitizing tablet should be configured with the following characteristics:

9,600 baud

Streaming mode

Binary data format

Fastest sampling rate at 9,600 baud

0.005 inch resolution

For the GTCO model 7 tablet, use the factory default, power-up/reset mode continuous-output mode:

5-bit binary

62 points/second

0.005 inch resolution

For the GTCO type 5 tablet, use the following switch settings (where 1 = on):

SW1	0	0	1	1	1	1	0	0
SW2	1	1	1	0	0	0	0	0
SW3	0	1	0	1	0	0	1	1

and use port J5 (71D4J5) with a null modem cable (i.e., pin 2—transmit, pin 3—receive, pin 7—ground).

For the Summa Graphics Bit Pad One, use the following switch settings:

Switch 1 position 7 on
position 8 off (n/a)
position 9 on

Switch 2 position 1 on
position 2 off
position 3 on
position 4 on
position 5 on (or off)

and use an output connector (J1) with a null modem (i.e., pin 2—transmit, pin 3—receive, pin 7—ground).

Note that the IBM PC serial port transmits data on pin 2, receives data on pin 3, and uses pin 7 as a ground.

If an IBM-compatible cable is not available with the digitizing tablet, then the following null modem cable may come in handy:

NULL MODEM

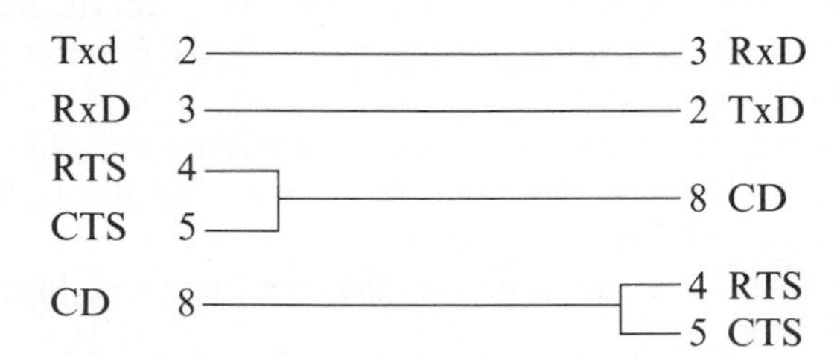

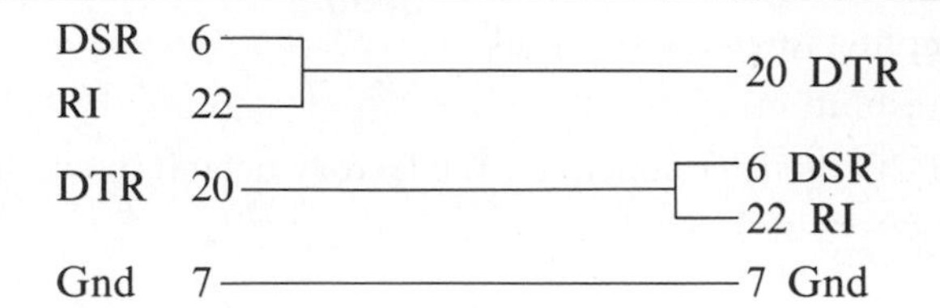

AIV.4 ▪ General Description of System Operation

The MICROCAD system was originally designed to work on an IBM Personal Computer or compatible computer with two floppy disk drives. The programs and data files were to be installed on a combination of the A: and B: floppy disk drives. On machines with hard disks, which are much more common, all of the programs and data files could be installed in a single subdirectory on the hard disk.

If the user wishes to use the software with a floppy disk–based machine, care must be exercised to put only a minimal amount of the DOS system software on the floppy disks. Early versions of DOS (e.g., DOS 2.10) may use less disk space and less computer memory. The user might start up with the A: floppy disk, issue a PATH command such as "PATH A:\", and then switch to the B: drive with a command of "B:".

When the execution of the MICROCAD system is carried out on a double-disk system in an IBM type of personal computer, three files on the B: drive diskette are created: an input file, an output file, and a result file. The input file contains the input data and can be modified to accommodate desirable variations of the original version. Input data can also be written to new files with different file names. The output file contains a complete listing of information, including the numerical values of the calculated stresses and displacements from the finite element analysis. The result file contains coded input for the plot routine, which provides graphic representations of the information listed in the output file. This file is in a compact form and therefore cannot be modified.

Because the B: drive diskette has a limited storage capacity, after each run of this program, the amount of space left on the diskette should be checked. This can be done by listing the file directory for the diskette. To do this, type "DIR B:". The three files described above and the number of bytes used by each are listed together with the total number of bytes remaining on the diskette.

Note: This diskette also requires extra space for temporary files used to produce the output and result files. If an error occurs and files TEMP1 and TEMP2 are found on this diskette, they must be erased.

In case more space is required on the diskette, this procedure may be followed:

Step 1 Any one of the above files that is not needed can be erased. (For example, you can delete an output file that duplicates an input file and the information plotted from the result file.) To ERASE a file, type "ERASE B: FILENAME.OUT" where FILENAME.OUT can be replaced with any file name including its own suffix.

Step 2 Any single file or group of files can be transferred to another diskette or set of diskettes, which may form a library of files capable of being transferred back to the B: drive diskette for re-use.

To transfer a file from the B: drive diskette to a new diskette, the new diskette must first be initialized. To INITIALIZE a new diskette (or erase an old one):

1. Exchange the B: drive diskette with the new one. The A: drive diskette must be in the left drive (or the upper drive in an IBM-PC portable model).
2. Type "A:FORMAT B:/S" and type ⟨cr⟩ when prompted. Be careful to type the above command correctly.

To COPY a file from the B: drive diskette, do the following:

1. Place the B: drive diskette in the right drive and the new (target) diskette in the left drive.
2. Type "COPY B:FILENAME.FOR A:" where FILENAME.FOR can be replaced with any file name including its own suffix.

To COPY a file from the library diskette to the B: drive diskette, place the diskettes in the same drives as above, and type "COPYA: FILENAME.FOR B:".

If an input file is to be revised and the new file given a different file name (both on the B: drive diskette), type "COPY OLDFILE.FOR NEWFILE.FOR". This creates two files with the same content so that the new file can be edited, leaving the old file unchanged.

AIV.5 ▪ Start-up Procedure

Step 1 After installation of the hardware, place the A: drive diskette in the left disk drive and the B: drive diskette in the right disk drive. (The A: drive will mean the upper disk drive in an IBM-PC portable model.)

Step 2 Turn on the system unit, display screen, and printer.

Step 3 After the system boots up, the user may be asked to enter the date as MONTH-DAY-YEAR, followed by ⟨cr⟩. Also enter the time as indicated, followed by ⟨cr⟩.

Step 4 The prompt "A>" will appear. If a digitizing tablet is being used, type "MODE COM1:9600, N,8,1" followed by ⟨cr⟩.

Step 5 Answer the prompts given by the program, and perform the operations described. If the display screen contains information that is difficult to remember, make hard copies for later reference by typing PrtSC, followed by ⟨cr⟩.

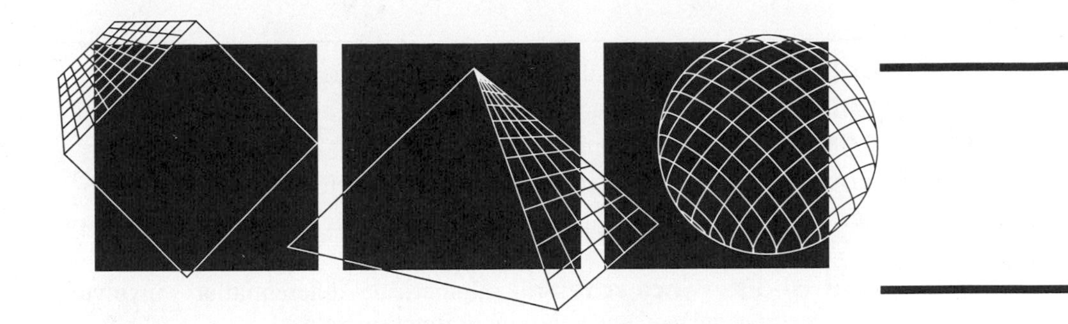

Glossary

application program A user-written program, or a commercial package, developed for solving specific problems.

artificial intelligence Software algorithms that provide expert knowledge and make logical decisions.

attribute The smallest information entry in a database or a characteristic of a display item in computer graphics.

axonometric projection A projection plan that is inclined to all the principal axes of an object.

backward difference scheme Numerical differentiation involving current and preceding steps.

banded matrix A sparse matrix in which all nonzero elements are clustered around the principal diagonal.

bandwidth The number of nonzero elements in each row of a banded matrix.

behavior constraint A constraint arising out of the performance requirements in an optimization problem.

binary tree A schematic representation of Boolean operations in constructive solid geometry; each node in the tree has two descendants.

bit map Information on the on/off status of pixels in a raster display.

bit plane An assembly of bit maps.

boundary representation (B-Rep) Similar to surface modeling, except that the interior (or exterior) of the object is defined at each surface.

B-spline A free-form curve obtained by combining a number of basis splines.

CAD Computer-aided design or computer-aided drafting.

CAE Computer-aided engineering.

CAM Computer-aided manufacturing.

center point A fixed pivot point about which a crank rotates.

center point curve The locus of the center points in four-position synthesis.

central difference scheme Numerical differentiation involving preceding, current, and subsequent function values.

Cholesky decomposition method A method used for solving simultaneous equations.

CIMS Computer integrated manufacturing system.

circle point The conjugate to a center point that moves in a circular arc about the center point.

circle point curve The locus of all possible circle points in four-position synthesis.

clipping The process of eliminating objects or parts of objects not contained within the specified window in two-dimensional viewing or view volume in three-dimensional viewing.

CNC Computer numerically controlled (machine).

coherence A feature in raster graphics whereby adjacent picture elements may be assumed to have similar display characteristics.

computer animation The production of continuous motion (motion pictures) through computer graphical displays.

concatenation of transformations The combining of the effects of more than one transformation.

constructive solid geometry A method of modeling solid objects by Boolean operations using three-dimensional graphics primitives.

convex constraint function A straight line joining two points on the constraint function lies entirely within the feasible region with respect to the particular constraint.

convex objective function Objective function with one minimum.

convex programming problem An optimization problem in which the objective and constraint function(s) are convex.

coupler A floating link that connect cranks.

coupler curve The curve traced by a point on a moving coupler.

CPU Central processing unit (in a digital computer).

CRT Cathode ray tube.

curve fitting The method of determining the functional relationship between given set of dependent and independent variables.

database A logical collection of related information.

data file A collection of relevant information.

diagonal matrix A matrix in which all elements except those at the main diagonal are zeros.

digitizer An input device for computer graphics.

discretization A process of subdividing a continuous domain into a finite number of subdomains.

DOF Degree of freedom; the number of unknown physical quantities in an analysis.

eigenvalue The parameters that result in unique solutions of certain types of simultaneous equations.

eigenvector The solution of simultaneous equations corresponding to an eigenvalue.

ELD Electro-luminescent display; an image display device based on light emission from gas-discharge elements with a high-voltage electric field.

expert system A software program or algorithm constructed to solve a special problem.

FDM Finite difference method, used to solve differential equations.

FEA Finite element analysis.

FEM Finite element method.

field The smallest information entry in a database.

forward difference scheme Numerical differentiation involving current and subsequent function value.

free-form curve A curve of arbitrary shape that may not be represented mathematically by a single equation.

free parameters A design parameter that can be assigned an arbitrary value.

free vibration Oscillation in a system, in the absence of external force, under the influence inherent in the system itself.

Gaussian elimination method A method for solving simultaneous equations.

Gaussian integration A numerical integration technique for definite integrals.

geometric modeling Defining an object from the point of view of its geometric properties.

graphics primitives Simple graphic shapes (such as lines, arcs, text strings, etc.) provided in graphics packages which act as building blocks for complex drawings.

hardware device One of the physical components in a computer system on which the software operates.

hexahedron element An element with eight nodes and six faces.

hierarchical data model A data model constructed in a tree structure with the most general information placed at the top of the tree.

high-level language A programming language far removed from machine language. Such a language is designed to be user-oriented rather than machine-oriented. A program written in a high-level language (FORTRAN, COBOL, ALGOL, Pascal, etc.) must be *compiled* into binary codes by a compiler before it can be run on a computer. A high-level language is independent of the architecture of the computer's CPU.

homogeneous coordinates A coordinate system used to represent n-space objects (namely, two- or three-dimensional objects) as $(n + 1)$-space objects. In a three-dimensional system, the coordinates therefore have four components. The first three components are the x, y, and z coordinates multiplied by a scale factor, and the last component is the scale factor itself. The homogeneous coordinates of a point with coordinates $\{x \quad y \quad z\}^T$ can be expressed as $\{wx \quad wy \quad wz \quad w\}^T$.

identity matrix A diagonal matrix in which each element in the principal diagonal is 1.

image space algorithm A hidden surface removal algorithm that operates on the projection of three-dimensional objects

inference engine A component of an expert system consisting of search and reasoning procedures which enable the system to offer optimum solutions with justification.

interactive graphics A graphics mode in which the user may change the display dynamically.

isometric view A graphic representation similar to axonometric view in which the principal axes of an object are equally inclined to the projection plane.

kinematic synthesis The act of designing mechanisms for specific tasks.

knowledge base A component of an expert system that stores information related to special fields of knowledge.

language binding The statement from a high-level language used to specify generic subroutines in such graphics languages as GKS and PHIGS.

LCD Liquid crystal display.

local minimum The minimum value of the objective function in the search direction relative to the initial value; not the absolute minimum.

low-level language A language closer to machine language than a high-level language. Such a language is difficult to use since a knowledge of the architecture of the CPU is essential. Low-level languages, however, provide for smaller programs and faster execution times. Low-level languages (such as BASIC Assembly Language) may be different for every machine, depending on the architecture.

MCU Machine control unit, which translates the instructions and converts them into appropriate signals for the cutting tool in a CNC machine.

metafile A GKS file used for the storage and transmittal of graphic data.

mirroring Constructing a geometric model by repeating part of the geometry in mirror image.

motion parameter A specification describing the motion of a mechanism.

network data model A data model with information networked together by common attributes.

normalized device coordinates (NDC) A normalized coordinate (0 to 1) system on the display surface.

null matrix A matrix with all zero elements.

NURBS Nonuniform rational B-spline; also called **rational B-spline.** Such a spline is produced by the projection of four-dimensional curves (in the homogeneous space) into three-dimensional space.

objective function A mathematical function defining the objective of an optimization process in terms of the design variables.

object space algorithm A hidden line removal algorithm that operates on the physical location of the objects in the world coordinate space.

path cognate An alternate mechanism with a coupler curve identical to that of a given mechanism.

penalty function A function added onto the objective function that increases the value of the objective function when constraints are violated.

perspective projection Lines of sight which are inclined at varying angles to projection planes.

picture element (pixel or pel) The smallest displayable element (or cell) on a display screen. On a CRT monochrome display, each phosphor dot acts as a pel. In color display on a CRT, a triad of phosphor dots (red, green, and blue) form a pel.

piecewise polynomial function A set of functions in which the relationships among the variables are described by a separate polynomial in different intervals.

plane stress The stress state in which there are stresses in only one plane.

polyline A GKS graphic primitive consisting of connected straight lines.

polymarker A GKS graphic primitive consisting of a series of similar markers placed at user-specified locations.

polynomial B-spline A B-spline in which the basis splines are polynomials.

precision point (or position) A point or position at which the synthesized mechanism meets the design motion requirement precisely.

preprocessor Software that facilitates automatic input to a finite element package.

principal diagonal The diagonal in a square matrix containing nonzero elements.

RAM Random access memory (in a digital computer).

raster A rectangular matrix of picture elements arranged in rows and columns.

raster-scan The process of scanning an image onto a display screen in a sequential manner—one *scan-line* after another. Each scan-line consists of a horizontal row of pixels. The intensity or color of each pixel is set to a predetermined value (during the scanning) so that the correct picture composition is produced.

regression A type of curve-fitting technique in which the functional relationship between the mean values of a random (dependent) variable and those of a dependent variable is sought.

relational model A collection of related information in tabulated form.

rendering The process of shading and shadowing in a picture composition.

ROM Read only memory (in a digital computer).

round-off error Error in numerical solutions due to the rounding off of decimal points beyond what the computer can accommodate.

segment A user-defined block of graphic primitives that may be treated as a customized primitive.

semantic model Descriptive information contained in a database.

side constraint A constraint that sets bounds on the design variables.

Simpson's rule A numerical integration scheme.

skew-symmetric matrix A matrix in which the elements are equal to the negative of the corresponding elements in the transposed matrix.

solid model A mathematical model of the surface and inside of three-dimensional objects.

sparse matrix A matrix with predominantly zero elements.

spatial subdivision method A method of solid modeling in which the object space is divided into small cells (voxels) which are either filled or empty.

strain energy energy induced in a solid due to deformation.

string a set of characters.

surface model A mathematical definition of a three-dimensional object in terms of its bounding surfaces.

surface patch A surface, generally of arbitrary shape, used for representing three-dimensional objects.

system softwareA program that handles the internal data management and operations of a computer.

tensor A physical quantity that can be defined by specifying magnitude, direction, and position.

tetrahedron element An element in the shape similar to a pyramid with four nodes and four faces.

torus element An element in the shape of a ring with a triangular or quadrilateral cross-section.

transformation In computer graphics, an operation that modifies the display by rotation, translation, or scaling.

transformation matrix A matrix that, when multiplied by the position vectors in a display, causes the required transformation.

transmission angle The angle between the coupler and the driven crank in a mechanism.

triangular matrix A matrix in which all the elements on one side of the main diagonal are zeros.

triangulation An algorithm that generates triangular elements.

truncation error An error in a numerical solution induced by omitting higher-order terms in the approximate numerical expressions of corresponding functions.

unary operation A transformation applied to an individual primitive in constructive solid geometry.

unconstrained optimization Optimization carried out in the absence of any constraints.

univariate search method A method of searching for the minima by changing one variable at a time.

vector display A method of producing images on a CRT monitor by analog voltage–driven deflected electron beams.

viewport The area on the display surface on which the contents of the window are displayed.

VLSI Very large system integration.

window A rectangular region in world coordinates from which the data for display are extracted in two-dimensional viewing.

work station A commercial term for a multiuser, multitasking personal computer. For an alternate definition, see GKS terminology in Appendix II.

world coordinates The coordinate system in which objects and their positions are actually defined.

zero-order method An optimization method in which the minima are searched using the value of the function itself, rather than its derivative.

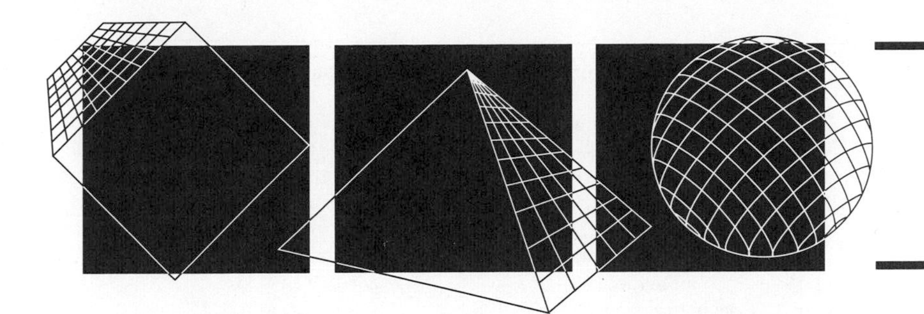

Index

AC-driven panels, 15
Association for Computer Machinery (ACM), 107
Addition and subtraction of matrices, 415–416
Adjoint matrices, 419
Algorithm
  area-coherence, 155–157
  Cohen-Sutherland, 171–172
  hidden line/surface removal, 147–157
  image space, 148
  object space, 148
  priority, 152–154
  for raster-scan graphics, 91–92
  for scan conversion, 92–94
  scan-line, 154
  shading, 73
  span-coherence, 154–155
  YX, 94
  Y-X, 94
  z-buffer, 148–149
American National Standard Institute (ANSI), 107
animation, 1, 5, 18–20
ANSYS, design case study using, 397–402
ANSYS program, 6, 281–285
  element library, 282, 283–284
  modeling capabilities, 281–282
  postprocessing capabilities, 285
  preprocessing capabilities, 282
  program capabilities, 281
  solution methods, 282
  special capabilities, 285
APPLICON, 373
Area-coherence algorithm, 155–157
  contained polygons, 155, 166
  disjoint polygons, 155–156
  intersecting polygons, 155, 156
  surrounding polygons, 155, 156–157
ASME Pressure Vessel Code, 25
Assembly model, 204
Atherton, P., 157
Attribute, 179
Augmented Lagrangian method, 323–325
  for equality constraints, 323–324
  for inequality constraints, 324–325
AutoCAD, 2, 378

Back-face test, 149–150
Backward difference method, 51
Backward substitution, 37
Banded matrix, 414
Bar elements, 216
Behavior constraint, 294
Beitz, W., 21, 405
Beta spline, 78
Bezier curves, 75–78
Bezier surface patch, 158–160
Binary tree, 134
Bit map, 13
Bit plane, 13
Boolean operations, 132
Boundary element method, 34
Boundary representation (B-Rep), 131–132
Bresenham, J.E., 92
B-splines
  nonuniform rational, 81
  polynomial, 75–81
B-spline surface patch, 160–161
Burmester, 353

CAD, 1–29
  benefits of, 6–7
  capabilities of, 4–50
    animation, 5
    engineering analysis, 5–6
    graphical representations, 4–5
  characteristics of, 7–8
  and computer algorithms, 7–8
  and finite element analysis, integration of, 235–239
    free node generator, 237–238
    mapped mesh generator, 238–239
  hardware requirements for, 8–15
  numerical techniques for, 33–67
  and software, 7
  and traditional design, 4–8
CADAM/CAEDS/CATIA, 373

CAD/CAM
commercial packages, 2
interface, 406–408
Cadkey, 2, 378, 381
CAD systems, commercial, major functions of
design analysis, 383
display control, 383–384
drafting, 383
editing/filing, 382
error messages and help, 384
geometric construction, 382–383
CAD systems, integrated, 373–381
CAD/CAM interface, 406–408
data management, storage, and transfer, 377
design analysis, 376–377
design case study using ANSYS, 397–402
engineering drafting, 377
expert systems for, 402–406
geometric modeling, 375–376
MICROCAD system, 384–397
See also CAD systems, commercial, major functions of.
CALMA, 2, 373, 378, 381
Cathode-ray tube (CRT), 11–13
CATIA, 134
Catmull, E., 148
Cayley, 361–362
Center of vision (CV), 144
Central difference method, 51
Central projection, 145
Central processing unit (CPU), 8–9
CGM, 113
Chebyshev spacing method, 348, 369
Cholesky LU decomposition method, 36, 39–42, 66
Circles, in mathematical formulations for graphics, 74–75
Clipping, 5, 170–172
Codd, E.F., 182
Cognate four-bar mechanisms, 360–362
Cohen, 171
Cohen-Sutherland algorithm, 171–172
Color coding, 5–6
Coherence
edge, 94
scan-line, 93
Column vector, 35
Computer-aided engineering (CAE), 1
Computer algorithms, 7
Computer graphics, 1
Computer graphics and design, 127–172
clipping, 170–172
computer simulation and animation, 165–167
geometric modeling, 127–128
geometric properties of graphics models, 161–165
hidden line/surface removal algorithms, 147–157
projections, mathematics of, 140–147
projections, principles of, 136–140
solid modeling, 131–135
surface modeling, 128–131
surface patches, 158–161
viewing in three dimensions, 135–136
windows, viewports, and viewing transformations, 167–170
Computer Graphics Metafile. *See* Graphics standards, CGM.
Computer graphics method, 299–300
Computer graphics, principles of, 73–122
algorithms for raster-scan graphics, 91–92
algorithms for scan conversion, 92–94
curve-fitting techniques, 82–90
GKS, 107–108
graphics packages and standards, 106–107, 112–113
graphics primitives in GKS, 108–112
image manipulation about arbitrary axes, 103–106
three-dimensional graphics, 114
three-dimensional transformations, 114–122
transformation of plane objects, 101–103
transformation of a point, 95–100
two-dimensional transformation, 94–95
Computer-integrated manufacturing system (CIMS), 1–4
Computer-numerically-controlled (CNC) machines, 17
Computer simulation and animation, 165–167
Concatenated transformation, in three-dimensional transformations, 116–117
Concatenation of transformations, of a point, 100
Constructive solid geometry (CSG), 132–133
Containment test, 152
Continuity, 78
Constrained minima, 297–299
Constrained optimization, direct method
method of feasible directions, 326–329
Constrained optimization, indirect methods, 316–225
augmented Lagrangian method, 323–325
extended interior penalty function method, 321–323
exterior penalty function method, 316–319
interior penalty function method, 319–321
Constrained optimization with one design variable, 302–303
Convex constraint function, 299
Convex objective function, 299
Convex programming problem, 299
Cook, R., 181
Coordinates
screen, 148
world, 148
CORE graphics system, 107
Cramer's rule, 36, 364
Crank-Nicholson scheme, 53
Creating images, methods for, 12
Cubic spline functions, 88–90
Curve-fitting techniques, 82–90
polynomial curve-fitting, 82–85
polynomial regression with least squares fit, 85–87
spline interpolation, 87–90
Curves, 75–81

Database, 177
Database program, FORTRAN 77, 183–188
Database management systems (DBMS), 179–180
data models for commercial packages, 180

Databases, design. *See* Design databases.
Databases, for integrated engineering systems, 202–204
 assembly model, 204
 material model, 203
 parts model, 203
 production model, 203
Data file, 180
 general, 192
 geometric, 193
 on previous projects, 194
 product structure, 193
 technical, 192–193
Data models, 180–192
 hierarchical, 188–189
 network, 189–192
 relational, 180–188
Davidon-Fletcher-Powell method, 315
DC-driven panels, 15
Deferral modes, in GKS, 423–424
Deflector system, 12
Deformation, 248
Degree of freedom (DOF), 211
 of a linkage system, 342
Derivation
 Galerkin weighted residual method, 223
 Rayleigh-Ritz method, 223
Desai, C.S., 215
Design constraints, 294–295
Design database management systems, basic requirements for, 201–202
Design databases, 177–205
 data files, 192–194
 data models, 180–192
 geometric databases for three-dimensional objects, 199–200
 geometric databases for two-dimensional objects, 194–199
 IGES standard, 200–201
 for integrated engineering systems, 202–204
Design optimization, 6, 293–334
 constrained, direct method, 326–329
 constrained, indirect methods, 316–325
 constrained, with one variable, 302–303
 constrained and unconstrained, 296–299
 finite element method in, 329–334
 objective function, 295–296
 unconstrained with multiple variables, 307–315
 unconstrained with one variable, 299–302
 unconstrained with two variables, 304–306
 variables, parameters, and constraints, 294–295
Design parameters, 294
Design variables, 294, 400
Determinants
 definitions, 418
 matrices and, 34–35
 properties of, 418–419
Diagonal equalization method, 315
Diagonal matrices, 414
Digital differential analyzer (DDA), 74
Digitizer tablet, 10
 electromagnetic, 10–11
 touch-sensitive, 11
Dimetric projections, 137
Direction cosines, 118
Direct methods, in optimization, 326
Discrete sets concept, 7
Discretization, 210–214
Discretized model, 34
Displacement, 250
Double-twist nematic (DTN), 14
DRAFT, 384

Edge coherence, 94
Eigenvalue problems, 36, 42–50
ELAS, 384
Elastic stress analysis by the finite element method, 247–286
 ANSYS program, 281–285
 finite element formulation, 255–259
 general purprose programs, 279–280
 linear elasticity theory, 247–255
 one-dimensional, of solids, 259–269
 two-dimensional, of solids, 269–279
Element displacement field, 270
Element equations, 223–224
Elements, in finite element analysis, 212
Ellipses, 75
Engineering analysis
 finite element method in, 214–216
 interpretation of results, 28
 mathematical modeling, 27–28
 mechanical, 24
 model analysis, 28
 physical conditions, 27
 performance, 24
Engineering design, general procedure for, 20–26
 conceptual geometry, 22–24
 design analysis, 24–25
 design synthesis, 21–22
 finalizing geometry, 25
 product specification, 20–21
Entities, 194, 201
Euler equation, 130
Expert system, 402–406
Explanatory interface, 406
Extended interior penalty function method, 321–323
Exterior penalty function method, 316–319

Feasible directions, method of, 326–329
Field, 179
Fill area primitive, in GKS, 109–110
Finite difference method, 24, 27, 28, 34, 53–57, 212
Finite element analysis (FEA), 2, 240
 integration of CAD and, 235–239
Finite element formulation, 255–259
Finite element method (FEM), 5, 24, 27, 28, 34, 209–240
 application of, in engineering analysis, 214–216
 automatic mesh generation, 229–235
 discretization, 210–214
 integration of CAD and, 235–239
 in optimization, 329–334
 steps in, 216–229
Finite element programs, general-purpose, 279–280
Fletcher, R., 314
Fletcher and Reeve's method, 314
Forward difference method, 51
FORTRAN 77 database program, 183–188
Free node generator, 237–238
Free vibration, 43

Galerkin residual method, 28, 223
Gas-plasma/electro-luminescent display, 11
Gaussian elimination, 36–39, 66, 274, 282
Gaussian integration scheme, 58, 59, 62–66, 266, 267
Geometric computations, 149–152
  back-face test, 149–150
  containment test, 152
  intersection test, 151–152
  minimax test, 150–151
Geometric databases
  for three-dimensional objects, 199–200
  for two-dimensional objects, 194–199
Geometric data file, sets of information in, 199–200
Geometric method of mechanism synthesis, 350–355
  five-position synthesis, 354–355
  four-position synthesis of a coupler, 353–354
  three-position synthesis of a coupler, 352–353
  two-position synthesis of a coupler, 350–352
Geometric modeling, 16–17, 127–128, 236
GEOMOD, 384
Global coordinate system, 214
Golden section method, 308–310
Gouraud, H., 92
Graphic input devices in GKS, 436–438
Graphical Kernal System (GKS), 107–112, 421–445
  accuracy levels, 427–428
  GKS-3D, 107–113
  input devices, graphic, 436–438
  input devices, initialization, 441–443
  input devices, modes of operation of, 438–441
  metafile, 443–444
  miscellaneous functions, 444–445
  segments, 430–436
  window, viewport, and clipping functions, 429
  workstation, 421–424
Graphical representation of analytical results, 17–18
Graphics models, geometric properties of, 161–165
  lengths of curves, 161–163
  surface area, 163
  volume, 164–165
Graphics packages, 106–107
Graphics standards, 106–107, 112–113
  Computer Graphics Metafile (CGM), 113
  Graphical Kernal System (GKS)-3D, 113
  North American Presentation Level Protocol (NAPLP), 113
  Programmer's Hierarchical Interactive Graphics System (PHIGS), 113
  Virtual Device Interface (VDI), 112–113
Groover, M. P., 406
Ground plane, 138
Grubler's equation, 342

Hidden line/surface removal algorithms, 73, 147–157
  area-coherence algorithm, 155–157
  geometric computations, 149–152
  priority algorithm, 152–154
  scan-line algorithms, 154
  span-coherence algorithm, 154–155
  z-buffer algorithm, 148–149
Hierarchical model, 188–189, 205
Homogeneous coordinates, 78, 122
Hooke's law 254, 278
Horizontal plane, 139
Hybrid solid modelers
  CATIA, 134
  UNISOLIDS, 134
Hyperpatch, 164

Identity matrix, 415
Image manipulation, 103–106
  reflection about an arbitrary axis, 104–106
  rotation about an arbitrary axis, 103–104
Image space algorithms; 148
Inference engine, 404, 405
Initial Geometric Exchange Specification (IGES) standard, 200–201
Input devices, in GKS
  initialization of, 441–443
  modes of operation of, 438–441
Input metafile, 444
Inside test, 93
Institute of Industrial Engineering (IIE), 2
Interactive computer graphics, 15–20
  geometric modeling, 16–17
  graphical representation of analytic results, 17–18
  kinematics and animation, 18–20
Interactive graphic communication, *See* Monitor.
Interior penalty function method, 319–321
Internal resistance, 248
International Standard Organization (ISO), 107
Intersection test, 151–152
Inverse matrices, 420
Isometric projections, 137
Isoparametric elements, 216

Joints, 78

Keyboard, 10
Keyframe, 166
Kinematics and animation, 18–20
Kinematic synthesis, 339
Knots, 78
  double, 79
  simple, 79
Knot sequence vector, 79–80
Knowledge base, 403
Krouse, J.K., 375

Language bindings, 107
Least squares method, 86–87
Lengths of curves, 161–163
Liquid crystal display (LCD), 11, 13–14
Liquid crystals (LCs), 13–14
  double-twist nematic (TN), 14
  supertwist nematic (STN), 14
  twisted nematic (TN), 14
Linear elasticity theory, 247–255
  fundamental assumptions, 248
  fundamental relationships, 252–255
  terminologies, 248–252
Lines, 74
Lines of sight, 139
Linkage synthesis on microcomputers, 339–369
  analytic method of, 355–360
  cognate four-bar mechanisms, 360–362

geometric method, 350–355
mechanism and linkage, 340–344
planar mechanisms, 344–346
precision points, 347–350
Linkage system, four-bar

Main diagonal, 35
Mapped mesh generator, 238–239
Marx, F.J., 331
Material model, 203
Mathematical formulations for graphics, 74–81
Circles, 74–75
Curves, 75–78
Ellipses, 75
Lines, 74
Mathematical modeling, 27–28
Matrices, adjoint and inverse, 419–420
Matrices and determinants, 34–35
Matrix
banded, 414
diagonal, 414
identity, 415
null, 415
skew-symmetric, 414
symmetric, 413
triangular, 414, 415
Matrix algebra, 34–35, 415–417
addition and subtraction, 415–416
multiplication, 416–417
multiplication by a scalar, 416
Matrix multiplication, 416–417
computer program for, 417
Matrix operations, review of, 413–420
adjoint and inverse matrices, 419–420
definitions, 413–415
determinants, 418–419
matrix algebra, 415–417
McAuto, 2
Mechanism and linkage, 340–344
Mechanism synthesis and analysis, 340. *See also* Geometric method of mechanism synthesis.
MEDUSA, 132, 373
Memory boards, 8
Memory disks, 9
Mesh generation, automatic, 229–235
by area density distribution, 232–234
by nodal density distribution, 230–232
by volume density distribution, 234–235
Metafile
input, 444
output, 443
Method of feasible directions, 326–329
MICROCAD, 384–397
case study using, 389–397
program structures, 385–389
MICROCAD system, computer configuration of, 467–472
hardware requirements, 468–470
null modem, 469–470
software requirements, 468
start-up procedure, 471–472
system operation, 470–471
Microprocessor, 8
Minima
constrained, 297–299
unconstrained, 296–297
Minimax test, 150–151
Mitchell, L.D., 20
Model analysis, 28
Modeled animation, 166
Monitor, 11–15
Cathod-ray tube (CRT) display, 11–13
Multiplication by a scalar, 416

Natural frequencies, 43
Nematic LCs, 14
Network model, 189–192, 205
Newell, M.E., 152
Newell, R.G., 152
Nodes, 212
Nonuniform rational B-splines. *See* NURBS.
Normal stress components, 250
Normalized device coordinates (NDC), 168
North American Presentation Level Protocol (NAPLP). *See* Graphics standards.
Null matrix, 415
Null modem, 469–470
Numerical analysis, 33
Numerical differentiation, 50–58
backward difference method, 51
central difference method, 51
Crank-Nicholson scheme, 53
forward difference method, 51
Runge-Kutta method, 53
See also Finite difference method.
Numerical integration, 58–66
Gaussian integration, 62–66
Simpson's one-third rule, 60–62
trapezoidal rule, 59–60
Numerical techniques for CAD, 33–37
eigenvalue problems, 42–50
matrices and determinants, 34–35
numerical differentiation, 50–58
numerical integration, 58–66
simultaneous linear equations, solution of, 35–42
Numerically controlled (NC), 19, 406
NURBS, 78, 81

Objective function, 295–296, 400
Object space algorithms, 148
One-dimensional stress analysis of solids, 259–269
Optimization, 25
by curve-fitting, 301–302
See also Constrained optimization; Unconstrained optimization.
Output attributes, in GKS, 424–428
Output metafile, 443

Pahl, G. 21, 405
Parts model, 203
Pel, 74
Perspective projection, 143–147
one-point, 138, 143
three-point, 138, 140, 143
two-point, 138, 140, 143, 144–147
PHIGS, 107, 113
Phong, Bui-Tuong, 92
Piecewise polynomial functions, 78–81
nonuniform rational B-splines, 81
polynomial B-splines, 78–81
Pixel, 74
Planar mechanisms, 344–346
Plane elements, 216
Plane stress, 269
Plotter, 15
Polygons
contained, 155–156
disjoint, 155–156
intersecting, 155–156
surrounding, 155, 156–157
Polyhedrons, 130
Polyline primitive in GKS, 108–109
Polymarker primitive in GKS, 109

Polynomial regression with least squares fit, 82, 85–87
Postmultiplier, 416
Powell's method, 311–313
Powell, M.J.D., 311
pp functions. *See* Piecewise polynomial functions.
Precision points, 347–350
Premultiplier, 416
Printer, 15
Priority algorithm, 152–154
Production model, 203
PROFILE, 197–199, 200, 447–466
  flowchart summary, 448
  program listing, 454–466
  subroutine flowchart
    arcs, 452
    create, 450
    curve, 453
    draw, 451
    main program, 449
  symbols, 447
Programmer's Hierarchical Interactive Graphics System. *See* PHIGS.
Projection, center of, 135
Projections
  isometric projection, 140–143
  parallel projections, 136–137
  perspective projections, 138–140, 143–147
Projectors, straight-line, 135
Pseudo-objective function, 316
Push-off factor, 328

Random access memory (RAM), 9
Random display system, 12
Raster display, 13
Raster line, 91
Raster-scan graphics, algorithms for, 91–92
Raster scan method, 13
Rayleigh-Ritz method, 223, 255
Read-only memory (ROM), 9
Record, 179
Reeves, C.M., 314
Reflection
  about an arbitrary axis, 104–106
  of a point, 99
  in three-dimensional transformations, 115–116
Relational file, 195
Relational model, 180–188, 205
Resonant vibration, 43
Roberts-Cheyshev theorem, 360, 369
ROMULUS, 132
Rotation
  of a point, 95–100
  in three-dimensional transformations, 115
Rotation about an arbitrary axis, 103–104
  in three-dimensional transformations, 118–122
Roundoff error, 39
Runge-Kutta method, 53

Sadowski, R.P., 2
Sancha, T.L., 152
Scalar matrix, 415
Scaling
  of a point, 98–99
  in three-dimensional transformations, 115
  of variables, 314–315
Scan conversion, algorithms for, 92–94
  polygons, 93–94
  rectangles, 92
  straight lines, 92
Scan-line algorithms, 154
Scan-line coherence property, 93
Screen coordinates, 144, 148
SDRC I-DEAS, 373, 378, 383
Segments, 430–436
Semantic models, 180
Shading algorithms, 73
Shear stress components, 250
Shigley, J.E., 20
Side constraints, 294
Simple window, 155
Simpson's one-third rule, 58, 60–62, 66
Skew-symmetric matrix, 414
Solid geometric modeling, 1
Solid model, 16
Solid modelers, commercial
  MEDUSA, 132
  ROMULUS, 132
Solid modeling, in computer graphics and design, 131–135, 236
  boundary representation, 131–132
  constructive solid geometry, 132–134
  sweeping, 134–135
Span-coherence algorithms, 154–155
Special Interest Group on Graphics (SIGGRAPH), 107
Spline interpolation, 82, 87–90
Stephenson mechanisms, 346
Stiffness equations, 224–225
Strains, 251
Strain-displacement relations, 252
Strain energy, 255
Stress, 249
Stress analysis of solids, one-dimensional, 259–269
Stress-strain relations, 254
Stroke-writing method, 12
Supertwist nematic, 14
Surface area, 163
Surface modeling, in computer graphics and design, 128–131
  explicit polygon method, 129
  method of explicit edges, 130
  method of polyhedrons, 130–131
  polygons by pointers, 130
Surface patches, 158–161
  Bezier, 158–160
  B-spline, 160–161
Sutherland, 171
Swanson, J.A., 331
Sweeping, 134–135
Symmetric digital differential analyzer (DDA), 74
Symmetric matrix, 413
Synthesis problems
  coplanar motion coordination, 367–368
  function generation, 362–366

Tensor quantity, 249
Thin geometry, 269
Three-dimensional graphics, 73, 114
Three-dimensional transformations
  concatenated transformation, 116–117
  reflection, 115–116
  rotation, 115
  rotation about an arbitrary axis, 118–122
  scaling, 115
  of solid objects, 117–118
Transformations
  of plane objects, 101–103
  of a point, 95–100
  two-dimensional, 94–95
  See also Three-dimensional transformations.

Translation
  of a point, 97–98
  in three-dimensional
    transformations, 114
Transpose, 413
Trapezoidal rule, 58, 59–60, 66
Triangularization method, 37
Triangular matrices, 414, 415
Triangulation, 232
Trimetric projections, 137
Truncation error, 60, 61
Turnkey systems, 373
Twisted nematic (TN), 14
Two-dimensional graphics, 73
Two-dimensional stress analysis of
    solids, 269–279

Unary operations, 133
Unconstrained minima, 296–297
Unconstrained optimization with
    multiple design variables
  Davidon-Fletcher-Powell method,
    315
  Fletcher and Reeve's method, 314
  Golden section method, 308–310
  Powell's method, 311–313
  scaling of the variables, 314–315
  univariate search method,
    307–308
Unconstrained optimization with
    one design variable
  computer graphics method
    299–300
  optimization by curve-fitting,
    301–302
Unconstrained optimization with
    two variables, 304–306
UNISOLIDS, 134
Univariate search method, 307–308
Usable feasible sector, 327
Usable sector, 327

Vanishing point, 139
Variable metric method, 315
Variational principle, 28
Vector display, 12
Versacad, 2
Very large systems integration
    (VLSI), 9
Viewing point, 139
Viewing in three dimensions,
    135–136
Viewing transformations, 168–170
Viewport, 382. *See also* Windows
    and viewports.
Virtual device, 107
Virtual Device Interface (VDI). *See*
    Graphics standards.
Volume, 164–165

Warnock, J.E., 155
WATT mechanisms, 346
Wave front, 282
Weiler, K., 157
Windowing, 5
Window, viewport, and clipping
    functions, 429
Windows and viewports, 167–168
Wire frame model, 16, 128
Work station, GKS
  input, 421
  outin, 421
  output, 421
Work station-dependent segment
    storage (WDSS), 430–433
Work station-independent segment
    storage (WISS), 430, 433–436
World coordinates, 148, 167

YX algorithm, 94
Y-X algorithm, 94

z-buffer algorithm, 148–149
Zienkiewicz, O.C., 215
Zooming, 5
Zero-order method, 311